Agroforestry and Ecosystem Services

Ranjith P. Udawatta • Shibu Jose
Editors

# Agroforestry and Ecosystem Services

Springer

*Editors*
Ranjith P. Udawatta
Department of Soil
University of Missouri
Columbia, MO, USA

Shibu Jose
College of Agriculture, Food & Natural Resources
University of Missouri
Columbia, MO, USA

ISBN 978-3-030-80059-8 ISBN 978-3-030-80060-4 (eBook)
https://doi.org/10.1007/978-3-030-80060-4



This Springer imprint is published by the registered company Springer Nature Switzerland AG
The registered company address is: Gewerbestrasse 11, 6330 Cham, Switzerland

# Contents

# Agroforestry for Ecosystem Services: An Introduction

**Shibu Jose and Ranjith P. Udawatta**

## Abbreviations

| | |
|---|---|
| AF | Agroforestry |
| AQ | Air quality |
| BD | Biodiversity |
| CS | Carbon sequestration |
| ES | Ecosystem services |
| NPSP | Nonpoint-source pollution |
| SH | Soil health |
| SOC | Soil organic carbon |
| SOM | Soil organic matter |
| UFF | Urban food forests |

## Introduction

In the last two decades, agroforestry (AF) has gained considerable attention as a land-use practice that can provide a suite of ecosystem services (ES). Advanced landscape-scale multifunctional agroforestry approaches can create stronger links between AF and ES. In AF, trees and/or livestock are intentionally integrated for increased benefits arising from interactions among components in the system (Gold and Garrett 2021). For example, trees, grasses, shrubs, and forbs within AF provide

S. Jose (✉)
College of Agriculture, Food and Natural Resources, University of Missouri,
Columbia, MO, USA
e-mail: joses@missouri.edu

R. P. Udawatta
The Center for Agroforestry, School of Natural Resources, University of Missouri,
Columbia, MO, USA
e-mail: udawattar@missouri.edu

 
R. P. Udawatta, S. Jose (eds.), *Agroforestry and Ecosystem Services*,
https://doi.org/10.1007/978-3-030-80060-4_1

canopy, understory, shrub layer, herbaceous layer, protection, and floor for structural and functional diversity that influence above- and belowground interactions. Additionally, careful selection of perennial species to meet landowner objectives and soil-site conditions along with strategic placement of each component on the AF landscape with advanced management protocols can yield numerous production, protection, and economic benefits.

The depth and breadth of available literature on ES of AF have expanded significantly over the past two decades (Jose 2009). Several recent studies have evaluated soil, water, and biodiversity benefits of AF. Additionally, several meta-analysis and reviews have evaluated regional and global scale benefits of AF (Bhagwat et al. 2008; De Beenhouwer et al. 2013; Kuyah et al. 2019). These studies, reviews, and meta-analysis have shown provisional, regulating, supporting, and cultural ES of AF at farm to regional scales. The objective of this introduction chapter is to summarize ES of AF and how AF can help alleviate some of the global concerns, primarily by providing an insight into the content of this edited volume.

## Agroforestry for Soil Ecosystem Services and Climate Resilience

In many regions, soil loss exceeds soil formation (FAO-TIPS 2015) and the projected 28–51% increase in rainfall intensity over the next century (Nearing 2001; Nearing et al. 2004) could further damage soil resources, thereby impacting soil health (SH). Changing climatic conditions could accelerate degradation of SH, particularly on farmlands with unsustainable agricultural practices. The world needs highly productive, sustainable farming systems with a lower environmental footprint (Ponisio et al. 2015). The current rate of environmental degradation necessitates sustainable soil management practices that are based on efficient nutrient cycling, and reduced degradation of natural resources while increasing productivity (Swift et al. 2004).

The role of AF in improving soil properties and ES has been reviewed by Young (1997), Buresh and Tian (1998), Udawatta et al. (2017, 2021), and Dollinger and Jose (2018) among others. Agroforestry-induced SH improvements can also contribute to sustainable production and other ES such as water quality (Dollinger and Jose 2018). Specifically, AF will help improve soil-related ES by enhancing soil carbon sequestration (CS; Nair et al. 2009a, b; Udawatta and Jose 2012; De Stefano and Jacobson 2018); soil physical, chemical, and biological properties (Tufekcioglu et al. 2003; Seobi et al. 2005; Kumar et al. 2008, 2012; Udawatta et al. 2008a, b, 2009; Chu et al. 2010; Adhikari et al. 2014; Akdemir et al. 2016); soil contaminant removal (Lin et al. 2010, 2011; Kaur et al. 2018); nutrient enrichment (Torralba et al. 2016); soil water dynamics (Anderson et al. 2009; Udawatta et al. 2011a, b; Sahin et al. 2016; Alagele et al. 2020); and soil biodiversity (BD; Bhagwat et al. 2008; Polglase et al. 2008; Jose 2012; Sistla et al. 2016; Udawatta et al. 2019).

Four chapters in this volume describe soil conservation, CS, SH, soil water buffering, and soil rehabilitation by AF. Sauer et al. (2021) evaluated soil services in Eurasia, Canada, USA, Africa, and Asia to describe soil CS, soil conservation, and SH benefits of AF. Their evaluation showed that soil with a high soil organic matter (SOM) content was a common characteristic of resilient production systems that also reduced wind and water erosion. The study further explained that these systems were less likely to experience soil degradation or nutrient depletion and more likely to successfully adapt to climate change. Their chapter provides detailed information from long-term studies in many countries on shelterbelts, silvopasture, and AF with rubber, cacao, and other combinations. Enhanced microclimates, reduced soil disturbance, and increased organic inputs from leaf litter and root, and manure (from grazing animals) decomposition, have each been shown to increase soil organic carbon (SOC) and enhance soil health. Their analysis highlights the importance of selection and integration of suitable components by region for optimum benefits while sustaining the land resource and supplying multiple ecosystem services.

The second chapter by Kremer (2021) describes the soil health ES, particularly from a soil biology perspective. Information on the impacts of AF on soil microbial community, and soil biological processes, is important for assessing and adjusting current management practices. According to the author, AF practices consistently promote the buildup and maintenance of SOC, soil biological activity, and increased microbial diversity that are critical components in improving SH, many of which are ecosystem functions necessary for the expression of ecosystem services. Soil health improvement is often an incidental benefit associated with functions such as nitrogen fixation by intercropped legumes that also increase SOM. This chapter also emphasizes plant resilience to environmental stress, including pests and diseases, as a result of enhanced soil microbiome diversity and function.

Favor and Udawatta et al. (2021) reviewed how AF can help reduce risks associated with viticulture, specifically on the belowground interactions between trees and grapevines and their effects on water availability, nutrient availability, and grapevine rooting patterns in vineyard AF systems. They found that trees in vineyards imparted a neutral to positive effect on parameters surrounding grapevine water status and stress. The review showed that trees improved overall SH, but trees had a slight negative effect on grapevine nutrient status near trees. Trees had induced greater rooting depth and density of grapevine due to improved soil structure. These authors concluded that the incorporation of trees in vineyards possibly created more resilient agroecosystems, and could improve certain grape quality and production parameters, increase farmer savings, and better the environment in numerous ways.

The case study by Rahayu et al. (2021) evaluated differences in soil properties among palm oil plantation (POP), traditional agriculture (TA), agroforestry homegarden (AHG), and natural forest (NF) land-use practices to determine agroforestry's potential for soil rehabilitation. They sampled soils from three villages of Kuala Pembuang, Muara Dua, and Telaga Pulang within the Rimba Raya Biodiversity Reserve located in Seruyan District, Central Kalimantan Province, Indonesia, and determined enzyme activity (cellulase, PMEase, and urease), C, and N. Enzyme activities were greater in NF, AHG, and TA treatments than POP. The lowest C and

N levels were found in POP. Authors recommended homegarden AF as a land-use system that could be an intermediary to rehabilitate and improve degraded soils.

## Agroforestry for Improved Water Quality

Globally hypoxia zones have expanded from less than 10 in 1910 to over 400 by 2010, a 400% increase in 100 years (http://www.vims.edu/research/topics). The hypoxia zone of the Gulf of Mexico expanded from 10,000 $km^2$ in 1985 to 21,000 $km^2$ in 2007 and the change is strongly correlated with the nutrient input and cropland acreage in the Midwest USA (Burkart and James 1999). Water pollution is a serious global issue affecting health and economies of every country. Emerging pollutants further strain already impacted water bodies in addition to sediment and nutrients originating from agricultural watersheds. Agroforestry practices including riparian buffers, upland buffers, alley cropping, windbreaks, and silvopasture have been confirmed to improve the quality of surface and groundwaters. Schultz et al. (2009, 2021) have summarized water quality benefits of AF as well as criteria for the design and maintenance of AF for water quality benefits. Others have explained water quality benefits from long-term studies (e.g., Udawatta et al. 2002, 2011b; Schultz et al. 2021) and model simulations (Senaviratne et al. 2013, 2014a, b, 2018). A meta-analysis and a review have shown that wider buffers were more effective in removing NPSP (Mayer et al. 2007; Liu et al. 2008) although wider buffers take row cropland out of production. This emphasizes the importance of buffer dimensions by soil type, landscape, and slope parameters within a watershed to determine optimum buffer parameters and species composition to maximize land productivity, farm economics, and water quality benefits.

Water quality improvement involves several changes and processes within the soil and AF vegetation. On the soil surface, litter material, roots, fallen branches, and vegetation provide a surface cover and reduce the impact of raindrops, thus reducing soil detachment, transport, flow velocity, and runoff volume (Schultz et al. 2009; Gantzer et al. 1987). Changes in soil parameters, including addition of carbon, improve water dynamics, infiltration, and soil water storage and reduce surface flow (Udawatta et al. 2017, 2021). Storage of nutrients in the perennial vegetation, removal of nutrient from shallow and deeper soil layers, and efficient recycling of nutrients help reduce nutrient losses from agricultural watersheds. Trees with deep-root systems can capture nutrients below the crop root zone and release to the soil surface. This "safety net" mechanism can help protect groundwater while improving nutrient-use efficiency (van Noordwijk et al. 1996; Allen et al. 2004; Nair and Graetz 2004; Nair et al. 2007). Attributes of perennial vegetation of AF such as longer life span, earlier leaf out, profile dewatering, and soil improvements can contribute to improved water quality on agricultural watersheds with AF (Udawatta et al. 2011b; Sahin et al. 2016; Alagele et al. 2020). In summary, changes induced by the combination of perennial and annual vegetation components of AF in soil parameters, and soil water and nutrient dynamics, help reduce nonpoint-source pollution (NPSP) from agricultural watersheds, thereby improving water quality.

In chapter "Water Quality and Quantity Benefits of Agroforestry and Processes: Long-Term Case Studies from Missouri, USA," Udawatta et al. (2021) present findings from two long-term case studies in Missouri, USA. Results showed that integration of AF into a corn (*Zea mays* L.)-soybean [*Glycine max* (L.) Merr.] rotation and a grazing management system had improved surface- and groundwater quality by reducing sediment, nitrogen, and phosphorus losses. Model simulations have further proven AF's water quality benefits on these watersheds. These water quality improvements of AF have been attributed to changes in soil properties, nutrient cycling, water use by the perennial vegetation, soil biodiversity, chemical-biological reactions, shallow-rooted buffer vegetation, deep-rooted trees, and microclimate modification. The interpretation of raw data and model simulations of this chapter highlight the importance of strategic placement of soil-site-climate-compatible AF on agricultural watersheds for enhanced water quality benefits and for other ES including pollination, aesthetic value, biodiversity, and diversified products.

## Agroforestry and Silvopasture

Silvopasture may become increasingly valuable to meet the demand for high-quality animal protein. Continuously increasing global population demands more food while the expansion of the middle class due to economic prosperity demands high-quality animal protein. As global population increases and land availability for production agriculture and forestry diminishes silvopasture could be promoted as a partial solution for providing food, animal protein, and fiber. Conventional animal operations generate large amounts of manure and disposal of nutrient-rich manure has become a serious environmental issue. Silvopasture is a land management practice that intentionally integrates trees, forages, and livestock into the same system. Silvopastoral systems are more intensive in nature, managed in space as well as in time, and those interactions arising from this combination are planned in such a way that it is mutually beneficial (Jose et al. 2019). In silvopasture, animals graze or browse under the shade of trees and both trees and forage or browse species use nutrients from manure and therefore contamination of surface- or groundwater is minimized.

Use of marginal and/or managed forest lands and conversion of grazing land to silvopasture could potentially address the land availability for high-value protein food while improving ES. A portion of forest land areas out of 250 million ha in the USA and 436 million ha in Canada can be potentially converted to silvopasture and managed intensively for various ES. Currently, 54 million ha of forest land is grazed in the USA (Sharrow et al. 2009). Protecting seedlings and saplings and even larger trees from animal damage should be given priority when conversion of a forest to silvopasture is considered (Jose and Dollinger 2019). In Spain and Portugal 6.3 million ha of land characterized by a savannah-like system is managed under dehesa raising pigs (Joffre et al. 1999). Silvopasture is a popular AF practice in South America. Globally, 3315 million ha pasture lands account for 30% of the global ice-free lands for livestock production and contribute 40% of the global agricultural production (Herrero et al. 2013). There is a greater potential for silvopasture to meet future animal protein demand while enhancing numerous ES. Silvopasture could be a more sustainable and resilient practice in a changing climate with continuously declining agricultural land base.

Gabe and Fike (2021) have explained provisional, supporting, regulating, and cultural services provided by silvopasture with numerous examples from all regions. They have also compared these services with conventional grazing. Although silvopasture is not widely adopted in the temperate region, conversion of marginal and unmanaged forests and pastures to silvopasture can increase these ES while increasing the land productivity. There are cultural barriers for the adoption of silvopasture—keeping cattle out of the woods has been a dominant legacy in forestry. However, a growing number of foresters and conservation consultants are recognizing the potential value of silvopasture on the landscape to bring more forest under management and as a way of additional income for the landowner. Gabe and Fike (2021) also have provided information on placement, design, and management considerations for silvopasture.

Dibala et al. (2021) have highlighted many contributions of silvopasture across the globe. They have focused on provisional services and detailed production potential of forage, meat, and milk in silvopastoral systems as direct indicators of food supply. Their study also emphasized animal welfare, pollination, CS, and BD benefits of silvopasture. Alternate food sources for animals are another aspect they have presented—using tree fodder and fruit.

Rivero and Dube (2021) carried out a study to classify a set of preliminary indicators that have the potential to identify sustainability trends and associated risks, and facilitate decision-making for the management of native forest ecosystems under silvopastoral conditions. South America is facing significant challenges and is expected to boost agricultural production while maintaining the forest cover. The region is trending towards sustainable management of agricultural, livestock, and especially forestry systems. However, indicators can provide the right direction for policy makers and federal and local governments. Their methodological triangulation (bibliography-experts-community) used 246 indicators and 50 of those showed most potential to measure sustainability. Three, nine, and ten indicators represented economic, environmental, and social dimensions, respectively. They have suggested to combine several methods to help simplify information, strengthen the validity of results, and reduce biases within the methodological framework. Their conclusion suggests that the method can be further improved to facilitate usefulness, relevance, ease of measurement, and representativeness of the reality of participating communities although the method is suitable for assessing sustainability.

## Agroforestry for Biodiversity Conservation and Enhanced Pollination

There is a strong link between BD conservation and ES (Hooper et al. 2005; Gallai et al. 2009; Rands et al. 2010). Biodiversity provides many essential services to the society including material and nonmaterial benefits and regulation of environmental functions (Rands et al. 2010). Biodiversity positively contributes to agriculture via pest control and pollination, and provides long-term resilience to disturbances and environmental changes (Hooper et al. 2005), thereby contributing substantially to economic and social development (Gallai et al. 2009). It also provides valuable ES and functions for agricultural production at genetic, species, and farming system levels (Thrupp 2000). Loss of BD reduces ecosystem functions and crop yields, and results in potential income loss while increasing health risks and malnutrition (Leakey 1999). Agroforestry can improve ES through BD conservation (Steffan-Dewenter et al. 2007; Jose 2012; Torralba et al. 2016). For example, increased mycorrhizal fungi associations of AF improve nutrient cycling and availability and water and nutrient uptake, and reduce erosion by increased aggregation and thereby increase land productivity compared to conventional monocropping (Bainard et al. 2011; Smith and Read 2008; Torralba et al. 2016). Diverse microbial communities of perennial vegetation degrade antibiotics and herbicides, thereby contributing to regulatory ES like soil health and water quality (Chu et al. 2010; Lin et al. 2010, 2011). In addition to soil health and water quality, AF-induced BD conservation contributes to numerous ES including air quality, pollination, pest control, and fire retardation as well as cultural services such as improvements in recreational, aesthetic, and cultural values (Schultz et al. 2009; Benayas and Bullock 2012; Torralba et al. 2016; Udawatta et al. 2011a, 2021; Bentrup et al. 2021). Numerous studies indicate that AF enhances BD, however, scientific articles are limited synthesizing positive and negative effects of AF on BD and ES (Swallow and Boffa 2006; Udawatta et al. 2019).

Various tropical AF systems with trees such as cacao (*Theobroma cacao*) and rubber (*Hevea brasiliensis*), homegardens, and banana-based AF have shown floral and faunal diversity than monocrop or disturbed forests (Kumar and Nair 2007; Kabir and Webb 2009). Greater insect, bird, and soil microbial diversity has been reported in tropical and temperate regions (Bugg et al. 1991; Smith et al. 1996; Stamps and Linit 1998; Söderström et al. 2001; Brandle et al. 2004; Harvey and Villalobos 2007). This greater diversity has been attributed to reduced stress, availability of food, safety, and spatial heterogeneity of the canopy and soils.

Udawatta et al. (2021) synthesized the most recent literature to explain the BD conservation value of AF. Findings of the chapter demonstrate that AF conserves faunal and floral diversity on agricultural farms. These authors emphasized the importance of strategic planning, selection of soil-site-climate-suitable combinations, integration of many species, and long-term maintenance of the practice for enhanced BD and ES. However, AF cannot be considered as the sole solution for

BD conservation as farmers and landowners usually integrate plants and animals that can contribute to production or income potential of the farm.

Pollination is a main benefit among many services of AF-induced BD conservation. Globally pollination service represents US$195–387 billion annual benefit for domesticated and wild plants (Costanza et al. 1997; Porto et al. 2020). In the USA, the direct and indirect values on pollination were $15.1 billion (2009) and $12 billion (2004) (Calderone 2012). Approximately 90% of flowering plants are pollinated by insects and over 75% of world's most important crops and 35% of food production depend on animal pollination (Kearns and Inouye 1997; Klein et al. 2007; Ollerton et al. 2011). Pollination is also needed for production of certain vitamins, minerals, as well as feed for livestock (Eilers et al. 2011). Climate change will likely have both direct and indirect negative impacts on crop pollination and there is a need to address this issue. Agroforestry practices increase pollinator diversity, which is essential for food production as well as maintenance of population levels of wild plants (Gallai et al. 2009; Varah et al. 2013). Agroforestry's plant diversity offers a variety of resources for increased pollination potential and may help lessen the effects of climate change. Pollination is a mutually beneficial interaction between plants and pollinators.

Bentrup et al. (2021) detailed pollination ES of AF in their chapter. By providing structural and functional diversity, AF influences pollinator diversity and populations. Agroforestry provides habitat, foraging resources, and nesting sites for pollinators. It also improves land connectivity and mitigates harmful effects of pesticides. Their review also emphasized the value of effective AF designs for enhanced pollination service.

## Agroforestry for Flood Reduction, Air Quality Improvement, and Peatland Protection

Globally, heavy rains and flooding events have increased in recent years and will increase by 5–51% according to climate models (Nearing 2001; Nearing et al. 2004). The reported and model-simulated predictions of global temperature surges (climate.nasa.gov/effects) will increase ocean temperatures and formation of heavy rains. Increasing number of rain events and their increased intensity have contributed to significant property and human losses in many parts of the world. According to Hirsh and Ryberg (2012) floods have contributed to 500,000 deaths globally, between 1980 and 2009.

Increasing temperature, certain anthropogenic activities, and natural causes affect air quality (AQ). Dust affects AQ of regions like the Middle East and Northern Africa. Agriculture and forestry can also deteriorate AQ if proper measures are not implemented. Concentrated animal operations are often scrutinized for causing poor AQ. Industry, transport, prescribed burning, waste incineration, and backyard burning can also deteriorate AQ. Particulate matter, gases, vapor, volatile organic compounds, metals, spores, and germs often degrade AQ. Poor AQ can aggravate asthma, emphysema, rashes, nausea, or headaches and increases breathing difficulties, hospitalization, premature birth, various other health issues, health cost, and deaths. For example, fine particles can reduce visibility by scatter and light absorption, and create a haze and may contribute to accidents and traffic fatalities.

Udawatta et al. (2021) have explained the role of AF in reducing flooding risks and improving AQ in their chapter. According to the literature, riparian buffers with a width of more than 100 m on both sides of the Missouri river can significantly reduce flooding, levee break, and sand deposition on land. Windbreaks, alley cropping, forest farming, and urban food forest (UFF) also reduce storm runoff, thus reducing flooding. The reduction of flooding by AF has been attributed to increased water storage and use by trees, and soil modification processes by the perennial nature of trees. The green vegetation of riparian buffers, alley cropping, windbreaks, and UFF improves AQ by removing particulate matter, gases, vapor, volatile organic chemicals, minerals, spores, and odor from the atmosphere. Agroforestry practices established for flood control and air cleaning improve CS, water quality, soil health, biodiversity, aesthetic value, and animal welfare and reduce health issues, noise pollution, and heating/cooling cost.

Dewantto et al. (2021) explained the role of agroforestry for protection of peatlands in Indonesia. Originally it was the indigenous people, who utilized peatland forests as a resource to produce traditional food crops, fruits, and spices; however, over the last five decades it was exploited for commercial oil plantations. Agroforestry practices have been proposed as an alternative livelihood to the rural communities that live near peatland ecosystems in Kalimantan (Borneo) and as a buffer to protect the peatland ecosystems. The team used readily available ecological data and the base map data gathered from participatory approach to determine the most suitable locations for buffer zones in the Rimba Raya Biodiversity Reserve. The participatory approach, communities, government, and private sector conducted the planning, surveying, and developing of suitability base maps and ArcGIS software was used to process the parameters. The southern area and some parts of the northern region of Rimba Raya Biodiversity Reserve were the most suitable locations to implement agroforestry as a buffer. The model has identified 9154 hectares for suitable buffers.

## Agroforestry and Cultural Ecosystem Services

The global land cover change has affected 70% of cultural ES to an unrepairable level. Agroforestry's trees, shrub, grass, and animal integration, which mimic forests, could help reverse some damage. Since cultural ES is intangible and difficult to measure, often proxy indicators are used as indirect CES assessments. Care must be taken when using proxy indicators. For example, a photograph can be used to assess cultural ES, but accessibility is not the same for each location when required.

Literature lacks scientific studies on cultural ES of AF although their benefits can change the adoptability of AF.

Falkowski and Diemont (2021) in this volume have summarized cultural ES of AF and how AF can improve those services. These intangible and nonmaterial benefits include aesthetic and spiritual appreciation, recreation, and education, which, in turn, enhance provisioning, supporting, and regulating of ecosystem services. They further explain how these services can affect societal structure by using four case studies from Mexico, Vietnam, Brazil, and Portugal. Findings of the study indicate that cultural ES is important but ignored in ES assessment because they are difficult to quantify. Their chapter also provides a framework for future studies with a list of recent citations and assessments.

## Country-Specific Examples, India and Australia

Each country has widely variable soil, climate conditions, vegetation types, and preferences of citizens in selecting vegetation that can affect the adopted tree-crop-animal combinations. Australia and India, the two countries we have selected for this volume, have diverse physiography, ecology, and climatic regions. This diversity contributes to greater biological diversity and a number of different land-use systems. For instance, there are 20 agroecological regions in India. Indeed, the biophysical heterogeneity and climatic variability of the country affect choice of tree and crop species and their productivity, implying profound variability in the nature and composition of agroforestry practices. Country-wide data on all AF practices can be useful for scientists, policy makers, and the general public to understand comparative benefits of these practices and for the development of national policies to promote selected practices. India launched a national initiative on AF research in 1983 and created the All India Coordinated Research Project on Agroforestry (AICRPA, Chinnamani 1993). The initiative like AICRPA can develop research, implementation, and national level policies to combat climate change, food insecurity, and other natural and man-made disasters.

The average decadal growth rate of $CO_2$, which was 2.0 ppm per year in the 2000s, had surged to 2.4 ppm per year during the 2010–2019 period (https://www.co2.earth/co2-acceleration). According to the United Nations Framework Convention on Climate Change (UNFCCC), CS is the process of removing C from the atmosphere and depositing it in a reservoir, or the transfer of atmospheric $CO_2$ to secure storage in long-lived pools (UNFCCC 2007). Green plants—especially woody perennials—play a central role in storing atmospheric C in the above- and belowground biomass as well as in the soils. As climate change begins to affect the planet in many ways, AF has been recognized as a potential solution to mitigate the negative effects of climate change.

Kumar and Kunhamu (2021) reviewed C sequestration potential of AF practices in both vegetation and soil for India in their chapter. The analysis showed that CS of aboveground ranged between 0.23 and 23.55 Mg C $ha^{-1}$ $year^{-1}$ and belowground (roots) varied from 0.03 to 5.08 Mg C $ha^{-1}$ $year^{-1}$. These wide ranges have been attributed to diverse range of ecoclimatic conditions, disparate array of AF, species diversity, and variations in management regimes. Similar to many other climatic regions, the Western Himalayan and the humid tropical AF systems were generally characterized by higher CSP than arid and semiarid regions. These authors have emphasized the importance of standard protocols for establishment, management, sampling, data collection, and data analysis so that various AF practices can be compared. They have stated that it is mandatory to develop a countrywide AF database for the development of national level C management strategies.

Nichols et al. (2021) described ES of AF in Australia. In recent years, Australia has experienced above-normal bushfires, heavy rains, and degraded AQ affecting the living standard of its citizens. This chapter explains how AF can help minimize these adverse effects. These AF-induced modifications help reduce health cost and improve food security and health of rural and indigenous cultures. Integration of diverse vegetation types including trees, shrubs, and crops has improved food diversity, nutrition status, and food security. Additionally, AF has helped improve BD, pollination, CS, and air quality in those areas as AF matures.

## Agroforestry Design and Economics

Climate change can negatively impact many communities, but rural communities and land could be more vulnerable (Lal et al. 2011). Therefore, it is imperative to develop resilient and productive ecosystems to reduce the vulnerability, especially for rural systems (Lin et al. 2008). A proper AF design that is suitable for soil, site, and landscape of the climatic region that also satisfies landowner objectives potentially can have greater success and longevity than an unsuitable design. These designs should also be based on scientific evidence as AF is an intensive land management practice where components are intentionally integrated for intended benefits arising from interactions among the components. However, the adoption is determined by economic returns; if the return of the new system is greater than the previous system there is a tendency to adopt the new system. There is a whole list of other factors that also affect the adoption including household preference, land tenure, market values, risk, etc.

Although AF provides many ES like CS, BD conservation, improved water and air quality, pollination, diverse food types, and cultural services to the society, these services may not improve the income of landowners/farmers directly. There is a need for methods that can easily value these services and assign a monetary value. However, literature is limited on simple and practical methods that can evaluate these services and demonstrate direct and indirect benefits to farmers and landowners. There is a need to compensate landowners/farmers for providing these services so that adoption of climate-resilient AF can be enhanced.

In their chapter, Lovell et al. (2021) offer a new perspective on the targeted placement of AF practices, based on a review of relevant literature. They have summarized placement of five main AF practices considering the landscape scale to whole farm. Their approach is based on AF placement in the landscape, considering contributions from landscape ecology, land suitability analysis, and landscape multifunctionality for optimization for suitability and the public good, while also matching the needs and preferences of the landowner. The final section proposes to use more advanced technologies in landscape modeling and visualization to further support new solutions.

Cai et al. (2021) examined the economic value of AF and concluded that AF generated significant economic value to the society. The economic value of ES provided by AF is measured by social utility change (in dollars) caused by a change in service from AF ecosystem. Their review indicates that economic valuation of ES of AF has received little attention. They have presented a detailed evaluation of economic value concepts for ES valuation approaches. The chapter suggests that certain ES evaluation methods may be more appropriate to value certain goods and services compared to other valuation methods. However, there are disadvantages and bias in those estimates and therefore there is a need to further improve these valuation methods. Most studies have evaluated economics of ES associated with riparian systems and windbreaks. Studies are limited on silvopasture and alley cropping while forest farming does not have economic evaluations in the literature. Authors have presented suggestions for future assessment of the economic value of ecosystem services of agroforestry.

## Conclusions

Various chapters, case studies, reviews, meta-analyses, and country examples of this book have indicated that AF improves provisional, regulating, supportive, and cultural ES and the environment in both the tropical and temperate regions. These chapters have clearly demonstrated the importance of AF integration into cropping and animal systems for improved soil conservation, water quality, BD conservation, land productivity, food security, pollination, flood control, air cleaning, cultural services, and greater economic return. Four chapters on soil services of AF have confirmed the importance of soil carbon for more resilient agroecosystems, especially to address climate threats and anomalies. Chapters on soil services, silvopasture, water quality, pollination services, and BD conservation demonstrate that AF can reverse some of the damages and improve the environment and agricultural sustainability. Rationally implemented and regularly maintained AF reduces flooding risks and improves air quality. Cultural benefits of AF play an important role in promoting AF adoption although sound scientific methods are required to quantify these benefits. Country-specific studies have reaffirmed the value of AF in terms of providing multiple ES that are beneficial to their citizens. Agroforestry design criteria and economic valuation, including that of ES, are important to promote AF adoption.

Many chapters emphasize the importance of long-term maintenance of AF systems and implementation of rational management practices on carefully selected and strategically placed species combinations on correct landscapes and soils for enhanced ES. Scientific evidence is clear, and AF can provide ES and improve the living standard. The solid scientific foundation suggests that federal, state, and regional governments, practitioners, farmers, and policy makers can now help increase AF adoption and thereby implement more resilient AF systems that can minimize the effects of climate change, diseases/pest infestation, food insecurity, and human health issues. Although numerous ES of AF has been documented, there is still room to further improve these benefits and identify site-specific systems for regional and local landscapes.

## References

Adhikari P, Udawatta RP, Anderson SH, Gantzer CJ (2014) Soil thermal properties under prairies, conservation buffers, and corn/soybean land use systems. Soil Sci Soc Am J 78:1977–1986

Akdemir E, Anderson SH, Udawatta RP (2016) Influence of agroforestry buffers on soil hydraulic properties relative to row crop management. Soil Sci 181:368–376. https://doi.org/10.1097/SS.0000000000000170

Alagele SM, Anderson SH, Udawatta RP, Veum KS, Rankoth LM (2020) Long-term perennial management and cropping effects on soil microbial biomass for claypan soils. Agron J 112:815–827. https://doi.org/10.1002/agj2.20116

Allen SC, Jose S, Nair PKR, Brecke BJ, Nkedi-Kizza P, Ramsey CL (2004) Safety-net role of tree roots: evidence from a pecan (Carya illinoensis K. Koch)–cotton (Gossypium hirsutum L.) alley cropping system in the southern United States. For Ecol Manag 192:395–407

Anderson SH, Udawatta RP, Seobi T, Garrett HE (2009) Soil water content and infiltration in agroforestry buffer strips. Agrofor Syst 75:5–16. https://doi.org/10.1007/s10457-008-9128-3

Bhagwat SA, Willis KJ, Birks HJB, Whittaker RJ (2008) Agroforestry: a refuge for tropical biodiversity? Trends Ecol Evol 23:261–267. https://doi.org/10.1016/j.tree.2008.01.005

Bainard LD, Kochb AM, Gordon AM, Newmaster SG, Thevathasan NV, Klironomosb JN (2011) Influence of trees on the spatial structure of arbuscular mycorrhizal communities in a temperate tree-based intercropping system. Agric Ecosyst Environ 144:13–20

Benayas JMR, Bullock JM (2012) Restoration of biodiversity and ecosystem services on agricultural land. Ecosystems 15:883–889

Brandle JR, Hodges L, Zhou XH (2004) Windbreaks in North American agricultural systems. Agrofor Syst 61:65–78

Bugg RL, Sarrantonio M, Dutcher JD, Phatak SC (1991) Understory cover crops in pecan orchards: possible management systems. Am J Alt Agric 6:50–62

Buresh RJ, Tian G (1998) Soil improvement by trees in sub-Saharan Africa. Agrofor Syst 38:51–76

Burkart MR, James DE (1999) Agricultural-nitrogen contributions to hypoxia in the Gulf of Mexico. J Environ Qual 28:850–859

Calderone NW (2012) Insect pollinated crops, insect pollinators and US agriculture: trend analysis of aggregate data for the period 1992–2009. PLoS One 7(5):e37235. https://doi.org/10.1371/journal.pone.0037235

Chinnamani S (1993) Agroforestry research in India: a brief review. Agroforest Syst 23: 253–259. https://doi.org/10.1007/BF00704919

Chu B, Goyne KW, Anderson SH, Lin CH, Udawatta RP (2010) Veterinary antibiotic sorption to agroforestry buffer, grass buffer, and cropland soils. Agrofor Syst 79:67–80

Costanza R, d'Arge R, de Groot R, Farber S, Grasso M, Hannon B, Limburg K, Naeem S, O'Neill RV, Paruelo J, Raskin RG, Sutton P, van den Belt M (1997) The value of the world's ecosystem services and natural capital. Nature 387:253–260

De Beenhouwer MM, Aerts R, Honnay O (2013) A global meta-analysis of the biodiversity and ecosystem service benefits of coffee and cacao agroforestry. Agric Ecosys Environ 175:1–7. https://doi.org/10.1016/j.agee.2013.05.003

De Stefano A, Jacobson MG (2018) Soil carbon sequestration in agroforestry systems: a meta-analysis. Agrofor Syst 92:285–299. https://doi.org/10.1007/s10457-017-0147-9

Dollinger J, Jose S (2018) Agroforestry for soil health. Agrofor Syst 92:213–219

Eilers EJ, Kremen C, Greenleaf SS, Garber AK, Klein AM (2011) Contribution of pollinator-mediated crops to nutrients in the human food supply. PLoS One 6(6):e21363. https://doi.org/10.1371/journal.pone.0021363

FAO and ITPS (2015) Status of the World's Soil Resources (SWSR)—Main Report. Food and Agriculture Organization of the United Nations and Intergovernmental Technical Panel on Soils, Rome, Italy

Gantzer CJ, Buyanovsky GA, Alberts EE, Remley PA (1987) Effects of soybean and corn residue decomposition on soil strength and splash detachment. Soil Sci Soc Am J 51:202–206

Gallai N, Salles JM, Settele J, Vaissière BE (2009) Economic valuation of the vulnerability of world agriculture confronted with pollinator decline. Ecol Econ 68(3):810–821

Gold MA, Garrett HE (2021) Agroforestry nomenclature, concepts, and practices. In: Garrett HE, Jose S, Gold M (eds) North American Agroforestry, 3rd edn. American Society of Agronomy, Madison, WI

Harvey CA, Villalobos JAG (2007) Agroforestry systems conserve species-rich but modified assemblages of tropical birds and bats. Biodivers Conserv 16:2257–2292. https://doi.org/10.1007/s10531-007-9194-2

Herrero M, Havlik P, Valin H et al (2013) Biomass use, production, feed efficiencies, and greenhouse gas emissions from global livestock systems. Proc Nat Acad Sci 110:20888–20893

Hooper DU, Chapin FS, Ewel JJ, Hector A, Inchausti P, Lavorel S, Lawton JH, Lodge DM, Loreau M, Naeem S, Schmid B, Setälä H, Symstad AJ, Vandermeer J, Wardle DA (2005) Effects of biodiversity on ecosystem functioning: a consensus of current knowledge. Ecol Monogr 75(1):3–35

Hirsh RM, Ryberg KR (2012) Has the magnitude of floods across the USA changed with global CO2 levels? Hydrological Sci J 57:1–9. https://doi.org/10.1080/02626667.2011.621895

Joffre R, Rambal S, Ratte JP (1999) The dehesa system of southern Spain and Portugal as a natural ecosystem mimic. Agrofor Syst 45:57–79

Jose S (2009) Agroforestry for ecosystem services and environmental benefits: an overview. Agrofor Syst 76:1–10

Jose S (2012) Agroforestry for conserving and enhancing biodiversity. Agrofor Syst 85:1–8

Jose S, Kumar BM, Walter D (2019) Ecological considerations in sustainable silvopasture design and management. Agrofor Syst 93:317–331. https://doi.org/10.1007/s10457-016-0065-2

Jose S, Dollinger J (2019) Silvopasture: a sustainable livestock production system. Agrofor Syst 93:1–9. https://doi.org/10.1007/s10457-019-00366-8

Kabir EM, Webb EL (2009) Can Homegardens Conserve Biodiversity in Bangladesh? Biotropica 40:95–103

Kaur B, Singh B, Kaur N, Singh D (2018) Phytoremediation of cadmium contaminated soil through multipurpose tree species. Agrofor Syst 92:473–448

Kearns CA, Inouye DW (1997) Pollinators, flowering plants, and conservation biology. Bioscience 47:297–307

Kumar BM, Nair PKR (2007) Tropical homegardens: a time-tested example of sustainable agroforestry. Springer

Klein AM, Vaissière BE, Cane JH, Steffan-Dewenter I, Cunningham SA, Kremen C, Tscharntke T (2007) Importance of pollinators in changing landscapes for world crops. Proc R Soc B Biol Sci 274:303–313

Kumar S, Anderson SH, Bricknell LG, Udawatta RP (2008) Soil hydraulic properties influenced by agroforestry and grass buffers for grazed pasture systems. J Soil Water Conserv 63:224–232

Kumar S, Anderson SH, Udawatta RP, Kallenbach RL (2012) Water infiltration influenced by agroforestry and grass buffers for a grazed pasture system. Agrofor Syst 84:325–335. https://doi.org/10.1007/s10457-011-9474-4

Kuyah S, Whitney CW, Jonsson M, Sileshi GW, Öborn I, Muthuri CW, Luedeling E (2019) Agroforestry delivers a win-win solution for ecosystem services in sub-Saharan Africa. A meta-analysis. Agron Sustain Dev 39:47

Lal R, Alavalapati J, Mercer E (2011) Socio-economic impacts of climate change on rural United States. Mitig Adapt Strateg Glob Chang 16:819–844

Leakey RRB (1999) Agroforestry for biodiversity in farming systems. In: Collins WW, Qualset CO (eds) Biodiversity in agroecosystems. CRC Press, New York, pp 127–145

Lin BB, Perfecto I, Vandermeer J (2008) Synergies between agricultural intensification and climate change could create surprising vulnerability for crops. Bioscience 58:847–854

Lin CH, Goyne KW, Kremer RJ, Lerch RN, Garrett HE (2010) Dissipation of sulfamethazine and tetracycline in the root zone of grass and tree species. J Environ Qual 39:1269–1278

Lin C-H, Lerch RN, Goyne KW, Garrett HE (2011) Reducing herbicides and veterinary antibiotic losses from agroecosystems using vegetative buffers. J Environ Qual 40:791–799

Liu X, Zhang X, Zhangm M (2008) Major factors influencing the efficacy of vegetated buffers on sediment trapping: a review and analysis. J Environ Qual 37:1667–1674

Mayer PM, Reynolds SK Jr, McCutchen MD, Canfield TJ (2007) Meta-analysis of nitrogen removal in riparian buffers. J Environ Qual 36:1172–1180

Nair VD, Graetz DA (2004) Agroforestry as an approach to minimizing nutrient loss from heavily fertilized soils: the Florida experience. Agrofor Syst 61:269–279

Nair VD, Nair PKR, Kalmbacher RS, Ezenwa IV (2007) Reducing nutrient loss from farms through silvopastoral practices in coarse-textured soils of Florida, USA. Ecol Eng 29:192–199

Nair PKR, Kumar BM, Nair VD (2009a) Agroforestry as a strategy for carbon sequestration. J Plant Nutr Soil Sci 172:10–23

Nair PKR, Nair VD, Gama-Rodriguez EF, Garciad R, Haile SG, Howlett DS, Kumar BM, Mosquera-Losada MR, Saha SK, Takimoto ANG, Tonucci RG (2009b) Soil carbon in agroforestry systems: an unexplored treasure? Nat Prec. https://doi.org/10.1038/npre.2009.4061.1

Nearing MA (2001) Potential changes in rainfall erosivity in the U.S. with climate change during the 21st century. J Soil Water Conserv 56:229–232

Nearing MA, Pruski FF, O'Neal MR (2004) Expected climate change impact on soil erosion rates: a review. J Soil Water Conserv 59:43–50

Ollerton J, Winfree R, Tarrant S (2011) How many flowering plants are pollinated by animals? Oikos 120:321–326

Polglase P, Paul K, Hawkins C, Siggins A, Turner J, Booth T, Crawford D, Jovanovic T, Hobbs T, Opic K, Almeida A, Carter J (2008) Regional opportunities for agroforestry systems in Australia. RIRDC/L and WA/FWPA/MDBC Joint Venture Agroforestry Program, RIRDC

Ponisio LC, M'Gonigle LK, Mace KC, Palomino J, de Valpine P, Kremen C (2015) Diversification practices reduce organic to conventional yield gap. Proc R Soc B 282:20141396

Porto RG, de Almeida RF, Cruz-Neto O (2020) Pollination ecosystem services: a comprehensive review of economic values, research funding and policy actions. Food Sec. https://doi.org/10.1007/s12571-020-01043-w

Rands MRW, Adams WM, Bennun L, Butchart SHM, Clements A, Coomes D, Entwistle A, Hodge I, Kapos V, Scharlemann JPW, Sutherland WJ, Vira B (2010) Biodiversity conservation: challenges beyond 2010. Science 329(5997):1298–1303

Sahin H, Anderson SH, Udawatta RP (2016) Water infiltration and soil water content in claypan soils influenced by agroforestry and grass buffers compared to row crop management. Agrofor Syst 90:839–860. https://doi.org/10.1007/s10457-016-9899-x

Sistla SA, Roddy AB, Williams NE, Kramer DB, Stevens K, Allison SD (2016) Agroforestry practices promote biodiversity and natural resource diversity in Atlantic Nicaragua. PLoS One 11(9):e0162529. https://doi.org/10.1371/journal.pone.0162529

Schultz RC, Isenhart TM, Colletti JP, Simpkins WW, Udawatta RP, Schultz PL (2009) Riparian and upland buffer practices. In: Garrett HE (ed) North American Agroforestry, an integrated science and practice, 2nd edn. American Society of Agronomy, Madison, WI, pp 163–218

Schultz RC, Udawatta RP, Isenhart TM, Colletti JP, Simpkins WW, Schultz PL (2021) Riparian and upland buffer practices. In: Garrett HE, Jose S, Gold MA (eds) North American Agroforestry, an integrated science and practice, 3rd edn. American Society of Agronomy, Madison, WI

Senaviratne GMMMA, Udawatta RP, Baffaut C, Anderson SH (2013) Agricultural policy environmental extender simulation of three adjacent row-crop watersheds in the claypan region. J Environ Qual 42:726–736. https://doi.org/10.2134/jeq2012.0241

Senaviratne GMMMA, Udawatta RP, Anderson SH, Baffaut C, Thompson A (2014a) Use of fuzzy rainfall-runoff predictions for claypan watersheds with conservation buffers. J Hydrol 507:1008–1018. https://doi.org/10.1016/j.jhydrol.2014.06.023

Senaviratne GMMMA, Udawatta RP, Baffaut C, Anderson SH (2014b) Evaluation of a stepwise, multiobjective, multivariable parameter optimization method for the APEX model. J Environ Qual 43:1381–1391. https://doi.org/10.2134/jeq2013.12.0509

Senaviratne GMMMA, Udawatta RP, Baffaut C, Lory J, Nelson NO, Bhandari A (2018) Evaluation of four parameterization strategies for the APEX model. Am Society Agricult Biol Eng 61:1603–1617. https://doi.org/10.13031/trans.12656

Seobi T, Anderson SH, Udawatta RP, Gantzer CJ (2005) Influences of grass and agroforestry buffer strips on soil hydraulic properties. Soil Sci Soc Am J 69:893–901

Sharrow S, Brauer D, Clason T (2009) Silvopastoral practices. In: Garrett HE (ed) North American Agroforestry: an integrated science and practice, 2nd edn. American Society of Agronomy, Madison, WI, pp 105–132

Smith SE, Read DJ (2008) Mycorrhizal symbiosis, 3rd edn. Academic Press, New York, NY

Smith MW, Arnold DC, Eikenbary RD, Rice NR, Shiferaw A, Cheary BS, Carroll BL (1996) Influence of ground cover on beneficial arthropods in pecan. Biol Control 6:164–176

Söderström B, Svensson B, Vessby K, Glimskär A (2001) Plants, insects and birds in semi-natural pastures in relation to local habitat and landscape factors. Biodivers Conserv 10(11):1839–1863

Stamps WT, Linit MJ (1998) Plant diversity and arthropod communities: implications for temperate agroforestry. Agrofor Syst 39:73–89

Steffan-Dewenter I, Kessler M, Barkmann J, Bos MM, Buchori D, Erasmi S, Faust H, Gerold G, Glenk K et al (2007) Tradeoffs between income, biodiversity, and ecosystem functioning during tropical rainforest conversion and agroforestry intensification. PNAS 104:4973–4978

Swallow B, Boffa J (2006) The potential for agroforestry to contribute to the conservation and enhancement of landscape biodiversity. In: Garrity D, Okono A, Grayson M, Parrott S (eds) World agroforestry into the future. World Agroforestry Centre, Nairobi, Kenya, pp 95–101

Swift MJ, Izac AMN, van Noordwijk M (2004) Biodiversity and ecosystem services in agricultural landscapes—are we asking the right questions? Agric Ecosyst Environ 104:113–134. https://doi.org/10.1016/j.agee.2004.01.013

Thrupp LA (2000) Linking agricultural biodiversity and food security: the valuable role of agrobiodiversity for sustainable agriculture. Int Aff 76(2):283–297

Torralba M, Fagerholm N, Burgess PJ, Moreno G, Plieninger T (2016) Do European agroforestry systems enhance biodiversity and ecosystem services? A meta-analysis. Agric Ecosyst Environ 230:150–161

Tufekcioglu A, Raich JW, Isenhart TM, Schultz RC (2003) Biomass, carbon and nitrogen dynamics of multi-species riparian buffers within an agricultural watershed in Iowa, USA. Agrofor Syst 57:187–198

Udawatta RP, Jose S (2012) Agroforestry strategies to sequester carbon in temperate North America. Agroforest Syst. https://doi.org/10.1007/s10457-012-9561-1

Udawatta RP, Krstansky JJ, Henderson GS, Garrett HE (2002) Agroforestry practices, runoff, and nutrient loss: a paired watershed comparison. J Environ Qual 31:1214–1225

Udawatta RP, Kremer RJ, Garrett HE, Anderson SH (2008a) Soil enzyme activities and physical properties in a watershed managed under agroforestry and row-crop system. Agric Ecosyst Environ 131:98–104. https://doi.org/10.1016/j.agee.2008.06.001

Udawatta RP, Gantzer CJ, Anderson SH, Garrett HE (2008b) Agroforestry and grass buffer effects on high resolution X-ray CT-measured pore characteristics. Soil Sci Soc Am J 72:295–304. https://doi.org/10.2136/sssaj2007.0057

Udawatta RP, Kremer RJ, Garrett HE, Anderson SH (2009) Soil enzyme activities and physical properties in a watershed managed under agroforestry and row-crop system. Agric Ecosyst Environ 131:98–104

Udawatta RP, Anderson SH, Motavalli PP, Garrett HE (2011a) Clay and temperature influences on sensor measured volumetric soil water content. Agrofor Syst 82:61–75

Udawatta RP, Garrett HE, Kallenbach RL (2011b) Agroforestry buffers for nonpoint source pollution reductions from agricultural watersheds. J Environ Qual 40:800–806

Udawatta RP, Gantzer CJ, Jose S (2017) Agroforestry practices and soil ecosystem services. In: Al-Kaisi MM, Lowery B (eds) Soil health and intensification of agroecosystems. Elsevier AP, pp 305–334

Udawatta RP, Rankoth LM, Jose S (2019) Agroforestry and biodiversity. Sustainability 11:02879. https://doi.org/10.3390/su11102879

Udawatta RP, Anderson SH, Kremer RJ, Garrett HE (2021) Soil health services of agroforestry. In: Garrett HE, Jose S, Gold MA (eds) North American Agroforestry, an integrated science and practice, 3rd edn. American Society of Agronomy, Madison, WI

UNFCCC (2007) Report of the conference of parties on its thirteenth session. Bali, Indonesia. In: United Nations framework convention on climate change. Geneva, Switzerland

van Noordwijk M, Lawson G, Soumare A, Groot JJR, Hairiah K (1996) Root distribution of trees and crops: competition and/or complementarity. In: Ong CK, Huxley P (eds) Tree-crop interactions. CAB International, Wallingford, UK, pp 319–364

Varah A, Jones H, Smith J, Potts SG (2013) Enhanced biodiversity and pollination in UK agroforestry systems. J Sci Food Agric 93(9):2073–2075

Young A (1997) Agroforestry for soil conservation. CAB International, Wallingford, UK, p 297

# Agroforestry Practices for Soil Conservation and Resilient Agriculture

**Thomas J. Sauer, Christian Dold, Amanda J. Ashworth, Christine C. Nieman, Guillermo Hernandez-Ramirez, Dirk Philipp, Alexander N. Gennadiev, and Yury G. Chendev**

## Abbreviations

| | |
|---|---|
| $^{137}Cs$ | Radioactive isotope of cesium traced in soil erosion studies |
| C | Carbon |
| $C_3$ | Plants with $C_3$ photosynthetic pathway, cool-season grasses |
| $C_4$ | Plants with $C_4$ photosynthetic pathway, warm-season grasses |
| Ca | Calcium |
| $CO_2$ | Carbon dioxide |
| DM | Dry matter |
| K | Potassium |
| MBC | Microbial carbon |
| MBN | Microbial nitrogen |
| Mg | Magnesium |
| N | Nitrogen |

T. J. Sauer (✉) · C. Dold
National Laboratory for Agriculture and the Environment, USDA-ARS, Ames, IA, USA
e-mail: tom.sauer@usda.gov

A. J. Ashworth
Poultry Production and Product Safety Research Unit, USDA-ARS, Fayetteville, AR, USA

C. C. Nieman
Dale Bumpers Small Farms Research Center, USDA-ARS, Booneville, AR, USA

G. Hernandez-Ramirez
Department of Renewable Resources, University of Alberta, Edmonton, AB, Canada

D. Philipp
Animal Science Department, University of Arkansas, Fayetteville, AR, USA

A. N. Gennadiev
Lomonosov Moscow State University, Moscow, Russia

Y. G. Chendev
Belgorod State University, Belgorod, Russia



R. P. Udawatta, S. Jose (eds.), *Agroforestry and Ecosystem Services*,
https://doi.org/10.1007/978-3-030-80060-4_2

| | |
|---|---|
| P | Phosphorous |
| PSFP | Prairie States Forestry Project |
| S | Sulfur |
| SIC | Soil inorganic carbon |
| SOC | Soil organic carbon |

## Introduction

Soil conservation has historically focused on the prevention of erosion by wind and water. A broader perspective is more common today and more relevant to a discussion of ecosystem services. This broader view of soil conservation includes preventing or reducing any type of soil degradation (physical, chemical, or biological) that limits the ability of land to be productive and provide ecosystem services. Recent assessments of soil degradation at the global scale present a sobering observation of the degree and extent of degradation by erosion, pollution, loss of organic matter, salinization, and desertification (Bai et al. 2008). Often multiple forms of degradation occur simultaneously or in sequence as the same process, e.g., erosion, may exacerbate other types of degradation. Thus, soil erosion remains a serious concern with regard to agricultural sustainability (Follett and Stewart 1985; Pimentel et al. 1995; Montgomery 2007) through many effects including loss of organic matter (Gregorich et al. 1998; Olson et al. 2016) and nutrient depletion (Cihacek et al. 1993; Wang et al. 2006; Leue and Lang 2012).

Healthy soils are needed to recycle nutrients through decomposition of organic inputs (leaves, roots, manures, etc.), processes that require and support a healthy soil biologic community (vertebrates, invertebrates, and microbial). Of all the indicators of soil health, the amount and quality of the soil organic matter are some of the most important (Bot and Benites 2005; Victoria et al. 2012). Not only do soils with substantial soil organic matter contain a reserve of available nutrients but also their active decomposer communities are better suited to remediate the soil after exposure to contamination. Soils with high organic matter also have better physical structure that avoids restrictive rooting and enables increased storage of plant available water (Hudson 1994; Saxton and Rawls 2006; Libohova et al. 2018).

## Agroforestry Systems and Soil Conservation

The standard agroforestry practices, windbreaks, riparian buffers, silvopasture, alley cropping, and forest farming have many variations due to climate, soil, and economic factors. The perennial woody vegetation of agroforestry practices contributes to soil conservation by providing year-round surface cover that protects the soil from water and wind erosion (Nair et al. 1995; McDonald et al. 2003). This stabilization of the surface soil enhances infiltration, which increases soil moisture in the surface layer and supports root exploration and biological activity (Sauer and Hernandez-Ramirez 2011). In temperate climates, tree windbreaks also capture drifting snow that, after spring snowmelt, contributes to an important source of additional water in semiarid areas.

In this chapter, we present the topic of soil conservation with agroforestry practices through focused regional perspectives. Although agroforestry practices can be modified to suit available socioeconomic and natural resources, several practices have evolved to become more common in particular regions. Here we focus on regions of high adoption or long history of practices, although concepts and findings are applicable to areas of similar climate and soils. The incorporation of trees into agroecosystems often addresses a particular resource limitation or environmental concern. In all cases, a healthy agricultural system is better able to compensate for market, climate, and other stresses. After the discussion of regional studies, a brief summary unifies the concepts presented and provides general conclusions.

## Shelterbelts in Eurasia and North America

### *Shelterbelts in Temperate Eurasia*

An understanding of the effects of agroforestry practices on soil conservation in Eurasia became possible only after the emergence of a new scientific discipline—soil science or pedology. In 1891–1892 under the guidance of the founder of genetic pedology, the Russian scientist Vasily Dokuchaev undertook expeditions to Kamennaya Steppe (modern Voronezh Oblast/state) and to an area of virgin grassland near the city of Starobelsk (modern Luhansk Oblast). Experimental areas were established and research projects implemented to study agricultural practices of the steppe territories and to evaluate a system of protective forest plantations for water and wind erosion control (Mil'kov et al. 1992). From this effort began the scientific period of protective afforestation of agricultural lands. The creation of scientific experiment stations in different geographic zones within Eurasia in the first half of the twentieth century facilitated the study of the influence of tree–shrub shelterbelts on landscapes, soils, and crop development.

In the more humid climate of the temperate belt of Eurasia (i.e., taiga landscapes), one of the significant roles of shelterbelts is to serve as biogeochemical barriers on the path of pollutant migration in agricultural lands. An example is the shallow groundwater remediation by windbreaks demonstrated at an experiment station in Turek (Poland). Groundwater at the site contained excessive concentrations of nutrients contributed by fertilizers applied to nearby fields that were transported through the soil to groundwater (Ryszkowski and Kedzior 2007). A significant remediation effect of tree root uptake of water and nutrients was observed up to 16 m from the edge of windbreaks but was detectable up to 45 m from the edge of windbreaks (Szajdak and Życzyńska-Bałoniak 2013).

At the same location, the stocks and composition of soil organic matter under young (14 years) and old (200 years) windbreaks and on the adjoining arable lands were studied. It was found that, under both young and old windbreaks, there is an increase in soil organic carbon (SOC) in comparison with the arable soils, with an obvious trend of SOC accumulation in the soil under the old windbreak. Also in the soil under the old windbreak, the maturation of soil organic matter is revealed by an increase in the content of aromatic carbon (Maryganova et al. 2010). Researchers found similar patterns of SOC change in the zone of broad-leaved forests of Eastern Europe (Novosil' Zonal Agroforestry Experimental Station, Tula Oblast of Russia; Kretinin 1996). Stressing the regional climatic differences in the process of organic matter accumulation in soils under windbreaks in the territory of Eastern Europe, Danilov and Lobanov (1973) noted that the accumulation of SOC is most intensively observed in forest-derived soils and less intensively in grassland soils (Chernozems or Mollisols).

## *Shelterbelts in Semiarid and Arid Eurasia*

A wide range of studies were conducted at research stations in the forest steppe and steppe zones of the East European Plain. In this region, shelterbelts improve the microclimate and water regime of soils under shelterbelts and in adjacent fields. They are a significant factor in the protection and transformation of soil cover (Fedorov and Gumerov 1990). The windbreaks influence the distribution of the snow cover, with more snow accumulation taking place near them than in the open (Erusalimskii 2007). A prolonged annual wetting of soil profiles by melting waters near the windbreaks caused leaching of carbonates and their migration down the soil profile. The depth of effervescence increases and the soil pH decreases markedly. As a result, soil conditions become more favorable for humus formation (Agroforestry and Soil Fertility 1991). The depth of the humus profile in chernozems under windbreaks is noticeably increased and becomes darker (Fedorov and Gumerov 1990; Agroforestry and Soil Fertility 1991). The number of water-stable aggregates in chernozems under windbreaks also increases (Danilov and Lobanov 1973).

In more arid landscapes of Eurasia (semidesert zone), afforestation of agricultural lands has also led to improvement of soil quality. Golubeva (1941) as cited in Van Eimern (1964) reported on the effectiveness of a shelterbelt in trapping windblown sediment during a dust storm in the North Caucasus (Fig. 1). At the Dzhanybek Research Station (Volgograd Oblast of Russia), in an arid climate and with saline soils, agroforestry as narrowband forest belts in the landscape reduced root mortality and the risk of secondary soil salinization (Sapanov 2016). At this location, scientists also developed a method for agricultural management of Solonetz soils (sodium-rich arid soils), which includes deep plowing to destroy the Solonetz horizon and adding gypsum to the Ap (surface plow layer) horizon. Simultaneously, single-row windbreaks were created for additional soil moisture storage by snow

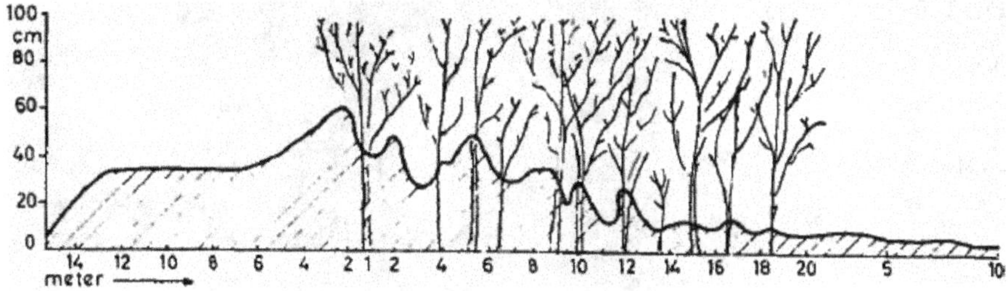


**Fig. 1** Depth of windblown sediment accumulation in a shelterbelt in the North Caucasus following a 3-day dust storm in March 1939. From Golubeva (1941) as cited in Van Eimern (1964)

retention. Consequently, the soil moisture regime changed from aridic (arid) to ustic (semiarid). The amount of soil moisture for plants increased, desalinization and desolonetzation of Solonetz took place, and a specific type of soil was formed that is characterized by favorable soil water physical properties and leaching of salts below 1.5 m. As a result of these changes, biological productivity of Solonetz and productivity of crops increased (Vomperskii et al. 2006).

## *Shelterbelts in the Semiarid US and Canadian Great Plains*

The long history of tree windbreaks on the Russian steppes (Mirov 1935; Vyssotsky 1935; Schroeder and Kort 1989) had a direct influence on the development of the largest agroforestry program in the US history. A period of above-normal rainfall in the early twentieth century led to cultivation of large areas of the Great Plains for the production of small grains. An extended, severe drought in the 1930s resulted in multiple years of extensive crop failure. The unprotected soil surface was left vulnerable to wind erosion, and dust clouds traveled across the Midwest to the east coast of the USA. During this "Dust Bowl" of the 1930s, the U.S. Forest Service coordinated the Prairies States Forestry Project (PSFP, Fig. 2) that included the planting of more than 200 million trees in 30,000 km of shelterbelts in six states from North Dakota to Texas (Droze 1977; Baer 1989). One of the key figures in the PSFP was a forester from Russia, Raphael Zon, who worked for the U.S. Forest Service at their Lake States Forest Experiment Station. Zon and others prepared a thorough feasibility analysis for the PSFP (U.S. Forest Service 1935) and advised field managers throughout the project, often drawing upon experiences and information from the Russian steppes.

Although multiple benefits of shelterbelt planting were recognized, the primary objective of the PSFP plantings was to reduce wind erosion and create a beneficial microclimate for crop growth (U.S. Forest Service 1935; Stoeckeler 1938). Most assessments of shelterbelt effectiveness in reducing erosion were not based on direct measurement but rather on visual observations or measurements of reduced wind speed and an inferred reduction in the ability to detach and transport soil particles.

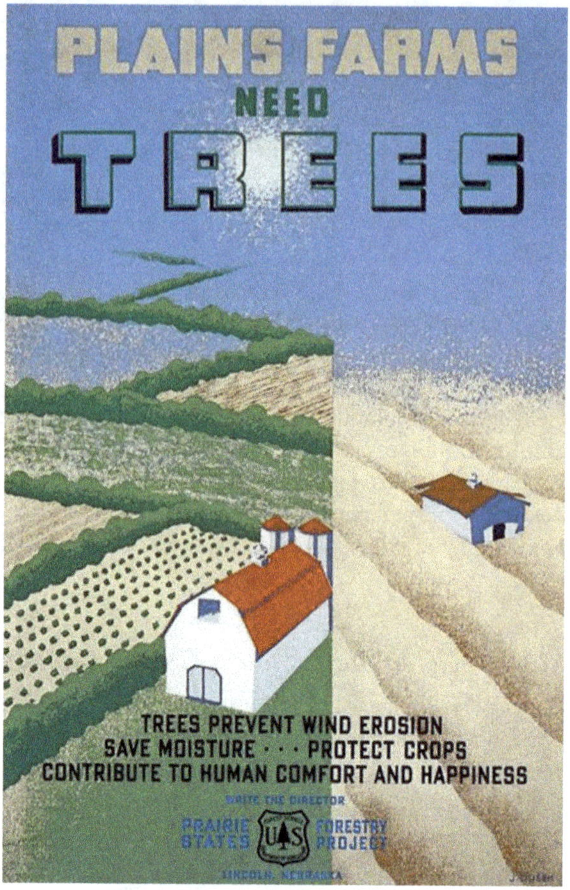


**Fig. 2** Poster promoting the Prairie States Forestry Project created for the Works Progress Administration by the artist Joseph Dusek

In recent decades, however, De Jong and Kowalchuk (1995), Kort (1987), and Van Pelt et al. (2007) used the distribution of $^{137}Cs$ to measure windblown sediment deposited in the lee of shelterbelts in Canada and Texas. De Jong and Kowalchuk (1995) found that 6-m-tall shelterbelts at 200 m spacing in Saskatchewan effectively reduced off-field soil loss but did not prevent some in-field soil redistribution. Sauer et al. (2007) found an ~5% increase in silt content in the surface soil on the leeward side of a 35-year-old red cedar–Scotch pine (*Juniperus virginiana–Pinus sylvestris*) shelterbelt in eastern Nebraska that was attributed to deposition by wind (Fig. 3). Simulation modeling and wind tunnel techniques have been used to investigate sediment movement around shelterbelts and to improve designs of tree windbreaks for optimal wind erosion control (Hagen 1976; Ticknor 1988; Raupach et al. 2001; Cornelis and Gabriels 2005).

A recent survey of trees outside of forests in the Great Plains estimated 458 million trees in the states of Kansas, Nebraska, North Dakota, and South Dakota (Meneguzzo et al. 2018) with 55% of those trees occurring in windbreaks covering an area of 450,000 ha (Fig. 4). The continued use of shelterbelts in the region and

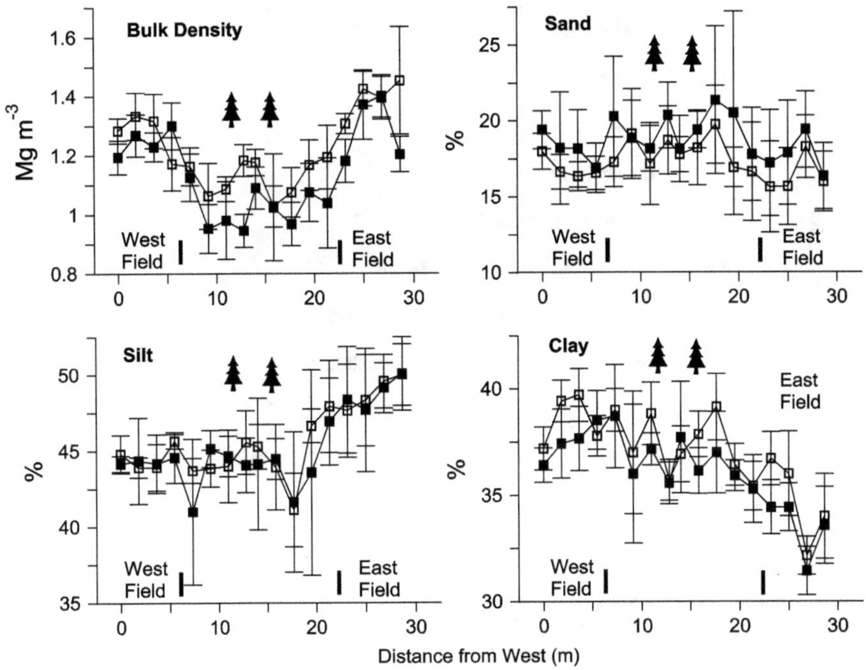


**Fig. 3** Transects of bulk density and soil particle size classes across a red cedar-Scotch pine windbreak near Mead, Nebraska (figure from Sauer et al. 2007). Closed and open squares are for 0–7.5 and 7.5–15 cm layers, respectively. Data indicate deposition of windblown silt on the leeward (east) side of the 35-year-old windbreak

survey data indicate that landowners and farm operators continue to give high importance to wind protection of crops, livestock, and homesteads by tree windbreaks (Hand et al. 2017).

The long history of shelterbelt planting in the Great Plains has also left a legacy of effects on soil properties. Dhillon and Van Rees (2017) found significantly greater SOC in the soil beneath six major shelterbelt tree species in Saskatchewan. The shelterbelt soils average 18.6 Mg C ha$^{-1}$ greater SOC in the surface 50 cm as compared to adjacent crop fields. Chendev et al. (2014, 2015a) measured changes in soil profile properties at windbreak sites in the U.S. Great Plains and Central Russian Upland. Greater SOC accumulation occurred at those U.S. and Russian windbreak locations with more cool and moist climatic conditions. For 55-year-old shelterbelts on chernozems in the European part of Russia, the average annual rate of SOC increase in the surface 1 m of soil varied from 0.7 to 1.5 Mg C ha$^{-1}$ at Streletskaya Steppe (leached chernozems), Yamskaya Steppe (typical chernozems), and Kamennaya Steppe (ordinary chernozems). By comparison, under a 19-year-old forest plantation at the Huron, South Dakota, the average annual rate of SOC increase was 1.9 Mg C ha$^{-1}$. Tree cover in all locations improved soil quality by increasing the surface horizon (A or A + AB) thickness, SOC content, and stocks (Chendev et al. 2014; Novykh and Chendev 2014).

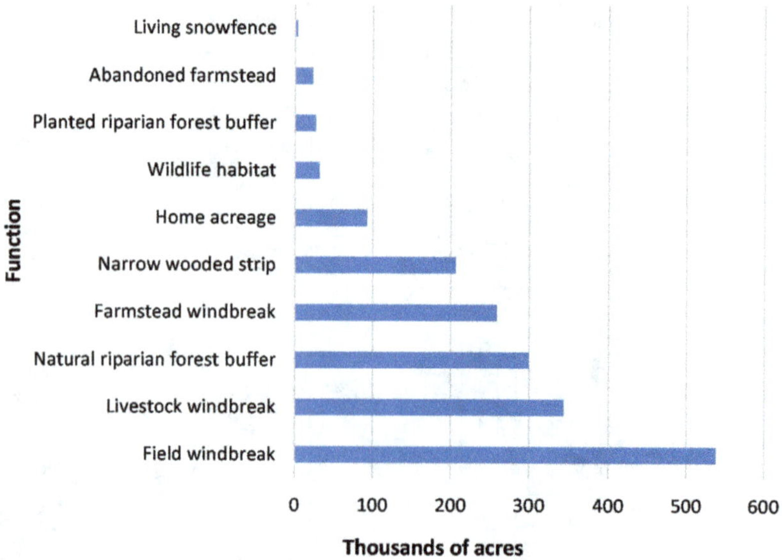


**Fig. 4** Area of tree windbreaks by function in the U.S. Great Plains states of Kansas, Nebraska, South Dakota, and North Dakota (from Meneguzzo et al. 2018)

A 10% increase (3.71 Mg C $ha^{-1}$) in SOC in the surface 15 cm of soil beneath a 35-year-old red cedar–Scotch pine shelterbelt in eastern Nebraska was attributed to lack of disturbance and increased organic inputs from litterfall and root decomposition (Sauer et al. 2007). Hernandez-Ramirez et al. (2011), using stable carbon isotope techniques, determined that 37.2% of the SOC in the surface 15 cm at this site was tree derived, with a mean residence time of 75 years. This site had lower pH and increased Ca content beneath the trees as compared to the adjacent crop fields. De Jong and Kowalchuk (1995) also observed changes in soil pH, bulk density, and water retention beneath shelterbelts in Saskatchewan.

## *Summary of Shelterbelt Contributions in Eurasia and North America*

Shelterbelts have become a common practice over the last century in temperate and semiarid regions of Eurasia and North America. This agroforestry practice is valued for reducing wind erosion, improving crop microclimate, and enhancing soil quality. Shelterbelts serve as a climate adaptation practice by increasing the resiliency of agroecosystems through drought mitigation and microclimate modification. Planted shelterbelts in the central forest steppe zone of European Russia and in the Great Plains in the USA also serve as C sinks by extracting C from the atmosphere and sequestering it in above- and belowground biomass and in the soil organic matter

(Chendev et al. 2015a, b). Shelterbelts can preserve existing soil quality and, when planted on marginal or degraded soils, improve soil health by adding SOC, which promotes multiple soil ecosystem services relating to nutrient cycling and water retention.

## Silvopasture Systems of the Southeastern USA

Silvopastoral systems include the integration of animals (small or large ruminant) in agroforestry landscapes. Higher trophic level production occurs in these integrated forage-livestock-forestry systems, as well as an increased system profitability due to risk management (Clason 1995). Silvopastures, among other agricultural production systems, emulate natural ecosystems more closely, thus resulting in greater provisioning of ecosystem services.

All states throughout the Mid-South USA sustain a climate suitable for establishing and maintaining silvopastoral systems because of sufficient rainfall (about 1100 mm $year^{-1}$) and an abundance of adapted forage species and varieties. A landscape of multiple land uses has been common since colonial times (Clason 1995), as before the 1940s the majority of forest range consisted of longleaf-slash pines (*Pinus elliottii*) and loblolly pines (*Pinus taeda*) (Barlow et al. 2016). After the 1940s, increased regulations on grazing caused ranchers to remove forested areas from the open cattle grazing territory. Current fire suppression in most areas allows for the intermittent reforestation of pastureland through succession (Barlow et al. 2016), resulting in an unintentional mix of open and forested pasture (e.g., savanna systems). However in this system, forest products are not harvested due to the volunteer nature of tree communities.

With trees reforesting open pasture and the strong likelihood of a more variable climate, higher temperatures, and greater variation in precipitation, increased drought and flooding occurrence and severity are likely. For these reasons it would be prudent for ranchers in the southeastern USA to transition to silvopastoral production. Trees are more tolerant to stochastic precipitation and variations in soil texture and elevations, and are more apt to persist in floodplains compared to perennial forages. Multilayer (tree and forage) canopies also allow for heightened capture of rain; reduced runoff velocity and soil erosion, particularly on sloped or already degraded environments; and better catchment of N and phosphorus (P) in runoff.

### *Soil-Based Ecosystem Services of Silvopasture Systems*

Silvopasture systems are one of the more complex agroforestry practices due to the interaction of trees, forage, and animals with landscape features and climate (Garrett et al. 2004). These interactions require greater management to achieve optimal production but also create multiple opportunities to protect and enhance soil resources.

Like shelterbelts, silvopasture systems have the ability to increase SOC and enhance the associated nutrient cycling and water retention properties (Haile et al. 2010; DeBruyne et al. 2011). Haile et al. (2010) used stable carbon isotope techniques at four silvopasture sites in Florida to determine that most of the SOC deeper in the soil profile and in the more stable <53 μm C fraction was tree derived. Increased SOC often results in improved soil structure and soil biological function. Karki et al. (2009) found lower penetration resistance and higher density of fungal biomass but lower aggregate stability at some positions in a longleaf pine (*Pinus palustris* Mill.)–Bahia grass (*Paspalum notatum* Flugge) silvopasture in Georgia compared to an open pasture. Some of the changes in these soil quality indicators were attributed to N source (fertilizer vs. legume) and microclimate variations.

More research has been devoted to nutrient cycling and especially nutrient losses from silvopasture systems. Michel et al. (2007), Nair et al. (2007), and Blazier et al. (2008) all found silvopasture systems to retain less P in the soil profile, thereby reducing the potential for runoff or leaching losses. Lower nutrient concentrations in the root zone are attributed to more extensive rooting and greater nutrient uptake by the combined stand of tree and forage as compared to forage or trees alone. Nair et al. (2007) also found higher concentrations of ammonium and nitrate N in the surface horizon below a treeless pasture in Florida as compared to a slash pine–Bahia grass and a native silvopasture. These changes in soil properties require time and are influenced by establishment practices and differences between thinning an existing tree stand and seeding forages or planting trees into an existing pasture.

Nyakatawa et al. (2012) found persistent low pH and soil C in silvopasture plots established from a thinned loblolly pine stand in Alabama. Soil disturbance for lime and fertilizer application and forage planting was identified as a likely cause for the lack of soil C increase. Adhikari et al. (2018) found that topography attributes had a significant effect on nutrient distribution in a multispecies silvopasture in Arkansas. Total N, S, and P were best predicted by the terrain analysis model, which was used to divide the 4.3 ha site into four topographic functional units with varying nutrient concentrations. Knowledge of topographic influences on nutrient distributions and dynamics enables fine-tuning of nutrient management to improve system efficiency.

Nutrient and soil biota dynamics are different under grazing management than hay production scenarios in agroforestry systems. Livestock grazing can increase concentrated nutrient cycling through animal excreta, increase overall net productivity through animal production, and may induce compensatory plant growth due to continuous defoliation. Consequently, grazed pasture systems may have greater SOC additions, productivity, and organic N and P levels compared to hay production systems (Franzluebbers et al. 2000). Furthermore, the addition of animal grazers enables greater energy flows by increasing trophic levels and providing manure (energy source) for soil microorganisms, thereby influencing soil community dynamics (Naiman 1988).

Soil health is a key indicator of the ability of soil to respond to agricultural management by maintaining both the agricultural production and the provision of other ecosystem services (Kibblewhite et al. 2008). There is growing evidence of the strong link between organisms above- and belowground (Wardle et al. 2004), suggesting that land management can largely drive soil health and soil ecosystem services. The long-term perenniality and extensive root structure of trees have a profound and interacting effect with both neighboring alley crops in silvopastoral systems and soil properties (physical, chemical, and biological), which ultimately influences soil health and the microclimate of alley fodder production. In addition, silvopastoral systems have been proposed as an approach to reduce pasture degradation particularly in tropical and subtropical pastures, which if left unchecked leads to a decline in the natural resource base (e.g., reduced carbohydrate storage and soil and water quality).

Forest systems are estimated to absorb up to 3 Pg of C annually (FAO 2003). Therefore, it is no surprise that a major contribution of silvopastoral systems is above- and belowground C substrate and nutrient requirements for promoting SOC formation and nutrient cycling. In addition to fine and course root turnover, leaf litter and tree biomass also serve as substrates for organic matter synthesis. These C transformations in turn facilitate enzymatic action by fungi and bacteria that are responsible for releasing nutrients to the soil matrix, regulating $CO_2$ gas exchange, improving microaggregation, and altering the biochemical soil environment (White and Rice 2007; Mikha and Rice 2004). In addition, such substrate formation facilitates soil macro- and mesofauna growth and diversity, which is a key element in the soil food web and impacts soil aeration and subsequent water infiltration.

## *Forage Considerations for Silvopasture Production*

The suitability of various forages, in terms of potential dry matter production and suitability in an agroforestry setting, has been well studied in the southeastern USA. Typically, there is some trade-off between forage production and tree growth and planting density owing to interspecific competition (Roth and Mitchell 1982; Alley et al. 1999; Bendfeldt et al. 2001). Roth and Mitchell (1982) noted that tall fescue reduced black walnut (*Juglans nigra*) tree growth by 45% and trunk diameter by 50%. As a result of shade and competition for water, cumulative forage production in silvopasture is often lower when compared to open pastures, especially as trees develop a dense canopy (Pearson and Whitaker 1974; Clary 1979; Sibbald et al. 1991; Silva-Pando et al. 2002). However forage quality may be greater in silvopasture systems (Kallenbach et al. 2006), as cool-season perennial forages have been shown to have higher crude protein (Kephart and Buxton 1993; Lin et al. 2001) and lower neutral detergent fiber (NDF) compared to open pastures (Kephart and Buxton 1993). However, reports of equal or higher acid detergent fiber (ADF) and NDF also exist (Lin et al. 2001).

Depending on agroforestry management objectives, there are several annual and perennial forages that can be grown in the southeastern USA. Sufficient dry matter (DM) production of any species will be challenged in more densely populated pine stands, owing to increased canopy cover (Lin et al. 1999) and lower soil moisture (Karki and Goodman 2015). Ultimately, high DM production will be determined by several factors, including soil type, water-holding capacity, microclimate, predominant tree and forage species type, tree age, and ability of the landowner to develop tree sward management that optimizes production while also implementing proper grazing for forage and animal production (Martsolf 1966). Below, we elaborate on possible alley forages based on production objectives.

## Perennial Cool-Season Forages

Tall fescue (*Lolium arundinaceum*) is by far the most common cool-season perennial forage in the southeastern USA, but this species will not grow well under intense summer heat. Fescue can be successfully integrated into silvopasture systems (Lehmkuhler et al. 1999) and tends to persist well even in minimally managed loblolly pine silvopasture (Burner and Brauer 2003). Orchardgrass (*Dactylis glomerata*), as the name implies, is also well suited for wooded areas, and has been shown to have greater forage production than tall fescue in a loblolly pine plantation (Burner 2003), but production capacity of each cool-season perennial is largely dependent on the level of shading.

Based on the reports by Lin et al. (1999), cool-season perennial grasses, Kentucky-31 tall fescue, and orchardgrass showed only a small reduction in DM production when grown under shade cloth between summer and fall. Early seasonal differences between shade treatments were more pronounced, pointing to the nature of perennial cool-season grasses to generate more DM in spring than fall and thus also being more sensitive to reduced solar radiation from their $C_3$ photosynthesis. Simulated shade treatments included 50 and 80% reduction in light, with the latter reducing DM production by 30–50%.

## Perennial Warm-Season Forages

Warm-season grasses ($C_4$) are less shade tolerant than their $C_3$ counterparts due to their requirement for high incident solar radiation (Kephart et al. 1992), thus requiring greater alley spacing between tree rows. In these systems, however, there are several options for warm-season forage integration in silvopastoral systems in the southeastern USA. Bermudagrass (*Cynodon dactylon*) is the most common perennial warm-season forage; it is highly competitive in open pasture, though DM production is greatly reduced in shaded environments. Not surprisingly, Lin et al. (1999) reported that the reduction in DM production in $C_4$ forages such as bermudagrass, big bluestem (*Andropogon gerardii*), and Indian grass (*Sorghastrum nutans*) was substantially reduced (30% reduction under 50% shade; 70% reduction under 80% shade) than in cool-seasonal grasses (<5% reduction under 50% shade; 30% reduction under 80% shade).

Such DM reductions do not mean that warm-season forages do not have a place in silvopastoral systems. Despite limited forage production of shaded warm-season perennials, species such as common bermudagrass, coastal bermudagrass, and Bahia grass (*Paspalum notatum*) have shown to be profitable in Louisiana loblolly pine plantations (Clason 1999).

### Annual Cool-Season Forages

Competition with trees for water resources puts many forage species at a great disadvantage, although cool-season annual forages can be used strategically during the year when precipitation is naturally high. Root pruning and barrier studies have shown that during summer months, water may be the greatest limiting factor (Jose et al. 2000; Gillespie et al. 2000; Sudmeyer et al. 2002; Burner et al. 2009). Root pruning and root barriers did not have an effect on annual ryegrass (*Lolium multiflorum*) (Burner et al. 2009), indicating that water competition is not a factor in early spring. Rye (*Secale cereal*) and ryegrass co-seeded in an open pasture and in a pine-walnut silvopasture in Missouri resulted in a 20% greater forage production in the open pasture; however, forage production was still high in silvopasture, averaging 5509 kg $ha^{-1}$ over 2 years (Kallenbach et al. 2006).

### Annual Warm-Season Forages

Warm-season annual forages are an additional option for silvopasture systems, though competition for water and light must also be considered. Pearl millet (*Pennisetum glaucum*) is a common warm-season annual in the southeastern USA and an option for cultivation in silvopasture, although research indicates that water competition is a significant factor for pearl millet growth in loblolly pine systems in Arkansas (Burner et al. 2009). Despite the competition for water, pearl millet provided a moderate amount of forage during the hottest part of summer when available biomass can be low, and livestock may benefit from shade in the humid southeastern USA.

## *Livestock Considerations in Silvopastoral Systems*

Since the purpose of a silvopastoral system is to establish pastures between tree plantings, grazing animas such as cattle and sheep are most commonplace. There are systems in other parts of the world where swine are produced in a woody environment, but market demands, supply chain requirements, geography, and climate in the USA lend to the use of small or large ruminants. Among large ruminant operations, cow-calf production is most common in the southeastern USA.

Managing cattle grazing in silvopastoral systems is relatively straightforward if some basic rules of animal husbandry are applied. Cattle are herd animals and as such require a minimum paddock size for efficient grazing and reasonable rotation cycles. In general, cattle should be stocked on silvopastures after tree growth reaches a height that cattle will not damage them through grazing. There are options to fence groups of trees or tree rows to prevent damage. This, however, is labor intensive and in many cases will prevent proper grazing as distances to water access points may be vastly increased and the extra fencing may be cost prohibitive. It is important to recognize that grazing management on silvopastures is similar to open pastures in terms of forage management. Standing forage will have to be grazed or harvested to maintain forage quality, kept weed free, and fertilized to promote growth.

Animal production in silvopasture *versus* open pasture varies but is generally equal or greater. Kallenbach (2009) showed 10% less winter weight loss, less calving difficulty, and heavier weaning weights for calves in a cow-calf silvopastoral system compared to open pasture. Kallenbach et al. (2006) detected no differences in average daily gain or gain $ha^{-1}$ for heifers on annual ryegrass/cereal rye mixture in a pine-walnut silvopasture. Despite greater forage production per area, gain per area was equal between the two treatments, which may be explained by the silvopasture microclimate that allowed for significantly greater forage production. Such positive impacts are expected from microclimates, owing to reduced solar radiation, lower temperatures, and provision of shade. Solar radiation is a major microclimatic parameter that is consistently lower, ranging from 14 to 58% lower in silvopasture compared to open pasture (Karki and Goodman 2010).

In terms of nutrient distribution, cattle behavior in silvopasture systems may better promote even forage utilization and nutrient deposition. Uneven grazing in pastures may reduce pasture utilization, and this is particularly an issue during warmer times of the year. In open pastures, cattle will spend more time lying or loafing in pasture while cattle in a silvopasture system spend more time grazing throughout the hottest parts of the day (Karki and Goodman 2010; Zuo and Miller-Goodman 2004). In a loblolly pine-Bahia grass silvopasture cattle spent 36–52% less time grazing in an open pasture compared to silvopasture (Karki and Goodman 2010). The shade protects animals from environmental stressors that can reduce animal production (Fike et al. 2004). Cattle also prefer shade from trees rather than artificial shade, likely owing to greater distribution of shade available from forested areas, opposed to structural shade (Zuo and Miller-Goodman 2004).

High grazing pressure may lead to soil erosion in pastures and silvopastoral systems, causing increased sediment delivery to waterways. For example, research has shown that various grazing practices and intensities can affect soil physical properties and increase runoff (Ludvikova et al. 2014; Russell and Bisinger 2015). In a 12-year grazing management study in Arkansas, authors found that the penetration resistance and bulk density of rotationally grazed watersheds were lower than continuously or overgrazed systems, which was also greater than ungrazed or hayed management (Pilon et al. 2017). Similarly, in this experiment, runoff volumes, sediment concentrations, and loads were lowest for the hayed and rotational grazed treatments and greatest for continuously grazed systems (Pilon et al. 2017). Soil in grazed Douglas fir [*Pseudotsuga menziesii* (Mirb.) Franco] silvopastures in Oregon sampled 11 years post-establishment had 13% higher bulk density and 7% lower total porosity, and water infiltration rate was 38% less compared to ungrazed Douglas fir forests (Sharrow 2007). However, after silvopastures were rested for 2 years without livestock grazing, soil bulk density, total porosity, and air-filled pore space were similar for forests and silvopastures, demonstrating that the effects on soil physical properties were quickly reversed (Sharrow 2007). Grazing effects on soil properties were shown to have little effect on silvopasture forage or tree production (Sharrow 2007), but further research is required to explore these relationships.

### *Summary of Silvopasture Considerations*

Compared across grazing practices, silvopastoral production systems are environmentally and economically beneficial alternatives to grazing of monoculture pastures. Specifically, integrating tree production in pastures has several positive effects on soil properties and nutrient cycling while creating more favorable microclimate for the animals and increasing overall system net primary productivity. However, given the degree of sensitivity of introduced or improved forages to shade, it is difficult to predict forage growth in wooded areas. Successful establishment of forages depends on a multitude of factors including tree and companion forage species, livestock type, and long-term system objectives. Silvopasture systems have also been found to increase SOC and other soil quality indicators and especially to improve the uptake of nutrients (P and N) that pose water quality risks.

## Agroforestry Practices for Soil Conservation and Agriculture Resilience in the Neotropics

### *Soil Quality*

In the neotropics, agroforestry systems have shown to sustain similar fertility and organic matter content as the natural forest ecosystems (Menezes et al. 2008). Comprehensive research conducted by Rousseau et al. (2012) in multistrata cacao (*Theobroma cacao* L.) agroforestry systems in Costa Rica and Panama revealed that critical abiotic indicators for good soil quality were low bulk density and high C storage as well as high-sum-of-bases calcium, magnesium, and potassium (Ca, Mg, K, respectively) and pH (slightly acid or near neutral). Based on a cluster classification analysis, these four indicators of good soil quality explained three-fourths of total data variance. These four measured soil properties were similar between the cacao agroforestry locations with good soil quality and their natural forests (replicated reference sites) (Rousseau et al. 2012). Moreover, although the cacao agroforestry systems had fewer tree species and lower densities than the forest, the multistrata canopy structures of these two types of ecosystems were very similar

(Guiracocha et al. 2001), supporting the beneficial role of tree inclusion and the associated canopy architecture in conserving soils and ecosystem functions.

Ilany et al. (2010) reported beneficial effects of agroforestry in maintaining soil quality when intercropping yerba mate (*Ilex paraguariensis*) with native araucaria trees (*Araucaria angustifolia*) in South America. They documented substantial tree effects such as sustaining soil organic matter contents as well as cation-exchange capacity (in particular available Ca) in their strongly acid soils (pH: 5–5.5). Similarly, in the highlands of Guatemala, coffee (*Coffea arabica*) crops under shading trees have been shown to store the same amounts of SOC as neighboring natural forests (Schmitt-Harsh et al. 2012). Researchers Schmitt-Harsh et al. (2012) further interpreted this agroforestry practice as a means for facilitation and maintenance of soil conservation in these tropical landscapes under a wide range of management intensities. In Chiapas, Mexico, Soto-Pinto et al. (2010) also quantified soil C density beneath agroforestry systems (e.g., coffee, silvopasture, living fences) in contrast to traditional maize crops and pastures with no trees. This study identified soil as the greatest C sink across all of their systems (e.g., agroforestry, monocrops, and pastures), and the agroforestry systems having overall higher soil C storages, in particular in the cases of the more permanent agroforestry land uses such as shaded coffee farms located in the highlands (>1000 m elevation) (Romero-Alvarado et al. 2002; Soto-Pinto et al. 2010).

## Soil Erosion

The presence of trees contributes to the reduction in soil-erosive processes in both humid and semiarid tropics by various mitigating mechanisms. As reported by Beliveau et al. (2017), a diverse agroforestry assemblage of fruit trees [e.g., Brazil nut (*Bertholletia excelsa*), mango (*Mangifera indica*), soursop (*Annona muricata*), Barbados cherry (*Malpighia glabra*), araza (*Eugenia stipitata*), and orange trees (*Citrus sinensis*)] in the Brazilian Amazon substantially reduced erosive runoff and sediment losses compared to a continuous cassava (*Manihot esculenta*) monocrop, which left most of the soil exposed to the intense rainfalls. Moreover, the assessed agroforestry system resulted in mitigation of soil erosion to a similar extent as the local natural forests, even after only 2 years of tree establishment. This response applied for both moderate terrain slopes and even more for steep slopes. This study emphasizes the consistent contribution of agroforestry practices to conserving, restoring, and maintaining soil integrity and nutrient retention in tropical regions with excessively wet climate (annual rainfall ranged from 2050 to 2720 mm). Likewise, after assessing C pools and dynamics in tropical semiarid environments in Northeast Brazil (annual rainfall of 822 mm), Maia et al. (2007) concluded that compared to intensive maize cultivation, a long-term silvopasture increases soil C accretion primarily by accumulating labile C fractions, and thus supporting improved soil quality as mentioned above. For the same comparison and biome, Aguiar et al. (2010) found that silvopasture reduces sediment load and erosive runoff, and these beneficial effects were attributable to the enhancements in key soil properties such as higher water conductivity and organic matter content, as well as increasing ground cover provided by the tree canopy. This study highlighted this agroforestry system as a clear example of a deployable management option leading to soil conservation.

Blanco Sepúlveda and Aguilar Carrillo (2015) demonstrated a robust relationship ($R^2 = 0.66$) between increasing ground coverage by a litter layer and reduction of soil erosion losses in coffee plantations including shading plants such as *Musa* spp. and *Inga* spp. in northern Nicaragua. The important role of litter in these agroforestry systems was highlighted even more in the case of pronounced terrain slopes (>50%) where at least 60% of ground litter cover is desirable to mitigate soil erosion by rainfall. Additionally, in such partially shaded coffee plantations, Soto-Pinto et al. (2000) established that a 38–48% shade optimized coffee productivity in Chiapas. Cusack and Montagnini (2004) reported that intermediate tree canopy density led to optimal understory regeneration in silvopastoral systems in Costa Rica.

Collectively, these studies point out the beneficial contributions of trees for developing sustainable land-use systems in tropical environments. Conversely, Meylan et al. (2017) found neutral effects of varying degrees of shading on productivity of coffee under *Erythrina* spp. trees. However, in keeping with Blanco Sepúlveda and Aguilar Carrillo (2015), Meylan et al. (2017) reported other benefits of the leguminous *Erythrina* spp. component such as increasing the litter ground cover and water infiltration in the soil profile as well as N fixation and transfer from *Erythrina* trees to coffee plants. They underscored these synergistic effects as important ecosystem services that reduce soil erosion and also support the overall conservation of the soil resource. Meylan et al. (2017) concluded that these tree contributions enabled resilience and sustainability of the production system as a whole, even under intensive management in the steep highlands of Costa Rica.

## *Multifunctionality and Adoption*

As a simple example of effective agroforestry practices in tropical landscapes, living fences (e.g., *Erythrina* spp.) have been found to contribute to soil C storage (Soto-Pinto et al. 2010), and serve as a biomass source (Budowski and Russo 1993) and as wildlife refugia (León and Harvey 2006). Likewise, as another agroforestry practice, fodder banks also provide multiple products and services (biomass, fuelwood, timber, fiber, etc.) (Montagnini et al. 2013). This multifunctionality of agroforestry systems substantiates their value and supports their adoption. However, implementation of agroforestry in tropical regions can be utterly constrained by a wide range of technical and socioeconomic limiting factors, including the availability of appropriate knowledge on how to establish and manage agroforestry systems. Following a survey evaluation of the key factors conditioning the adoption of alley cropping in Haiti, Bayard et al. (2007) found that training land managers on the topic of soil conservation practices greatly favors the implementation and maintenance of alley cropping systems. As alley cropping can mitigate soil erosive losses and land degradation, training on this underlying aspect resulted in decisive engagement by land managers in Haiti. This training seemed to trigger the establishment and retention of alley cropping systems in their farms. Bayard et al. (2007) also pointed to socioeconomic factors that drive agroforestry adoption such as gender, age, income, and farm size.

## Agroforestry in Africa and Asia

### *Soil Conservation and Agroforestry in African Drylands*

The desertification of soils, that is, the degradation of soil in semiarid, arid, and subhumid climates (drylands), is of special concern in many African countries. Desertification comprises sand encroachment and loss of topsoil by wind and water erosion as well as salinity, acidification, and pollution of soils by irrigation and fertilizer mismanagement (UNCCD 2017). Agroforestry has been recognized as one tool to combat desertification, and many large-scale agroforestry projects have been developed in African drylands for this purpose. In 2002, the Pan African Agency for the Great Green Wall was founded, with the initial aim to plant a 15-km-wide shelterbelt of trees on 8000 km across the Sahel to combat desertification. Approximately 20 Mha of land has reportedly been restored, and millions of trees were planted in Ethiopia, Senegal, Nigeria, Sudan, Burkina Faso, Mali, and Niger. There are attempts to modify the Great Green Wall by including shrubs, silvopastoral agroforestry systems, and apiculture in conjunction with shelterbelts (O'Connor and Ford 2014; UNCCD 2017). Another example is the tree planting efforts of small-scale farmers in Niger, who fostered natural regeneration of 200 million trees on 5 Mha of agricultural land (Sendzimir et al. 2011; Reij and Garrity 2016). Similar attempts of natural regeneration have been reported in Ethiopia and Senegal (Reij and Garrity 2016). Including trees for soil conservation in West African drylands reportedly increased crop yields, especially on marginal soils and under low-precipitation regimes (Bayala et al. 2012). Agroforestry also improved eroded terraces in Uganda, where improved fallows are planted on eroded terraces (Siriri et al. 2013).

Despite these efforts, there is an ongoing debate about whether or not agroforestry systems have beneficial or adverse effects on soils in African drylands. For example, trees in drylands may reduce soil moisture, lower groundwater tables, and reduce recharge rates owing to the high water consumption of trees, deep-ranging taproots, and fine lateral roots at the soil surface. In contrast, including trees in the landscape may increase water-use efficiency of the agroecosystem owing to reduced crop water loss by evaporation. Tree planting can also increase preferential flow and infiltration while decreasing surface runoff (Ong et al. 2006, 2007; Bargués Tobella et al. 2014; Mwangi et al. 2016). The success or failure of soil conservation by agroforestry in African drylands depends on design, density, species and phenology, age, and pruning management of the trees. This complexity is further increased by climate and available soil and water resources, which may differ on a regional scale (Mwangi et al. 2016). For example, soil moisture was not severely affected in corn-pigeon pea-*Gliricidia sepium* systems in Malawi (Chirwa et al. 2007), and in young sorghum/sesame-*Acacia senegal* systems in Sudan (Raddad and Luukkanen 2007). However, Eucalyptus plantings reportedly reduced streamflow in semiarid areas in Africa (Ong et al. 2006). Tree species differ in their capability for soil conservation benefits such as soil water recharge, erosion control, and nutrient or C storage (Sinare and Gordon 2015; Bargués Tobella et al. 2017; Siriri et al. 2013). Again, regional differences have to be considered: *Grevillea robusta* is a promising species for subhumid climates in East African highlands but did not perform well in semiarid areas owing to its evergreen foliage (Ong et al. 2006). In addition, dense tree plantations are more likely to deplete groundwater reservoirs than sparsely planted parklands (Bargués Tobella et al. 2014; Ilstedt et al. 2016). Deciduous trees with leaf fall during the dry season may have less negative impact on the water budget than evergreen tree species. Also, shoot and root pruning can decrease competition to available water resources (Ong et al. 2006, 2007). The establishment of shelterbelts, windbreaks, shelter forests, and hedges has shown to reduce wind speed and sand encroachment while increasing soil moisture and decreasing soil temperature in the drylands of Sudan, Nigeria, and Kenya. However, the wrong design of the shelterbelts can increase wind speed (Stigter et al. 2002).

Recent studies on SOC show no clear trend on how agroforestry systems impact soil C and nutrients in African dryland soils. In general, agroforestry systems exhibit somewhat higher soil SOC, but the difference between treatments is often not significant owing to high variability. Gelaw et al. (2015) compared agrisilvicultural and silvopastoral agroforestry systems to pasture, rainfed, and irrigated production in Northern Ethiopia. Although rainfed agriculture showed lowest SOC values after 50 years of use, and agrisilvicultural SOC stocks were higher, the difference among systems was not significant. Demessie et al. (2013) found a massive reduction of SOC stocks after 50 years of land conversion from natural forests in both agroforestry and farmland in Southern Ethiopia. At the same time, soil pH and bulk density were not significantly different between agroforestry and monoculture. Marone et al. (2017) found declining SOC stocks in the order fallow > rangeland > parkland in Senegal, indicating that the integration of a few trees in crop production (i.e., parkland) may not be sufficient to increase or stabilize SOC. Many studies do not consider soil inorganic carbon (SIC) in agroforestry systems, which may be a substantial C pool in drylands. Carmi et al. (2017) found that $CO_2$ respired by tree roots was incorporated in soils as SIC in two Israeli forests. Such an effect may also be detectable in agroforestry systems in African drylands.

## *Agroforestry and Soil Conservation in Humid Zones of Africa*

Soils in humid areas of Africa are threatened by soil degradation such as soil erosion, nutrient depletion, and other factors. Agroforestry systems in humid zones in Africa are often considered to have beneficial effect on a range of soil physical and chemical parameters. Ketema and Yimer (2014) found higher soil moisture contents, porosity, and infiltration rates, and lower bulk density in agroforestry systems than in monoculture in Ethiopia. Zake et al. (2015) reported higher SOC and N in coffee-banana-agroforestry systems compared to banana monoculture in Uganda, while P, K, and pH were significantly lower. Tumwebaze and Byakagaba (2016) found higher SOC in coffee agroforestry systems than in coffee monoculture in Uganda. Kassa et al. (2017, 2018) reported significantly higher C, N, P, and K contents, as well as higher cation-exchange capacity and base saturation in agroforestry systems and natural forests than in monoculture in Ethiopia. Bajigo et al. (2015) found no difference in SOC among woodlot, home gardens, and parkland systems in Ethiopia, but did not compare these systems to monoculture. Lagerlöf et al. (2014) investigated the microbial carbon (MBC) and nitrogen (MBN) between monoculture and agricultural systems and found significantly higher values in woodlots in Kenya. However, there was no significant difference between agrisilvicultural and monocultural systems in MBC, MBN, and microbial community composition after 20–30 years of agrisilvicultural land use. Plant available phosphorus (Olsen P) was significantly higher in agroforestry systems, and soil pH indicated significantly higher acidity in monoculture than in woodlots. In addition, Negash and Starr (2015) found no significant differences between Enset-coffee, fruit tree-coffee agroforestry systems, and Enset monoculture. Dawoe et al. (2014) investigated the impact of forest conversion to cocoa agroforestry systems on soil health in humid lowlands in Ghana. While soil C and N concentration significantly decreased after 30 years of cultivation, P, K, Mg, and Ca concentration as well as pH did not change compared to natural forests. Similar to African drylands, research on the impact of agroforestry on soil conservation in humid areas has shown beneficial, no, or even adverse effects.

## *Agroforestry and Soil Conservation in Asia*

Agroforestry systems have a long tradition in Asia, and they have been recognized for the ability to enhance or protect soil resources. The conversion from monoculture to *Ginkgo biloba* agroforestry systems resulted in significantly higher SOC and organic carbon fractions than the monoculture pedants in subtropical China (Wang et al. 2015). Similar results have been reported from Nepal, where the conversion from monoculture to agroforestry significantly increased soil quality parameters, such as soil C, N, pH, base saturation, and cation-exchange capacity (Schwab et al. 2015). This is in accordance to Baral et al. (2013), who reported increased SOC in agroforestry systems than monoculture in Nepal. Agroforestry has also been promoted to improve the soil quality of rubber (*Hevea brasiliensis*) monocultures. The diversification of rubber by introducing other tree species has significantly increased SOC and N contents (Chen et al. 2017). However, the increase in SOC in agroforestry systems is often due to an increase in the labile fraction of soil C. For example, the introduction of 7-year-old N-fixing alder trees (*Alnus nepalensis*) in monoculture tea (*Camellia sinensis*) plantations in China increased the C content and the amount of biomass fungi and soil bacteria, which is an indicator of an increase in the labile C fraction (Mortimer et al. 2015). Agroforestry systems increased SOC, especially in the topsoil, in India after 26 years of cultivation and in comparison to monoculture. This increase was mainly connected to an incremented contents of different labile C pools (Ramesh et al. 2015). The results are in accordance with another study in Northern India, where the non-labile C fraction of poplar (*Populus deltoids*) and wheat (*Triticum aestivum*) agroforestry systems was lower than that in monoculture and natural vegetation. In addition, the total C pool was significantly lower than in uncultivated land, and not significantly different to monoculture (Benbi et al. 2015). This indicates that agroforestry systems cannot stabilize the soil quality parameters of uncultivated land, and that the increased SOC pool is susceptible to mineralization in these systems. Also, Chen et al. (2017) reported an increase in C and N bound in macroaggregates in rubber-based agroforestry systems compared to rubber monoculture, which suggests an increase of labile C. However, the authors also detected an increase in microaggregates, hence non-labile C, which is an indicator that there is a difference in the impact on labile C either among agroforestry systems or among monoculture cropping systems (i.e., control plots).

Agroforestry has also been used as a strategy to combat soil degradation induced by erosion. For example, the Great Green Wall project in China, which started in 1978 with the anticipated end in 2050, constructed a shelterbelt of 4.1 M $km^2$ with the objective to combat encroachment of the Gobi Desert, and reduce dust storms. Indeed, dust storm intensity significantly declined, and vegetation improved in the area (Tan and Li 2015). Agroforestry systems can also diminish the negative impact of soil erosion connected to rainfall events. Liu et al. (2016) investigated soil erosion induced by raindrops in different tea and rubber agroforestry systems in China and found that rubber agroforestry systems with low sub-canopies reduced splash erosion compared to open environments, while high canopies like rubber monocultures increased splash erosion. These rubber agroforestry systems also had a higher amount of water-stable aggregates, which indicates a reduced susceptibility to soil erosion (Chen et al. 2017). Agroforestry hedgerows in Indian uplands also successfully reduced soil erosion compared to the traditional monoculture system. The runoff and soil loss were lower in these hedgerow systems with a positive effect on soil C, P, and K (Jakhar et al. 2017).

## *Summary of Agroforestry Contributions in Neotropical Africa and Asia*

The ability of agroforestry systems in Africa and Asia to contribute to soil conservation depends on many factors, and agroforestry can have beneficial, no, or even adverse effect on different soil health parameters. Agroforestry systems can reduce the risk of wind and water erosion and reduce soil loss on steep upland soils and terraces. Agroforestry has been reportedly beneficial on soil chemical and physical parameters on degraded and/or marginal soils. In addition, the transition to agroforestry-based systems may increase soil health compared to highly unsustainable monoculture systems. In general, tropical and silvopastoral agroforestry systems have the highest C sequestration rates compared to other agroforestry systems and climate regimes (Feliciano et al. 2018). However, the increase in SOC is often connected to an increase in the labile C fraction, which may be susceptible to mineralization.

The precondition for the success of agroforestry systems in terms of soil health is the careful design of the system, including tree species, tree density, management, and cultural settings among others. However, it should be noted that, according to recent studies, agroforestry systems may not be able to compensate for the soil quality loss in natural vegetation. In addition, there is still a lack of scientific understanding under which circumstances agroforestry systems improve soils in terms of crop yield (Bayala et al. 2012), C sequestration (Luedeling et al. 2011; Nair and Nair 2014), and other soil chemical, physical, and biological parameters, especially on the African continent. There are even agroforestry systems that have not been fully investigated yet (Nair et al. 2017). The few meta-analysis and modeling approaches available are often limited by the high data variation and the lack of standardization in case studies (Luedeling et al. 2011; Feliciano et al. 2018; Nair and Nair 2014). This is further complicated by the nature of agroforestry systems of being highly variable, and the control plots used (in comparison to natural vegetation, and monoculture, or among agroforestry systems).

## Conclusions Regarding Agroforestry Contributions to Soil Conservation

Protection from erosion and soils with a high soil organic matter content are common characteristics of resilient production systems. These systems are less likely to experience soil degradation or nutrient depletion and more likely to successfully adapt to climate change. All agroforestry systems, by nature of their combination of perennial woody and nonwoody plants, provide increased protective cover for the soil surface that reduces the risk of soil loss through wind and water erosion. Enhanced microclimates, reduced soil disturbance, and increased organic inputs from leaf litter, root, and grazing animal manure decomposition have each been shown to increase SOC and enhance soil health.

There are many variations of the standard agroforestry systems, often including unique features designed to address specific priorities or regional concerns. While efficient production of food, fuel, and fiber is the primary motivation for adopting any agroforestry practice, the soil conservation benefits are also recognized as an important benefit of these systems. Whether seeking to restore productivity to degraded lands, improve productivity of marginal lands, or increase the resiliency of established systems, agroforestry practices provide multiple means to protect and enhance soil quality. Land managers have the opportunity to optimize and integrate tree, crop, and livestock production with regard to soil and climate variation to achieve their production goals while sustaining the land resource and supplying multiple ecosystem services.

**Acknowledgement** Portions of the sections "Shelterbelts in Temperate Eurasia" and "Shelterbelts in Semiarid and Arid Eurasia" were completed under support of the Russian Science Foundation, grant no. 19-17-00056.

## References

Adhikari K, Owens PR, Ashworth AJ, Sauer TJ, Libohova Z, Richter JL, Miller DM (2018) Topographic controls on soil nutrient variations in a silvopasture system. Agrosyst Geosci Environ. https://doi.org/10.2134/age2018.04.0008

Agroforestry and Soil Fertility (1991) Edit by ES Pavlovskii. Agropromizdat, Moscow. 288p (In Russian)

Aguiar MI, Maia SMF, Xavier FAS, Mendonza ES, Filho JAA, Oliveira TS (2010) Sediment, nutrient and water losses by water erosion under agroforestry systems in the semi-arid region in northeastern Brazil. Agrofor Syst 79:277–289. https://doi.org/10.1007/s10457-010-9310-2

Alley JL, Garrett HE, McGraw RL, Dwyer JP, Blanche CA (1999) Forage legumes as living mulches for trees in agroforestry practices—preliminary results. Agrofor Syst 44:281–291

Baer NW (1989) Shelterbelts and windbreaks in the Great Plains. J Forestry 87:32–36

Bai ZG, Dent DL, Olsson L, Schaepman ME (2008) Proxy global assessment of land degradation. Soil Use Manag 24:223–234

Bajigo ATM, Moges Y, Anjulo A (2015) Estimation of carbon stored in agroforestry practices in Gununo Watershed, Wolayitta Zone, Ethiopia. J Ecosyst Ecography 5.1

Baral S, Malla R, Khanal S, Shakya R (2013) Trees on farms: diversity, carbon pool and contribution to rural livelihoods in Kanchanpur District of Nepal. Banko Janakari, 23(1), 3–11. https://doi.org/10.3126/banko.v23i1.9462

Bargués Tobella A, Reese H, Almaw A, Bayala J, Malmer A, Laudon H, Ilstedt U (2014) The effect of trees on preferential flow and soil infiltrability in an agroforestry parkland in semiarid Burkina Faso. Water Resour Res 50:3342–3354

Bargués Tobella A, Hasselquist NJ, Bazié HR, Nyberg G, Laudon H, Bayala J, Ilstedt U (2017) Strategies trees use to overcome seasonal water limitation in an agroforestry system in semiarid West Africa. Ecohydrology 10

Barlow RJ, Hunt S, Kush JS (2016) The silviculture of silvopasture. In: Proceedings of the 18th biennial southern silvicultural research conference. USDA For. Serv. South Research Stn. Asheville, North Carolina. e-Gen. Tech. Rep. SRS-212. 614, p 285–287

Bayala J, Sileshi GW, Coe R, Kalinganire A, Tchoundjeu Z, Sinclair F, Garrity D (2012) Cereal yield response to conservation agriculture practices in drylands of West Africa: a quantitative synthesis. J Arid Environ 78:13–25
Bayard B, Jolly CM, Shannon DA (2007) The economics of adoption and management of alley cropping in Haiti. J. Environ. Manage. 84:62–70
Beliveau A, Lucotte M, Davidson R, Paquet S, Mertens F, Passos CJ, Romana CA (2017) Reduction of soil erosion and mercury losses in agroforestry systems compared to forests and cultivated fields in the Brazilian Amazon. J Environ Manag 203:522–532
Bendfeldt ES, Feldhake CM, Burger JA (2001) Establishing trees in Appalachian silvopasture: response to shelters, grass control, mulch, and fertilization. Agrofor Syst 53:291–295
Blanco Sepúlveda R, Aguilar Carrillo A (2015) Soil erosion and erosion thresholds in an agroforestry system of coffee (Coffea arabica) and mixed shade trees (Inga spp and Musa spp) in Northern Nicaragua. Agric Ecosyst Environ 210:25–35
Blazier MA, Gaston LA, Clason TR, Farrish KW, Oswald BP, Evans HA (2008) Nutrient dynamics and tree growth of silvopastoral systems: impact of poultry litter. J Environ Qual 37:1546–1558
Bot A, Benites J (2005) The importance of soil organic matter—key to drought resistant soil and sustained food production. Food and Agriculture Organization of the United Nations, Rome. 78p
Budowski G, Russo RO (1993) Live fence posts in Costa Rica: a compilation of the farmer's beliefs and technologies. J Sustain Agric 3:65–87
Burner DM (2003) Influence of alley crop environment on orchardgrass and tall fescue herbage. Agron J 95:1163–1171
Burner DM, Brauer DK (2003) Herbage response to spacing of loblolly pine trees in a minimal management silvopasture in southeastern USA. Agrofor Syst 57:69–77
Burner DM, Pote DH, Belesky DP (2009) Effect of loblolly pine root pruning on alley cropped herbage production and tree growth. Agron J 101:184–192
Benbi DK, Brar K, Toor AS, Singh P (2015) Total and labile pools of soil organic carbon in cultivated and undisturbed soils in northern India. Geoderma, 237(Supplement C), 149-158. https://www.sciencedirect.com/science/article/pii/S0016706114003346
Carmi I, Kronfeld J, Moinester M (2017) Sequestration of atmospheric carbon dioxide as inorganic carbon in the unsaturated zone under semi-arid forests. arXiv preprint arXiv:1702.05249
Chen C, Liu W, Jiang X, Wu J (2017) Effects of rubber-based agroforestry systems on soil aggregation and associated soil organic carbon: implications for land use. Geoderma 299:13–24
Chendev YuG, Novykh LL, Sauer TJ, Petin AN, Zazdravnykh EA, Burras CL (2014) Evolution of soil carbon storage and morphometric properties of afforested soils in the U.S. Great Plains. In: Soil carbon. Progress in soil science. Springer International Publishing, New York, p 475–482
Chendev YG, Sauer TJ, Gennadiev AN, Novykh LL, Petin AN, Petina VI, Zazdravnykh EA, Burras CL (2015a) Accumulation of organic carbon in Chernozems (Mollisols) under shelterbelts in Russia and the United States. Eurasian Soil Sci 48:43–53
Chendev YG, Sauer TJ, Ramirez GH, Burras CL (2015b) History of East European Chernozem soil degradation; protection and restoration by tree windbreaks in the Russian Steppe. Sustainability 7:705–724
Chirwa PW, Ong CK, Maghembe J, Black CR (2007) Soil water dynamics in cropping systems containing Gliricidia sepium, pigeon pea and maize in southern Malawi. Agrofor Syst 69:29–43
Cihacek LJ, Sweeney MD, Deibert EJ (1993) Characterization of wind erosion sediments in the Red River Valley of North Dakota. J Environ Qual 22:305–310
Clary WP (1979) Grazing and overstory effects on rotationally burned slash pine plantation ranges. J Range Manag 32:264–266
Clason TR (1995) Economic implications of silvopasture on southern pine plantations. Agrofor Syst 29:227–238
Clason TR (1999) Silvopastoral practices sustain timber and forage production in commercial loblolly pine plantations of northwest Louisiana, USA. Agrofor Syst 44:293–303

Cornelis WM, Gabriels D (2005) Optimal windbreak design for wind-erosion control. J Arid Environ 61:315–332

Cusack D, Montagnini F (2004) The role of native species plantations in recovery of understory woody diversity in degraded pasturelands of Costa Rica. For Ecol Manag 188:1–15

Danilov GG, Lobanov DA (1973) Agroforestry in forest-steppe zone. Lesnaya promyshlennost' Press, Moscow. 125p. (In Russian)

Dawoe EK, Quashie-Sam JS, Oppong SK (2014) Effect of land-use conversion from forest to cocoa agroforest on soil characteristics and quality of a Ferric Lixisol in lowland humid Ghana. Agrofor Syst 88:87–99

De Jong E, Kowalchuk TE (1995) The effect of shelterbelts on erosion and soil properties. Soil Sci 159:337–345

DeBruyne SA, Feldhake CM, Burger JA, Fike JH (2011) Tree effects on forage growth and soil water in an Appalachian silvopasture. Agrofor Syst 83:189–200

Demessie A, Singh BR, Lal R (2013) Soil carbon and nitrogen stocks under chronosequence of farm and traditional agroforestry land uses in Gambo District, Southern Ethiopia. Nutr Cycl Agroecosyst 95:365–375

Dhillon GS, Van Rees KCJ (2017) Soil organic carbon sequestration by shelterbelt agroforestry systems in Saskatchewan. Can J Soil Sci 97:394–409

DordrechtMaia SMF, Xavier FAS, Oliveira TS, Mendonza ES, Filho JAA (2007) Organic carbon pools in a Luvisol under agroforestry and conventional farming systems in the semi-arid region of Ceara, Brazil. Agrofor Syst 71:127–138. https://doi.org/10.1007/s10457-007-9063-8

Droze WH (1977) Trees, prairies, and people. Texas Woman's University, Denton, TX. 313p

Erusalimskii VI (2007) Influence of forest melioration on microclimate, soil fertility and yield in conditions of the Non-Chernozem Region South. Bulletin VV Dokuchaev Soil Science Institute 60:75–81. (In Russian)

FAO (2003) State of the World's Forests 2003. Rome, p 126

Fedorov SI, Gumerov FR (1990) Influence of windbreaks to morphologic and agrochemical properties of chernozems. In: Crop rotation, soil tillage and fertilization for cultivation of crops by intensive technology. Ufa, p 16–20 (In Russian)

Feliciano D, Ledo A, Hillier J, Nayak DR (2018) Which agroforestry options give the greatest soil and above ground carbon benefits in different world regions? Agric Ecosyst Environ 254:117–129

Fike JH, Buergler AL, Burger JA, Kallenbach RL (2004) Considerations for establishing and managing silvopastures. Forage and Grazinglands. https://doi.org/10.1094/FG-2004-1209-01-RV

Follett RF, Stewart BA (1985) Soil erosion and crop productivity. American Society of Agronomy, Madison, WI, 533p

Franzluebbers AJ, Stuedemann JA, Schomberg HH, Wilkinson SR (2000) Soil organic C and N pools under long-term pasture management in the Southern Piedmont USA. Soil Biol and Biochem 32:469–478

Garrett HE, Kerley MS, Ladyman KP, Walter WD, Godsey LD, Van Sambeek JW, Brauer DK (2004) Hardwood silvopasture management in North America. Agrofor Syst 61:21–33

Gelaw AM, Singh BR, Lal R (2015) Organic carbon and nitrogen associated with soil aggregates and particle sizes under different land uses in Tigray, Northern Ethiopia. Land Degrad Develop 26:690–700

Gillespie AR, Jose S, Mengel DB, Hoover WL, Pope PE, Seifert JR, Biehle DJ, Stall T, Benjamin TJ (2000) Defining competition vectors in a temperate alley cropping system in the Midwestern USA. 1. Production physiology. Agrofor Syst 48:25–40

Golubeva LA (1941) The influence of forest shelterbelts of different design on microclimate and snow accumulation. VNIALMI, Moscow (as cited in Van Eimern, 1964)

Gregorich EG, Greer KJ, Anderson DW, Liang BC (1998) Carbon distribution and losses: erosion and deposition effects. Soil Tillage Res 47:291–302

Guiracocha G, Harvey C, Somarriba E, Krauss U, Carrillo E (2001) Conservacion de la biodiversidad en sistemas agroforestales con cacao y banano en Talamanca Costa Rica. Agroforesteria en las Americas 8:7–11

Hagen LJ (1976) Windbreak design for optimum wind erosion control. In: Tinus RW (ed) Shelterbelts on the great plains, Great Plains Agricultural Council Publ. No. 78, p 31–36, 218

Haile SG, Nair VD, Nari PKR (2010) Contribution of trees to carbon storage in soils of silvopastoral systems in Florida. USA Global Change Biology 16:427–438

Hand AM, Bowman T, Tyndall JC (2017) Influences on farmer and rancher interest in supplying woody biomass for energy in the US Northern Great Plains. Agrofor Syst. https://doi.org/10.1007/s10457-017-0170-x

Hernandez-Ramirez G, Sauer TJ, Cambardella CA, Brandle JR, James DE (2011) Carbon sources and dynamics in afforested and cultivated corn belt soils. Soil Sci Soc Am J 75:216–225

Hudson BD (1994) Soil organic matter and available water capacity. J Soil Water Conserv 49:189–194

Ilany T, Ashton MS, Montagnini F, Martinez C (2010) Using agroforestry to improve soil fertility: effects of intercropping on Ilex paraguariensis (yerba mate) plantations with Araucaria angustifolia. Agrofor Syst 80:399–409. https://doi.org/10.1007/s10457-010-9317-8

Ilstedt U, Tobella AB, Bazié H, Bayala J, Verbeeten E, Nyberg G, Sanou J, Benegas L, Murdiyarso D, Laudon H (2016) Intermediate tree cover can maximize groundwater recharge in the seasonally dry tropics. Sci Rep 6:21930

Jakhar P, Dass A, Adhikary PP, Sudhishri S, Naik BS, Hombegowda HC, Madhu M, Lenka NK, Chaudhary PR, Panda RK (2017) Multitier agroforestry system for integrated resource conservation on uplands of Eastern Ghats region in India. Agrofor Syst 91:697–712

Jose S, Gillespie AR, Seifert JR, Biehle DJ (2000) Defining competition vectors in a temperate alley cropping system in the Midwestern USA. 2. Competition for water. Agrofor Syst 48:41–59

Kallenbach RL (2009) Integrating silvopastures into current forage-livestock systems. In: agroforestry comes of age: putting science into practice. Proceedings of the 11th North American Agroforestry Conference, Columbia, MO, 31 May–3 June, p 455–461

Kallenbach RL, Kerley MS, Bishop-Hurley GJ (2006) Cumulative forage production, forage quality and livestock performance from an annual ryegrass and cereal rye mixture in a pine walnut silvopasture. Agrofor Syst 66:43–53

Karki U, Goodman MS (2010) Cattle distribution and behavior in southern-pine silvopasture versus open-pasture. Agrofor Syst 78:159–168

Karki U, Goodman MS (2015) Microclimatic differences between mature loblolly-pine silvopasture and open-pasture. Agrofor Syst 89:319–325

Karki U, Goodman MS, Sladden SE (2009) Nitrogen source influences on forage and soil in young southern-pine silvopasture. Agric Ecosyst Environ 131:70–76

Kassa H, Dondeyne S, Poesen J, Frankl A, Nyssen J (2017) Impact of deforestation on soil fertility, soil carbon and nitrogen stocks: the case of the Gacheb catchment in the White Nile Basin, Ethiopia. Agric Ecosyst Environ 247:273–282

Kassa H, Dondeyne S, Poesen J, Frankl A, Nyssen J (2018) Agro-ecological implications of forest and agroforestry systems conversion to cereal-based farming systems in the White Nile Basin, Ethiopia. Agroecol Sustain Food Syst 42:149–168

Kephart KD, Buxton DR (1993) Forage quality responses of C3 and C4 perennial grasses to shade. Crop Sci 33:831–837

Kephart KD, Buxton DR, Taylor ES (1992) Growth of C3 and C4 perennial grasses under reduced irradiance. Crop Sci 32:1033–1038

Ketema H, Yimer F (2014) Soil property variation under agroforestry based conservation tillage and maize based conventional tillage in Southern Ethiopia. Soil Tillage Res 141:25–31

Kibblewhite MG, Ritz K, Swift MJ (2008) Soil health in agricultural systems. Philos Trans Royal Soc B 363:685–701

Kort J (1987) Effect of shelterbelts on soil erosion as measured by cesium-137 levels. 1987 report-PFRA Shelterbelt Centre. Agriculture and Agri-Food Canada, Indian Head, Saskatchewan, p 37
Kretinin VM (1996) Forest melioration in agrolandscape of Novosil experimental station. In: collection of scientific works. 1(107). All-Russia Scientific Institute of Agroforestry, Volgograd, p 47–56 (In Russian)
Lagerlöf J, Adolfsson L, Börjesson G, Ehlers K, Vinyoles GP, Sundh I (2014) Land-use intensification and agroforestry in the Kenyan highland: impacts on soil microbial community composition and functional capacity. Appl Soil Ecol 82:93–99
Lehmkuhler JW, Kerley MS, Garrett HE, Cutter BE, McGraw RL (1999) Comparison of continuous and rotational silvopastoral systems for established walnut plantations in southwest Missouri, USA. Agrofor Syst 44:267–279
León MC, Harvey CA (2006) Live fences and landscape connectivity in a neotropical agricultural landscape. Agrofor Syst 68:15–26
Leue M, Lang F (2012) Recycling soil nutrients by using channel deposits as fertilizers? Nutr Cycl Agroecosyst 93:75–88
Li Y, Chen Y, Wang X, Niu Y, Lian J (2017) Improvements in soil carbon and nitrogen capacities after shrub planting to stabilize sand dunes in China's Horqin Sandy Land. Sustainability 9:662
Libohova Z, Seybold C, Wysocki D, Wills S, Schoeneberger P, Williams C, Lindbo D, Stott D, Owens PR (2018) Reevaluating the effects of soil organic matter and other properties on available water-holding capacity using the National Cooperative Soil Survey Characterization Database. J Soil Water Conserv 73:411–421
Lin CH, McGraw RL, George MR, Garrett HE (1999) Shade effects on forage crops with potential in temperate agroforestry practices. Agrofor Syst 44:109–119
Lin CH, McGraw RL, George MF, Garrett HE (2001) Nutritive quality and morphological development under partial shade of some forage species with agroforestry potential. Agrofor Syst 53:269–281
Liu W, Zhu C, Wu J, Chen C (2016) Are rubber-based agroforestry systems effective in controlling rain splash erosion? Catena 147:16–24
Ludvikova V, Pavlu VV, Gaisler J, Hejcman M, Pavlu L (2014) Long term defoliation by cattle grazing with and without trampling differently affects soil penetration resistance and plant species composition in Agrostis capillaris grassland. Agric Ecosyst Environ 197:204–211
Luedeling ESG, Beedy T, Dietz J (2011) Carbon sequestration potential of agroforestry systems in Africa. In: Kumar B, Mohan N, Ramachandran PK (eds) Carbon sequestration potential of agroforestry systems. Advances in agroforestry. Springer
Maia SM, Xavier FAZ, Oliveira TS, Mendonça ES, Araújo Filho JA. (2007). Organic carbon pools in a Luvisol under agroforestry and conventional farming systems in the semi-arid region of Ceará, Brazil. Agrofor. Syst. 71:127–138. https://doi.org/10.1007/s10457-007-9063-8
Marone D, Poirier V, Coyea M, Olivier A, Munson AD (2017) Carbon storage in agroforestry systems in the semi-arid zone of Niayes, Senegal. Agrofor Syst 91:941–954
Martsolf JD (1966) Microclimatic modification through shade induced changes in net radiation. Dissertation, University of Missouri in Columbia
Maryganova V, Szajdak LW, Tychinskaya L, Parmon S (2010) Chemical composition and hydrophobic-hydrophilic properties of humic acids from soils under shelterbelts of different age. In: Szajdak LW, Karabano AK (eds) Physical, chemical and biological processes in soils. Wydawnictwo-Drukarnia "Prodruk", Poznan, pp 359–372
McDonald MA, Lawrence A, Shrestha PK (2003) Soil erosion. In: Schroth G, Sinclair FL (eds) Trees, crops and soil fertility. CABI Publishing, Wallingford, UK, pp 325–343
Meneguzzo DM, Lister AJ, Sullivan C (2018) Summary of findings from the great plains tree and forest invasives initiative. Gen. Tech. Rep. NRS-GTR-177, U.S. Department of Agriculture, Forest Service, Northern Research Station, Newtown Square, PA. 24p. https://doi.org/10.2737/NRS-GTR-177
Menezes JMT, van Leeuwen J, Valeri SV, da Cruz MCP, Leandro RC (2008) Comparison of soils used for agroforestry and of remaining forests, in Northern Rondonia State, Brazil. Revista Brasileira de Ciencia do Solo 32:893–898. https://doi.org/10.1590/S0100-06832008000200043

Meylan L, Gary C, Allinne C, Ortiz J, Jackson L, Rapidel B (2017) Evaluating the effect of shade trees on provision of ecosystem services in intensively managed coffee plantations Agric. Ecosyst. Environ. 245 (2017) 32–42. https://doi.org/10.1016/j.agee.2017.05.005
Michel G-A, Nair VD, Nair PKR (2007) Silvopasture for reducing phosphorus loss from subtropical sandy soils. Plant Soil 297:267–276
Mikha M, Rice CW (2004) Tillage and manure effects on soil and aggregate associated carbon and nitrogen. Soil Sci Soc Am J 68:809–816
Mil'kov FN, Nesterov AI, Petrov PG, Skachkov BI (1992) Kamennaya steppe: agroforestry landscapes. Voronezh State University Press, Voronezh, 224p (In Russian)
Mirov NT (1935) Two centuries of afforestation and shelterbelt planting on the Russian steppes. J Forestry 33:971–973
Montagnini F, Ibrahim M, Murgueitio Restrepo E (2013) Silvopastoral systems and climate change mitigation in Latin America. Bois et Forêts des Tropiques 316:3–16
Montgomery DR (2007) Soil erosion and agricultural sustainability. Proc Natl Acad Sci 104:13268–13272
Mortimer PE, Gui H, Xu J, Zhang C, Barrios E, Hyde KD (2015) Alder trees enhance crop productivity and soil microbial biomass in tea plantations. Appl Soil Ecol 96:25–32
Mwangi HM, Julich S, Patil SD, McDonald MA, Feger K-H (2016) Modelling the impact of agroforestry on hydrology of Mara River Basin in East Africa. Hydrol Process 30:3139–3155
Naiman RJ (1988) Animal influences on ecosystem dynamics. Bioscience 38:750–752
Nair PKR, Nair VD (2014) 'Solid–fluid–gas': the state of knowledge on carbon-sequestration potential of agroforestry systems in Africa. Curr Opinion Environ Sustain 6:22–27
Nair PKR, Kang BT, Kass DCL (1995) Nutrient cycling and soil-erosion control in agroforestry systems. In: Juo ASR, Freed RD (eds) Agriculture and environment: bridging food production and environmental protection in developing countries. ASA Special Publ No 60, Madison, WI, p 117–138
Nair VD, Nair PKR, Kalmbacher RS, Ezenwa IV (2007) Reducing nutrient loss from farms through silvopastoral practices in coarse-textured soils of Florida, USA. Ecol Eng 29:192–199
Nair PKR, Viswanath S, Lubina PA (2017) Cinderella agroforestry systems. Agrofor Syst 91:901–917
Negash M, Starr M (2015) Biomass and soil carbon stocks of indigenous agroforestry systems on the south-eastern Rift Valley escarpment, Ethiopia. Plant Soil 393:95–107
Novykh LL, Chendev YG (2014) Change in the morphological properties of chernozems in an agrosilvicultural landscape. Arid Ecosyst 4:6–10
Nyakatawa EZ, Mays DA, Naka K, Bukenya JO (2012) Carbon, nitrogen, and phosphorus dynamics in a loblolly pine-goat silvopasture system in the Southeast USA. Agrofor Syst 86:129–140
O'Connor D, Ford J (2014) Increasing the effectiveness of the "Great Green Wall" as an adaptation to the effects of climate change and desertification in the Sahel. Sustainability 6:7142–7154
Oldeman LR (1994) The global extent of soil degradation. In: Greenland DJ, Szabolcs I (eds) Soil resilience and sustainable land use. CAB International, p 99–118
Olson KR, Al-Kaisi M, Lal R, Cihacek L (2016) Impact of soil erosion on soil organic carbon stocks. J Soil Water Conserv 71:61A–67A
Ong C, Black C, Muthuri C (2006) Modifying forestry and agroforestry to increase water productivity in the semi-arid tropics. CAB reviews: perspectives in agriculture, veterinary science. Nutr Nat Resour 1:1–19
Ong C, Anyango S, Muthuri C, Black C (2007) Water use and water productivity of agroforestry systems in the semi-arid tropics. Ann Arid Zone 46:255–284
Pearson HA, Whitaker LB (1974) Forage and cattle responses to different grazing intensities on southern pine ridge. J Range Manag 27:444–446
Pilon C, Moore PA Jr, Pote DH, Pennington JH, Martin JW, Brauer DK, Raper RL, Dabney SM, Lee J (2017) Long-term effects of grazing management and buffer strips on soil erosion from pastures. J Environ Qual 46:364–372
Pimentel D, Harvey C, Resosudarmo P, Sinclair K, Kurz D, McNair M, Crist S, Shpritz L, Fitton L, Saffouri R, Blair R (1995) Environmental and economic costs of soil erosion and conservation benefits. Science 267:1117–1123

Raddad EY, Luukkanen O (2007) The influence of different Acacia senegal agroforestry systems on soil water and crop yields in clay soils of the Blue Nile region, Sudan. Agric Water Manag 87:61–72

Ramesh T, Manjaiah KM, Mohopatra KP, Rajasekar K, Ngachan SV (2015) Assessment of soil organic carbon stocks and fractions under different agroforestry systems in subtropical hill agroecosystems of north-east India. Agrofor Syst 89:677–690

Raupach MR, Woods N, Dorr G, Leys JF, Cleugh HA (2001) The entrapment of particles by windbreaks. Atmos Environ 35:3373–3383

Reij C, Garrity D (2016) Scaling up farmer-managed natural regeneration in Africa to restore degraded landscapes. Biotropica 48:834–843

Romero-Alvarado Y, Soto-Pinto L, García-Barrios L, Barrera-Gaytán JF (2002) Coffee yields and soil nutrients under the shades of Inga sp. vs. multiple species in Chiapas. Mexico Agrofor Syst 54:215–224

Roth PL, Mitchell RJ (1982) Effects of selected cover crops on the growth of black walnut. In: Black Walnut for the Future. USDA For. Serv. North Central Forest Exp Stn St. Paul, MN. Gen. Tech. Rep. NC-74, p 110–113

Rousseau GX, Deheuvels O, Rodriguez Arias I, Somarriba E (2012) Indicating soil quality in cacao-based agroforestry systems and old-growth forests: the potential of soil macrofaunal assemblage. Ecol Indic 23:535–543

Russell JR, Bisinger JJ (2015) Grazing system effects on soil compaction in Southern Iowa pastures. Animal Industry report: AS 661, ASL R2987. Iowa State Univ., Ames

Ryszkowski L, Kedzior A (2007) Modification of water flows and nitrogen fluxes in shelterbelts. Ecol Engin 29:388–400

Sapanov MK (2016) Results of studies of soil moisture at the Dzhanybek station. Bull VV Dokuchaev Soil Sci Inst 83:22–39. (In Russian)

Sauer TJ Hernandez-Ramirez G (2011) Agroforestry. In: Hatfield JL, Sauer TJ (eds) Soil management: building a stable base for agriculture. Soil Science Society of America, Madison, WI, p 351–370, 409

Sauer TJ, Cambardella CA, Brandle JR (2007) Soil carbon and tree litter dynamics in a red-cedar-scotch pine shelterbelt. Agrofor Syst 71:163–174

Saxton KE, Rawls WJ (2006) Soil water characteristic estimates by texture and organic matter for hydrologic solutions. Soil Sci Soc Am J 70:1569–1578

Schmitt-Harsh M, Evans TP, Castellanos E, Randolph JC (2012) Carbon stocks in coffee agroforests and mixed dry tropical forests in the western highlands of Guatemala. Agrofor Syst 86:141–157. https://doi.org/10.1007/s10457-012-9549-x

Schroeder WR, Kort J (1989) Shelterbelts in the Soviet Union. J Soil Water Conserv 44:130–134

Schwab N, Schickhoff U, Fischer E (2015) Transition to agroforestry significantly improves soil quality: a case study in the central mid-hills of Nepal. Agric Ecosyst Environ 205:57–69

Sendzimir J, Reij C, Magnuszewski P (2011) Rebuilding resilience in the Sahel: regreening in the Maradi and Zinder regions of Niger. Ecol Soc 16

Sharrow SH (2007) Soil compaction by grazing livestock in silvopastures as evidenced by changes in soil physical properties. Agrofor Syst 71:215–223

Sibbald AR, Griffiths JH, Elston DA (1991) The effects of the presence of widely spaced conifers on under-storey herbage production in the U.K. For Ecol Manag 45:71–77

Silva-Pando FJ, Gonzalez-Hernandez MP, Rozados-Lorenzo MJ (2002) Pasture production in a silvopastoral system in relation with microclimate variables in the Atlantic coast of Spain. Agrofor Syst 56:203–211

Sinare H, Gordon LJ (2015) Ecosystem services from woody vegetation on agricultural lands in Sudano-Sahelian West Africa. Agric Ecosyst Environ 200:186–199

Siriri D, Wilson J, Coe R, Tenywa MM, Bekunda MA, Ong CK, Black CR (2013) Trees improve water storage and reduce soil evaporation in agroforestry systems on bench terraces in SW Uganda. Agrofor Syst 87:45–58

Soto-Pinto L, Perfecto I, Castillo-Hernandez J, Caballero-Nieto J (2000) Shade effect on coffee production at the northern Tzeltal zone of the state of Chiapas, Mexico. Agric Ecosyst Environ 80:61–69

Soto-Pinto L, Anzueto M, Mendoza J, Jimenez Ferrer G, de Jong B (2010) Carbon sequestration through agroforestry in indigenous communities of Chiapas, Mexico. Agrofor Syst 78:39–51. https://doi.org/10.1007/s10457-009-9247-5

Stigter CJ, Mohammed AE, Nasr Al-amin NK, Onyewotu LOZ, Oteng'i SBB, Kainkwa RMR (2002) Agroforestry solutions to some African wind problems. J Wind Engin Indust Aerodynam 90:1101–1114

Stoeckeler JH (1938) Shelterbelt planting reduces wind erosion damage in western Oklahoma. J Am Soc Agron 30:923–931

Sudmeyer RA, Hall DJM, Eastham J, Adams MA (2002) The tree-crop interface: the effects of root pruning in south-western Australia. Aust J Exp Agric 42:763–772

Szajdak LW, Życzyńska-Bałoniak I (2013) Effectiveness of a shelterbelt in decreasing the level of inorganic elements in agricultural landscape. Estonian J Ecol 62:24–34

Tan M, Li X (2015) Does the Green Great Wall effectively decrease dust storm intensity in China? A study based on NOAA NDVI and weather station data. Land Use Policy 43:42–47

Ticknor KA (1988) Design and use of field windbreaks in wind erosion control systems. Agric Ecosys Environ 22(23):123–132

Tumwebaze SB, Byakagaba P (2016) Soil organic carbon stocks under coffee agroforestry systems and coffee monoculture in Uganda. Agric Ecosyst Environ 216:188–193

UNCCD (2017) The Great Green Wall Initiative. http://www2.unccd.int/actions/great-green-wall-initiative

US Forest Service (1935) Possibilities of shelterbelt planting in the plains region. US Government Printing Office, Washington, DC. 201p

Van Eimern J (1964) Windbreaks and shelterbelts. World Meteorological Organization Technical Note No. 59, Geneva, Switzerland, p 188

Van Pelt RS, Zobeck TM, Ritchie JC, Gill TE (2007) Validating the use of $^{137}Cs$ measurements to estimate rates of soil redistribution by wind. Catena 70:455–464

Victoria R, Banwart S, Black H, Ingram J, Joosten H, Milne E, Noellemeyer E (2012) The benefits of soil carbon. In: United Nations Environmental Program (UNEP) Year Book 2012, p 19–33

Vomperskii SE, Dobrovol'skii GB, Sapanov MK, Sizemskaya ML, Sokolova TA (2006) A man-made forest oasis in a semi-desert. Her Russ Acad Sci 76:798–804. (In Russian)

Vyssotsky GN (1935) Shelterbelts in the steppes of Russia. J Forestry 33:781–787

Wang X, Oenema O, Hoogmoed WB, Perdok UD, Cai D (2006) Dust storm erosion and its impact on soil carbon and nitrogen losses in northern China. Catena 66:221–227

Wang G, Welham C, Feng C, Chen L, Cao F (2015) Enhanced soil carbon storage under agroforestry and afforestation in subtropical China. Forests 6:2307

Wardle DA, Bardgett RD, Klironomos JN, Setälä H, Van Der Putten WH, Wall D H (2004) Ecological linkages between aboveground and belowground biota. Science, 304(5677), 1629–1633

White PM, Rice CW (2007) Tillage effects on microbial and carbon dynamics during plant residue decomposition. Soil Sci Soc Am J 73:138–145

Zake J, Pietsch SA, Friedel JK, Zechmeister-Boltenstern S (2015) Can agroforestry improve soil fertility and carbon storage in smallholder banana farming systems? J Plant Nutr Soil Sci 178:237–249

Zhou X, Guan D, Wu J, Yang T, Yuan F, Musa A, Jin C, Wang A, Zhang Y (2017) Quantitative investigations of water balances of a dune-interdune landscape during the growing season in the Horqin Sandy Land, Northeastern China. Sustainability 9:1058

Zuo H, Miller-Goodman MS (2004) Landscape use by cattle affected by pasture developments and season. J Range Manag 57:426–434

# Soil Health Ecosystem Services of Agroforestry

**Robert J. Kremer**

## Abbreviations

| | |
|---|---|
| AMF | Arbuscular mycorrhizal fungi |
| PLFA | Phospholipid fatty acids |
| SOC | Soil organic carbon |
| SOM | Soil organic matter |
| WSA | Water-stable aggregation |

## Introduction

The modern era of agricultural production that is dependent on chemical fertilizer and synthetic pesticide inputs and high-yield crop varieties has contributed greatly to soil degradation, decreased soil organic matter, decreased water quality, increased greenhouse gas emissions, depletion of stratospheric ozone, and increased water use (Khan et al. 2007; Motavalli et al. 2008; Rani and Goel 2012). The continued excessive use of tillage under these production conditions leads to deterioration of the physical, chemical, and biological health of soil; potential decreases in crop productivity; as well as environmental degradation. More efforts are needed to optimize soil productivity in an environmentally sustainable approach that preserves the capacity of soil to function as a healthy system while protecting ecosystems.

Agroforestry offers complex ecosystems in contrast to monocultures of trees, single-species grasslands (i.e., pastures or cultivated forage crops), and agricultural and horticultural crops. Multiple provisioning services are accommodated by the tree crop and intercropped grain, food, forage, or pasture crops, with or without livestock. Agroforestry practices as integrated systems for efficient production of forestry, horticulture, agronomic, and/or livestock-related goods have evolved to

R. J. Kremer (✉)
University of Missouri, Columbia, MO, USA
e-mail: KremerR@missouri.edu



R. P. Udawatta, S. Jose (eds.), *Agroforestry and Ecosystem Services*,
https://doi.org/10.1007/978-3-030-80060-4_3

include and emphasize major roles of promoting soil conservation and soil health and providing ecosystem services. Soil health benefits and ecosystem services associated with agroforestry have received more attention recently as established systems consistently demonstrated long-term crop and soil productivity (Dollinger and Jose 2018).

The objective of this chapter is to summarize recent studies on agroforestry systems in which soil health parameters were assessed and related to ecosystem function and suggest adjustments in management that have potential to further improve the provision of ecosystem services.

## Ecosystem Services

Ecosystem services are benefits that society derive from ecosystems and are manifested as supporting, provisioning, regulating, and cultural services connected to human well-being and sustainability (Daily et al. 2009). Polasky et al. (2011) suggest that ecosystem services include carbon sequestration due to its impact on climate regulation; nutrient retention due to impact on water quality; water flow and infiltration because of roles in plant uptake, flood, and drought mitigation; and provision for agricultural production (e.g., soil productivity, pollination). Ensuring availability of robust ecosystem services under current land-use practices is critical for sustainable crop, forestry, and livestock production; developing bioenergy production systems; and maintaining or improving soil, water, and environmental quality. Changing land use or land management including agricultural and forestry practices may lead to changes in provision and value of ecosystem services (Polasky et al. 2011). Changes in land use or management often increase availability and value of some services but decrease others. For example, land management decisions for maximizing single outputs such as crop yields or timber production likely generate a simultaneous decline in provision of other services. Ecosystem services are generated and delivered through different ecosystem functions (Brussaard 2012), which are processes mediated primarily by the soil microbiome (Table 1). Maximum production of ecosystem services is associated with non-stressed agroecosystems, in which an undisturbed soil habitat supports a diversity of plants that provide ample C substrates to sustain the soil microbiome.

In addition to its status as a key soil health indicator, SOC plays a dynamic role in influencing the terrestrial climate and aquatic environment while providing essential ecosystem services (Brussaard 2012). Indeed, the capacity of soil to deliver a range of agricultural and ecosystem services including fertility maintenance and mitigation of atmospheric C emissions is dependent on SOC content (Feller et al. 2006). Thus, the loss of SOC through land degradation is a major factor in the deterioration of ecosystem services, which must be addressed with appropriate management practices for conservation of the critical ecosystem services provided by this important soil component (Feller et al. 2006).

**Table 1** Ecosystem services of agroforestry relative to soil health

| Ecosystem service | Soil health function | Ecosystem benefit | Comments |
|---|---|---|---|
| Provisioning; human acquisition | Productivity (yield) of agro- or natural ecosystem | Food, fiber, biofuel, potable water | Ecosystem management affects soil health and services |
| Erosion management | Stable soil aggregation | Maintained landscapes; viable aquatic habitats; clean water | Tillage, inorganic fertilizer use; clear-cutting lead to erosion |
| Soil aeration; water infiltration; water quality | Soil aggregation; bio-pore formation (porosity) | Water retention; maintain soil biological processes | Balanced fungi-bacteria communities and meso- and macrofauna |
| Carbon sequestration | Adequately distributed C pools within soil (SOC components) | Storage in both aboveground biomass and soil; soil water, air movement; low GHG emissions | Soil organic matter and microbial biomass buildup and maintenance are critical |
| Nutrient cycling | Mineralization via soil enzymatic activity; solubilization of minerals via microbial acid production | Nutrients recycled from organic residues and soil minerals; preserves water quality | High microbial diversity necessary; reduces/avoids inputs of chemical fertilizers |
| Xenobiotic degradation | Neutralizes nontarget effects of pesticides | Avoids harmful effects on soil organisms and functions | High microbial diversity necessary |
| Sustain biodiversity | Microbiome abundance and functional diversity; plant growth promotion | Plant and soil microbial diversity in balance for productivity and health | Diverse plant stands required to sustain diverse microbiome and functions |
| Adapting to climate change | C sequestration; presence of microbes that transform and cycle C and N | Reduced GHG emissions; C and N transformation products remain within ecosystem | Diverse microbiome to include C and N transforming bacteria |

Information based on Broadhead (2015), Jose (2009), Udawatta et al. (2017)
*GHG* greenhouse gas, *SOC* soil organic carbon

## *Agroforestry and Ecosystem Services*

Development of modern agroforestry as a production system for practical use began in the 1970s (Udawatta et al. 2017). The practice has increasingly gained importance as beneficial impacts on ecosystem services were recognized for crop, tree, and livestock production and soil and environmental quality. Agroforestry provides a variety of spatial and temporal configurations of trees on the landscape that contribute to maximum production of ecosystem services. The great variety of agroforestry systems established within any geographic region makes it nearly impossible to assign general ecological principles that mediate the provision of environmental services (Gordon et al. 2009) or soil health. Despite differences in the nature and types of ecological services rendered by various agroforestry systems, some broadly defined benefits are common with all (Table 1). These include general soil improvement including productivity (due to incorporation of annual vegetation including tree litterfall) and carbon (C) storage, reduced soil erosion, improved soil structure due to presence of perennial root systems, expansion of biodiversity (both above- and belowground), improved water quality, shelter in livestock systems, enhanced nutrient cycling and nutrient-use efficiency, and potential for integrated pest management with reduced pesticide inputs (Gordon et al. 2009). The integration of trees, agricultural crops, and/or animals into an agroforestry system has the potential to enhance soil fertility, reduce erosion, improve water quality, enhance biodiversity, and sequester carbon (Garrett and McGraw 2000; Garrity 2004; Nair et al. 2009).

Agroforestry, as a land-use management practice based on integration of trees and/or shrubs with crops and/or livestock to achieve economic, conservation, and ecological goals (Gold and Garrett 2009), is considered to play a critical role in providing and maintaining ecosystem services and benefits ranging from farm to global scales (Jose 2009). Integrating multifunctional agroforestry practices with other cropping systems within a region has the potential to improve ecosystem services such as food, feed, fiber, and fuel production; C and nutrient cycling; and soil, water, and air quality. Production practices that promote ecosystem services are particularly important with increasing concerns of challenges to agriculture and forestry including high costs of production, environmental degradation, food security, and climate change.

Jose (2009) specifically classified the major ecosystem services of agroforestry as carbon sequestration, soil enrichment, biodiversity conservation, and air and water quality. Trees or shrub components of agroforestry systems increase carbon sequestration compared to crop monocultures or single-species pastures by storing significant amounts of C in aboveground biomass as well as belowground in SOM and root systems (Jose 2009). Agroforestry enhances and maintains long-term soil productivity and sustainability by enriching soil N stores using N-fixing shrubs/trees and alley crops and increasing other important nutrients especially by plant species associated with N-fixing bacteria and/or mycorrhizae. Biodiversity is conserved in agroforestry by providing habitats for species able to tolerate minimal disturbance; preserving germplasm of sensitive species; reducing the conversion of natural habitat by providing a sustainable alternative to traditional agricultural systems; providing other ecosystem services such as erosion control and water recharge, which prevents degradation and loss of surrounding habitat; and improving air and water quality through establishment of windbreaks and shelterbelts that limit the effects of wind and riparian buffers that minimize or eliminate the movement of soil particles and fertilizer and pesticide contaminants in runoff, helping to ensure clean water. A more detailed description of the ecosystem services provided by agroforestry that includes excellent supporting examples is presented by Udawatta et al. (2017). Agroforestry is recognized as a viable strategy to restore and sustain soil health. Considerable evidence has accumulated over the past 20 years showing positive impacts on SOC enrichment, plant nutrient availability, and enhanced soil microbial abundance and activity (Dollinger and Jose 2018).

## Soil Health

The concept of soil health was proposed by Doran (2002) as the capacity of soil to function, within ecosystem boundaries, in sustaining plant and animal productivity, maintaining or enhancing water and air quality, and promoting plant and animal health. Coleman et al. (1998) distinguish soil health from soil quality by suggesting that the health and balanced activity of all groups of organisms within an ecosystem are implicit and should be specifically noted as components of soil health. Lehman et al. (2015) noted the importance of microbial diversity and activity as the basis for soil function because soil health relies on diverse soil biological communities that support high levels of critical environmental services. Kibblewhite et al. (2008) considered soil health within the context of sustainable agriculture, which assures that agricultural production does not outweigh the provision of ecosystem services. Thus a "healthy agricultural soil is capable of supporting the production of food and fiber to a level and with quality sufficient to meet human requirements, *and* deliver ecosystem services essential to maintain environmental quality, quality of life for humans, animals, plants and conservation of biodiversity" (Kibblewhite et al. 2008). Soil health functions can be assigned for crop production (i.e., water infiltration capacity, storage/release of nutrients, disease suppression) but functions necessary for important ecosystem services (C sequestration, water quality maintenance, biodiversity enhancement) must also be provided (Stirling et al. 2016). Optimal soil health requires a balance between soil functions for productivity, environmental quality, and plant and animal health, all of which are greatly affected by management and land-use decisions. Good management practices that consider soil health must consider all functions, rather than focus on single functions, such as crop productivity (Doran 2002). In summary, soil health focuses on the living, dynamic nature of soil that incorporates the biological attributes of biodiversity, soil food web structure, ecosystem functioning, and intimate relationships of soil microorganisms with plants and animals (Kremer 2016).

Soil health is still an evolving concept based primarily on numerous indicators of chemical and physical properties but relatively few biological indicators for assessment of management impacts on our soil resource. Soil organic C (SOC) as part of soil organic matter (SOM) is considered the key soil health indicator based on its multifunctional nature serving as a reservoir of plant nutrients, formation of stable fractions contributing to soil aeration and water infiltration, and as a source of readily decomposable substrate for the majority of the soil microbial community (Doran and Smith 1987; Brussaard 2012). Although many of the crop production functions and most ecosystem services are driven by biological processes (Kibblewhite et al. 2008), microbial diversity and microbial functional groups are not used as standard soil health indicators in assessment models due to lack of sufficient databases and due to difficulty in devising in-field sampling methods that maintain in situ conditions (Lehman et al. 2015). Several soil health assessment models are available to evaluate the effects of land management on the soil resource (Morrow et al. 2016; Stott et al. 2013). However, these models are considered deficient in indicators representing important measures of soil properties such as soil loss (Morrow et al. 2016), disease suppression (Van Bruggen and Semenov 2000), and microbial diversity and biological/biochemical activity (Stott et al. 2013).

Analysis of supplemental biological indicators, or "bioindicators," is often included with currently available soil health assessments. Proposed bioindicators include soil C mineralization, active C (AC or permanganate-oxidizable C), water-extractable (soluble) C, soil enzymes, soil microbial community structure and biodiversity components, soil fauna (i.e., earthworms), and plant disease criteria (Killham and Staddon 2002; Morrow et al. 2016; Stott et al. 2013; Van Bruggen and Semenov 2000). Furthermore, relationships of soil microorganisms formed by intimate associations with plants (i.e., mycorrhizal symbioses) strongly suggest that they are also major contributors to soil health. Reference soils with similar inherent soil characteristics are analyzed concurrently to gauge potential changes in soil health due to different management systems (Zuber et al. 2020), which should include conventional agricultural sites for comparative assessment of impacts of agroforestry practices (Sparling 1997). Functioning soil microbiomes in healthy soils depend on interactions with properly functioning soil chemical and physical factors (Stirling et al. 2016); thus these factors must be considered in describing the impacts of agroforestry on soil health. The soil microbiome is the critical component in soil health and ecosystem services due to mediation of decomposition, formation of soil OM, and nutrient transformations, which influence soil chemical and physical properties that contribute to soil fertility, productivity, and sustainability.

## *Agroforestry and Soil Health*

The benefits of agroforestry practices on soil health are realized through an emphasis on improved productivity on a land-equivalent basis that provides opportunities for sustainable pest management and weed control, which result in added vegetative production for soil conservation and/or forage for livestock. Soil health improvement is often an incidental benefit associated, for example, with nitrogen fixation by intercropped legumes that also increase soil organic matter (Udawatta et al. 2017). Quantitative assessments of soil health indicators to support these benefits are reported for agroforestry practices that vary in types and number of both tree and associated crops, pasture, or other interplanting. Udawatta and Jose (2012) estimated C sequestration potential for four main agroforestry practices (silvopasture, alley cropping, windbreaks, and riparian buffers) in North America at 530 Tg C year$^{-1}$, which is about 2.65 times the amount sequestered by croplands.

The high input of C into soil is a primary means of enhancing soil health by influencing other indicators including soil organic matter quality, nutrient cycling, structural and functional characteristics of the soil microbiome, and sustaining ecosystem services.

In an agroforestry practice with a corn-soybean rotation no-tilled in the alleys between either grass or grass plus pin oak (*Quercus palustris*) buffer strips in northeast Missouri, the soil health indicators water-stable aggregation (WSA), SOC, total soil N, and selected soil enzymes were significantly higher in both grass and grass plus tree strips than in the continuously cropped alleys (Udawatta et al. 2008). Assessments of soil enzyme activities in ecosystems aid in quantifying specific soil biological processes. A follow-up at the same site revealed that soil physical properties including bulk density were improved under agroforestry buffer management, demonstrating the linkage of biological, chemical, and physical indicators in overall soil health (Udawatta et al. 2009). An alley cropping agroforestry practice with eastern cottonwood trees (*Populus deltoides*) plus tall fescue grass (*Schedonorus arundinaceus*) buffers, tall fescue grass buffers, and permanent pasture alleys of tall fescue plus forage legumes, which were subjected to intermittent grazing by cattle, and a row-cropping (corn-soybean rotation) system were evaluated for comparative effects on soil health (Paudel et al. 2011, 2012). All soil health indicators and biological activities were highest in the alleys with perennial vegetation relative to the row crop that resulted from accumulation of greater SOC and total N, which was strongly correlated with soil enzymatic activities. High enzymatic activities in the grazed pasture alleys (up to 2.5 times higher for soil dehydrogenase activity) demonstrated the beneficial impacts of including livestock in the silvopasture component of this agroforestry practice.

Alley cropping agroforestry with perennial alley species is most effective in restoring and maintaining soil health compared to similar systems in which alleys are planted with row crops, especially when tillage and chemical inputs are used and no cover crop (vegetation established as a conservation cover after "cash crops" are harvested) is planted. In a pecan (*Carya illinoinensis*) agroforestry practice with alleys intercropped to the perennial legume, kura clover (*Trifolium ambiguum*), soil health indicators including SOC content, soil enzyme activities, and stable soil aggregates improved compared with no or unmanaged vegetation in alleys of adjacent trees (Kremer and Kussman 2011). A 10-year-old fruit tree alley cropping system with multiple perennial native grasses and forbs revealed increases in SOC, aggregate stability, and soil enzyme activities by 30%, 70%, and 30%, respectively, over a 6-year period compared with an adjacent orchard with alleys of tall fescue (Kremer and Hezel 2013; Kremer et al. 2015). This ecologically managed agroforestry practice demonstrated how the use of native perennial vegetation optimizes horticultural crop production while promoting soil health and soil conservation and many other ecosystem services including C sequestration and habitats for pollinating arthropods without inputs of inorganic fertilizers or synthetic pesticides.

A limited number of studies show that microbial diversity essential to soil health and ecosystem services are promoted by agroforestry. Generally, practices resulting in increased microbial diversity also show similar trends for increased soil microbial activity. Mycorrhizae, detected using phospholipid fatty acid (PLFA) analysis of soils in an oak (*Quercus* spp.) agroforestry system, accumulated similar biomass densities in both the grass alleys and within the tree rows but were significantly higher by 33% than the adjacent cultivated row-crop field (Unger et al. 2013). Similar trends were found for dehydrogenase and fluorescein diacetate hydrolase, enzymes associated with soil microbial activity and indicative of the size of the microbial population. Soil microbial components (bacteria, fungi, protozoa) were 12–35% greater for the tree rows and grass alleys relative to the row-crop field. The long-term fruit tree with perennial native vegetation system revealed higher soil microbial biomass and increased selected microbial groups including gram-negative bacteria, important for plant growth promotion, and arbuscular mycorrhizal fungi (AMF) compared with trees and tall fescue alleys (Kremer et al. 2015). Zhang et al. (2018) investigated the impact of plant-tree associations on the soil fungal community using molecular techniques. In three production systems, one with barley (*Hordeum vulgare*) only and those with barley and tree species, *Populus euramevicana* or *Taxodium distichum*, fungal diversity was greater in the rhizosphere compared to bulk soil, but no difference in the fungal diversity between the different systems.

Soils of Canadian hedgerows (windbreaks) and woodlands exhibited significantly greater bacterial abundance, measured by 16S rRNA gene sequencing, than in adjacent annual row-crop fields (Banerjee et al. 2016). Bacterial community composition also differed among land-use practices, and the sites with trees increased all soil C fractions suggesting greater species richness and diversity. Dobo et al. (2018) compared AMF diversity and spore density in the agroforestry practice of forest farming in Ethiopia based on three economically important tree species interplanted with either or both coffee (*Coffea arabica*) and ensete (Ethiopian banana; *Ensete ventricosum*). A slight effect of system design on both density and diversity of AMF was detected, with tree-ensete having greater AMF diversity and density than tree-coffee and multiple-species cropping systems. Forested riparian buffers in the Sacramento Valley of California exhibited twice as much soil organic C compared to adjacent cultivated systems (Young-Matthews et al. 2010). As plant productivity increased in the riparian buffers, the soil microbiome including beneficial nematodes increased, which correlated with higher soil health scores and associated with more ecosystem functions including C sequestration, water quality, and belowground biodiversity.

The foregoing studies demonstrate that most agroforestry practices consistently promote the buildup and maintenance of SOC, soil biological activity, and increased microbial diversity that are critical components in improving soil health, many of which are ecosystem functions necessary for the expression of ecosystem services.

## Agroforestry Management to Sustain and Improve Soil Health and Ecosystem Services

Concerns about degradation of environmental quality and decreased food quality and safety associated with current industrial agricultural production have prompted many farmers to develop more ecologically based production systems to provide greater ecosystem services while economically producing nutrient-dense foods. Ecologically based farming conserves and improves the soil resource and protects environmental quality by using natural resources with little or no synthetic chemicals to minimize ecosystem disturbance while providing multiple ecosystem services (Daily et al. 2009). Regenerative agriculture is a component of ecological agriculture, defined as a biological system for growing food and restoring degraded land by increasing soil health (LaCanne and Lundgren 2018). Agroforestry falls well within ecologically based production systems based on concepts of restoring and enhancing biological diversity, soil conservation, effective productive land areas, and others (Gold and Garrett 2009). However, for some specific practices, better agroforestry design planning could further improve soil health and ecosystem services (Dollinger and Jose 2018). For example, expansion of temperate alley cropping systems to combine multiple tree and shrub species in a "woody polyculture" approach to include tree crops that produce food or fodder has been proposed to reduce climate change impacts by reducing land area required for production, and enhance C sequestration and other ecosystem services (Wolz et al. 2018). This section offers some potential management suggestions that may achieve better soil health and ecosystem services in selected agroforestry practices based on insights gained from previously reported research.

Alley cropping systems offer the greatest opportunity for overall improved soil health and ecosystem functioning. This is especially applicable where alleys are managed as conventional row cropping that use some form of tillage and are fallowed between harvest and next season's planting. Indeed, designation of "silvoarable" systems by some practitioners suggests that conventional production is the standard for use in parallel tree rows with annual crops (Beuschel et al. 2018; Cardinael et al. 2018). Thus, it is not surprising that alley cropping systems such as hybrid black walnut (*Juglans regia* X *nigra*) and durum wheat (*Triticum turgidum*), "cultivated annually," attain higher SOC and increased stable aggregates within the tree row compared with the alley (Cardinael et al. 2018). Similarly, poplar (*Populus* spp.) alley practices cropped with several cereal crops under "reduced tillage" using chisel and rotary harrow showed, not surprisingly, that standard soil health indicators and fungal abundance were higher in tree rows compared with intercropped soils, which was correctly attributed to the lack of tillage and higher soil porosity under trees (Beuschel et al. 2018). In some systems trees appear to enhance soil microbial activity leading to improved soil productivity by building residual C through additions of litter and fine roots during the early phases of agroforestry establishment. This was the case for a pecan (*Carya illinoinensis*)–cotton (*Gossypium hirsutum*) system compared to a pecan orchard or cotton monoculture;

however as the system matured, SOC levels decreased likely due to continuous annual tillage for cotton production that accelerated microbial respiration and subsequent SOM breakdown (Lee and Jose 2003).

No-tillage practices established for production of alley crops have decreased soil disturbance yet soil health and biological indicators have not greatly exceeded those in cultivated sites. For example, soil bacterial diversity measured in a 21-year-old agroforestry system of silver maple (*Acer saccharinum*) tree rows and row-cropped alleys in Missouri did not differ significantly between tree row and alley cropped soils (Bardhan et al. 2013). Even though alleys were established with a corn-wheat rotation no-till, tree roots overlapped into soils of alleys after 20 years causing a more uniform soil microbiome possibly due to increasing of the variety of carbon exudates from both trees and crops. However, the absence of vegetation (i.e., cover crops) between cropping seasons and use of inorganic fertilizers and pesticides during crop production likely suppressed the development of a soil bacterial community different from the tree row. Veum et al. (2012) found that SOC fractions preferentially metabolized by the soil microbiome and aggregate stability were enhanced in grass and tree plus grass buffers in an alley cropping practice compared to the no-till alley cropped to a corn-soybean rotation. They suggested that silt loam soils at this northeast Missouri location under no-till were more susceptible to erosion and SOM turnover compared with the grass and tree plus grass buffer strips.

Even though no-till was deployed at these and other sites (Bardhan et al. 2013; Udawatta et al. 2008), the apparent lack of cover crop integration with the alley crops, also absent at sites with conventional tillage (Lee and Jose 2003; Paudel et al. 2011), hindered improvement in many soil health and biological functions in an agroforestry practice expected to optimize multiple ecosystem benefits from the biophysical interactions provided by the tree and crop combinations (Gold and Garrett 2009). However, several practices with positive impacts on agronomic production can be integrated, especially in alley cropping, to complement the benefits of agroforestry and contribute to greater sustainability of the overall system (Table 2).

Soil disturbance caused by tillage degrades soil structure, disrupts the hyphal network and nutrient absorption mediated by AMF (Evans and Miller 1988), reduces microbial diversity and biomass (Lehman et al. 2015), and detrimentally affects biological functional activity that leads to reduced SOM and C sequestration (Mehra et al. 2018). Shifting to no-till in alley crop production offsets the negative effects of conventional tillage and, when combined with cover crops, not only maintains the advantages of annual crop production but also improves ecosystem services and soil health (Blanco-Canqui et al. 2015) without dramatically altering established agroforestry practices (Table 2). The addition of cover crops enhances the multifunctionality of agroforestry. For example, cover crops can sustain or increase crop yields and provide forage for livestock grazing and, in alleys established with perennial species, forage and/or seed may be harvested as supplemental income before the tree component becomes productive (Kremer and Kussman 2011).

Deliberate use of organic amendments such as manures and composts in crop production alleys provides available carbon substrates to optimize soil microbial

**Table 2** Management for improving soil health and maximizing ecosystem services in agroforestry

| Agroforestry practice | Integrated management | Enhanced ecosystem and soil health benefits |
|---|---|---|
| Alley cropping | No tillage | Reduce soil erosion; improve soil structure; increase soil microbial diversity, activity |
| Alley cropping | Integrate cover crops in postharvest to planting period of annual crops | Soil cover reduces erosion; increases water infiltration; increases microbial abundance, diversity; increases SOM; reduces fertilizer, pesticide inputs; provides pollinator habitat |
| Alley cropping | Apply organic amendments (manure, compost, biochar); maintain crop residue cover | Increase SOM; increase microbial abundance, diversity; plant nutrient source; cover stem erosion; protect/enhance beneficial insects |
| Alley cropping | Plant non-transgenic (non-GM) corn, soybean, cotton in annual cropping systems | Reduce potential effects of GM products (i.e., Bt) and glyphosate on soil microbiome |
| Alley cropping | Introduce perennial crops into rotation; establish woody polycultures | Increase SOM, soil aggregation, soil microbial diversity and activity, soil fertility |
| Alley cropping and silvopasture | Introduce perennial native grasses and forbs | Increase SOM, soil aggregation, soil microbial diversity and activity, soil structure and productivity due to deep-rooting systems |
| Alley cropping and silvopasture | Prescribed periodic burning of established native vegetation | Stimulate new root growth and exudation; increase microbial activity; reduce weeds, pests; reduce chemical inputs |
| Riparian, upland buffer strips | Integrate native, i.e., warm-season grasses, to increase plant diversity | Improve rhizosphere microbial activity; stimulate microbial degradation of pesticides in runoff trapped by dense root systems; increase C sequestration and microbiome and nematode diversity |
| Riparian, upland buffer strips | Amend grass vegetation with selected microbial inoculants | Native soil microbiome supplemented with selected microbes degrades pesticides/herbicides in runoff/sediments |
| Silvopasture, riparian buffers, windbreaks | Integration of livestock grazing | Increase C sequestration; increase SOM quantity and quality; enhance nutrient cycling; enhance soil microbiome diversity |

*GM* genetically modified, *SOM* soil organic matter, *SOC* soil organic carbon

functions and reduce or eliminate inorganic fertilizer inputs (Hatfield and Walthall 2015). Addition of small amounts of biochar to perennial alley vegetation may stimulate specific microbial groups and increase microbial biomass and biological activity (Kremer et al. 2015). Retention of crop residues linked with no-tillage in cropped alleys provides a physical barrier against soil erosion, a source of photosynthesized carbon for SOM formation, increased soil biological activities leading to improved soil structure, water infiltration, and soil tilth.

Selection of specific crops/cultivars, along with traits for aboveground production and belowground soil biodiversity, contributes to a full suite of ecosystem services and products in agroforestry systems. A preference for non-transgenic crops that avoids introduction of Bt toxins detrimental to nontarget insects' important soil food web function (Velmourougane and Blaise 2017) and that eliminates weed management dependent on glyphosate, which disrupts the beneficial rhizosphere microbiome (Kremer and Means 2009), will optimize the value of harvested annual crops and deliver improved ecosystem goods and services (Brussaard 2012). Use of selected native grass species in vegetative buffer strips can enhance the breakdown of herbicides such as atrazine by stimulating soil microbiome components capable of biodegradation (Lin et al. 2011). Recently introduced neonicotinoid insecticides for soil, seed, and foliar insect pests of many annual crops should be used with caution in alley cropping. Neonicotinoid residues persist in soils and detrimentally affect a broad range of nontarget soil and aquatic invertebrates, which potentially alters the diversity and activity of trophic groups within the soil food web, with ultimate consequences for overall ecosystem functioning (Kremer 2018).

## Conclusions

This chapter described how the traits of agroforestry as a sustainable land management and production practice result in considerably greater soil health and ecosystem services relative to the industrial agricultural production model. Successful practice of agroforestry culminates in the expression of high levels of soil health and biological indicators that coincide with many desirable ecosystem functions. Despite the unique compatibility of agroforestry within the landscapes of various ecosystems, improved and sustainable management within agroforestry practices can further mediate improvements in soil health and ecosystem services. Most of the resulting improvements center on increased abundance, diversity, and capability of the soil microbiome to release nutrients, increase nutrient bioavailability, improve plant resilience to environmental stress and disease, and improve various chemical and physical properties of soils (Brussaard 2012). Eventually the adaptation of new microbiological and molecular methods for screening soils and rhizospheres will allow better insight into mechanisms at the molecular level and enhance our understanding of agroforestry production efficacy and ecosystem preservation for use in future management decisions.

As for estimating the benefits and soil health provided by agroforestry systems, models such as InVEST could be applied to calculate the provision and value of ecosystem services and species habitat under alternative land-use scenarios (Polasky et al. 2011). To attain optimum levels of soil health and ecosystem services and overall sustainability, a full range of soil management elements (LaCanne and Lundgren 2018; Stirling et al. 2016), detailed in the previous section, must be integrated and followed judiciously within agroforestry farming systems:

- Continuous inputs of organic matter; maintain living roots in soil all year
- Permanent plant residue cover on soil surface

- Build diversity; diverse soil microbiome and diverse crop rotation sequence
- Limit soil disturbance; no or minimum tillage
- Integration of livestock

In summary and in agreement with Udawatta and Jose (2012) and Wolz et al. (2018), future field research should involve design criteria for appropriate configuration, species selection for highly productive tree crops and planting densities including woody polyculture and complementary crop combinations, and efficient management for the various agroforestry practices to optimize overall soil health and ecosystem services.

## References


Banerjee S, Baah-Acheamfour M, Carlyle CM, Bissett A, Richardon AE, Siddique T, Bork EW, Chang SX (2016) Determinants of bacterial communities in Canadian agroforestry systems. Environ Microbiol 18:1805–1816

Bardhan S, Jose S, Udawatta RP, Fritschi F (2013) Microbial community diversity in a 21-year-old temperate alley cropping system. Agrofor Syst 87:1031–1041

Beuschel R, Piepho H-P, Joergensen RG, Wachendorf C (2018) Similar spatial patterns of soil quality indicators in three poplar-based silvo-arable alley cropping systems in Germany. Biol Fert Soils. https://doi.org/10.1007/s00374-018-1324-3

Blanco-Canqui H, Shaver TM, Lindquist JL, Shapiro CA, Elmore RW, Francis CA, Hergert GW (2015) Cover crops and ecosystem services: insights from studies in temperate soils. Agron J 107:2449–2474

Broadhead J (2015) Competition and phenology in agroforestry. In: Ong CK, Black CR, Wilson J (eds) Tree-crop interactions, 2nd edn. CABI Publishers, Oxfordshire, UK, pp 191–220

Brussaard L (2012) Ecosystem services provided by the soil biota. In: Wall DH (ed) Soil ecology and ecosystem services. Oxford University Press, Oxford, UK, pp 45–58

Cardinael R, Guenet B, Chevallier T, Dupraz C, Cozzi T, Chenu C (2018) High organic inputs explain shallow and deep SOC storage in long-term agroforestry system—combining experimental and modelling approaches. Biogeosciences 15:297–317

Coleman DC, Hendrix PF, Odum EP (1998) Ecosystem health: an overview. In: Huang PM (ed) Soil Chemistry and Ecosystem Health. SSSA Spec. Publ. 52. Soil Science Society of America, Madison, WI, pp 1–20

Daily G, Polasky S, Goldstein J, Kareiva PM, Mooney HA, Pejchar L, Ricketts TH, Salzman J, Shallenberger R (2009) Ecosystem services in decision-making: time to delivery. Front Ecol Environ 7:21–28

Dobo B, Asefa F, Asfaw Z (2018) Effect of tree-enset-coffee based agro-forestry practices on arbuscular mycorrhizal fungi (AMF) species diversity and spore density. Agrofor Syst 92:525–540

Dollinger J, Jose S (2018) Agroforestry for soil health. Agrofor Syst 92:213–219

Doran JW, Smith MS (1987) Organic matter management and utilization of soil and fertilizer nutrients. In: Follet RF (ed) Soil fertility and organic matter as critical components of production systems. Soil Science Society of America, Madison, WI, pp 53–72

Doran JW (2002) Soil health and global sustainability: translating science into practice. Agric Ecosyst Environ 88:119–127

Evans DG, Miller MH (1988) Vesicular-arbuscular mycorrhizas and the soil-disturbance-induced reduction of nutrient absorption in maize. I Causal relations. New Phytol 110:67–74

Feller C, Manlay RJ, Swift MJ, Bernoux M (2006) Functions, services and value of soil organic matter for human societies and the environment: historical perspective. In: Frossard E, Blum

WE, Warkentin BP (eds) Function of soils for human societies and the environment. The Geological Society of London, London, UK, pp 9–22
Garrett HE, McGraw RL (2000) Alley cropping practices. In: Garrett HE, Rietveld WJ, Fisher RF (eds) North American Agroforestry: an integrated science and practice, 1st edn. American Society of Agronomy, Madison, WI, pp 149–188
Garrity DP (2004) Agroforestry and the achievement of the millennium development goals. Agrofor Syst 61:5–17
Gold MA, Garrett HE (2009) Agroforestry nomenclature, concepts, and practices. In: Garrett HE (ed) North American Agroforestry: an integrated science and practice. American Society of Agronomy, Madison, WI, pp 45–56
Gordon AM, Thevathasan NV, Nair PK (2009) An agroecological foundation for temperate agroforestry. In: Garrett HE (ed) North American Agroforestry: an integrated science and practice. American Society of Agronomy, Madison, WI, pp 25–41
Hatfield JL, Walthall CL (2015) Soil biological fertility: foundation for the next revolution in agriculture? Comm Soil Sci Plant Anal 46:753–762
Jose S (2009) Agroforestry for ecosystem services and environmental benefits: an overview. Agrofor Syst 76:1–10
Khan SA, Mulvaney RL, Ellsworth TR, Boast CW (2007) The myth of nitrogen fertilizer for soil carbon sequestration. J Environ Qual 36:1821–1832
Kibblewhite MG, Ritz K, Swift MJ (2008) Soil health in agricultural systems. Phil Trans R Soc Biol Sci 363:685–701
Killham K, Staddon WJ (2002) Bioindicators and sensors of soil health and the application of geostatistics. In: Burns RG, Dick RP (eds) Enzymes in the environment: activity, ecology, and applications. Marcel Dekker, New York, pp 391–405
Kremer RJ (2018) Pesticide use and biodiversity in soils. In: Reicosky D (ed) Managing soil health for sustainable agriculture, Monitoring and management, vol 2. Burleigh Dodds Science Publishing, Cambridge, UK, pp 247–263
Kremer RJ (2016) Soil health, quality & microbial diversity. Acres USA 46(12):38–43
Kremer RJ, Hezel LF, Veum KS (2015) Soil health improvement in an organic orchard production system in northwest Missouri. In: Proceedings of the organic agricultural research symposium, LaCrosse, WI, p 1–4
Kremer RJ, Hezel LF (2013) Soil quality improvement under an ecologically based farming system in Northwest Missouri. Renew Agric Food Syst 28:245–254
Kremer RJ, Kussman RD (2011) Soil quality in a pecan—kura clover alley cropping system in the Midwestern USA. Agrofor Syst 83:213–223
Kremer RJ, Means NE (2009) Glyphosate and glyphosate-resistant crop interactions with soil and rhizosphere microorganisms. Eur J Agron 31:153–161
LaCanne CE, Lundgren JG (2018) Regenerative agriculture: merging farming and natural resource conservation profitably. PeerJ 6:e4428. https://doi.org/10.7717/peerj.4428
Lee K-H, Jose S (2003) Soil respiration and microbial biomass in a pecan-cotton alley cropping system in Southern USA. Agrofor Syst 58:45–54
Lehman RM, Cambardella CA, Stott DE, Acosta-Martinez V, Manter DK, Buyer JS, Maul J, Smith JE, Collins HP, Halvorson JJ, Kremer RJ, Lundgren JG, Ducey TF, Jin VL, Karlen DL (2015) Understanding and enhancing soil health: the solution for reversing soil degradation. Sustainability 7:988–1027
Lin C-H, Lerch RN, Kremer RJ, Garrett HE (2011) Stimulated rhizodegradation of atrazine by selected plant species. J Environ Qual 40:1113–1121
Mehra P, Baker J, Sojka RE, Boan N, Desbiolles J, Kirkhan MB, Ross C, Gupta R (2018) A review of tillage practices and their potential to impact soil carbon dynamics. Adv Agron 150:185–228
Morrow JG, Huggins DR, Carpenter-Boggs LA, Reganold JP (2016) Evaluating measures to assess soil health in long-term agroecosystem trials. Soil Sci Soc Am J 80:450–462
Motavalli PP, Goyne KW, Udawatta RP (2008) Environmental impacts of enhanced-efficiency nitrogen fertilizers. Crop Manage. https://doi.org/10.1094/CM-2008-0730-02-RV

Nair PKR, Kumar BM, Nair VD (2009) Agroforestry as a strategy for carbon sequestration. J Plant Nutr Soil Sci 172:10–23
Paudel BR, Udawatta R, Anderson SH (2011) Agroforestry and grass buffer effects on soil quality parameters for grazed pasture and row-crop systems. Appl Soil Ecol 48:125–132
Paudel BR, Udawatta R, Anderson SH, Kremer RJ (2012) Soil quality indicator responses to row crop, grazed pasture, and agroforestry buffer management. Agrofor Syst 84:311–323
Polasky S, Nelson E, Pennington D, Johnson KA (2011) The impact of land-use change on ecosystem services, biodiversity and returns to landowners: a case study in the state of Minnesota. Environ Resour Econ 48:219–242
Rani A, Goel R (2012) Role of PGPR under different agroclimatic conditions. In: Maheshwari DK (ed) Bacteria in agrobiology: plant probiotics. Springer, Heidelberg, pp 169–184
Sparling GP (1997) Soil microbial biomass, activity and nutrient cycling as indicators of soil health. In: Pankhurst CE, Doube BM, Gupta VVSR (eds) Biological Indicators of Soil Health. CAB International, Wallingford, UK, pp 97–119
Stirling G, Hayden H, Pattison T, Stirling M (2016) Soil health, soil biology, soilborne diseases and sustainable agriculture. CSIRO Publishing, Clayton South, VIC, Australia
Stott DE, Cambardella CA, Karlen DL, Harmel RD (2013) A soil quality and metabolic activity assessment after fifty-seven years of agricultural management. Soil Sci Soc Am J 77:903–913
Udawatta RP, Jose S (2012) Agroforestry strategies to sequester carbon in temperate North America. Agrofor Syst 86:225–242
Udawatta RP, Gantzer CJ, Jose S (2017) Agroforestry practices and soil ecosystem services. In: Al-Kaisi MM, Lowery B (eds) Soil health and intensification of agroecosystems. Elsevier-Academic Press, San Diego, CA, pp 305–333
Udawatta RP, Kremer RJ, Garrett HE, Anderson SH (2009) Agroforestry buffer effects on soil physical properties and enzyme activity on a row-cropped watershed. Agric Ecosyst Environ 131:98–104
Udawatta RP, Kremer RJ, Adamson BW, Anderson SH (2008) Variations in soil aggregate stability and enzyme activities in a temperate agroforestry practice. Appl Soil Ecol 39:153–160
Unger IM, Goyne KW, Kremer RJ, Kennedy AC (2013) Microbial community diversity in agroforestry and grass vegetative filter strips. Agrofor Syst 87:395–402
Van Bruggen AHC, Semenov AM (2000) In search of biological indicators of soil health and disease suppression. Appl Soil Ecol 15:13–24
Velmourougane K, Blaise D (2017) Impact of transgenic Bt cotton on soil health. CAB Rev 12(046):1–8
Veum KS, Goyne KW, Kremer RJ, Motavalli PP (2012) Relationships among water stable aggregates and organic matter fractions under conservation management. Soil Sci Soc Am J 76:2143–2153
Wolz KJ, Lovell ST, Branham BE, Eddy WC, Keeley K, Revord RS, Wander MM, Yang WH, DeLucia EH (2018) Frontiers in alley cropping: transformative solutions for temperate agriculture. Glob Change Biol 24:883–894
Young-Matthews A, Culman SW, Sánchez-Moreno S, O'Geen AT, Ferris H, Hollander AD, Jackson LE (2010) Plant-soil biodiversity relationships and nutrient retention in agricultural riparian zones of the Sacramento Valley. California Agroforest Syst 80:41–60
Zhang Q, Zhang M, Zhou P, Fang Y, Ji Y (2018) Impact of tree species on barley rhizosphere-associated fungi in an agroforestry ecosystem as revealed by 18S rDNA PCRDGGE. Agrofor Syst 92:541–554
Zuber SM, Veum KS, Myers RL, Kitchen NR, Anderson SH (2020) Role of inherent soil characteristics in assessing soil health across Missouri. Agric Environ Lett 5:e20021. https://doi.org/10.1002/ael2.200210

# Belowground Services in Vineyard Agroforestry Systems

**Katherine Favor and Ranjith P. Udawatta**

## Abbreviations

| | |
|---|---|
| AGF | Agroforestry |
| CC | Cover crops |
| CEC | Cation-exchange capacity |
| CWSI | Crop Water Stress Index |
| K | Potassium |
| Ksat | Saturated hydraulic conductivity |
| N | Nitrogen |
| OM | Organic matter |
| P | Phosphorus |
| PAM | Plant available moisture |
| PD | Plant density |
| RD | Root density |
| VP | Vegetative potential |

## Introduction

Although many agroforestry (AGF) systems such as silvopasture, alley cropping, and windbreaks are increasingly being utilized, AGF's applications in viticulture have been severely overlooked. Modern viticulture is a largely monocultural practice, and as such, it relies heavily on the use of chemical fertilizers, herbicides, and pesticides (Dupraz et al. 2009). With climate change and environmental degradation on the rise, conventional vineyards, now more than ever, face the threats of reduced

K. Favor (✉)
National Center for Appropriate Technology, TS Glide Ranch, Davis, CA, USA

R. P. Udawatta
University of Missouri, Columbia, MO, USA
e-mail: udawattar@missouri.edu



R. P. Udawatta, S. Jose (eds.), *Agroforestry and Ecosystem Services*,
https://doi.org/10.1007/978-3-030-80060-4_4

soil fertility, heat stress, water scarcity, extreme climate events, and increased pest and disease pressure. Agroforestry, defined as the intentional cultivation of trees and crops in the same system (Nair 1993a), is one solution that can create more sustainable viticulture systems while simultaneously providing numerous other ecosystem services. Agroforestry has been practiced in vineyards for thousands of years and is the traditional method of cultivating grapevines; only recently, in the nineteenth century with the rise of mechanization, did vineyards shift to monocultures (Dupraz et al. 2009; Fabre 2014). Despite AGF's historical presence in vineyards, modern vineyards have not typically employed AGF practices due in part to anecdotal fears of above- and belowground competition between trees and vines (Trambouze and Goma-Fortin 2013; Zelba et al. 2016; Dupraz et al. 2018; Grimaldi 2018). Other than some vineyards in Argentina, Portugal, Spain, Nepal, Italy, Iran, and Greece, the practice is not very common (Amouretti 1988; Bartolucci and Dhakal 1999; Altieri and Nicholls 2002; Raj and Lal 2014; Wezel et al. 2014; Gholami et al. 2018; NPCS Board of Consultants and Engineers n.d.). However, today, AGF in vineyards is being looked to once again as a way to reduce pest pressure, boost soil fertility and quality, withstand the stresses of climate change, and reduce farmer reliance on agrochemicals while simultaneously benefiting the environment in various ways. This review examines specifically the belowground interactions that occur between trees and vines in vineyard AGF systems and the ways that they can address some of the problems facing modern viticulture.

## Problems in Modern Viticulture

Farming industries worldwide, including conventional viticulture, experience a range of problems that can jeopardize both the production of crops and the health of the environment as a whole. Globally, soil erosion is increasing at epidemic rates of 2.5% per year—a rate 10–40 times faster than the rate of soil renewal (Pimentel 2006; Borrelli et al. 2013). Viticulture is not immune to these losses; vineyards in the Bairrada wine region of Portugal have been shown to experience sediment loss at alarmingly high rates, up to 29 Mg ha$^{-1}$ year$^{-1}$, with total nitrogen (N) losses of up to 20 kg ha$^{-1}$ year$^{-1}$ (Ferreira et al. 2018). Similarly, bare-soil vineyards in an 8-year study in Tuscany, Italy, experienced N losses of 12.5 kg ha$^{-1}$ year$^{-1}$ and phosphorus (P) losses of 5 kg ha$^{-1}$ year$^{-1}$ (Napoli et al. 2017). Soil erosion results in the loss of soil organic carbon as well; in a study on vineyards in Sicily, Novara et al. (2018) found that soil organic carbon was lost at a rate of 0.20 Mg ha$^{-1}$ year$^{-1}$, and that total sediment loss was 16 Mg ha$^{-1}$ year$^{-1}$. Fertility losses such as these result in the need to apply high amounts of fertilizers and can cause real economic losses for farmers (Novara et al. 2018). Using data from vineyards in Northeastern Spain, economists estimated that the amount of N lost by normal, bare-soiled vineyards each year amounts to 2.4% of a vineyard's annual income, and that the amount of P lost each year amounts to 1.2% of annual income (Martínez-Casasnovas and Ramos 2006).

Heat stress—in particular, droughts and extreme heat —is another source of stress for vineyards and its prevalence is predicted to increase in the coming years with climate change (Pachauri and Meyer 2015). Heat stress has been shown to negatively affect grapevines and to result in significant yield reduction. High temperatures early in the season can cause reduced numbers of inflorescences and reduced fruit set (Dunn and Martin 2008; Pagay and Collins 2017) and high temperatures late in the season can cause fruit abscission (Stephenson 1981). A study by Greer and Weston (2009) found that vines which were exposed to daily daytime temperatures of 40 °C at flowering, fruit set, and veraison experienced significant flower abscission and a 35% reduction in photosynthesis. After exposure to temperatures above 40.6 °C for 3–4 days, grapes experienced delayed ripening as well (Dokoozlian 2016).

At the other extreme, abnormal climate change patterns can also cause grape losses due to unseasonal frost. Erratic warm temperatures earlier in the season cause grapes to exit dormancy before the last frost hits, which can result in desiccation of that year's crop completely (Gosme et al. 2019). This is increasingly becoming a problem for vineyards around the globe, and farmers are resorting to various measures to prevent frost damage, including lighting fires in their vineyards at night, burning straw to produce smoke and prevent radiative cooling, and using helicopters to invert warm air down onto vines (Gosme et al. 2019). Farmers who cannot afford these measures simply lose that year's crop.

In the coming years, climate change predictions estimate that both periods of drought and periods of extreme precipitation will increase (Di Carlo et al. 2019). Although it is difficult for drought to kill grapevines outright, drought can stunt vegetative growth, reduce fruit quality, and even suppress fruit production completely (Medrano et al. 2003; Charrier et al. 2018). In areas where vines are irrigated, excess drought can result in expensive water bills for farmers and even the drying up of groundwater (Cooley et al. 2015). Conversely, increased precipitation can also have negative impacts on the quality of wine (Di Carlo et al. 2019).

Monoculture vineyards, which make up the majority of vineyards in today's world, are extremely vulnerable to pests and plagues, and they are less able to recover from pest damage following disturbance (Francis et al. 2004). Without diversified habitat, diverse populations of natural pest predators cannot be supported; as such, vineyards experience high pest pressure. With increased pest pressure, farmers either experience reduced yield or are forced to rely heavily on the use of chemical pesticides, which are known to cause a variety of environmental and human health risks (Altieri et al. 2005; Nicholls et al. 2008; Nicolopoulou-Stamati et al. 2016; California Department of Pesticide Regulation 2017). Pesticides kill beneficial insects and natural pest enemies in addition to targeted pests, and pesticide application over time can produce resistant pests, resulting in an increased dependence on pesticides in the long run (Mahmood et al. 2015). With increased reliance on chemical pesticides, grape growers are forced to spend large parts of their income on external inputs, and thus, it is difficult to generate a sustainable profit at a small scale (Sellers and Alampi-Sottini 2016). For small wine grape farmers, the high-input demands of monoculture viticulture can be a major inhibitor to profit.

With growing concern over climate change, the environmental impacts of conventional viticulture, and rising production costs, sustainable viticulture solutions are needed now more than ever. Modern viticulture is both affected by and simultaneously contributes to environmental and economic problems. Viticultural practices must become more sustainable in order for the wine grape industry to continue to thrive in the coming years in the face of environmental changes.

## Agroforestry in Vineyards as a Sustainable Solution

Agroforestry is an integrated land management system that sequesters carbon, conserves biodiversity, contributes to increased air and water quality, alleviates poverty, reduces pest pressure, and increases food production, all while enriching and protecting the soil (Root 1973; Jose 2009). In terms of soil health, AGF systems as a whole have been proven to increase soil porosity; enhance microbial activity; enrich soil organic matter (OM); increase water infiltration, cycle nutrients, and buffer pH; reduce bulk density; reduce leaching; resist erosion; reduce farmers' reliance on chemical fertilizer; and improve other soil physical and chemical processes, resulting in an overall Soil Quality Index measurement higher than that of monoculture systems (Amacher et al. 2007; Thomazini et al. 2015; Udawatta et al. 2020).

Agroforestry has been shown to have positive applications specifically in vineyards as well. In terms of aboveground interactions between trees and grapevines in vineyard AGF systems, there are many proven advantages. Trees have been shown to reduce farmers' reliance on chemical pesticides by attracting beneficial insects and reducing pest populations in vineyards (Wilson et al. 2017). The windbreak effect of rows of trees in and around vineyards has been shown to cause vines to photosynthesize more, not only due to increased stomatal opening, but also due to the greater leaf area production that occurs when vines are sheltered from wind (Pienaar 2005). In an extensive study on vineyard AGF systems at the Restinclières research site in Montpellier, France, the shading effect from trees was found to significantly reduce vine heat stress without significantly reducing photosynthetically active radiation (Grimaldi et al. 2017; Grimaldi et al. 2019). On this same research site, trees were also found to reduce vine cold stress; it was discovered that trees create a "night mask" effect that reduces radiative cooling and shelters vines from frost (Gosme et al. 2019). In addition to all of these aboveground benefits, AGF systems also provide various other ecosystem services including purifying water, mitigating pollution, sequestering carbon, conserving biodiversity, and maintaining a beautiful landscape aesthetic (Garcia et al. 2018).

The benefits of AGF on the belowground parameters of water, nutrition, and rooting patterns in vineyard soils have also been studied, as this chapter reviews. What makes a soil suitable for growing grapes is dependent on many factors, including soil structure, available water-holding capacity, nutrient availability, OM quantity, bulk density, porosity, and pH (Thomazini et al. 2015). Trees have been proven to improve many of these belowground soil quality parameters in vineyards and, although more research on the belowground interactions between trees and grapevines must be done, there is growing evidence that incorporating trees into vineyards could play a valuable role in the future of viticulture in the coming years.

## The Effect of Trees on Soil Water Parameters in Vineyards

Premier wine grape production typically takes place in semiarid climates that receive little rainfall, most commonly in Mediterranean, maritime, and continental climate regions (Stevenson 2005). On average, vineyards consume 300–700 mm of water annually, which is higher than the annual precipitation in many of the areas where viticulture is practiced (Medrano et al. 2015). Although grapevines themselves are a drought-resistant species, because of the low rainfall that wine-growing regions tend to receive, conserving moisture is still of the greatest priority in most vineyards (Charrier et al. 2018). Little research has been done on the effect that trees have on soil water parameters in vineyards in particular; however, there is much research on the impacts of trees on soil water parameters in other agroecosystems.

### *Increased Water Conservation in Agroforestry Systems*

While competition for water between tree and crop species does occur, trees also conserve soil moisture in AGF systems through a variety of mechanisms. Shade from trees conserves soil moisture by decreasing temperature and solar irradiance levels, which results in decreased evaporation (Lin 2007). The mulching effect from tree litterfall and pruning also reduces evaporation by covering soil and reducing soil temperatures (Riha and McIntyre 1999). Both the mulching effect of trees and the simple presence of lateral tree roots reduce runoff as well, slowing the flow of water and resulting in higher infiltration rates (Riha and McIntyre 1999). The mulching effect of trees also reduces kinetic impact from rain; reduced kinetic impact from rain maintains surface soil structure intact and therefore sustains high water infiltration rates (Lanyon et al. 2004). Water infiltration is also influenced by the amount of macropores in the soil. Trees increase the quantity of macropores in soil by breaking up compacted soils with their roots and leaving behind old root channels that serve as passages for increased water infiltration (Young 1989a).

Trees increase soil water-holding capacity as well by improving overall soil structure. Trees improve soil structure by boosting both OM and microbial populations, each of which leads to the formation of water-stable aggregates that create micro- and mesopores in the soil, capable of holding increased amounts of water (Lal 1989). Agroforestry systems can increase OM by up to 100%, and on average, every 1% increase in OM increases soil available water-holding capacity by 1.9 mm 100 mm$^{-1}$, or 1.9% (Young 1989b; Minasny and McBratney 2015). Udawatta et al. (2011a) reported greater soil water recharge capability in AGF systems than in corn

(*Zea mays* L.)-soybean [*Glycine max* (L.) Merr.] rotations during rain events as well (Fig. 1). All in all, AGF systems are able to significantly increase soil moisture, water infiltration rates, water recharge, and water-holding capacity (Young 1989b), which in turn results in greater drought resistance and less reliance on irrigation (Shantz 1927).

## Competition Between Trees and Grapevines for Water

Despite the increased infiltration rates, increased water-holding capacity, reduced runoff, and reduced evapotranspiration due to the incorporation of trees in cropping systems, some competition for water between trees and crops in AGF systems is inevitable (Udawatta et al. 2011b, 2014, 2016). Although little research has been done on competition for water between trees and grapevines specifically, there is an abundance of research that has shown that competition for water between grapevines and other crops, including cover crops (CC), does exist, and that this competition can result in varying degrees of water stress (Celette and Gary 2013).

Excess competition can result in high levels of water stress, which, if great enough, can reduce both the number of bunches per vine, berry weight, and total yield per vine (McCarthy et al. 1983). Various studies have confirmed that excessive water stress reduces photosynthesis, because of both reduced leaf area and increased stomatal closure, which results in lower berry sugar levels (Winkel and Rambal 1993; Gómez-del-Campo et al. 2002; Schultz 2003). In a study on the effect of different irrigation treatments on Colombard grapevines, both fruit growth and vegetative growth were found to be inversely correlated with increases in water stress (Stevens et al. 1995). Additionally, when grapes experience significant water stress, sugar metabolism and flavor development are negatively affected as well (Jones and Webb 2010; Bondada and Keller 2012).

## Striking Water Stress Balance in Grapevines

Although excess competition can cause undesirable levels of water stress in grapevines, some water stress actually is desirable for high-quality wine grape growing. High water availability is considered undesirable when growing grapes because it promotes excess vigor in grapevines and diversion of resources from developing fruit to shoot tips (Wheeler and Pickering 2005). As stated by Lanyon et al. (2004), "Optimum berry quality is seldom achieved if vines are excessively vigorous," due to a number of reasons. Excess vigor manifests as higher leaf area, greater trunk growth, and excessive shoot growth rates (Wheeler and Pickering 2003). Excessive shoot growth rates subsequently cause high in-canopy shading, which can cause a

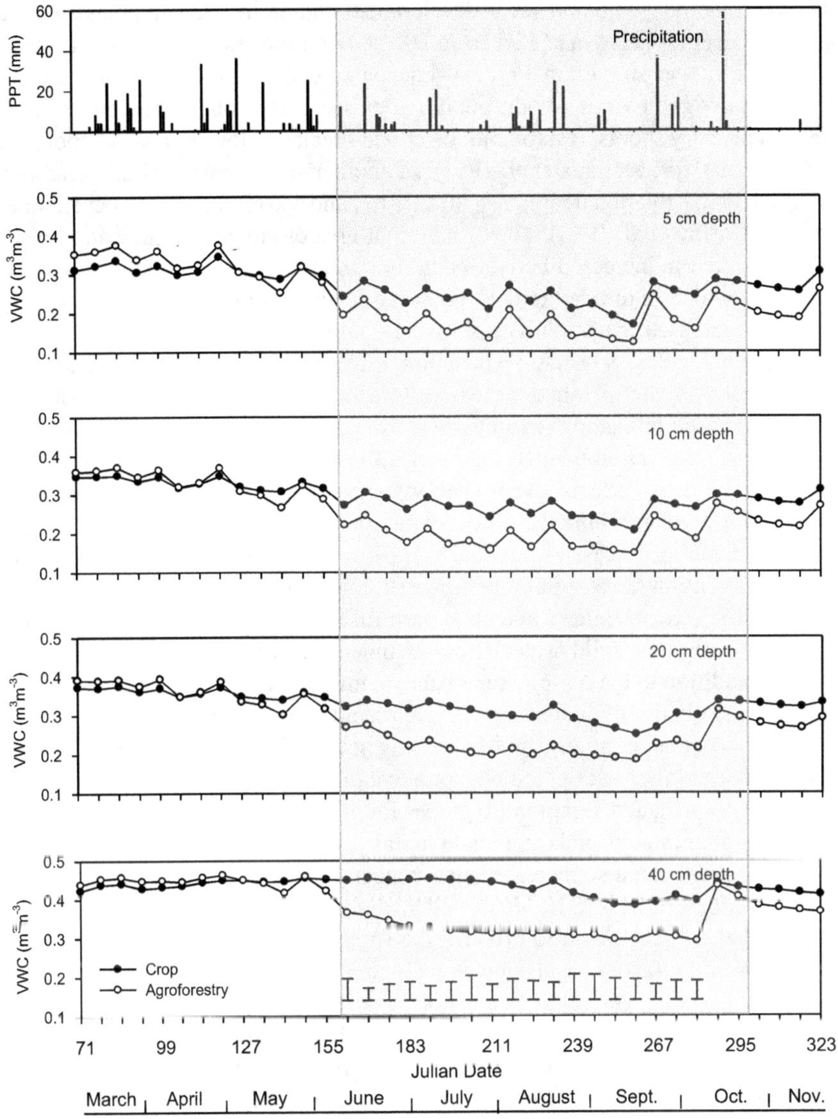


**Fig. 1** Daily precipitation and volumetric soil water content at 12:00 noon ($n = 4$) for crop and agroforestry (AGF) treatments for 5, 10, 20, and 40 cm depths during 2007 at the Greenley Research Center, University of Missouri, USA. Bars on the 40 cm depth graph indicate LSD values for significant differences in water content between crop and AGF treatments at the $\alpha = 0.05$ level. Source: Udawatta et al. (2011a) (reproduced with permission)

reduction in anthocyanin and sugar development and an increase in potassium (K) content and pH (Wheeler and Pickering 2005). High moisture levels affect overall yield as well; two studies in France—one on Grenache vines and the other on Cabernet Sauvignon vines—both found that excess water inhibits the bud burst of basal and primary shoots, resulting in lower bud break and lower yield (Carbonneau and Casteran 1979; Mériaux et al. 1981). In another study in Australia, researchers compared three irrigation treatments: 40%, 20%, and 0% replacement of evaporated water (McCarthy et al. 1983). They found that greater amounts of irrigation water applied resulted in increased berry weight, due to an increased amount of water in the berries, which in turn led to delayed sugar accumulation and diluted flavors. In this study, increased irrigation reduced wine quality as well; highly irrigated vines produced wine with less brilliant wine color, lower amounts of anthocyanins, lower total phenolics, higher pH, and increased K, which are all indicators of poor wine quality. Increased in-canopy shading—which was caused by increased vegetation, which was in turn caused by increased irrigation—was not the only culprit of these adverse wine quality effects; even when vigor was controlled for by applying the plant growth regulator, *ethephon*, poor wine quality was still observed with high levels of irrigation. Excessive vegetative growth can also indirectly reduce grape yield and quality by creating microclimatic humidity that causes vines to be more vulnerable to powdery mildew and other harmful fungi (Smart and Robinson 1991).

For these reasons, mild water stress is indeed desirable when growing wine grapes. In addition to the reasons stated above, mild water stress has been shown to improve wine quality by increasing the sugar:acid ratio, lowering malate and total titratable acid concentrations, and increasing total soluble solids (Vanzyl 1984). Mild water stress increases grape phenological profiles as well; a study comparing irrigated to nonirrigated Tempranillo grapes found that nonirrigated grapes had significantly higher total phenols and total tannins in grape skins (Esteban et al. 2001). Mild water stress can also increase sugar concentration in berries. In a study comparing the effects of 25%, 50%, 70%, and 90% soil moisture regimes, soil moisture regimes of 25% were found to produce the smallest berries and subsequently the highest concentrations of sugars and phenological compounds (Fig. 2).

Grape yield and wine quality are not negatively affected by moderate water stress, but they can be affected by the time at which water stress occurs. Water stress that occurs at certain periods within a vine's growth cycle can positively affect vines, while water stress that occurs at other periods can affect vines negatively (Vanzyl 1984). Mild water stress during the period from bud burst to flowering, for instance, can suppress shoot growth, which results in less vegetative growth and potential for higher wine quality (Vanzyl 1984). During flowering and phase I of berry development, however, grapes are very susceptible to water stress, and water stress can cause stunts in cell division, lower fruit set, and desiccation of clusters (Hardie and Considine 1976; Vanzyl 1984). After veraison, when cell division is no longer occurring, berry mass is not as sensitive to water stress (Vanzyl 1984), although extreme water stress can still result in failure of fruit to mature (Hardie and Considine 1976). In general, neither water stress nor water excesses after the period of veraison impact berry sugar accumulation. Sugar concentration might be

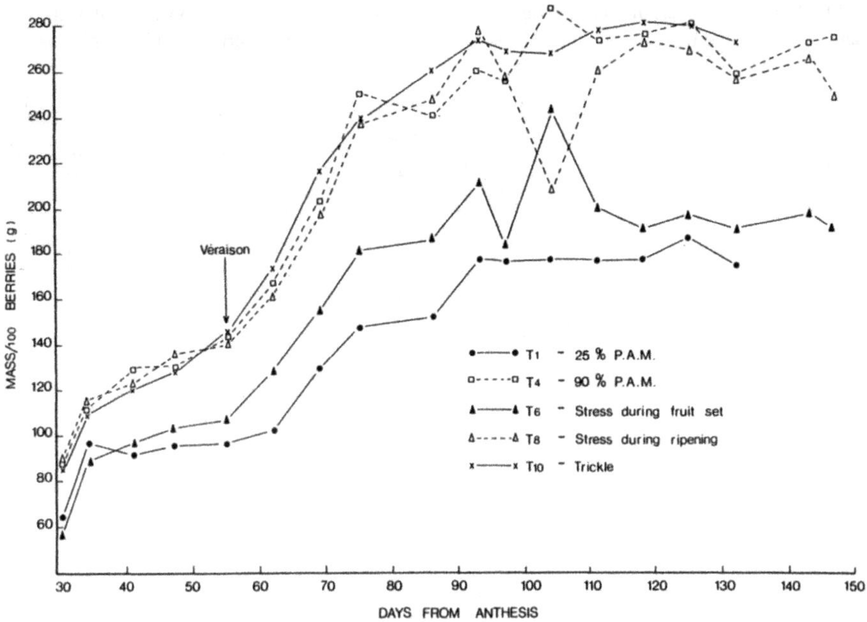


**Fig. 2** Effect of irrigation treatments of 25% plant available moisture (PAM), 90% PAM, 25% PAM stress during fruit set alone, 25% PAM stress during ripening alone, and trickle irrigation (concentrated irrigation) at 90% PAM on the cumulative berry mass of Colombard grapes during the 1979/1980 season in South Africa. Source: Vanzyl (1984) (reproduced with permission)

increased by water stress during the ripening period due to berry shrinkage, but actual sugar accumulation is affected neither by water deficiencies nor by excesses during this period in the grapevine growth cycle (Hunter et al. 2014).

With grapevine growth, striking the balance between too much water and too little water is of utmost importance. Vines must receive sufficient water at the right times in order to produce the minimum amount of vegetative growth that is needed to support fruit development and ripening, and in order to support cell development for sufficient yield. However, vines also must experience slight water stress so as to prevent excessive vegetative growth and so as to not divert nutrient sources away from fruit production (Wheeler and Pickering 2003). To illustrate, a study on Cabernet Sauvignon vines in California applied water in increasing amounts in four different treatments and found that vines receiving high amounts of water experienced delayed maturity and lower yield compared to vines receiving moderate amounts of water. However, vines receiving low amounts of water and vines receiving no water also had lower yields than the "moderate water" treatment (Neja et al. 1977). This study reflects the importance of balancing water stress in grapevines; some competition is a good thing, but too much competition can be detrimental.

For these reasons, viticulturists often employ techniques to actually cut back water to ideal stress levels and to induce slight water competition (Wheeler and Pickering 2005). Such soil-water-reducing techniques include regulated deficit

**Table 1** Effect of partial root drying on yield, water use, and fruit composition of Cabernet Sauvignon grafted to Ramsey. Source Wheeler and Pickering 2005 (reprinted with permission)

| Parameter | Control (fully irrigated vines) | Treatment (vines irrigated with partial root-zone drying) | Significance |
|---|---|---|---|
| Yield (kg vine$^{-1}$) | 4.73 | 4.88 | ns |
| Water-use efficiency (g fruit L irrigation$^{-1}$) | 4.9 | 7.2 | <0.01 |
| Total soluble solids (°Brix) | 22.8 | 22.9 | ns |
| pH | 3.44 | 3.26 | <0.05 |
| Titratable acidity (g L$^{-1}$) | 5.8 | 8.4 | <0.05 |
| Color (mg g fruit weight$^{-1}$) | 1.19 | 1.72 | <0.05 |
| Glycosyl-glucose (mol g fruit weight$^{-1}$) | 2.64 | 3.75 | <0.05 |

irrigation, partial root-zone drying, root pruning, high-density vine planting, and CC-induced competition (Wheeler and Pickering 2003; Wheeler and Pickering 2005). Competition for water and nutrients from tree roots in vineyard AGF systems is also speculated to be a valuable technique for inducing desirable levels of water stress. Such stress-inducing techniques result in higher water-use efficiency, more balanced acidity, more brilliant color (mg g fruit weight$^{-1}$), higher glycosyl-glucose (mol g fruit weight$^{-1}$), and increased perception of ripeness of aroma and flavor (Dry et al. 1996; Wheeler and Pickering 2003; Wheeler and Pickering 2005). Benefits of these techniques can be summarized in Table 1 and Figs. 3 and 4.

More research is needed to determine whether the competition for water between trees and vines in vineyard agroforestry systems would result in overall positive or negative effects for grapevines. This is a determination that would of course depend on the species of trees being intercropped, the soil available water, the architecture of both species' root systems (which is dependent on both species type and management practices), as well as the amount and timing of transpiration from each species. The amount of competition would also depend on management practices; trees that are pruned and/or root pruned and systems that are irrigated more would experience less competition (Sudmeyer and Flugge 2004).

Although much research has yet to be done in this area, in the gray literature there does exist an extensive study at the Restincliéres AGF site in Montpellier, France, in which competitive effects between certain types of trees and vines were quantified. In this study, Syrah and Grenache vines were intercropped with sorb (*Sorbus domestica* L.) and stone pine (*Pinus pinea* L.) in both N/S and E/W orientations, at both high (15 m × 2.5 m) and low (15 m × 3.75 m) tree densities. Early grapevine water stress was estimated using the apex method and late-stage soil water stress was quantified using environmental isotope hydrology. Between all treatments, all tree planting densities, and all row orientations, no negative effects from competition for water were observed between trees and grapevines (Trambouze and Goma-Fortin 2013). In a similar unpublished study, GreenSeeker technology was used to measure the Normalized Difference Vegetation Index of vines at different distances from fruit trees. No significant differences in vegetative growth were

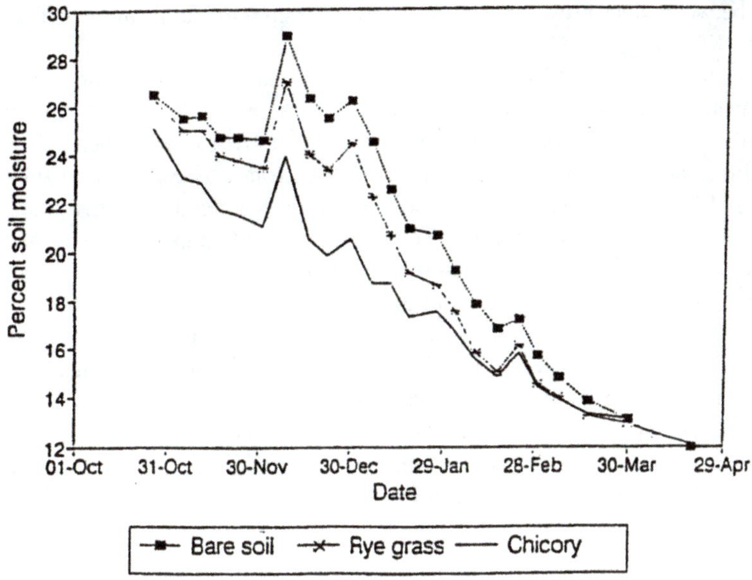


**Fig. 3** Effect of vineyard floor management on soil moisture levels in Cabernet Sauvignon vineyard in Hawke's Bay, New Zealand. Chicory and ryegrass (*Lolium perenne* L.) cover crop treatment resulted in lower soil moisture. Source: Wheeler and Pickering 2005 (reproduced with permission)

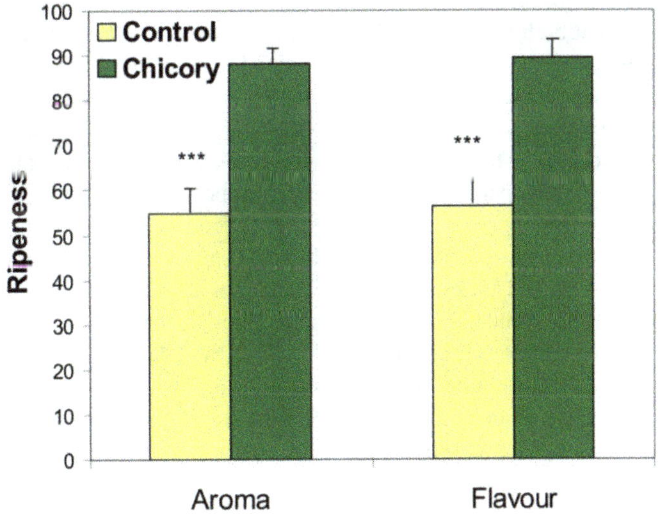


**Fig. 4** Effect of chicory cover crop on perceived ripeness of aroma and flavor in 4-year-old Cabernet Sauvignon wine in Hawke's Bay, New Zealand. Data shown are mean values [$n = 44$] + std. error; *** indicates that treatments are significantly different at $p < 0.001$. Source: Wheeler and Pickering 2005 (reproduced with permission)

noted between vines growing near trees and vines growing far from trees (Dufourcq et al. 2017).

In a similar yet different study at the Restinclières experimental site, data on the Crop Water Stress Index (CWSI) of vines in a vineyard AGF system was collected using thermal infrared imagery. Results showed that, overall, there were not significant differences in CWSI at different distances from a tree hedgerow (Grimaldi et al. 2017). Yet another study on the Restinclières viticulture experimental site found that competition for water between trees and grapevines was minimal, although competition for N was significant (Trambouze et al. 2017). Available literature suggests that the low amount of water stress in vineyard agroforestry systems may be due to trees' ability to redistribute water from deep in the ground through the process of hydraulic lift, reduce evaporative losses from the soil by modifying the climatic demand, reduce transpiration losses by creating a cooler microclimate, and increase water storage capacity by increasing soil OM and porosity (Smart et al. 2005; Trambouze et al. 2017; Grimaldi 2018). When competition for water does occur, it is slight, and can be remedied by root pruning with a subsoiler (Dupraz et al. 2009).

There is evidence that both tree roots and grapevine roots exhibit hydraulic redistribution, defined as the transfer of water from deep edaphic sources to drier soils (Smart et al. 2005). In both trees and grapevines, this process occurs both vertically (roots draw water up from deep profiles into shallower ones) and laterally (roots draw water from irrigated areas to nonirrigated areas) (Smart et al. 2005). This phenomenon allows both tree and vine roots to expand to unirrigated parts of the soil, allowing them to absorb nutrients and maintain strong anchorage across a broader area. The fact that both grapes and trees have the capacity for hydraulic redistribution is hypothesized as one of the reasons why low competition appears to exist between grapevines and trees in vineyard AGF systems (Smart et al. 2005; Grimaldi 2018).

Overall, the existing studies on AGF in vineyards suggest that trees have a neutral to positive effect on parameters surrounding grapevine water status. Grapevines are a drought-tolerant species which are capable of producing higher quality berries and higher yields under slight water stress (Carbonneau and Casteran 1979; Mériaux et al. 1981; Wheeler and Pickering 2005; Charrier et al. 2018). Although trees and grapevines do impart some levels of water stress through competition and root niche overlap, trees can also conserve water in vineyards by reducing evapotranspiration through increased shade and mulch, by increasing water infiltration through improvements in soil structure and water-holding capacity, and by distributing water from wet to dry zones through hydraulic distribution (Young 1989a, b; Morlat and Jacquet 1984; Riha and McIntyre 1999; Lanyon et al. 2004; Smart et al. 2005; Kailis and Harris 2007; Lin 2007; Bhadha et al. 2018). Given all trade-offs, research findings suggest that tree intercropping in vineyards would not induce damagingly high levels of water stress through competition for water (Dupraz et al. 2009; Grimaldi 2018). Studies have confirmed that the presence of trees in vineyards does not increase CWSI in vines (Grimaldi et al. 2017) and that competition for water is not responsible for reductions in fruit quality, vegetative growth, nor yield

(Trambouze et al. 2017). More studies on the effects of tree/vine competition for water are needed, especially ones that examine different tree species, grape trellis systems, row orientations, and layouts, in order to definitively quantify the effects of trees on grapevine water status.

## The Effect of Trees on Vineyard Nutrition Parameters

Conventional vineyards commonly face nutritional issues in soil due to low organic carbon levels, high levels of erosion, low microbial activity, and compaction (Pool et al. 1990; Garcia et al. 2018). More than other land-use systems, conventionally cultivated vineyards are considered one of the most erosion-prone land-use practices because of the lack of ground cover, high rates of tillage, and high levels of compaction associated with traditional management practices (Coll et al. 2011). Conventional vineyard floor management generally leads to impaired soil structure and reduced soil water-holding capacity (Biddoccu et al. 2017; Rodrigo-Comino et al. 2018), which in turn results in reduced biological activity and consequently diminishing levels of OM and nutrients over time (Pool et al. 1990). Many studies have proven the importance of incorporating CC and other service crops into vineyards to address these issues (Garcia et al. 2018), but few studies have evaluated the incorporation of trees specifically. The use of trees in AGF systems in general can have dichotomous effects on crop nutrition; trees can both cause nutrient stress due to increased competition between species and can increase nutrient availability through a variety of mechanisms.

### *Increased Nutrient Availability in Vineyard Agroforestry Systems*

Trees increase the soil nutrients available for crop uptake by increasing OM, cycling nutrients from deep soil profiles to shallow ones, fixing N (in the case of leguminous trees), and transforming nutrients into a more plant-absorbable form through increased microbial activity (Young 1989b). Although vineyards do not necessarily require high levels of N, they do perform better when there are adequate levels of soil OM and nutrients (Pool et al. 1990). Research suggests that the increased nutrient availability imparted by trees in vineyard AGF systems may balance out some of the competition for nutrients that occurs in these systems.

Agroforestry systems have the potential to increase soil OM by 50–100% (Young 1989b). They have been shown to return an average of 7.4 tons of OM per hectare per year in the form of pruning alone, and they also produce OM through litterfall, root slough, and root exudates (Nair 1993b; Schroeder 1993; Thevathasan and Gordon 2004). Nutrients that take the form of OM are released slowly at rates comparable to rates of plant absorption, and they are in a stable molecular form that is resistant to leaching (Young 1989b). Organic matter produced by trees serves as a source of food for microbes, which results in increased microbial populations; indeed, trees in AGF systems have been shown to increase soil microbiological activity by up to 30% (Young 1989a). Microbes excrete enzymes that mineralize nutrients, that stabilize carbon and N in the soil, and that decompose OM into simple, plant-available forms, resulting in higher plant nutrient uptake (Paudel et al. 2011; Adetunji et al. 2017). Increased OM in AGF systems also results in increased cation-exchange capacity (CEC), which translates to a greater ability of soil to hold onto exchangeable cations. This results in better retention of applied nutrients and resistance from nutrient leaching (Young 1989b; Maher et al. 2008).

Trees increase nutrient cycling in AGF systems as well by drawing nutrients up from deep in the ground, converting them into plant tissue and OM, dropping OM to the ground in the form of leaf litter and aboveground debris, and thereby releasing nutrients into the upper soil profiles, making them available for other crops to take up (Ramachandran et al. 1999). N-fixing trees are capable of cycling N from the atmosphere into the soil as well, through the process of N fixation. Depending on the species, trees can fix N at average rates of 40–200 kg N $ha^{-1}$ $year^{-1}$ (Nair 1993a).

## *Reduced Nutrient Losses in Vineyard Agroforestry Systems*

Trees also allow more nutrients to remain in cropping systems by reducing nutrient losses from leaching, erosion, and runoff. There is much evidence supporting the use of vegetative ground cover in general in vineyards to reduce such nutrient losses due to leaching and erosion. A study in Italy compared the erosion rates of conventionally tilled vineyards to those of vineyards with a grass CC by measuring infiltration rates, runoff discharge, and sediment yield at various rainfall intensities in each system. In the summer after high rainfall events, grass-covered vineyards experienced 83% less mean annual soil loss than conventionally tilled vineyards (Bagagiolo et al. 2017). Another study in Germany compared the erosion rates between a bare-soil vineyard and a grass-covered vineyard and found that soil losses and runoff rates were significantly higher in the bare-soil vineyards (Kirchhoff et al. 2017). Other studies have measured the erosion rates of bare-soil vineyards as well, and they support the conclusion that bare soils are one of the greatest determining causes of soil erosion in vineyards (Cerdà and Rodrigo-Comino 2018; Rodrigo-Comino et al. 2018). These studies have suggested the use of tree hedgerows, a type of AGF system, as a possible solution for halting erosion in vineyards (Cerdà and Rodrigo-Comino 2018).

Although there is little research on erosion reduction specifically in vineyard AGF systems, other studies have shown that AGF in general reduces soil erosion levels. As mentioned above, AGF systems have been shown to increase soil OM by up to 100% (Young 1989b), and just a 10% increase in OM results in a decrease in soil erodibility by roughly 13–23% (Young 1989c). Litterfall from trees in AGF systems translates to increased ground cover, which also results in reduced surface runoff and thus reduced erosion (Kimmins 1997; Pimentel 2006). While bare soil is exposed to the kinetic force of rain, which "seals the surface" of soils, breaks down soil structure where impact has occurred, dislodges soil particles, and reduces infiltration rates, AGF systems have a layer of surface mulch that protects soil from kinetic impact (Riha and McIntyre 1999; Cerdà and Rodrigo-Comino 2018). Indeed, studies comparing hedgerow intercropped AGF systems to monoculture systems found that the AGF systems in question had saturated hydraulic conductivity (Ksat) rates of 50 cm $h^{-1}$, while the monoculture systems had rates of only 18.5 cm $h^{-1}$ (Riha and McIntyre 1999). In a study comparing silvopasture AGF systems to treeless pastures in Missouri, Kumar et al. (2012) saw 31 times greater quasi-steady-state infiltration (qs) and 46 times greater saturated hydraulic conductivity (Ksat) in the AGF systems than in the treeless pastures. Similarly, Seobi et al. (2005) observed 14 times greater Ksat in grass and AGF buffers compared to a corn-soybean rotation in Missouri. Increased infiltration results in reduced runoff, which results in fewer nutrients that are carried out of the system (Seobi et al. 2005). Agroforestry reduces erosion potential by reducing compaction as well (Seobi et al. 2005; Kumar et al. 2012; Udawatta et al. 2011a).

Because nutrient loss due to soil erosion is such a large problem for vineyards, addressing soil erosion can result in significant farmer savings on fertilizer inputs. Depending on a number of factors such as vineyard size, slope, and soil type, among others, it is estimated that, on average, European viticulturists could save up to 1088 Euros $ha^{-1}$ annually by planting vegetative cover in vineyards, due to the increased nutrient retention that vegetative cover provides (Galati et al. 2015). It is speculated that AGF systems could be one such vegetative cover that is suitable for addressing erosion issues and maintaining nutrients within the cropping system (Cerdà and Rodrigo-Comino 2018).

## *Competition Between Trees and Grapevines for Nutrients*

Despite the increased nutrient availability that trees provide to crops, trees do compete with crops for nutrients. In general, competition for belowground nutrients is more of a limiting factor for crop growth in AGF systems than even light is (Gillespie et al. 2000) and this pattern extends to vineyard AGF systems as well (Dupraz et al. 2009; Grimaldi 2018). In a study on a 12-year-old vineyard AGF system in France in which grapevines were intercropped with stone pine (*Pinus pinea* L.), and service tree (*Sorbus domestica* L.), at densities of 222 trees $ha^{-1}$, data on vine nutrient status, vigor parameters, yield, berry quality, and soil electromagnetic conductivity was collected. Results showed that beyond 4 m from tree rows, no negative effects on grapevine yield due to competition for nutrients were experienced. However, at distances of 2.5–3.23 m from tree rows, high levels of competition for nutrients, especially N, were experienced. These negative effects manifested as reductions in vine vigor and yield; however, no reductions in berry quality were observed. Very few negative effects from water competition were experienced; nutrients were determined to be the limiting factor (Dupraz et al. 2009; Trambouze and Goma-Fortin 2013; Grimaldi 2018).

In line with these results, another study examining competition between vines and CC also discovered that vines are sensitive to N competition in particular, more so than other factors. In a study comparing five vineyard floor management treatments – bare soil without tillage, bare soil with tillage, sawdust mulch, chicory CC without tillage, and permanent chicory CC- researchers found that vines which received the bare-soil treatment (no competition) had the highest petiole nitrate concentration. Vines receiving CC treatments (both with tillage and without), on the other hand, had lower tissue N content, lower shoot growth, and lower pruning weights, showing that the presence of CC in vineyards does indeed result in competition for nutrients (Wheeler et al. 2005). A similar experiment comparing clean cultivation to CC treatments echoed these findings and found that, while cover crops increased water infiltration and did not compete excessively with vines for water, they did cause a significant decrease in the tissue N status of grapevines (Saayman and Huyssteen 1983). All of these findings point to the conclusion that nutrients, rather than water, are most likely the limiting factor for grapevine growth.

## *Striking Nutritional Balance in Grapevines*

Agroforestry's applications in vineyards can negatively affect grapevine nutrient status (Dupraz et al. 2009; Trambouze et al. 2017). However, in instances when vines are excessively vigorous, some competition for N can be beneficial. High soil fertility does not necessarily equate to higher yield or higher quality wine grapes, and in grapevines there exists a fine balance between healthy competition and excessive competition for nutrients (Wheeler and Pickering 2003). Too little N can result in severe stress, reduced yields, and decreased bud fertility, but too much N can result in reduced fruit set, excess allocation of resources to vegetative growth, increased in-canopy shading, and poor fruit quality (Wheeler and Pickering 2003). Vegetative imbalance from excessive N can delay crop maturation, prevent berry sugar accumulation, reduce phenolic concentration, and increase susceptibility to diseases such as powdery mildew and *Botrytis cinerea* (Wheeler and Pickering 2005). Additionally, excess vegetative growth leads to increased production costs from an increased need for spraying, trimming, leaf pulling, and thinning (Smart and Smith 1988). In general, grapevines have lower N requirements than many other crops, and they can maintain high yields and high-quality production in soils that are slightly deficient in N (Smart and Smith 1988; Martison 2010).

In wine grape growing, striking the balance between excessive and healthy levels of competition for nutrients is of utmost importance (Smart and Smith 1988). Nutrients are more of a limiting factor for grapevines than is water (Ussahatanonta et al. 2008) and nutritional balance can be difficult to achieve (Smart and Smith 1988). Although grapevines thrive under levels of slight nutrient deficiency, both nutrient surpluses and extreme nutrient deficits negatively impact vine growth, grape quality, and yield (Wheeler and Pickering 2005). In vineyard AGF systems, trees provide many nutritional benefits to the soil by increasing OM, cycling nutrients from deep soil profiles to shallow ones, fixing N, supporting microbial activity, increasing CEC, resisting nutrient loss from leaching and erosion, and increasing plant absorbability of nutrients (Young 1989a, b; Nair 1993b; Schroeder 1993; Ramachandran et al. 1999; Thevathasan and Gordon 2004; Paudel et al. 2011; Adetunji et al. 2017). These positive benefits may balance out some of the negative effects on nutrient status caused by competition between trees and vines. However, the current literature reveals that, overall, trees do cause negative effects on grapevine yield and growth within 4 m of tree hedgerows due to competition for N (Trambouze et al. 2017). Thus, it is most likely that the negative effects that trees have on vine nutritional parameters outweigh their positive benefits within 4 m of trees (Trambouze et al. 2017). Beyond 4 m of distance, there does not seem to be any effect, neither positive nor negative, but further evidence is needed to confirm the current studies' findings (Trambouze et al. 2017).

## The Effect of Trees on Vine Root Systems

Competition for nutrients and competition for water are two limiting factors that can hinder grapevine production in vineyard AGF systems when not managed correctly. Even though grapevines can, as proven, thrive at some levels of competition with other deep-rooted plants, excessive competition can be damaging. However, much of the ability of vines to absorb both water and nutrients in the face of competition depends on the health and spatial distribution of the vine's root system (Morlat and Jacquet 1984). Yield and overall quality of grapes are largely dependent on the ability of a vine's root system to exploit soil resources, and as such, it is important to examine the effects that interspecific interactions have on the roots of vines in specific (Morlat and Jacquet 1984). Research suggests that the depth and expansion of grapevine roots are highly dependent on soil structure and permeability, even more so than genotype (Smart et al. 2006), and that grapevine root plasticity is also influenced by planting density and competition (Hidalgo 1968).

### *Improved Soil Structure in Vineyard Agroforestry Systems*

Although trees in vineyard AGF systems can compete with vines for nutrients and water, these negative effects can be balanced by the positive influences that trees have on soil structure and quality (Smart et al. 2006). Soil structure consists of the spatial arrangement of individual soil particles, their aggregates, and the pore space that is formed between them (Lanyon et al. 2004). Soil structure affects soil strength, water-holding capacity, nutrient retention, aeration, friability, erodibility, plant root movement, and biological activity (Lanyon et al. 2004). High-quality soil structure allows for deeper and stronger vine root systems that are better able to exploit soil resources (Smart et al. 2006), and thus, it allows for higher grape production and quality despite competition.

According to Northcote (1988), soil porosity, and the increased aeration and water-holding capacity that come with it, is an even more important determinant of quality wine grape production than nutrient availability is. High aggregate stability (which leads to high porosity, high levels of water infiltration, low bulk density, and consequently greater root expansion) was also found to be a major wine grape quality determinant (Oliver et al. 2013). Soil penetrability, also determined by soil structure, is an important determinant of grapevine yield and quality as well (Henry 1993). An experiment was conducted in which grapevines were grown in soils with varying compaction levels. Researchers found that both size and depth of grapevine root systems decreased with increasing bulk density, and that grapevine roots did not occupy pores <200μm in diameter (Henry 1993). In a large-scale study across a variety of soil conditions throughout Australia, Myburgh et al. (1998) found that compacted soils with higher bulk density, greater incidence of cemented hardpans, and lower porosity were correlated with higher levels of grapevine root restriction and subsequent reduced yield and fruit quality.

Agroforestry has been shown to improve soil structure—including soil porosity, penetrability, aggregate stability, water-holding capacity, and strength—through a variety of mechanisms; as such, it has the potential to improve grapevine rooting potential, and consequentially, production and fruit quality (Young 1989a, b). Depending on what kinds of trees are used in vineyard AGF systems, a mulching effect from litterfall and pruning materials can occur that can have considerable beneficial effects on soil structure (Riha and McIntyre 1999). Soil cover improves soil structure by reducing raindrop and irrigation impact, which leads to conserved surface macroporosity, which leads to greater water infiltration rates and resultingly improved soil penetration (Lanyon et al. 2004). An 8-year study on different ground cover treatments including red fescue sod (*Festuca rubra* L.), postemergence herbicides, preemergence herbicides, and mulch confirmed that mulch lowers bulk density, decreases compaction, increases soil porosity, and increases water infiltration compared to bare-soil treatments and even the CC treatment (Oliveira and Merwin 2001).

Agroforestry systems also improve soil structure through the high amounts of root biomass that trees produce (Seobi et al. 2005). Tree roots in AGF systems improve soil macroporosity by breaking up compacted soils and leaving behind old root channels that grapevine roots are able to occupy for greater rooting depth capability (Mckenry 1984; Young 1989a). Finer roots also contribute to improved soil structure. In a study on AGF buffers in corn-soybean systems, researchers observed that tree buffer treatments produced higher porosity, increased coarse mesoporosity, and improved soil structure, most likely due to the increased root development in the tree buffer treatments (Seobi et al. 2005). In the case of vineyards, improved soil quality such as seen in this experiment would result in greater vine rooting capacity (Henry 1993).

The increased OM content that AGF systems impart to soil also improves soil structure. Soil structure is largely influenced by the amount of OM in soil (Young 1989b). In addition to increasing water infiltration and fertility, as mentioned previously, higher levels of OM translate to higher aggregate stability and overall improved structure (Balesdent et al. 2000). Organic matter contains sticky substances from bacterial exudates, organic gels, fungal hyphae, and excretions from fauna, and is able to "glue" soil particles together, thereby creating stable soil pores (Rashid et al. 2016). Agroforestry systems have been proven to increase soil OM by 50–100%, thus increasing porosity, reducing bulk density, and increasing soil water-holding capacity (Young 1989b). It can be speculated that because of this increase in OM, AGF's applications in vineyards would result in improved soil structure, resulting in increased rooting capability and subsequent higher yields and higher quality fruit production (Henry 1993).

## *Soil Niche Competition Between Tree and Grapevine Roots*

Grapevine productivity in vineyard AGF systems is also dependent on the measured extent of associated tree roots (Chirko et al. 1996). In order to avoid competition within AGF systems in general, it is important to take into consideration crop and associated tree root distribution patterns. Within the top 30 cm of any intercropped system there is typically intense competition between roots for nutrients and water, which results in lower yields and lower plant biomass production (Jose et al. 2009). However, belowground competition can be tempered through spatial separation of tree and crop roots, for example by combining deep-rooted trees with short-rooted crops (Lott et al. 1995). In the case of AGF's applications in vineyards, both tree roots and vine roots can be very long. Although the majority of grapevine roots are found in the top 1–2 m of soil, their roots, like trees, can reach greater deep depths (Smart et al. 2006). It is estimated that 63% of grapevine roots are found in the upper 60 cm of soil, 80% of grapevine roots are found in the upper 1.0 m of soil, and the remaining roots can extend to depths of 12 m (Lavee 2000; Smart et al. 2006). In contrast, 77% of coniferous forest tree roots are found in the upper 60 cm of soil, and 91% of coniferous tree roots are found in the upper 1.0 m of soil, revealing that grapevines might have a higher concentration of roots at deeper soil profiles than even some trees do (Jackson et al. 1996). Laterally, grapevine roots can spread outwards from the vine trunk up to 10 m (Smart et al. 2006).

The amount of overlapping soil niche occupation between grapevine roots and tree roots depends on the kind of trees utilized in the vineyard AGF system. Of the few vineyard AGF systems in existence today, most consist of grapes intercropped with olives (*Olea europaea* L.), Portuguese oak (*Quercus lusitanica* Lam.), elm (Ulmus sp.), poplar (*Populus* sp.), and wild cherry (*Prunus* sp.) (Altieri and Nicholls 2002; Mantzanas et al. 2016). Due to a lack of research, it is not known which trees might be most compatibly grown with grapevines (Lanyon et al. 2004). In the case of olive trees, lateral tree roots generally extend up to 12 m, and vertical roots grow

**Table 2** Average root length density per unit area and per unit of volume for different crops under field conditions. Source: Smart and Coombe (1983) (reprinted with permission)

| | Root length ( $cm_{root} \cdot cm_{soil}^{-1}$ ) | Root density ( $cm_{root} \cdot cm_{soil}^{-3}$ ) |
|---|---|---|
| Grapevines | 0.9–4 | 0.002–0.03 |
| Apple trees | 0.8–24 | 0.01–0.2 |
| Pear trees | 7–69 | 0.12–0.56 |
| Prune trees | 16–68 | 0-52–0.59 |
| Conifers | 5–126 | 0.5–0.69 |
| Cereals | 100–4000 | |

even deeper (Kailis and Harris 2007). Like grapevines, the majority of nutrient uptake occurs in the top 1 m of soil, and water uptake occurs within the top 1.2–1.7 m of soil (Morlat and Jacquet 1984; Kailis and Harris 2007). A study on an 11-year-old *Sorbus domestica* L./grapevine AGF system found that tree roots and vine roots occupied the same soil profile at distances of up to 8 m from the tree rows (Trambouze and Goma-Fortin 2013).

However, even though tree and vine roots occupy similar soil profiles, findings surrounding grapevine root morphology propone that there are still sufficient morphological and physiological differences between tree roots and vine roots to allow water and nutrient capture in different areas (Grimaldi 2018). Grapevine roots have been shown to exploit biopores left behind by dead tree roots and have been known to occupy "fracture lines" created by tree roots as well (Mckenry 1984). Because of this phenomenon, research suggests that grapevine roots evolved in competition with trees, and that it is possible for tree roots and grapevine roots to occupy different niches, even though they might exist within the same soil profile (Mckenry 1984).

The extent of interspecific competition for nutrients also depends on the root density per unit area volume of the competing species. The absorption rate of nutrients in plants is dependent on root length density and thus higher root length density in competing species can result in higher rates of competition for N (Fargione and Tilman 2006). Average root length density per unit area and per unit volume varies by species (Table 2). More research must be done to determine which tree species are most compatible with grapevines, both in terms of spatial root distribution and in terms of nutrient and water absorption potential (Jonsson et al. 1988).

## *Balancing Competition Through Root Plasticity*

Inferences about how grapevine roots will perform in AGF systems can be drawn based on the evidence of how grapevine roots perform when in competition with CC and with other vines in high-density plantings. Archer and Strauss (1985), in a study on grapevine root distributions at varying planting densities, found that vineyards with narrower spacings were able to utilize soil more efficiently and exploit more

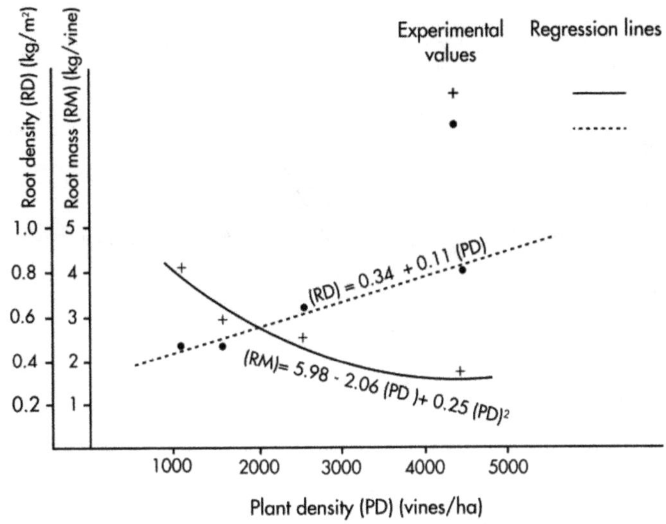


**Fig. 5** Relationship between vine root density (RD) ($kg/m^2$), vine room mass (RM) (kg/vine), and plant density (PD) (number of vines/ha). Source: Archer and Saayman 2018 (reproduced with permission)

nutrients and water while occupying a smaller space. This study found that when grapevine roots compete with other vines for water at higher planting densities, the horizontal space occupied by roots decreases, but root density increases, showing that grapevine root morphology can be modified to better exploit a smaller area, if necessary. Similarly, Hidalgo (1968) also found that throughout the entire soil profile, as plant density increased, root mass per vine decreased, but that root density increased (Fig. 5). Root density is positively correlated with vine vigor (Fig. 6). These findings point to the possibility that reduced nutrient availability from competition might be at least partially compensated for by increased root plasticity due to competition. Branas and Vergnes (1957) also found that as vine planting density increased, the quantity of shallow roots (25–45 cm) decreased, while the quantity of deep roots (65+ cm) increased, showing better utilization of soil volume in response to competition. It can be speculated that grapevines experiencing competition from tree roots rather than other vine roots would exhibit similar root density distribution patterns and exhaustive exploitation of soil resources.

Grapevine root plasticity can be induced by competition for water as well; the available soil water supply can determine the quantity of roots and the vertical distribution of roots (Morlat and Jacquet 1984). Studies have shown that more grapevine roots are produced under dry irrigation regimes than wet irrigation regimes, demonstrating that grapevine roots can indeed exhibit plasticity in response to resource scarcity as well (Freeman and Smart 1976; Freeman et al. 1982).

The positive effects that trees impart on soil structure, soil quality, and root plasticity allow for deeper and stronger grapevine root systems that can better absorb nutrients and water despite competition from trees (Smart et al. 2006). Trees

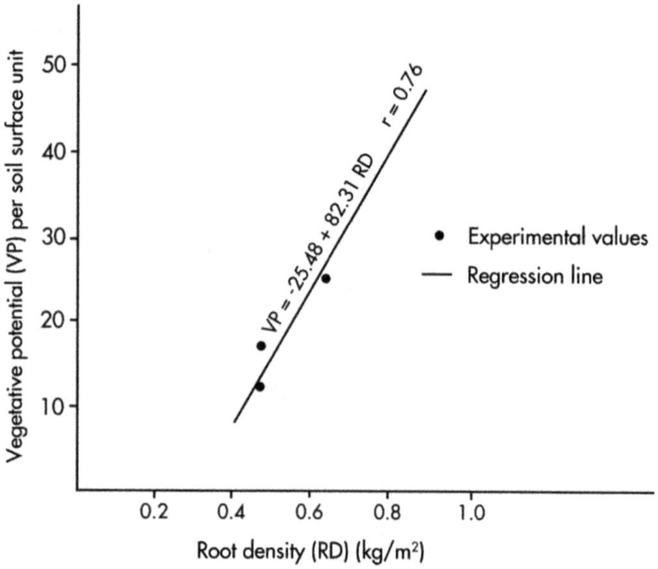


**Fig. 6** Relationship between root density (RD) (kg/m$^2$) and vegetative potential (vigor or shoot mass, VP) per square m of surface unit. Source: Archer and Saayman (2018) (reproduced with permission)

and vines are both perennial species with roots that occupy many of the same soil niches, which can result in high levels of competition (Morlat and Jacquet 1984; Kailis and Harris 2007). However, trees increase organic matter, aggregate stability, macroporosity, mesoporosity, water infiltration, water-holding capacity, penetrability, and overall quality in soils, and they decrease bulk density, incidence of hardpans, and irrigation impact, which all contribute to a soil environment that allows grapevine roots to grow more deeply (Young 1989a, b; Henry 1993; Lanyon et al. 2004; Seobi et al. 2005). Fracture lines left behind by tree roots also allow opportunities for grapevines to grow even deeper than they otherwise would have (Mckenry 1984). Additionally, competition from tree roots can trigger grapevine root plasticity, which results in increased root length density and increased nutrient and water absorption capacity per cm of soil (Branas and Vergnes 1957; Hidalgo 1968; Freeman et al. 1982; Fargione and Tilman 2006). Tree roots and grapevine roots are indeed able to adapt to competition and thrive despite occupying overlapping niches.

## Practical Implications

As with any farming technology, the use of trees in vineyards comes with trade-offs, and if vines are to be successfully intercropped with trees, vineyard AGF systems must be designed strategically. More important than high yields or a high growth rate in vineyards is the concept of growing a "balanced" vine—one which produces sufficient yield to be economically viable and which has sufficient vegetative growth to produce quality fruit (Wheeler and Pickering 2005). In vineyards, this balance is typically struck by allowing slight nutrient and water stress (Wheeler and Pickering 2005), and, in the case of vineyard AGF systems, trees could complete this function through regulated competition. However, excessive competition between trees and vines for nutrients and water is damaging and must be prevented when designing vineyard AGF systems (McCarthy et al. 1983; Giese et al. 2014).

In vineyard AGF systems, competition for N can be addressed by planting leguminous cover crops, applying higher rates of fertilizer in vine rows closer to tree rows, or selecting N-fixing trees for intercropping (Nair 1993c, d). Competition for water has been shown to be less of an issue in vineyard agroforestry systems, but during drought years, competition for water can be addressed through management practices such as root pruning, branch pruning, and tree thinning (Peter and Lehmann 2000; Reynolds et al. 2007; Dupraz et al. 2009; Senaviratne et al. 2012; Trambouze et al. 2017; Grimaldi 2018). Competition for both water and N can be addressed by combining vines with tree species whose roots occupy different soil niches than grapevine roots, or by spacing trees more widely (Nair 1993e). To minimize competition, grapes can be intercropped with trees that have lower root length densities, such as apples, pears, and plums (Smart and Coombe 1983). Preservation of soil structure and quality and reductions in erosion can be achieved by choosing trees with high litterfall production (Nair 1993a; Oliveira and Merwin 2001). If grapevines are intercropped with trees that also have economic value, such as nuts, fruits, and timber, it may be possible to produce more income per hectare through the increased use of vertical space (Nair 1993e). By designing vineyard agroforestry systems with strategic species combinations, cultural practices, and spacing, grapevines and trees can be integrated into a holistic system that is resilient against climate change, pests, and plagues; that produces high yields and high-quality wine; that improves soil fertility and quality; that reduces farmer reliance on agrochemicals; that is economically sustainable; and that betters the environment.

## Conclusion

Trees that are grown in association with grapevines both positively and negatively influence belowground soil parameters such as vine water status, vine nutrient status, and rooting patterns. The use of trees in vineyards results in greater water infiltration and water-holding capacity, greater nutrient availability, better soil quality, and more efficient vine rooting patterns while simultaneously causing greater competition for N and water. Studies have shown that trees do slightly negatively impact the growth, quality, and yield of grapevines within 4 m of trees, potentially due to competition for N; however, the positive ecological and cost-saving benefits that trees impart to a viticultural ecosystem as a whole might very well balance out these negative effects. In short, there are positives and negatives to intercropping grapevines with trees. There is sufficient scientific evidence that AGF has great potentials in viticulture, especially in the face of extreme temperatures, pests, plagues, and weather events that are predicted to come with climate change. This is a determination that is, of course, up to each individual viticulturist to decide, determinations which may be informed by more research on vineyard AGF systems done for all major wine-producing regions and tree-vine combinations.

## References

Adetunji AT, Lewu FB, Mulidzi R, Ncube B (2017) The biological activities of β-glucosidase, phosphatase and urease as soil quality indicators: a review. J Soil Sci Plant Nutr 17:794–807

Altieri MA, Nicholls CI (2002) The simplification of traditional vineyard based agroforests in northwestern Portugal: some ecological implications. Agrofor Syst 56:185–191

Altieri MA, Ponti L, Nicholls CI, York A (2005) Designing biodiverse, pest-resilient vineyards through habitat management. Practical Winery and Vineyard. https://agroecology.berkeley.edu/resources/Altieri_2005_Designing_biodiverse_pestresilient_vineyards.pdf. Accessed 14 Jan 2020

Amacher MC, O'Neill KP, Perry CH (2007) Soil Vital Signs: a New Soil Quality Index (SQI) for Assessing Forest Soil Health. U.S. Department of Agriculture, Forest Service, Rocky Mountain Research Station. Fort Collins, Colorado: U.S. Department of Agriculture

Amouretti MC (1988) La viticulture antique: contraintes et choix techniques. Revue des Études Anciennes 90:5–17

Archer E, Saayman D (2018) Vine roots. The Institute for Grape and Wine Sciences (IGWS), Stellenbosch University, South Africa

Archer E, Strauss HC (1985) Effect of plant density on root distribution of three-year-old grafted 99 Richter grapevines. South African J Enol Vitic 6:25–30

Bagagiolo G, Biddoccu M, Rabino D, Cavallo E (2017) Monitoring erosion in sloping vineyards: effectiveness of grass covering in different periods. In: Conference Proceedings of Convegno AIAM - SIA 2017, September 2017, Milan, Italy

Balesdent J, Chenu C, Balabane M (2000) Relationship of soil organic matter dynamics to physical protection and tillage. Soil Tillage Res 53:215–230

Bartolucci P, Dhakal BR (1999) Prospects for olive growing in Nepal. Food and Agriculture Organization of the United Nations, Department of Agriculture, Fruit Development Division Olive Production Development Project, TCP/NEP/6713, Kathmandu

Bhadha JH, Capasso J, Khatiwada R, Swanson S, Laborde C (2018) Raising soil organic matter content to improve water holding capacity. University of Florida, IFAS Extension

Biddoccu M, Pitacco A, Ferraris S, Cavallo E (2017) Temporal variability of soil management effects on soil hydrological properties, runoff and erosion at the field scale in a hillslope vineyard, North-West Italy. Soil Tillage Res 165:46–58

Bondada BR, Keller M (2012) Not All Shrivels Are Created Equal—Morpho-Anatomical and Compositional Characteristics Differ among Different Shrivel Types That Develop during Ripening of Grape (Vitis vinifera L.) Berries. Am J Plant Sci 3:879–898

Borrelli P, Robinson DA, Fleischer LR, Lugato E, Ballabio C, Alewell C, Meusburger K, Modugno S, Schutt B, Ferro V, Bagarello V, Van Oost K, Montanarella L, Panagos P (2013) An assessment of the global impact of 21st century land use change on soil erosion. Nat Commun 8:1–13

Branas J, Vergnes A (1957) Morphologie du système radiculaire de la vigne. Prog Agric Vitic 74:1–47

California Department of Pesticide Regulation (2017) 2017 Pesticide Use Report Highlights. California Department of Pesticide Regulation

Carbonneau A, Casteran P (1979) Irrigation-depressing effect on floral initiation of cabernet sauvignon grapevines in Bordeaux area. Am J Enol Vitic 30:3–7

Celette F, Gary C (2013) Dynamics of water and nitrogen stress along the grapevine cycle as affected by cover cropping. Eur J Agron 45:142–152

Cerdà A, Rodrigo-Comino J (2018) Is the hillslope position relevant for runoff and soil loss activation under high rainfall conditions in vineyards? Ecohydrol Hydrobiol 255:1–14

Charrier G, Delzon S, Domec JC, Zhang L, Delmas CEL, Merlin I, Corso D, King A, Ojeda H, Ollat N, Prieto JA, Scholach T, Skinner P, van Leeuwen C, Gambetta GA (2018) Drought will not leave your glass empty: low risk of hydraulic failure revealed by long-term drought observations in world's top wine regions. Sci Adv 4:1–9

Chirko CP, Gold MA, Nguyen PV, Jiang JP (1996) Influence of orientation on wheat yield and photosynthetic photon flux density (Qp) at the tree and crop interface in a Paulownia-wheat intercropping system. For Ecol Manag 89:149–156

Coll P, Le Cadre E, Blanchart E, Hinsinger P, Villenave C (2011) Organic viticulture and soil quality: a long-term study in Southern France. Appl Soil Ecol 50:37–44

Cooley H, Donnelly K, Phurisamban R, Subramanian M (2015) Impacts of California's ongoing drought: agriculture. Pacific Institute, Oakland, CA

Di Carlo P, Aruffo E, Brune WH (2019) Precipitation intensity under a warming climate is threatening some Italian premium wines. Sci Total Environ 685:508–513

Dokoozlian NK (2016) Grape berry growth and development. In: The grapevine. UC Davis, Davis, CA

Dry PR, Loveys BR, Botting D, During H (1996) Effects of partial root–zone drying on grapevine vigour, yield, composition of fruit and use of water. In: Proceedings of the Ninth Australian Wine Industry Technical Conference, 16–19 July 1995, Adelaide, South Australia

Dufourcq T, Lopez F, Vergnes M, Gontier L (2017) Use of proxy sensors to characterize spatial variability on agroforestry vineyards. Group of International Experts for Cooperation on Vitivinicultural Systems GiESCO. Conference Paper, 20th International GIESCO Symposium, Mendoza, Argentina

Dunn GM, Martin SR (2008) Do temperature conditions at budburst affect flower number in Vitis vinifera L. cv. Cabernet Sauvignon? Aust J Grape Wine Res 6(2):116–124

Dupraz C, Goma-Fortin N, Trambouze W, Kreiter S, Tixier MS, Gary C (2009) Vineyard agroforestry: a new concept for a sustainable vineyard production. In: Proceedings from the 2nd World Congress of Agroforestry (WCA2) August 2009, Ken, France. (hal-02758349)

Dupraz C, Blitz-Frayret C, Lecomte I, Molto Q, Reyes F, Gosme M (2018) Influence of latitude on the light availability for intercrops in agroforestry alley-cropping system. Agrofor Syst 92:1019–1033

Esteban MA, Villanueva MJ, Lissarrague JR (2001) Effect of irrigation on changes in the anthocyanin composition of the skin of cv Tempranillo (Vitis vinifera L) grape berries during ripening. J Sci Food Agric 81(4):409–420

Fabre E (2014) Des vignes entre champs et bois. Le cas de la marge pyrénéenne du vignoble languedocien (milieu XVIIIe-milieu XIXe siècle). In: Proceedings of the Bordeaux conference. Association for the History of Rural Societies, Caen, France, pp 373–386

Fargione J, Tilman D (2006) Plant Species Traits and Capacity for Resource Reduction Predict Yield and Abundance under Competition in Nitrogen-Limited Grassland. Funct Ecol 20:533–540

Ferreira CS, Keizera JJ, Santosa LM, Serpaa D, Silvaa V, Cerqueira M, Ferreira AJD, Abrantes N (2018) Runoff, sediment and nutrient exports from a Mediterranean vineyard under integrated production: an experiment at plot scale. Agric Ecosyst Environ 256:184–193

Francis CA, Altieri MA, Nicholls CI (2004) Designing species-rich, pest-suppressive agroecosystems through habitat management. In: Rickerl D, Farncis C (eds) Agroecosystems analysis. Madison, WI, pp 49–62

Freeman BM, Smart RE (1976) A root observation laboratory for studies with grapevines. Am J Enol Vitic 27:36–39

Freeman BM, Kliewer WM, Stern P (1982) Influence of windbreaks and climatic region on diurnal fluctuation of leaf water potential, stomatal conductance, and leaf temperature of grapevines. Am J Enol Vitic 33:233–236

Galati A, Gristina L, Crescimanno M, Barone E, Novara A (2015) Towards more efficient incentives for agri-environment measures in degraded and eroded vineyards. Land Degrad Dev 26:557–564

Garcia L, Celette F, Gary C, Ripoche A, Valdés-Gómez H, Metay A (2018) Management of service crops for the provision of ecosystem services in vineyards: a review. Agric Ecosyst Environ 251:158–170

Gholami R, Hadjiamiri A, Ghanbari F (2018) Effect of plant density and grape training method on pomological characteristics and yield of olive and grape in intercropping system. J Sustain Agri Prod Sci 28:15–24

Giese G, Velasco-Cruz C, Roberts L, Heitman J, Wolf TK (2014) Complete vineyard floor cover crops favorably limit grapevine vegetative growth. Sci Hortic 170:256–266

Gillespie AR, Jose S, Mengel DB, Hoover WL, Pope PE, Seifert JR, Biehle DJ, Stall T, Benjamin TJ (2000) Defining competition vectors in a temperate alley cropping system in the midwestern USA: 1. Production physiology. Agrofor Syst 48:25–40

Gómez-del-Campo M, Ruiz C, Lissarrague JS (2002) Effect of water stress on leaf area development, photosynthesis, and productivity in Chardonnay and Airén grapevines. Am J Enol Vitic 53:138–143

Gosme M, Delmotte S, Grimaldi J, Trambouze W (2019) Diachronic study of the effect of growing trees on grapevine yield: 24 years of experience in the South of France. In: 4th World Congress on Agroforestry, INRA SYSTEM. Conference Paper, Montpellier, France

Greer DH, Weston C (2009) Heat stress affects flowering, berry growth, sugar accumulation and photosynthesis of Vitis vinifera cv. Semillon grapevines grown in a controlled environment. Funct Plant Biol 37:206–214

Grimaldi J (2018) Impacts of agroforestry on microclimate for grape and wine production: assessment in Southern France. Dissertation, University of Toulouse, Toulouse, France

Grimaldi J, Trambouze W, Dufourcq T, Vergnes M, Pelletier C, Helen F, Fieuzal R, Houet T, Bustillo V (2017) Can intercropped trees mitigate heat and drought effects on grapevines? A study of microclimate patterns in agroforestry vineyards, Southern France. In: Proceedings of the IUFRO Landscape Ecology Conference, 24–29 September 2017, Halle, Germany

Grimaldi J, Wang Y, Gastellu-Etchegorry JP, Chavanon E, Lauret N, Guilleux J, Bustillo V, Houet T (2019) What are the impacts of tree shade on the absorption of light by grapevine in alley-cropped vineyards? In: Proceedings of the 4th World Congress on Agroforestry, 20–22 May 2019, Montpellier, France

Hardie WJ, Considine JA (1976) Response of grapes to water-deficit stress in particular stages of development. Am J Enol Vitic 27:55–61

Henry CC (1993) The effect of container volume and pore diameter on the growth of grapevines (Vitis vinifera L.). Dissertation, Lincoln University

Hidalgo L (1968) Contribuţii la studiul densităţii radiculare la viţa de vie. Revista de Horticultură şi Viticultură 7-8:116–121

Hunter JJ, Volschenk CG, Novello V, Strever AE, Fouché GW (2014) Integrative effects of vine water relations and grape ripeness level of Vitis vinifera L. cv. Shiraz/Richter 99. I. Physiological changes and vegetative-reproductive growth balances. South African J Enol Vitic 35:332–358

Jackson RB, Canadell J, Ehleringer JR, Mooney HA, Sala OE, Schulze ED (1996) A global analysis of root distributions for terrestrial biomes. Oecologia 108:389–411

Jones GV, Webb LB (2010) Climate change, viticulture, and wine: challenges and opportunities. J Wine Res 21:103–106

Jonsson K, Fidjeland L, Maghembe JA, Högberg P (1988) The vertical distribution of fine roots of five tree species and maize in Morogoro, Tanzania. Agrofor Syst 6:63–69

Jose S (2009) Agroforestry for ecosystem services and environmental benefits: an overview. Agrofor Syst 76:1–10

Jose S, Holzmueller EJ, Gillespie AR (2009) Tree–crop interactions in temperate Agroforestry. In: Garrett HE (ed) North American Agroforestry: an integrated science and practice, 2nd edn. American Society of Agronomy, Madison, WI, pp 57–74

Kailis S, Harris DJ (2007) Producing table olives. Landlinks Press, Collingwood, VIC, Australia

Kimmins JP (1997) Forest ecology: a foundation for sustainble management. University of Minnesota, United States, Prentice Hall

Kirchhoff M, Rodrigo-Comino J, Seeger M, Ries JB (2017) Soil erosion in sloping vineyards under conventional and organic land use managements (Saar-Mosel Valley, Germany). Geograph Res Lett 43:119–140

Kumar S, Anderson SH, Udawatta RP, Kallenbach RL (2012) Water infiltration influenced by agroforestry and grass buffers for a grazed pasture system. Agrofor Syst 84:325–335

Lal R (1989) Agroforestry systems and soil surface management of a tropical alfisol: V. Agrofor Syst 8:217–238

Lanyon DM, Cass A, Hansen D (2004) The effect of soil properties on vine performance. Technical Report, CSIRO Land and Water, South Australia

Lavee S (2000) Grapevine (Vitis Vinifera) growth and performance in warm climates. In: Erez A (ed) Temperate fruit crops in warm climates. Bet Dagen, Israel, pp 343–366

Lin BB (2007) Agroforestry management as an adaptive strategy against potential microclimate extremes in coffee agriculture. Agric For Meteorol 144:85–94

Lott JE, Khan AA, Ong CK, Black CR (1995) Sap flow measurements of lateral tree roots in agroforestry systems. Tree Physiol 16:95–1001

Maher M, Prasad M, Raviv M (2008) 11—Organic soilless media components. In: Raviv M, Lieth JH (eds) Soilless culture: theory and practice, pp 459–504

Mahmood I, Imadi SR, Shazadi K, Gul A, Hakeem KR (2015) Effects of pesticides on environment. In: Hakeem K, Akhtar M, Abdullah S (eds) Plant, soil and microbes. Springer International Publishing, Switzerland, pp 253–269

Mantzanas K, Pantera A, Delapre L, Koutsoulis D, Papadopoulos A, Fotiadis G, Sidiropoulou A, Papanastasis VP (2016) Olive Trees Intercropped with Cereals and Legumes in Kassandra Peninsula, Northern Greece. In: Proceedings of the 3rd European Agroforestry Conference, 23–25 May 2016, Montpellier, France. Abstract Number 94, p 248–250

Martínez-Casasnovas JA, Ramos MC (2006) The cost of soil erosion in vineyard fields in Penedès–Anoia Region (NE Spain). Catena 68:194–199

Martison T (2010) Sources and sinks: allocation of photosynthates during the Growing Season. Appellation Cornell (4). Cornell College of Agriculture and Life Sciences. https://grapesandwine.cals.cornell.edu/newsletters/appellation-cornell/2010-newsletters/issue-4/sources-and-sinks-allocation-photosynthates/. Accessed 15 Jan 2020

McCarthy MG, Cirami RM, McCloud P (1983) Vine and fruit responses to supplementary irrigation and canopy management. South African J Enol Vitic 4:67–76

Mckenry MV (1984) Grape root phenology relative to control of parasitic nematodes. Am J Enol Vitic 35:206–211

Medrano H, Escalona JM, Cifre J, Bota J, Flexas J (2003) A ten-year study on the physiology of two Spanish grapevine cultivars under field conditions: effects of water availability from leaf photosynthesis to grape yield and quality. Funct Plant Biol 30:607–619

Medrano H, Tomás M, Mortorell S, Escalona JM, Pou A, Fuentes S, Flexas J, Bota J (2015) Improving water use efficiency of vineyards in semi-arid regions. A review. Agron Sustain Dev 35:499–517

Mériaux S, Rollin H, Rutten P, Lessut J, Lallemand R (1981) Effets de la sécheresse sur la vigne (Vitis vinifera L.) II.—Etudes sur "Grenache". Agronomie 1:375–382
Minasny B, McBratney AB (2015) Limited effect of organic matter on soil available water capacity. Eur J Soil Sci 69:39–47
Morlat R, Remoué M, Pinet P (1984) Influence de la densité de plantation et du mode d'entretien du sol sur l'enracinement d'un peuplement de vigne planté en sol favorable. Agronomie 4:485–491
Myburgh P, Cass A, Clingeleffer P (1998) Root systems and soils in Australian vineyards and orchards: an assessment: 1996 Barossa Valley Rotary Foundation Fellowship report. Cooperative Research Centre for Soil and Land Management, South Australia
Nair PK (1993a) An introduction to agroforestry. Klgwer Academic Publishers, Dordrecht, The Netherlands
Nair PK (1993b) Nutrient cycling and soil organic matter. In: Nair PK (author) An introduction to agroforestry. Kluwer Academic Publishers, Dordrecht, The Netherlands, pp 277–306
Nair PK (1993c) Nitrogen fixation. In: Nair PK (author) An introduction to agroforestry. Kluwer Academic Publishers, Dordrecht, The Netherlands, pp 307–323
Nair PK (1993d) Alley cropping. In: Nair PK (author) An introduction to agroforestry. Kluwer Academic Publishers, Dordrecht, The Netherlands, pp 123–139
Nair PK (1993e) Component interactions. In: Nair PK (author) An introduction to agroforestry. Kluwer Academic Publishers, Dordrecht, The Netherlands, pp 243–258
Napoli M, Dalla Marta A, Zanchi CA, Orlandini S (2017) Assessment of soil and nutrient losses by runoff under different soil management practices in an Italian hilly vineyard. Soil Tillage Res 168:71–80
Neja RA, Wildman WE, Ayers RS, Kasimatis AN (1977) Grapevine response to irrigation and trellis treatments in the Salinas Valley. Am J Enol Vitic 28:16–26
Nicholls CI, Ponti L, Altieri MA (2008) Enhancing plant diversity for improved insect pest management in Northern California organic vineyards. Acta Hortic 785:263–278
Nicolopoulou-Stamati P, Maipas S, Kotampasi C, Stamatis P, Hens L (2016) Chemical pesticides and human health: the urgent need for a new concept in agriculture. Front Public Health 4:148
Northcote KH (1988) Soil and Australian viticulture. In: Coombe BG, Dry PR (eds) Viticulture: Volume 1 – Resources, 2nd edn. Adelaide, Australia, pp 61–90
Novara A, Pisciotta A, Minacapilli M, Maltese A, Capodici F, Cerdà A, Gristina L (2018) The impact of soil erosion on soil fertility and vine vigor. A multidisciplinary approach based on field, laboratory and remote sensing approaches. Sci Total Environ 622-623:474–480
NPCS Board of Consultants and Engineers (n.d.) Handbook on citrus fruits: cultivation and oil extraction. Asia Pacific Business Press Inc, Kamla, Nagar, Delhi, India
Oliveira MT, Merwin IA (2001) Soil physical conditions in a New York orchard after eight years under different groundcover management systems. Plant Soil 234:233–237
Oliver DP, Bramley RG, Riches D, Porter IJ, Edwards J (2013) Review: soil physical and chemical properties as indicators of soil quality in Australian viticulture. Aust J Grape Wine Res 19:129–140
Pachauri RK, Meyer L (2015) Climate Change 2014 Synthesis Report. Intergovernmental Panel on Climate Change, World Meteorological Organization (WMO). Report, Geneva, Switzerland
Pagay V, Collins C (2017) Effects of timing and intensity of elevated temperatures on reproductive development of field-grown Shiraz grapevines. OENO One 51(4)
Paudel B, Udawatta RP, Kremer RJ, Anderson SH (2011) Soil quality indicator responses to row crop, grazed pasture, and agroforestry buffer management. Agrofor Syst 82:311–323
Peter I, Lehmann J (2000) Pruning effects on root distribution and nutrient dynamics in an acacia hedgerow planting in northern Kenya. Agrofor Syst 50:59–75
Pienaar JW (2005) The effect of wind on the performance of the grapevine. Thesis, Stellenbosch University
Pimentel D (2006) Soil erosion: a food and environmental threat. Environ Dev Sustain 8:119–137
Pool RM, Dunst RM, Lakso AN (1990) Comparison of sod, mulch, cultivation, and herbicide floor management practices for grape production in nonirrigated vineyards. J Am Soc Hortic Sci 115:872–877

Raj AJ, Lal SB (2014) Agroforestry theory and practices. Scientific Publishers, Jodhpur, India
Ramachandran NP, Buresh RJ, Mugendi DN, Latt CR (1999) Nutrient cycling in tropical agroforestry systems: myths and science. In: Buck LE, Lassoie JP, Fernandes EC (eds) Agroforestry in sustainable agricultural systems. CRC Press
Rashid MI, Mujawar LH, Shahzad T, Almeelbi T, Ismail IM, Oves M (2016) Bacteria and fungi can contribute to nutrients bioavailability and aggregate formation in degraded soils. Microbiol Res 183:26–41
Reynolds PE, Simpson JA, Thevathasan NV, Gordon AM (2007) Effects of tree competition on corn and soybean photosynthesis, growth, and yield in a temperate tree-based agroforestry intercropping system in southern Ontario, Canada. Ecol Eng 29:362–371
Riha SJ, McIntyre BD (1999) Water management with hedgerow agroforestry systems. In: Buck LE, Lassoie JP, EDM F (eds) Agroforestry in sustainable agricultural systems. Lewis Publishers, pp 47–67
Rodrigo-Comino J, Cerdà A, Keesstra SD (2018) Soil erosion as an environmental concern in vineyards. The case study of Celler del Roure, Eastern Spain, by Means of Rainfall Simulation Experiments. Beverages 4:1–11
Root RB (1973) Organization of a Plant-Arthropod Association in Simple and Diverse Habitats: the Fauna of Collards (Brassica oleracea). Ecol Monogr 43:95–124
Saayman D, Huyssteen LV (1983) Preliminary studies on the effect of a permanent cover crop and root pruning on an irrigated Colombar vineyard. South African J Enol Vitic 4:7–12
Schroeder P (1993) Organic matter cycling by tropical agroforestry systems: a review. J Trop For Sci 7:462–474
Schultz HR (2003) Differences in hydraulic architecture account for near-isohydric and anisohydric behaviour of two field-grown Vitis vinifera L. cultivars during drought. Plant Cell Environ 26:1393–1405
Sellers R, Alampi-Sottini V (2016) The influence of size on winery performance: evidence from Italy. Wine Econ Policy 5:33–41
Senaviratne A, Udawatta RP, Nelson KA, Shannon KK, Jose S (2012) Temporal and spatial influence of perennial upland buffers of corn and soybean yields. Semigroup Forum 104:1356–1362
Seobi T, Anderson SH, Udawatta RP, Gantzer C (2005) Influence of grass and agroforestry buffer strips on soil hydraulic properties for an Albaqualf. Soil Sci Soc Am J 69:893–901
Shantz HI (1927) Drought resistance and soil moisture. Ecology 8:145–157
Smart DR, Carlisle E, Goebel M, Núnez BA (2005) Transverse hydraulic redistribution by a grapevine. Plant Cell Environ 28:157–166
Smart DR, Schwass E, Lakso A, Morano L (2006) Grapevine rooting patterns: a comprehensive analysis and a review. Am J Enol Vitic 57:89–104
Smart RE, Coombe BG (1983) Water relations of grapevines. In: Kozlowski TT (ed) Water deficits and plant growth, Additional woody crop plants, vol VII. Academic Press, New York, pp 137–196
Smart RE, Robinson M (1991) Sunlight into wine: a handbook for winegrape canopy management. Winetitles, Adelaide, Australia
Smart RE, Smith SM (1988) Canopy management: identifying the problems and practical solutions. In: Proceedings of the 2nd International Symposium on Cool Climate Viticulture and Oenology, January 1988, New Zealand Society for Viticulture and Oenology, Auckland, NZ
Stephenson AG (1981) Flower and fruit abortion: proximate causes and ultimate functions. Annu Rev Ecol Syst 12:253–279
Stevens RM, Harvey G, Aspinall D (1995) Grapevine growth of shoots and fruit linearly correlate with water stress indices based on root-weighted soil matric potential. Aust J Grape Wine Res 1:58–66
Stevenson T (2005) The Sotheby's wine encyclopedia, 4th edn. Penguin Group, London, New York, Melbourne, Munich, Delhi
Sudmeyer R, Flugge F (2004) The economics of managing tree–crop competition in windbreak and alley systems. Aust J Exp Agric 45:1403–1414
Thevathasan N, Gordon AM (2004) Ecology of tree intercropping systems in the North temperate region: experiences from Southern Ontario, Canada. Agrofor Syst 61:257–268

Thomazini A, Mendonça ES, Cardoso IM, Garbin ML (2015) SOC dynamics and soil quality index of agroforestry systems in the Atlantic rainforest of Brazil. Geoderma Reg 5:15–24
Trambouze W, Goma-Fortin N (2013) Agroforesterie viticole: résultats de 11 ans d'étude sur la production et la vigueur des vignes. Chambre d'agriculture de l'Hérault, Portugal
Trambouze W, Gouttesoulard D, Saubion C (2017) Agroforesterie viticole: 20 ans de complantation arbres/vignes, pour quels résultats agronomiques? Report. RMT Agroforesteries Chambre D' Agriculture Hérault, Paris, France
Udawatta RP, Anderson SH, Motavalli PP, Garrett HE (2011a) Calibration of a water content reflectometer and soil water dynamics for an agroforestry practice. Agrofor Syst 82:61–75
Udawatta RP, Garrett HE, Kallenbach RL (2011b) Agroforestry buffers for non-point source pollution reductions from agricultural watersheds. J Environ Qual 40:800–806
Udawatta RP, Nelson KA, Jose S, Motavalli PP (2014) Temporal and spatial differences in crop yields of a mature silver maple alley cropping system. Agron J 106:407–415
Udawatta RP, Gantzer CJ, Reinbott TM, Wright RL, Pierce RA II (2016) Yield differences influenced by distance from riparian buffers and CRP. Agron J 108:647–655
Udawatta RP, Anderson SH, Kremer RJ (2020) Soil benefits of agroforestry. In: Garrett HE, Jose S, Gold M (eds) North American Agroforestry: an integrated science and practice, 3rd edn. American Society of Agronomy-Crop Science Society of America-Soil Science Society of America (ASA-CSSA-SSSA), Madison, WI
Ussahatanonta S, Jackson DI, Rowe RN (2008) Effects of nutrient and water stress on vegetative and reproductive growth in Vitis vinifera L. Aust J Grape Wine Res 2:64–69
Vanzyl JL (1984) Response of Colombar grapevines to irrigation as regards quality aspects and growth. South African J Enol Vitic 5:19–28
Wezel A, David C, Ferrer A, Letort A, Feret S, Peigné J, Vian JF, Celette F (2014) Agroecological practices supporting the provision of goods and services in agriculture. Examples from France and Europe. ISARA Lyon, France and the Food and Agriculture Organization, Rome, Italy. https://www.researchgate.net/publication/266143585_Agroecological_practices_supporting_the_provision_of_goods_and_services_in_agriculture_Examples_from_France_and_Europe. Accessed 16 Jan 2020
Wheeler SJ, Pickering GJ (2003) Optimizing grape quality through soil management practices. J Food Agric Environ 1:190–197
Wheeler SJ, Pickering GJ (2005) Effects of soil management techniques on grape and wine quality. In: Dris R (author) Fruits: growth, nutrition, and quality. WFL Publisher, Helsinki, Finland, pp 195–208
Wheeler SJ, Black AS, Pickering GJ (2005) Vineyard floor management improves wine quality in highly vigorous Vitis vinifera 'Cabernet Sauvignon' in New Zealand. N Z J Crop Hortic Sci 33:317–328
Wilson H, Miles AF, Daane KM, Altieri MA (2017) Vineyard proximity to riparian habitat influences Western grape leafhopper (Erythroneura elegantula Osborn) populations. Agric Ecosyst Environ 211:43–50
Winkel T, Rambal S (1993) Influence of water stress on grapevines growing in the field: from leaf to whole-plant response. Aust J Plant Physiol 20:143–157
Young A (1989a) Effects of trees on soils. In: Young A (author) Agroforestry for soil conservation. CAB International, Wallingford, UK, pp 93–103
Young A (1989b) Soil organic matter. In: Young A (author) Agroforestry for soil conservation. CAB International, Wallingford, UK, pp 105–128
Young A (1989c) Trends in soil-conservation research and policy. In: Young A (author) Agroforestry for soil conservation. CAB International, Wallingford, UK, pp 17–52
Zelba O, Lojka B, Houška J, Kotrba R, Martiník A, Weger J (2016) Inventory of Agroforestry in Czech Republic. In: Proceedings of the 3rd European Agroforestry Conference, 23–25 May 2016, Montpellier, France. Abstract Number 144.5, p 211–213

# Selected Soil Properties Among Agroforestry, Natural Forest, Traditional Agriculture, and Palm Oil Land Uses in Central Kalimantan

**Kania Dewi Rahayu, Ranjith P. Udawatta, Shibu Jose, Manjula Nathan, Yves Laumonier, and Sarjiya Antonius**

## Introduction

Increasing human activities and also severe natural disasters have affected tropical forests in Indonesia and have turned some of these forests into degraded lands. The deforestation in Indonesia has begun in the twentieth century due to the expansion of agricultural lands and advancement in agriculture mechanization (Food and Agriculture Organization 2016). The uncontrolled and illegal land conversions have caused significant damages such as degradation of forest functions, soil health, and food insecurity (Iqbal and Sumaryanto 2007).

Palm oil is one of the world's most rapidly expanding plantations. Based on Wicke et al. (2011), the projection of additional land for palm oil production will reach 1–28 Mha in Indonesia based on the analyses of national level data on land-use change by 2020. The expansion of palm oil plantations contributes to deforestation in four ways; that is, (1) palm oil is the primary reason for forest clearing, (2) the most common opening procedure for palm oil plantations is logging and/or fire, (3) there is high demand of palm oil for economic reasons, and (4) there are indirect effects such as forest clearing for roads for access to plantation sites (Fitzherbert

K. D. Rahayu (✉) · R. P. Udawatta · S. Jose
The Center for Agroforestry, School of Natural Resources, University of Missouri, Columbia, MO, USA
e-mail: udawattar@missouri.edu; joses@missouri.edu

M. Nathan
Division of Plant Sciences, University of Missouri, Columbia, MO, USA

Y. Laumonier
Center for International Forestry Research (CIFOR), Bogor, West Java, Indonesia

S. Antonius
Research Center for Biology, Indonesian Institute of Sciences (LIPI), Bogor, West Java, Indonesia



R. P. Udawatta, S. Jose (eds.), *Agroforestry and Ecosystem Services*,
https://doi.org/10.1007/978-3-030-80060-4_5

et al. 2008). Based on Vijay et al. (2016), in 2013, Indonesia ranked number one in the world on harvested palm oil volume and its volume can reach between 11 and 50% of the world production.

The negative impacts of palm oil plantations are habitat fragmentation, land cover change, species loss, and greenhouse gas emission (Fitzherbert et al. 2008). The land conversion from natural forest to palm oil plantation is also reducing ecosystem services including soil health and fertility. A study conducted by Allen et al. (2015) reported that the conversion from forest to palm oil and rubber plantation can decrease soil $NH_4^+$ transformation rates and microbial biomass. According to Sitorus and Pravitasari (2017) about 3 Mha of land in East Kalimantan, West Kalimantan, North Sumatera, and Riau and Central Kalimantan has recorded the greatest degradation.

Soil quality is affected by land changes and extensive land conversions. The indicators of soil quality include physical, chemical, and biological characteristics. Based on USDA (2018), there are several characteristics that must be evaluated to rank soils. Those indicators are aggregate stability, available water capacity, bulk density, infiltration, slaking, soil crusts, soil structure, chemical indicators, reactive carbon, soil electrical conductivity, soil nitrate, soil pH, earthworms, particulate organic matter, potentially mineralizable nitrogen, soil enzymes, soil respiration, and total organic carbon. Soil quality indicators also help understand soil conditions and can be used to compare historical management conditions (Yakovchenko et al. 1998). Biological soil indicators respond immediately to management practices and thus can be used to quantify changes in soil parameters (Dick 1994; Bandick and Dick 1999).

Unsustainable agricultural practices affect the environment. Soil health is one of the aspects that can be affected by poor land-use practices. For example, clearing the forests and converting forests into other land-use systems can reduce soil fertility, disturb soils, and reduce the quality of soil physical and biological parameters (Doran and Zeiss 2000). Agroforestry has a great potential to stabilize the sloping lands and forest reserve by establishing buffer zones (Kang and Akinifesi 2000). These integrated management practices can improve soil health, fauna, and microflora, and land productivity (Schroth et al. 2001).

Most of the soil quality indicators are greater in agroforestry practices than in grazed pastures and row-crop system (Jose 2009; Dollinger and Jose 2018). Due to trees, soil enzyme activities like dehydrogenase, ß-glucosidase, and microbial biomass were greater under the perennial vegetation than grazing areas (Paudel et al. 2012). Furthermore, agroforestry practices can also help the economy of smallholder agriculture with their produce from multi-strata vegetation structures. The combination of food crops, cash crops, and trees can provide biological and economic benefits to landowners.

Literature is limited on studies evaluating peat swamp forest ecosystems especially comparing with agroforestry practices in peat areas and other land-use practices on enzyme activities, soil carbon, nitrogen, and soil respiration. This research examined soil health indicators in multiple land-use systems in Kalimantan, Indonesia, with enzyme activities and carbon measurements as major variables to determine the effects of land management on soil properties. The secondary objective was to understand the potency of homegarden agroforestry as an alternative land use in the restoration area of PT. Rimba Raya Biodiversity Reserve.

# Materials and Methods

## *Study Site*

This research was conducted from June 2017 to August 2017 in Rimba Raya Biodiversity Reserve located in Seruyan District in the Province of Central Kalimantan, Indonesia, for the fieldwork and Indonesian Institute of Sciences Agriculture Microbiology Laboratory, Bogor, for the laboratory work. The Rimba Raya Biodiversity Reserve is 64,500 ha in extent along the Tanjung Puting National Park (Fig. 1). The project site is bounded by Tanjung Puting National Park to the west, the Java Sea to the south, and the Seruyan River to the east. The eastern border of Rimba Raya coincides with the Seruyan River for almost 100 km. Rimba Raya concession area consists of several habitat types such as small amount of lowland

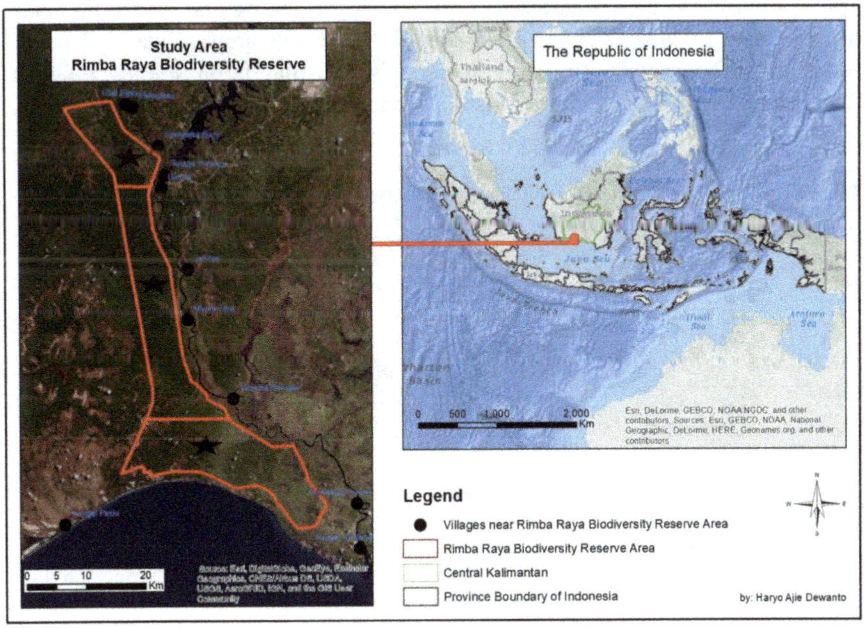


**Fig. 1** Study locations (stars) in Rimba Raya Biodiversity Reserve Concession Area, Seruyan Regency, Central Kalimantan (Rimba Raya Biodiversity 2016). The inset map shows approximate study areas in Indonesia

dipterocarp, *kerangas*, peat swamp, and riverine forests, which provides a strong conservation rationale for the border location.

Rimba Raya Biodiversity Reserve project initiated by InfiniteEARTH Rimba Raya was the first forest-carbon project validated under the Verified Carbon Standard (VCS) and also the first REDD+ (reducing emissions from deforestation and forest degradation plus conservation and sustainable development) project in the world that has received triple-gold validation under the Climate Community and Biodiversity Alliance Standard (CCBA) in 2010 (Lemons et al. 2011).

## Soil Types

Codominant soil types in Rimba Raya Concession Area are derived from peat and riverine alluvium. Sediment-derived soils especially in *kerangas* vegetation area have coarse-textured material with poor drainage (Lemons et al. 2011). Soil types in Rimba Raya area are grouped into five mapping units (Table 1).

In the soil mapping unit 3, the most dominant soils are Haplohemist and Sulfihemists with organic materials as the parent material. In soil mapping unit 14 the dominant soils are Endoaquepts, Sulfaquents, and Alluvium as the parent material. Endoaquepts and Dystrudepts are described as the saturated inceptisols and these are dominant soils in the soil mapping unit 20. Some soils in the area have developed from Alluvium parent materials. In soil mapping unit 20 the dominant soil types are Endoaquepts and Dystrudepts. These soils are saturated inceptisols and acidic inceptisols and the parent material is Alluvium. In the soil mapping unit 52 the dominant soils are Quartzipsamments and Durorthods in alluvial floodplain

**Table 1** Soil types in Rimba Raya Biodiversity Reserve (Lemons et al. 2011)

| Soil mapping unit | Dominant soils | General description | Parent material | Sub-landform | Relief |
|---|---|---|---|---|---|
| 3 | Haplohemist, Sulfihemists | Moderately decomposed peat soils, some of which are sulfic | Organic | Peat dome | Flat |
| 14 | Endoaquepts, Sulfaquents | Saturated inceptisols and saturated sulfic entisols | Alluvium | Delta or estuary | Flat |
| 20 | Endoaquepts, Dystrudepts | Saturated inceptisols and acidic inceptisols | Alluvium | Alluvial flood plane | Flat |
| 52 | Quartzipsamments, Durorthods | Quartzic entisols and spodosols with a cemented hardpan | Sediment | Terraces | Flat rolling |
| 61 | Haplorthods, Palehumults | Free-draining spodosols and humus-rich ultisols | Sediment | Terraces | Flat rolling |

and Alluvium is the parent material. The last soil mapping unit (61) has Haplorthods and Palehumults as the dominant soils. The parent material is sediment. These soil types are free-draining Spodosols and humus rich.

## *Experimental Design and Sampling*

Completely randomized block design with three replications was used to collect soil samples. The management treatments were palm oil plantation (POP), traditional agriculture (TA), agroforestry homegarden (AHG), and natural forest (NF). Soil was collected from three villages: Kuala Pembuang, Muara Dua, and Telaga Pulang for each treatment. Natural forest was the control. Soil sampling was conducted from June to July 2017 from two depths, i.e., 0–15 cm and 15–30 cm. Soil was stored at 4 °C (Parham and Deng 2000) before taken to the Indonesian Institute of Sciences (LIPI) laboratory.

## *Cellulase Activity*

Cellulase activity was determined according to Deng (1994) and Kanazawa and Miyashita (2012). Five grams of soil and 0.2 mL of toluene were added to a test tube. The mixture was allowed to stand for 10 min. After 10 min, 25 mL of 0.2 M acetate buffer at pH 5.5 that contained 10 mg of CMC (carboxymethyl cellulose sodium salt) was added to the test tube and the test tube was sealed using parafilm. The sealed tubes were incubated at 50 °C for 4 h.

After the incubation, 2 mL of sample solution was centrifuged for 10 min at 8000 rpm to remove colored humus. After centrifugation, the supernatant was treated with the Prussian blue and ammonium molybdate (Somogyi-Nelson). The Prussian blue was used to develop substrate color and Somogyi-Nelson reagent was used to reduce sugars (as the substrate). These reducing sugars were produced while the sample was incubated by carboxymethyl cellulose (CMC). The concentration of reduced sugars by Prussian blue and molybdate was measured using spectrophotometer at 710 nm with two replications. Those replications consisted of soil sample with substrate and soil sample without substrate. The following formula was used to determine the potential activity of cellulase:

$$\frac{(S*C)*100}{a*b*\mathrm{CF}}$$

where *S*: sample concentration (μg $mL^{-1}$ glucose), *C*: control concentration (μg $mL^{-1}$ glucose), *a*: sample weight (g), *b*: incubation period (h), and CF: correction factor.

## *PMEase Activity*

PMEase activity was determined according to Eivazi and Tabatabai (1977), and the measurement of PMEase activity was based on p-nitrophenol (pNPP) release. Five grams of soil and 1 mL p-nitrophenol as the substrate were added to a test tube. After that, 4 mL of universal buffer at pH 6.5 was added to the test tube and homogenized using vortex. Samples were incubated in a water bath incubator at 38 °C for 2 h. After the incubation, 1 mL of 0.5 M $CaCl_2$ was added into the test tube and homogenized using vortex. After samples were homogenized, 4 mL of 0.5 M NaOH was added to the test tube and homogenized.

Two mL of sample from each reaction tube was transferred to a microtube and centrifuged at 8000 rpm for 10 min. The concentration of pNPP in the supernatant was measured by using a spectrophotometer at 400 nm with two replications. Those two replications consist of samples with substrate and samples without substrate. The following formula was used to determine the potential activity of PMEase:

$$\frac{\left(\left(S-C\right)*\mathrm{df}\right)*100\%\mathrm{dm}}{a*b}$$

where *S*: sample concentration (μg pNPP), *C*: control concentration (μg pNPP), df: dilution factor, 100%dm: factor for soil dry weight, *a*: molecular weight of pNPP, and *b*: incubation period.

## *Urease Activity*

Urease activity was determined according to Tabatabai (1994). Five grams of soil sample and 2.5 mL of urea substrate were added to a 100 mL flask. Samples were incubated at 37 °C for 2 h. After incubation, 50 mL of 1 M KCl was added to the flask and then samples were mixed by using an orbital shaker for 30 min. After 30 min of incubation in an orbital shaker, 2 mL of sample was added to microtubes and centrifuged at 8000 rpm for 10 min. After centrifugation, 1 mL of sample was dissolved in 9 mL aquades and 0.1 mL Nessler B was added. Samples were homogenized by using vortex. The mixture was allowed to stand for 10 min and samples' absorbance was measured at 420 nm using a spectrophotometer. The absorbance result was compared with $NH_4Cl$ absorbency as the standard. The following formula was used to determine the potential activity of urease:

$$\frac{\left(\left(S-C\right)*\mathrm{df}\right)*100\%\mathrm{dm}}{B*a*b}$$

where *S*: sample concentration (μg pNPP), *C*: control concentration (μg pNPP), df: dilution factor, 100%dm: factor for soil dry weight, *B*: soil weight (g), *a*: molecular weight of $NH^{4+}$ (g $mol^{-1}$) (18), and *b*: incubation period (h).

### C, N, and C:N Ratio

The concentrations of carbon (C), nitrogen (N), and carbon:nitrogen ratio (C:N) were determined by dry combustion using an elemental analyzer (Applied Spectra, USA). Soil samples were air-dried and sieved in a 2 mm sieve before the analysis. The samples were oxidized at a 950 °C for 2 min in a dry combustion tube. Three different volumes of hippuric acid (5 mg, 10 mg, and 20 mg) were added to each 40 g of soil. Hippuric acid was used to reduce the $N_2O$ fluxes and emission during the analysis (Kool et al. 2006; Tivet et al. 2011). The combustion method detected C as $CO_2$ (Dieckow et al. 2007).

### Statistical Analysis

The data were analyzed as a completely randomized block design using Proc GLM in SAS version 9.2 (SAS). ANOVA (analysis of variance) was used to determine the significance level for each treatment and soil depths. Differences were declared at the 5% level of significance ($p \geq 0.05$).

## Results and Discussion

### Soil Carbon

The highest soil carbon concentration was observed at the 0–15 cm soil depth (36.8%) of the NF treatment (Fig. 2). The lowest soil carbon level was found at the 0–15 cm soil depth (1.3%) of POP treatment. The second highest of soil carbon level was recorded for 0–15 cm of TA and 15–30 cm of NF (25.6% and 25.3%) and it was followed by TA 15–30 cm (23.8%) and AHG 0–15 cm soil depths (21.7%). The carbon concentration of POP at 15–30 cm was significantly lower than NF at 15–30 cm soil depth (Fig. 2).

The data showed that soil carbon percentage varied by land-use type and soil depth. The variation of soil carbon can be attributed to land-use management such as deforestation and afforestation which can alter carbon pool levels in the soil (Jandi et al. 2007). The conversion from forest to cultivated land can cause a carbon loss. The loss of soil carbon by land conversion can be up to 30% (Murty et al. 2002). All of the palm oil areas in the concession area were converted from peat forest to palm oil plantations. Most of the agriculture sites such as in Kuala Pembuang were also converted from forest to paddy fields. These land conversions may have reduced soil carbon in TA and POP treatments in Rimba Raya area.

A study conducted by Saha et al. (2009) also showed similar results that soil carbon was higher in AHG than in monoculture managements. In our study, soil

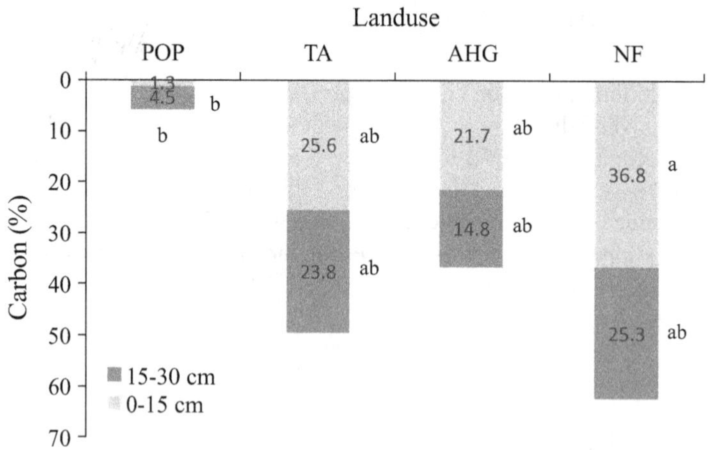


**Fig. 2** The mean soil carbon percentage for palm oil plantation (POP), traditional agriculture (TA), agroforestry homegarden (AHG), and natural forest (NF) treatments at 0–15 cm and 15–30 cm soil depth in Kuala Pembuang, Muara Dua, and Telaga Pulang Sites, Indonesia. Columns with different letters are significantly different at $\alpha = 0.05$

carbon level in AHG was higher than in POP (monoculture). This is because AHG consists of several species including perennials and annuals. However, our results disagree with Saha et al. (2009) because we found that soil carbon level in TA (monoculture) was higher than in AHG. This can be due to integration of crop residue in the TA treatment and addition of manure and other fertilizers for increased crop production (Kong et al. 2004).

Disturbances in tropical peat forest such as fire, drought, and also deforestation may have altered the amount of carbon in the soil (Saha et al. 2009). In Rimba Raya working area, all of the natural forests and agriculture sites have experienced severe logging events 10–15 years ago before the restoration effort in early 2008. Fire has become one of the major disturbances that have happened at the concession area of Rimba Raya including these study sites. Severe fire damages also occurred in 2015 and 2016 and the most severe one occurred at the peat forest in Kuala Pembuang. Almost all of the agriculture activities of the local people were performed on the previously fire-damaged sites.

## *Soil Nitrogen Concentration*

The highest level of N in the soil was observed for NF at 15–30 cm depth (1.2%) followed by TA at 0–15 cm (1.1%). The nitrogen for AHG was 0.61% at 15–30 cm depth and 0.6% at 0–15 cm soil depth. The lowest N level was observed for POP soils (0.4% and 0.2%) at both soil depth. Nitrogen concentration of NF (0–15 cm and 15–30 cm) and TA at a depth of 0–15 cm was significantly greater than POP at

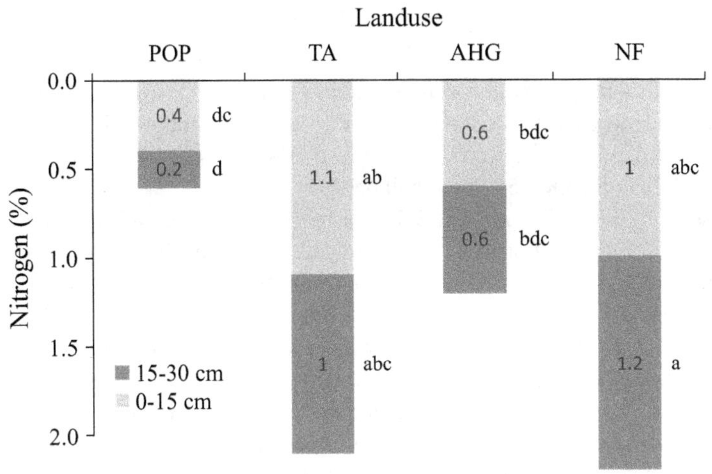


**Fig. 3** The mean soil nitrogen percentage for palm oil plantation (POP), traditional agriculture (TA), agroforestry homegarden (AHG), and natural forest (NF) treatments at 0–15 cm and 15–30 cm soil depth in Kuala Pembuang, Muara Dua, and Telaga Pulang sites, Indonesia. Columns with different letters are significantly different at $\alpha = 0.05$

15–30 cm soil depth and NF at 0–15 cm was significantly different with AHG at both soil depths and POP at 0–15 cm (Fig. 3).

Soil microbial biomass contributes to affecting the amount of soil N. Soil biomass affects the soil nutrient cycle that can influence the soil microbial symbiosis. Due to its competitive ability, soil microbial biomass is considered as the source for essential nutrients including N (Horwarth 2003). These conditions can explain why the natural forest has higher soil N concentration. We can assume that due to its natural condition, natural forest has higher number of soil microbial biomass (Jackson et al. 2000). The results also indicate that the amount of N concentration in AHG treatment was quite high and has a potential to store nitrogen.

Homegarden system, as the alternative land use, also showed better N levels because tree-based land use in tropical areas offers a greater potential to increase N level in the soil and prevents N loss from soil (Pandey et al. 2010). In general, AHG sites have several species of legume plants for human and animal use and these plants can help improve soil N status. Thus these systems have several ecological benefits. One of those benefits is the effective and efficient nutrient cycling. The abundance of plants in homegarden system results in continuous recycling of soil matter (Galhena et al. 2013).

Similar to our result, lower N concentration was observed by a study conducted by Allen et al. (2015). In their study, the forest that has been converted to palm oil plantations had lower N concentrations than the natural forest. Their study indicated that the proportion of clay in palm oil soil can affect the microbial activity and subsequent N immobilization can contribute to reducing N concentrations in the soil.

## *C:N Ratio*

The highest C:N ratio of 36.3 was recorded in TA treatment in Telaga Pulang for the 0–15 cm soil depth followed by 34.66 for TA treatment in Muara Dua within 0–15 cm soil depth. The lowest C:N ratio was observed (0.00) for a POP treatment in Telaga Pulang for the 15–30 cm soil depth.

The second lowest C:N ratio of 2.51 was recorded for Muara Dua POP treatment for 0–15 cm soil depth. For most treatments, C:N ratio for the 15–30 cm depth was higher than at 0–15 cm soil depth except for POP treatment in Kuala Pembuang (Table 2). Factors that can affect or lower the soil C:N ratio include N-rich soils which may cause the decrease of C:N ratio. Additionally, soils that are fertilized regularly and rapid decomposition of organic material cause lower C:N ratios (Ernfors et al. 2007). POP treatment in Muara Dua and Telaga Pulang has been established 10 years ago. Those sites are frequently fertilized; thus those sites have high fertilizer accumulation, especially urea. The management system of palm oil

**Table 2** C:N ratio for palm oil plantation (POP), traditional agriculture (TA), agroforestry homegarden (AHG), and natural forest (NF) treatments at 0–15 cm and 15–30 cm soil depth in Kuala Pembuang, Muara Dua, and Telaga Pulang sites, Indonesia

| Location | Land use | Depth (cm) | Mean C:N ratio |
|---|---|---|---|
| Kuala Pembuang | Natural forest | 0–15 | 10.74 |
| | | 15–30 | 17.37 |
| | Agriculture | 0–15 | 15.10 |
| | | 15–30 | 15.10 |
| | Homegarden | 0–15 | 7.49 |
| | | 15–30 | 10.15 |
| | Palm oil | 0–15 | 13.85 |
| | | 15–30 | 3.10 |
| Muara Dua | Natural forest | 0–15 | 26.83 |
| | | 15–30 | 29.85 |
| | Agriculture | 0–15 | 19.45 |
| | | 15–30 | 34.66 |
| | Homegarden | 0–1515 | 6.25 |
| | | 15–30 | 9.78 |
| | Palm oil | 0–15 | 2.51 |
| | | 15–30 | 2.66 |
| Telaga Pulang | Natural forest | 0–15 | 25.26 |
| | | 15–30 | 28.15 |
| | Agriculture | 0–15 | 36.30 |
| | | 15–30 | 31.49 |
| | Homegarden | 0–15 | 31.95 |
| | | 15–30 | 17.83 |
| | Palm oil | 0–15 | 3.91 |
| | | 15–30 | 0.00 |

plantation can also affect the decomposition rate, soil moisture, and soil temperature. Land conversion from natural peat forest to palm oil plantation can lower C:N ratio due to the increasing of the soil temperature and decomposition rate (Melling et al. 2013).

## *Urease Activity*

Urease activity ranged from 137μg $N\text{-}NH_4\ g^{-1}\ h^{-1}$ to 569μg $N\text{-}NH_4\ g^{-1}\ h^{-1}$ among the study treatments (Fig. 4). The highest activity of urease was 569.7μg $N\text{-}NH_4\ g^{-1}\ h^{-1}$ and it was observed in the NF at 0–15 cm depth and followed by 515.6μg $N\text{-}NH_4\ g^{-1}\ h^{-1}$ in AHG at 0–15 cm depth and TA treatment 473.8μg $N\text{-}NH_4\ g^{-1}\ h^{-1}$ at 15–30 cm soil depth (Fig. 4). The lowest activity of urease was 137.7μg $N\text{-}NH_4\ g^{-1}\ h^{-1}$ and 155.2μg $N\text{-}NH_4\ g^{-1}\ h^{-1}$ for 0–15 cm and 15–30 cm depths in POP (Fig. 4). The urease activities in TA treatment at 15–30 cm, NF treatment at 0–15 cm, and AHG at 0–15 cm were significantly higher than POP at 0–15 cm and 15–30 cm (Fig. 4).

The 0–15 cm and 15–30 cm depths are classified as the surface soil. On these depths the highest and most active microorganisms occur and high activity of microorganisms can increase enzyme activity including urease (Taylor et al. 2002). Based

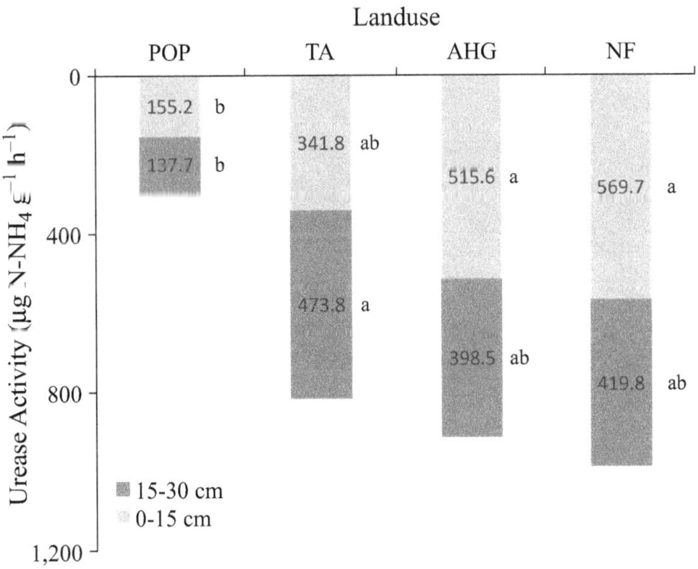


**Fig. 4** The mean urease activity for palm oil plantation (POP), traditional agriculture (TA), agroforestry homegarden (AHG), and natural forest (NF) treatments at 0–15 cm and 15–30 cm soil depth in Kuala Pembuang, Muara Dua, and Telaga Pulang sites, Indonesia. Columns with different letters are significantly different at $\alpha = 0.05$

on the result, urease activity was highest in the 0–15 cm depth because this depth is still in the surface soil layer where carbon and microbial activity were higher than those in deeper soil layers.

Urease activity was positively correlated with C level (Fig. 5). The correlation between soil carbon and urease activity was significant ($p = 0.0002$, $\alpha = 0.05$).

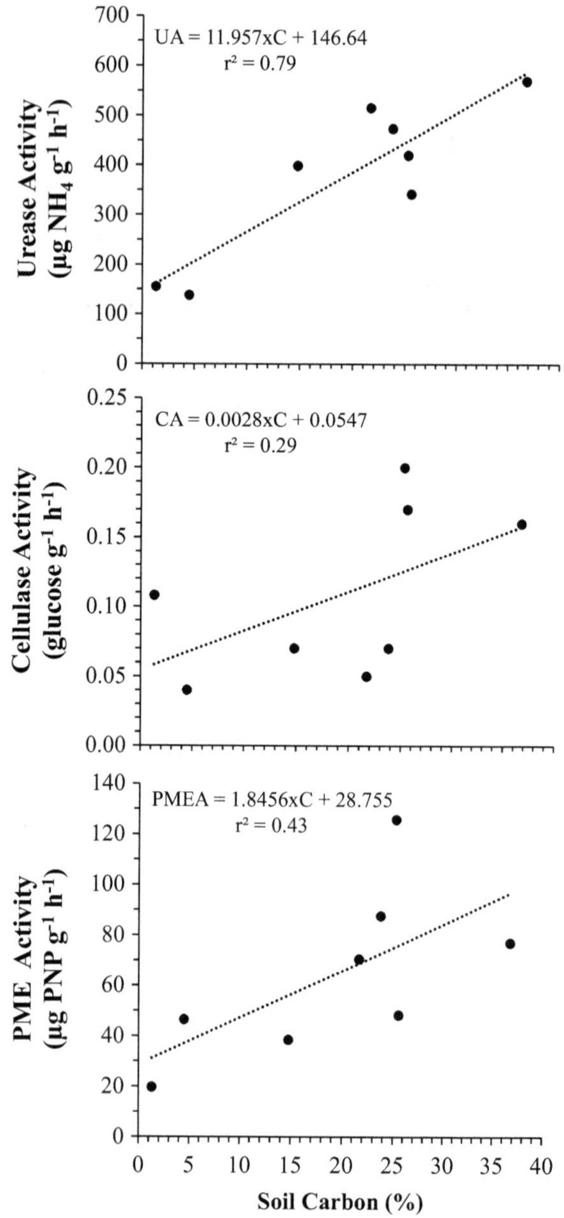


**Fig. 5** Linear regression for the correlation between urease (UA), cellulase (CA), and PME activity (PMEA) and soil carbon for palm oil plantation (POP), traditional agriculture (TA), agroforestry homegarden (AHG), and natural forest (NF) treatments at 0–15 cm and 15–30 cm soil depth in Kuala Pembuang, Muara Dua, and Telaga Pulang sites

Findings of our study also agree with a study conducted by Blonska (2010). Their study showed that urease activity in NF soils was positively correlated with C content, C:N ratio, and soil moisture. The activity of soil enzyme including urease is related with soil chemical properties such as pH, EC, available nutrients, soil texture, soil carbon, and soil nitrogen. A study conducted by Kumari et al. (2017) showed results similar to our study that urease activity has a positive correlation with soil carbon.

## *Cellulase Activity*

The highest activity of the cellulase enzyme was 0.17μg glucose $g^{-1}$ $h^{-1}$ in TA treatment at 0–15 cm soil depth. It was followed by the activity of cellulase enzyme in NF at 0–15 cm depth (0.16μg glucose $g^{-1}$ $h^{-1}$). The cellulase activity of AHG was 0.05μg glucose $g^{-1}$ $h^{-1}$ at 0–15 cm and 0.07μg glucose $g^{-1}$ $h^{-1}$ at 15–30 cm depth while the cellulase activity in POP treatment was 0.108μg glucose $g^{-1}$ $h^{-1}$ at 0–15 cm and 0.04μg glucose $g^{-1}$ $h^{-1}$ at 15–30 cm (Fig. 6). However, cellulase activity was not significantly different among the treatments and depths. Soil carbon and cellulase activity were weaker than with urease (Fig. 5). Soil C explained only 29% of the urease activity for these soils.

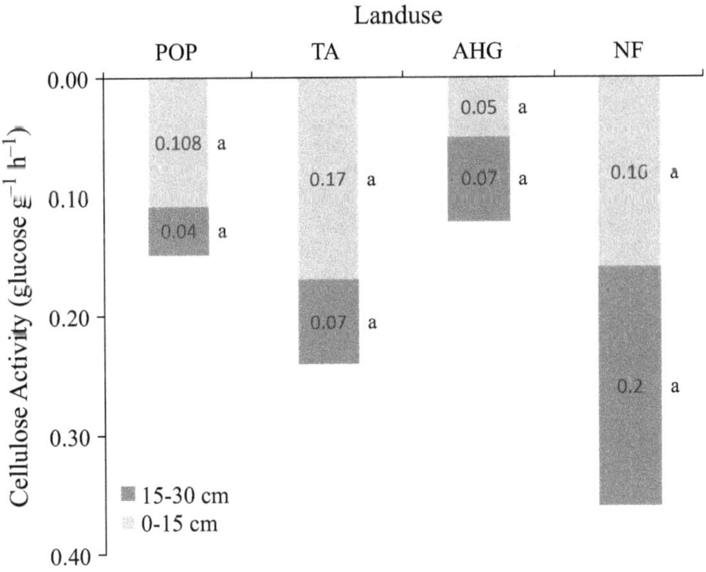


**Fig. 6** The mean cellulase activity for palm oil plantation (POP), traditional agriculture (TA), agroforestry homegarden (AHG), and natural forest (NF) treatments at 0–15 cm and 15–30 cm soil depth in Kuala Pembuang, Muara Dua, and Telaga Pulang sites. Columns with different letters are significantly different at $\alpha = 0.05$

Higher cellulase activity observed in the TA treatment was also similar to a study conducted by Lupicka et al. (2016). In their study, TA treatment had high cellulase activity especially the site with roundup treatment. They found that agriculture sites had high cellulase activity because the other additives such as urea phosphate stimulate the activity of the cellulolytic microflora.

The activity of cellulase in the NF treatment can be attributed to the diverse vegetation including woody plants, shrubs, and grasses. Diverse plant populations, litter material, as well as their interactions can increase the activity of cellulase (Chaer et al. 2009). The high soil C percentage can increase the activity of cellulase enzyme because there was an enrichment of the hydrolytic enzymes including cellulase in organic matter and maintain a high metabolic activity (Cepeda et al. 2008).

## *PMEase Activity*

The highest activity of PMEase was 125.8μg PNP $g^{-1}$ $h^{-1}$ and it was observed at 15–30 cm depth of NF (Fig. 7). This was followed by PMEase activity at TA at 15–30 cm (87.6 μPNP $g^{-1}$ $h^{-1}$), NF at 0–15 cm (77μg PNP $g^{-1}$ $h^{-1}$), and AHG at 0–15 cm (70.5μg PNP $g^{-1}$ $h^{-1}$). The lowest PMEase activity was 19.7μg PNP $g^{-1}$ $h^{-1}$ in POP treatment at 0–15 cm (Fig. 7). However, PMEase activities were similar for

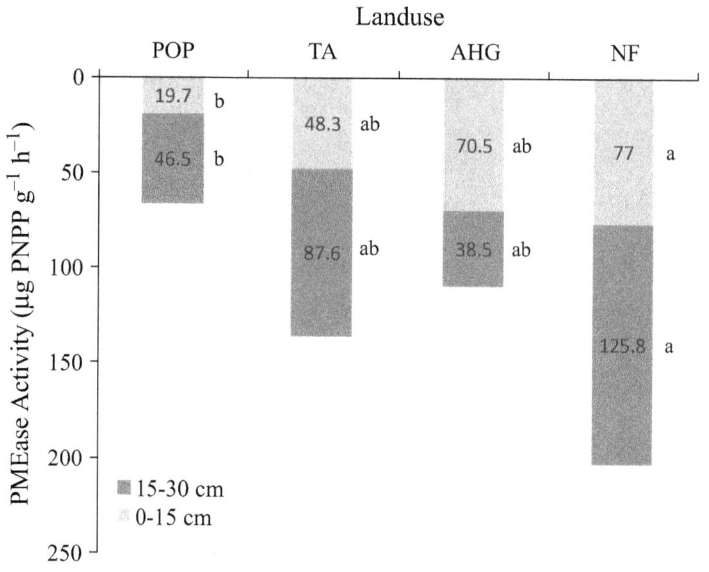


**Fig. 7** The mean PMEase activity for palm oil plantation (POP), traditional agriculture (TA), agroforestry homegarden (AHG), and natural forest (NF) treatments at 0–15 cm and 15–30 cm soil depth in Kuala Pembuang, Muara Dua, and Telaga Pulang sites, Indonesia. Columns with different letters are significantly different at $\alpha = 0.05$

15–30 cm depth of POP (46.5µg PNP $g^{-1}$ $h^{-1}$) and 0–15 cm depth of TA (48.3µg PNP $g^{-1}$ $h^{-1}$). The activity of NF at 0–15 cm and 15–30 cm was significantly higher than POP at both soil depths (Fig. 7).

## Conclusions

This study evaluated several soil health parameters to understand the effects of land management on these parameters. The study intended to understand the role of sustainable land management practices like agroforestry in improving soil health. The highest activity of those studied enzymes was mostly found in NF, AHG, and TA treatments. Among the studied enzymes, urease showed the highest activities on all four treatments. Palm oil plantation treatment had the lowest enzyme activities, and carbon and nitrogen concentration, and also the lowest C:N ratio. The 0–15 cm indicated greater activities and concentrations of all measured variables than the 15–30 cm soil depth. Based on our results AHG can be considered as the alternative land-use system for PT. Rimba Raya Area to improve soil health indicators.

**Acknowledgements** This work was funded through the USAID-CIFOR Project. We would like to express our thanks to the Center for Agroforestry, University of Missouri, for the support. We would like to express our sincere thanks to PT. Rimba Raya for their support and guidance during the fieldwork. We would also like to express our thanks Mr. Heru, Mr. Sodik, Wuryanto, and the team of Microbiology and Ecology Laboratory, Center for Biology, Indonesian Institute of Sciences (LIPI), for their support and guidance during the laboratory work. We would like to thank Dr. Mark Ellersieck for his guidance and advice on data analysis.

## References

Allen K, Corre MD, Tjoa A, Veldkamp E (2015) Soil nitrogen-cycling responses to conversion of lowland forests to oil palm and rubber plantations in Sumatera, Indonesia. PLoS One 10(7):1–21

Bandick AK, Dick RP (1999) Field management effects on soil enzyme activities. Soil Biol Biochem 31:1471–1479

Blonska E (2010) Enzyme activity in forest peat soils. FFP 52(1):20–25

Cepeda CT, Leiros MC, Sotres FG (2008) Hydrolitic enzyme activities in agricultural and forest soils. Some implications for their use as indicators of soil quality. Soil Biol Biochem 20:1–10

Chaer G, Fernandes M, Myrold D, Bottomley P (2009) Comparative resistance and resilience of soil microbial communities and enzyme activities in adjacent native forest and agricultural soils. Microb Ecol 58:414–424

Deng S (1994) Cellulase activity of soils and the effect of tillage management on enzyme activities in soil. Doctoral dissertation, Iowa State University

Dick RP (1994) Soil enzyme activities as indicators of soil quality. In: Doran JV, Coleman DC, Bezdicek DF, Stewart BE (eds) Defining soil quality for sustainable environment. SSSA-ASA, Madison, pp 107–124

Dieckow J, Mielniczuk J, Knicker H, Bayer C, Dick DP, Kögel-Knabner I (2007) Comparison of carbon and nitrogen determination methods for samples of a paleudult subjected to no-till cropping systems. Sci Agric 64(5):532–540
Dollinger J, Jose S (2018) Agroforestry for soil health. Agrofor Syst 92:213–219
Doran JW, Zeiss MR (2000) Soil health sustainability: managing the biotic component of soil quality. Appl Soil Ecol 15:3–11
Eivazi F, Tabatabai MA (1977) Phosphates in soils. Soil Biol Biochem 9:167–172
Ernfors M, von Arnold K, Stendahl J et al (2007) Nitrous oxide emissions from drained organic forest soils - an up scaling based on C:N ratios. Biogeochemistry 84:219–231
Fitzherbert EB, Struebig MJ, Morel A et al (2008) How will oil palm expansion affect biodiversity? Trends Ecol Evol 23(10):538–545
Food and Agriculture Organization (2016) State of the world's forest-forest and agriculture: land use challenges and opportunities. FAO, Rome
Galhena DH, Freed R, Maredia KM (2013) Home gardens: a promising approach to enhance household food security and well-being. Ag Food Sec 2:1–13
Horwarth W (2003) Biomass: soil microbial biomass. In: Encyclopedia of agrochemicals. https://doi.org/10.1002/047126363X.agr362
Iqbal M, Sumaryanto (2007) Strategi pengendalian alih fungsi lahan pertanian bertumpu pada partisipasi masyarakat. AKP 5(2):167–182
Jandi R, Lindner M, Vesterdal L et al (2007) How strongly can forest management influence soil carbon sequestration? Geoderma 137:253–268
Jackson CR, Liew KC, Yule CM (2008) Structural and functional change with depth in microbial communities in a tropical Malaysian peat swamp forest. Microb Ecol 57:402–412
Jose S (2009) Agroforestry for ecosystem services and environmental benefit: an overview. Agrofor Syst 76:1–10
Kanazawa S, Miyashita K (2012) Cellulase activity in forest soils. J Soil Sci Plant Nutr 33(3):399–406
Kang BT, Akinifesi FK (2000) Agroforestry as alternative land-use production systems for the tropics. Nat Res For 24:137–151
Kong AYY, Six J, Bryant DC, Denison RF, van Kessel C (2004) The relationship between carbon input, aggregation, and soil organic carbon stabilization in sustainable cropping system. Soil Sci Soc Am J 69(4):1078–1085
Kool DM, Hoffland E, Abrahamse SPA, van Groenigen JW (2006) What artificial urine composition is adequate for simulating soil $N_2O$ fluxes and mineral N dynamics. Soil Biol Biochem 38(7):1757–1763
Kumari JA, Rao PC, Padmaja G, Madhavi M (2017) Effect of physico-chemical properties on soil enzyme urease activity in some soils of Ranga Reddy District of Telangana State, India. Int J Curr Microbiol App Sci 6(11):1708–1714
Lemons T, Reece BL, Procanik J et al (2011) The project design document: the Rimba Raya Biodiversity Reserve Project. PT. Rimba Raya Conservation, Jakarta
Lupicka TK, Krecidlo L, Krecidlo M (2016) The comparison of cellulolytic activity of the modified soil treated with roundup. Chem Didact Ecol Metrol 21:133–139
Melling L, Hatano R, Goh KJ (2013) Methane fluxes from three ecosystems in tropical peatland of Sarawak, Malaysia. Soil Biol Biochem 37:1445–1453
Murty D, Kirschbaum MUF, Mcmurtrie RE, Mcgilvray H (2002) Does conversion of forest to agricultural land change soil carbon and nitrogen? Glob Chang Biol 8(2):105–123
Pandey CB, Singh GB, Singh SK, Singh RK (2010) Soil nitrogen and microbial biomass carbon dynamics in native forests and derived agricultural land use in humid tropical climate of India. Plant Soil 333:453–467
Parham JA, Deng SP (2000) Detection, quantification and characterization of β-glucosaminidase activity in soil. Soil Biol Biochem 32:1183–1190

Paudel B, Udawatta RP, Kremer RJ, Anderson SH (2012) Soil quality indicator response to row crop, grazed pasture, and agroforestry buffer management. Agrofor Syst 84:311–323

Saha SK, Nair PKR, Nair VD, Kumar BM (2009) Soil carbon stock in relation to plant diversity of homegardens in Kerala, India. Agrofor Syst 76:53–65

SAS 9.2 Software (2008) https://support.sas.com/software/92/

Schroth G, Lehmann J, Rodrigues MRL, Barros E, Macêdo JLV (2001) Plant-soil interactions in multistrata agroforestry in the humid tropics. Agrofor Syst 53:85–102

Sitorus SRP, Pravitasari AE (2017) Land degradation and land slide in Indonesia. SJDGGE 1(2):61–71

Tabatabai MA (1994) Soil enzymes. In: Weaver RW, Angle JS, Bottomley PS (eds) Methods of soil analysis part 2: microbiological and biochemical properties. SSSA Book, Madison

Taylor JP, Wilson B, Mills MS, Burns RG (2002) Comparison of microbial numbers and enzymatic activities in surface soils and subsoils using various techniques. Soil Biol Biochem 34:387–401

Tivet F, de Moraes Sa JC, Borszowskei PR, Letourmy P, Briedis C, Ferreira AO, dos Santos JB, Inagaki TM (2011) Soil carbon inventory by wet oxidation and dry combustion methods: effects of land use, soil texture, gradients, and sampling depth on the linear model of c-equivalent correction factor. Soil Sci Soc Am J 76:1048–1059

USDA (2018) USDA website. https://www.nrcs.usda.gov/Internet/FSE_DOCUMENTS/nrcseprd331820.pdf/. Accessed 16 Apr 2018

Vijay V, Pimm SL, Jenkins CN, Smith SJ (2016) The impacts of oil palm on recent deforestation and biodiversity loss. PLoS One 11(7):e0159668. https://doi.org/10.1371/journal.pone.0159668:1-19

Wicke B, Sikkema R, Dornburg V, Faajii A (2011) Exploring land use changes and the role of palm oil production in Indonesia and Malaysia. Land Use Policy 28:193–206

Yakovchenko VP, Sikora LJ, Millner PD (1998) Carbon and nitrogen mineralization of added particulate and microorganic matter. Soil Biol Biochem 30(14):2139–2146

# Water Quality and Quantity Benefits of Agroforestry and Processes: Long-Term Case Studies from Missouri, USA

**Ranjith P. Udawatta, Harold E. Garrett, Shibu Jose, and Sarah T. Lovell**

## Abbreviations

| | |
|---|---|
| AF | Agroforestry |
| CT | Computed tomography |
| GAF | Grazing AF watershed study |
| GB | Grass buffer |
| HARC | Horticulture Agroforestry Research Center |
| Ksat | Saturated hydraulic conductivity |
| MLRA | Major land resource areas |
| N | Nitrogen |
| NPSP | Nonpoint source pollution |
| P | Phosphorus |
| RAF | Row crop AF watershed study |
| SOM | Soil organic matter |

## Introduction

Technological advances, rising demand for food, and land tenure have tremendously changed the US agricultural landscape during the past century. Currently, US farms are more mechanized, mostly mono cropped, less diverse, and large, and require high inputs as compared to six or seven decades ago (NRC 2010). Cereal yields have almost doubled over the years due to these changes; however, some environmental indicators have been negatively impacted by these changes. For instance, 44% of rivers and streams and 64% of lakes, reservoirs, and ponds in the USA are impaired and unsuitable for human use and recreational activities (USEPA 2017). In

R. P. Udawatta (✉) · H. E. Garrett · S. Jose · S. T. Lovell
The Center for Agroforestry, School of Natural Resources, University of Missouri, Columbia, MO, USA
e-mail: UdawattaR@missouri.edu; joses@missouri.edu; slovell@missouri.edu

 
R. P. Udawatta, S. Jose (eds.), *Agroforestry and Ecosystem Services*,
https://doi.org/10.1007/978-3-030-80060-4_6

Missouri, 54% of rivers and streams and 35% of lakes do not meet water quality standards (MDNR 2012). As of 2000, 62% of the cropland in Missouri was losing soil above the tolerance level of 7.6 Mg $ha^{-1}$ (USDA-NRCS 2012). The main sources of impairment of rivers and streams were sediments, nutrients, pathogens, and chemicals originating from agricultural lands. Sediment in runoff water is the largest contaminant of surface water by weight and volume (Koltun et al. 1997), and is identified by the USA as the leading pollutant in rivers and streams . Annually, 75 billion metric tons of soil are being eroded from croplands worldwide (http://www.fao.org/global-soil-partnership/resources/highlights/detail/en/c/416516 accessesed 11/02/2019), but in the USA, a significant reduction of soil erosion has been reported due to adoption of conservation practices (USDA-NRCS 2007a).

Other pollutants including nutrients, herbicides, pathogens, and various antibiotics from agricultural practices impact water, soil, biodiversity, and the environment. The total nitrogen (all forms), phosphate ($P_2O_5$), and potash ($K_2O$) fertilizer usage has increased by 34%, 40%, and 45%, respectively, in 2017, compared to 2002 (FAO 2019). Furthermore, Asia and America account for more than half of the total global fertilizer usage. Nutrients lost from farmland in the corn-belt region have been strongly associated with the expansion of the hypoxia zone in the Gulf of Mexico and the reduction of economies of seven surrounding states . USDA-NRCS (2013) reported that the Lower Mississippi River Basin discharged 30.26 kg N $ha^{-1}$ $y^{-1}$ and 2.91 kg P $ha^{-1}$ $y^{-1}$ between 2003 and 2006 to water bodies. From 1990 to 2016, the global pesticide usage increased from 1.5 $kgha^{-1}$ to 2.57 $kgha^{-1}$, and the pesticide usage in the USA increased from 1.59 $kgha^{-1}$ to 3.39 $kgha^{-1}$ (http://www.fao.org/faostat/en/#data/EP/visualize, accessed on 11/02/2019). Antibiotic overuse for prophylaxis, metaphylaxis, and therapeutic treatments in livestock is another major concern, as up to 90% of the antibiotic parent compounds are directly excreted (Massé et al. 2014; Sarmah et al. 2006; Kumar et al. 2005).

Although agricultural practices are often criticized for impacting water bodies with nonpoint source pollution (NPSP), reductions in NPSP have been seen in the last several decades, due to various conservation practices, government support, and also implementation of the Clean Water Act in the 1970s (Udawatta et al. 2004, 2006a). During the last 50 years, soil erosion in the USA by water on cropland has been reduced from 1.68 billion tons in 1982 to 960 million tons in 2007, a 43% reduction, due to the adoption of soil conservation practices (USDA-NRCS, 2007b). Although there are improvements due to conservation practices, 28% of the cropland still had erosion rates higher than the tolerance level in 2007 as compared to 40% in 1982 (USDA-NRCS 2012). In Missouri, soil loss has been halved from 11 tons $ha^{-1}$ in 1987 to 2013 (MDNR 2017). Many other states have also reported similar reductions in soil and nutrient losses from agricultural watersheds due to the adoption of various conservation practices. Combined, this evidence suggests that conservation practices such as minimum tillage, cover crops, crop rotation, buffers, and nutrient management can be an effective strategy, although the mechanisms and processes for these improvements need further study in order to continue this positive trend.

Sustainable agriculture and ecosystem services are also impacted by monocropping and shrinking land area available for agriculture. Monocropping and heavy chemical use have been strongly connected to declining environmental quality and biodiversity in many regions of the world (Udawatta et al. 2019). Globally, the agricultural land base has declined at a significant rate due to the expansion of urban, industrial, and commercial areas. For instance, in the USA, croplands decreased by 12.5 million ha between 1992 and 2012, and from 63% to 51% of the total land area from 1949 to 2007 (USDA-ERS 2012). The predicted increase in soil erosivity between 16 and 58% due to climate change could further impact land productivity, water, soil, and the environment (Nearing 2001). This emphasizes the importance of site-suitable sustainable land management practices that also produce more food from a more diverse set of crops, thus minimizing environmental degradation while satisfying dietary needs.

Agroforestry (AF) is a complex land management practice where trees and/or shrubs are intentionally integrated into row-crop and/or livestock practices for various production, environmental, and economic benefits (Gold and Garrett 2009). These benefits are the result of biophysical interactions among the components in these systems (Garrett et al. 1994; Jose 2019). Therefore, AF is known to address issues on sustainability and ecosystem services including water and soil quality, and biodiversity (Udawatta et al. 2002, 2009, 2019, 2020; Schultz et al. 2009; Jose 2009). A recent meta-analysis from Europe has shown overall positive effects of AF on biodiversity (Torralba et al. 2016). Riparian buffers, alley cropping, shelter belts, silvopasture, forest farming, and urban food forests are the main AF practices usually practiced in temperate zones. However, agroforestry buffers differ from vegetative or grassed filter strips in that they consist of a viable forest ecosystem (Schultz et al. 2009). This chapter synthesizes water quality and quantity benefits of AF from the long-term Greenley paired row crop AF watershed study (RAF) located in northeast Missouri and the grazing AF watershed study (GAF) at the Horticulture and Agroforestry Research Center (HARC) in Central Missouri, USA.

## *The Two Study Sites (RAF and GAF)*

The Greenley paired row crop AF watershed study (RAF) and the grazing AF watershed study (GAF) both have contrasting land management practices, soils, experimental designs, and statistical data analysis procedures. The three adjacent row crop watersheds at the Greenley Center were in a corn (*Zea mays* L.)-soybean [*Glycine max* (L.) Merr.] rotation while the six watersheds at HARC were in cattle grazing (Fig. 1). The three row crop watersheds are located in the claypan region, Major Land Resource Area (MLRA) 113. Soils in these watersheds have a shrink-swell smectite clay-rich horizon with very low hydraulic conductivity and thus runoff potential is great on these soils. The six cattle grazing watersheds are located on loess-derived deep Menfro soils (MLRA 107B) with 12–15% slopes; thus runoff

**Fig. 1** Alley cropping agroforestry with row crops and biomass crops (RAF) at the Greenley Research Center (left) and riparian agroforestry with cattle grazing (GAF) at the Horticulture and Agroforestry Research Center (right), Missouri, USA (photo Ranjith Udawatta)

potential is significantly lower compared to the RAF watersheds. Both study areas have grass buffers (GB) and no-buffer control treatments.

Six upland tree+grass buffers at RAF represent alley cropping AF practices. Cottonwood+grass buffers at GAF represent riparian buffers at the bottom of watersheds along a waterbody. The three row crop watersheds used the paired watershed approach (USEPA 1993), while grazing watersheds used a completely randomized block design for statistical analysis.

Three adjacent north-facing RAF watersheds at the Greenley Research Center were instrumented with H-flumes, approach sections, water samplers, and flow meters in 1991 to collect water samples (Fig. 2; 40°01' N, 92°11' W; Udawatta et al. 2002). Watersheds were on a no-till corn-soybean rotation and drained by fescue grass [*Schedonorus phoenix* (Scop.) Holub] waterways. The east watershed is 1.65 ha, the center watershed is 4.44 ha, and the west watershed is 3.16 ha. Grass legume buffers of redtop (*Agrostis gigantea* Roth), smooth bromegrass (*Bromus inermis* L.), and birdsfoot trefoil (*Lotus corniculatus* L.) were established on contours on the west and center watersheds in 1997 after a 7-year calibration period. Bur oak (*Quercus macrocarpa* Michx.), pin oak (*Q. palustris* Muenchh.), and swamp white oak (*Q. bicolor* Willd.) were alternately planted at a 3-m spacing in the center of the grass-legume buffer strips of the center watershed to create AF buffers. Crop areas on the AF and grass buffer watersheds were converted to biomass crops consisting of switch grass (*Panicum virgatum* L.), big blue stem (*Andropogon gerardi* Vitman.), Indian grass (*Sorghastrum nutans* [L.J. Nash]), Illinois bundleflower (*Desmanthus illinoensis* (Michx.) MacMill. ex B.L. Rob. & Fernald), and partridge pea (*Chamaecrista fasciculata* (Michx.) Greene) in 2012. More details on management, soils, and weather can be found in Udawatta et al. (2002, 2011a) and Alagele et al. (2020a).

Six mini watersheds of GAF were established and instrumented with equipment similar to RAF in 2000 to collect water samples (Fig. 2; 39°02'N, 92°46'W, 195 m amsl; Udawatta et al. 2010). Treatments were AF buffers, GB, and control with two

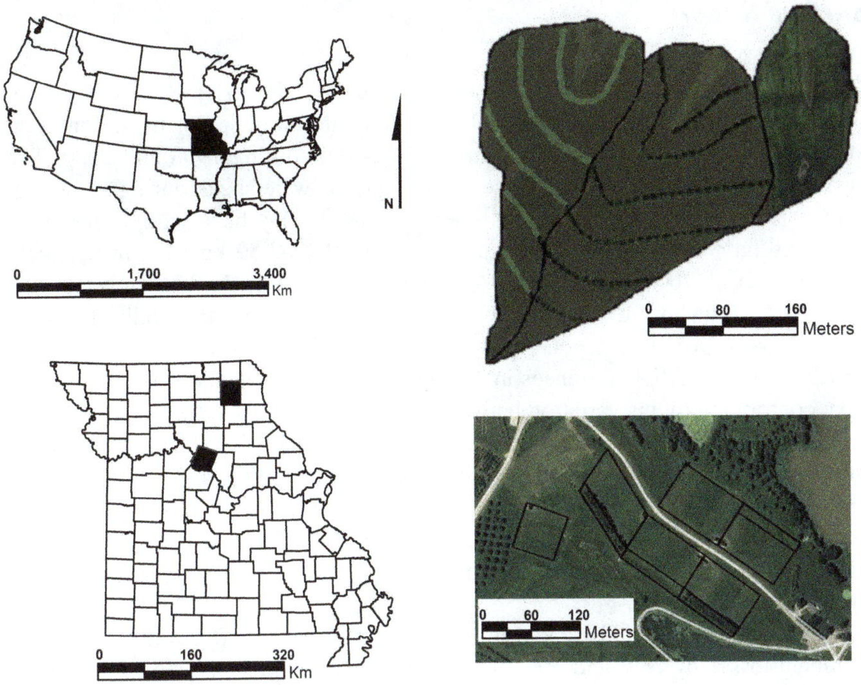


**Fig. 2** Left: Location of three row crop watersheds (RAF) of the paired watershed study at Greenley Research Center, Novelty, and six grazing watersheds (GAF) at the Horticulture and Agroforestry Research Center (HARC), New Franklin, Howard County, Missouri. Top right: Wide light green bands represent grass buffers on the contour grass strip watershed, dark dotted lines represent agroforestry buffers (tree+grass) on the agroforestry watershed, green areas represent row crop areas, and black bands represent watershed boundary. Bottom right: Narrow dark strips on two watersheds represent agroforestry buffers (tree+grass), open strips represent grass buffers, and black bands represent watershed boundaries

replications. All grazing and buffer areas were seeded with tall fescue (*Festuca arundinacea* Schreb; Kentucky 31) in 2000. The grazing areas were reseeded with a mixture of fescue, red clover (*Trifolium pretense* L.), and lespedeza (*Kummerowia stipulacea* Maxim.) in 2003 (Kumar et al. 2008). Buffers located at the lower landscape positions on AF and GB watersheds are 15 m wide and 60 m long. Four rows of eastern cottonwood (*Populus deltoides* Bortr. ex Marsh.) trees at 3 m within- and between-row spacing were established on the two AF watersheds. Four-wire permanent fences between the buffers and grazing areas prevented cattle access to buffer areas. Grazing areas were divided into six paddocks and each paddock was grazed for 3.5 days and rested for 17.5 days. More details on management, soils, and weather data can be found in the study of Kumar et al. (2008) and Udawatta et al. (2010, 2011a).

## *Surface Runoff*

The RAF at the University of Missouri Greenley Memorial Research Center (Udawatta et al. 2002, 2011a) showed reductions in runoff, sediment, and nutrient losses by agroforestry (tree+grass) and grass buffers during 1998 and 2000. The mean annual N and P loss on these three adjacent watersheds during the initial 7-year (1991–1997, no treatments) period was 16 kg ha$^{-1}$ with a range of 13–19 kg ha$^{-1}$ and 1.36 kg ha$^{-1}$ with a range of 0.29–3.59 kg ha$^{-1}$, respectively (Udawatta et al. 2004, 2006a). USDA-NRCS (2013) reported 30.26 kg N ha$^{-1}$ y$^{-1}$ and 2.91 kg P ha$^{-1}$ y$^{-1}$ discharge between 2003 and 2006 to water bodies from the Lower Mississippi River Basin. These losses are 1.9–2.1 times larger than the losses observed at the RAF. Differences in management, soils, weather, and regions may have contributed to this variation between the RAF and USDA-NRCS (2013) findings. During the 1998–2000 treatment period, the AF and GB treatments reduced runoff by 1–10%, total nitrogen (TN) loads by 21% and 20%, and total phosphorus (TP) loads by 4–26%. Using the mean reductions for N (20.5%) and P (15%), watersheds lost 12.8 kg N ha$^{-1}$ y$^{-1}$ and 1.2 kg P ha$^{-1}$ y$^{-1}$ between 1998 and 2000. These reductions were observed during a 3-year period soon after AF and GB treatments were established.

From 2004 to 2008, GB and AF treatments reduced runoff by 23% and 15%, sediment losses by 28% and 30%, TN losses by 13% and 11%, and TP losses by 22% and 26%, respectively (Udawatta et al. 2011a, Fig. 3). The sediment losses on the control, AF, and GB watersheds were 101, 76, and 82 kg ha$^{-1}$. The control watershed lost 4 and 3.2 times more TN than the AF and GB watersheds, respectively. A comparison between the two treatment periods indicated improved reductions in losses with time. Runoff volume and P loss were lower during 2004–2008 than 1998–2000. Losses were larger during years with more rain than drought years. This suggests the importance of buffers or conservation practices for reduction of

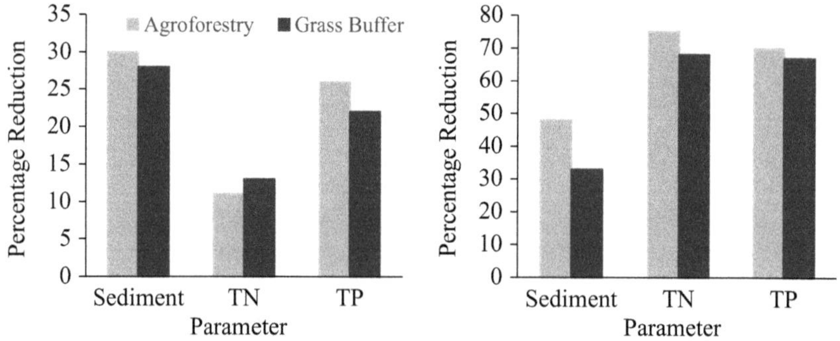


**Fig. 3** Percent reduction of sediment, total nitrogen (TN), and total phosphorus (TP) on row crop agroforestry watersheds (RAF) at Greenley Research Center (left) and grazing agroforestry (GAF) watersheds (right) at the Horticulture and Agroforestry Research Center (adapted from Udawatta et al. 2011a)

NPSP from agricultural watersheds and for other benefits like C sequestration, biodiversity, and reduction of evaporation from the soil surface.

Crop areas of AF and GB watersheds were converted to biomass crops in 2012. The region had 40% lower precipitation in 2012 than the long-term mean, which resulted in poor establishment. Biomass crops were reseeded in 2013 with better success. The two biomass watersheds with buffers had significantly lower runoff, sediment, N, and P losses during the measurement period than the crop watershed without buffers. The crop watershed lost 12 and 1.2 kg $ha^{-1}$ $y^{-1}$ N and P during 2014 and 2018. On average the two biomass watersheds lost only 3 and 0.2 kg N and P during this period. Total N loss was significantly reduced by buffer practices with biomass crops during 2012–2018 than for the two other study periods. On the same watersheds, N and P losses before establishment of the buffers were 16 and 1.29 kg $ha^{-1}y^{-1}$. After buffers were established these losses were significantly reduced as compared to the no-buffer watershed. Biomass crops have helped further reduce these losses. Additionally, harvested biomass also removes nutrients from these agricultural watersheds, offering further protection of water bodies for years to come.

During the study period GAF watersheds at HARC also showed significant improvements in water quality by AF and GB as compared to the control with no conservation practices (Fig. 3). These watersheds showed greater improvements in water quality than row crop watersheds in the claypan region. These differences can be attributed to soil types, vegetation characteristics, and management. RAF watersheds have grass waterways while cattle can access near flume areas at the grazing watersheds and these differences may have caused differences in water quality among watersheds.

Simulating APEX (Agricultural Policy Environmental eXtender) and fuzzy logic models, Senaviratne et al. (2013, 2014a, b, 2018) proved the long-term water quality benefits of AF. These authors calibrated and validated the model using long-term water quality and crop yield data and met the model performance criteria for all measured water quality parameters. These simulations evaluated water quality benefits associated with changes in buffer position, shape, and number of buffers for 5, 10, 20, and 30 years. Results showed that upland and edge-of-field buffers are required for water quality benefits. The most effective system consisted of an upland buffer and an edge-of-field buffer. These two buffers were located in the mid slope (2–5% slope) and footslope areas (5–9% slope) of the watershed. Buffers located in upland areas and near water bodies help retain soil and nutrients within the watershed while reducing the slope length and steepness. The edge-of-field buffer is similar to a riparian buffer and serves as the final defense mechanism.

## Groundwater

The effects of AF and GB on groundwater nitrate ($NO_3$-N) and TN concentrations were studied on AF and GB watersheds at the HARC in the GAF study (Wickramaratne 2017; Figs. 2 and 4). Two of the watersheds each had a transect of three wells installed at summit, backslope, and footslope positions. Well depth ranged from 12 m at the summit to 4 m at the footslope positions. These two footslope wells are located near the bottom of the AF and GB. Weekly groundwater samples were analyzed from December 2014 to December 2016 for $NO_3$-N and TN to understand buffer type and landscape effects on these N species. Both $NO_3$-N and TN were significantly lower at the footslope than at the summit and backslope ($p < 0.001$). The median concentrations of $NO_3$-N and TN were 0 mg $L^{-1}$ and 0.24 mg $L^{-1}$ at the footslope, compared to median concentrations >1.4 mg $L^{-1}$ at the summit and backslope. Across both transects, footslope $NO_3$-N concentrations were 30–400 times lower than the backslope and summit, demonstrating almost complete removal of $NO_3^-$ from groundwater. In this study, 97% of $NO_3$-N in groundwater was removed as water flowed from the summit to the footslope (Fig. 4). This effective $NO_3$-N removal occurred over a short flow path (40 m horizontal distance) with significant $NO_3^-$ inputs from the upper landscape, indicating that the collective

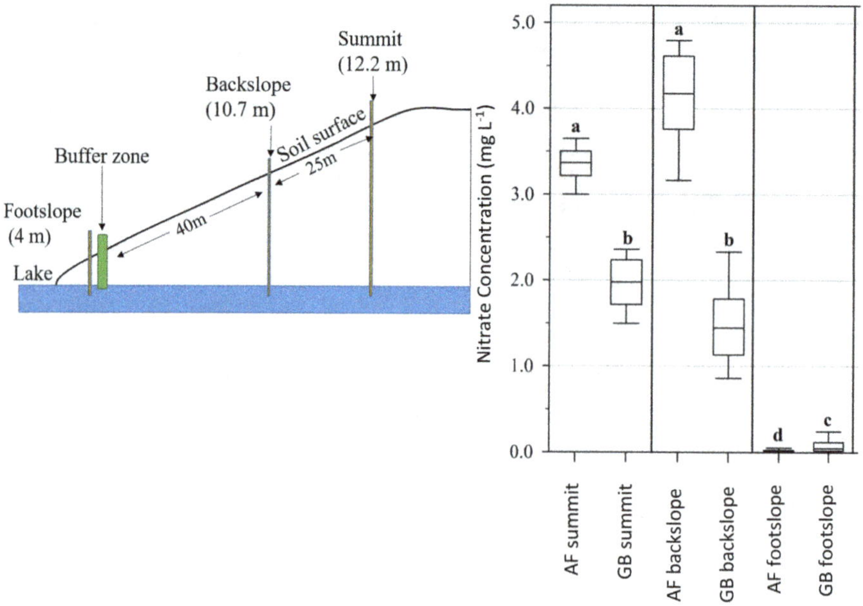


**Fig. 4** Schematic of well distribution and 2-year (December 2014–2016) mean concentrations of weekly nitrate samples at summit, backslope, and footslope wells of agroforestry (AF) and grass buffer (GB) transects at the grazing watersheds of the GAF study at the Horticulture and Agroforestry Research Center, New Franklin, Missouri (source: Wickramaratne 2017)

biological processes occurring within the riparian zones can almost eliminate $NO_3$-N in groundwater.

The $NO_3$-N/$Cl^-$ ratio was significantly lower at the footslope than at the summit and backslope, suggesting stronger denitrification at the footslope. The study also reported greater reduction of nitrate and TN concentrations by AF buffers than by GB. In spite of greater $NO_3$-N ($p < 0.001$) and TN ($p < 0.001$) concentrations at the summit and backslope positions of the AF watershed as compared to the GB watershed, the median $NO_3$-N concentration was lower at the footslope of the AF watershed than of the GB watershed (0 mg $L^{-1}$ in AF and 0.05 mg $L^{-1}$ in GB; $p < 0.001$).

At the footslope of both the AF and GB treatments, the combination of very low $NO_3$-N/$Cl^-$ ratios, shallow water table, high organic C in surface sediment, and frequent reducing conditions demonstrated that denitrification was the primary process causing $NO_3^-$ reductions in groundwater. Results indicate that proper vegetation management in lower landscape positions (and riparian areas) adjacent to cattle pastures can essentially eliminate $NO_3^-$ in shallow groundwater, and thus reduce its transport to streams and groundwater aquifers. These findings have a greater implication for public health as groundwater provides more than 33% of the water used for public drinking water supplies in the USA (Kenny et al. 2009).

## *Water Quality*

Integration of perennial vegetation changes soil parameters, watershed hydrology, water dynamics, and nutrients. Additionally, trees influence wind, heat, radiation, and humidity of these watersheds. Undisturbed soils, shade, and organic material influence various chemical reactions and biological activities as well as flora and fauna diversity. Some of these processes are not well understood and need further studies to explain how these changes influence water quality and quantity. For instance, there is only one study on AF effects on soil thermal properties and this study was conducted under laboratory conditions (Adhikari et al. 2014). Similarly, studies on activities and processes related to microbiology, rhizosphere, degradation of chemicals, and interactions among components of AF are limited in the literature. The next section summarizes the effects of soil, water dynamics, vegetation, and chemical reactions on NPSP reduction from these two watershed projects.

### Soil Physical Properties

Perennial vegetation of AF changes soil physical, chemical, and biological properties (Dollinger and Jose 2018). These changes influence water quality and quantity and thus help improve water quality and water dynamics. Stems, branches, leaves, roots, and various other types of residue on the surface can stop soil and nutrients from leaving watersheds. The simplest explanation is that litter and live materials from AF protect soils and nutrients. At the terminal velocity, a single raindrop has

2.01 mJ of kinetic energy, which can dislodge soil to produce splash erosion (Gantzer et al. 1987). The denser grass vegetation of buffers acts as a barrier imposing greater resistance to flow and enhancing sedimentation within the buffer areas. Trees, shrubs, grasses, and litter materials of buffers reduce soil bulk density and improve aggregate stability, porosity, water-holding capacity, infiltration, soil water dynamics, soil thermal properties, and other soil physical properties (Bharati et al. 2002; Seobi et al. 2005; Udawatta et al. 2006b, 2009, 2011a; Sauer et al. 2007; Kumar et al. 2010) that can help reduce NPSP.

Litter material, exudates, fungi, and detritus provide organic matter and strengthen bonding between soil particles, improving soil aggregate stability (Kremer and Kussman 2011). Studies conducted at the RAF and GAF sites have shown significantly greater aggregate stability in AF systems (Udawatta et al. 2008; Paudel et al. 2011, 2012; Alagele et al. 2019a). Other studies have found similar results: according to Alagele et al. (2019a), row crops had 2.6–3.4 times fewer aggregates than perennial vegetation of AF. Similarly, Paudel et al. (2011) reported three times more aggregates in AF and GB soils than crop soils in Central Missouri. These authors have attributed the increased aggregate stability to soil organic matter (SOM) accumulation, increased tree roots, and reduced soil disturbance. As SOM increases, a concomitant increase in aggregate stability occurs, reducing soil erodibility (Udawatta et al. 2009; Paudel et al. 2011, 2012; Al-Kaisi et al. 2014; Alagele et al. 2019a).

Seobi et al. (2005) sampled 0–40 cm depth soils in 10 cm increments to assess soil bulk density and porosity 6 years after the establishment of AF buffers on the row crop watersheds. Results from the study showed that bulk density was reduced by 2.3% and porosity was increased by 3% compared to the row crop treatments. On the same watersheds, Akdemir et al. (2016) reported 8.5% lower bulk density in the 0–10 cm soil depth of AF and GB soils than in the crop soils 17 years after the establishment of buffers (Table 1). Soil porosity was 4.7% greater in the AF and GB compared to the row crop treatment. Authors reported that most of the improvement occurred during the first 6 years. Using a similar sampling strategy, Kumar et al. (2008) showed that soil bulk density was 13% higher for rotational and continuously grazed areas compared to AF and GB 6 years after buffers were established. Soil porosity was greater in buffers relative to grazing areas.

**Table 1** Effects of agroforestry, grass buffers, and row crops on soil bulk density and saturated hydraulic conductivity (Ksat) at the Greenley paired watershed study (Adapted from Akdemir et al. 2016)

| | Bulk density | | | | Hydraulic conductivity | | | |
|---|---|---|---|---|---|---|---|---|
| Depth cm | 5 | 15 | 25 | 35 | 5 | 15 | 25 | 35 |
| Treatment | mg $m^{-3}$ | | | | mm $h^{-1}$ | | | |
| Agroforestry | 1.28 | 1.36 | 1.36 | 1.28 | 40 | 25 | 15 | 30 |
| Grass buffer | 1.20 | 1.42 | 1.37 | 1.24 | 12 | 15 | 10 | 11 |
| Row crop | 1.36 | 1.46 | 1.37 | 1.28 | 12 | 20 | 8 | 1 |

Agroforestry practices also improve soil pore size distribution measured by water retention and computed tomography (CT). Seobi et al. (2005) noticed 33% and 3% greater total porosity for grass and AF buffers relative to row crops in four, 10-cm-depth increments 6 years after buffers were placed in Northeast Missouri. In a recent study on the same watersheds, Akdemir et al. (2016) confirmed the above findings. Sampling loess soils in Central Missouri, Kumar et al. (2008) showed 5.7, 4.5, and 3.9 times greater soil macroporosity in AF, GB, and rotationally grazed paddocks relative to a continuously grazed treatment. Using CT, Udawatta et al. (2006b, 2008), Udawatta and Anderson (2008), and Kumar et al. (2010) showed greater porosity in undisturbed grass and AF relative to row crop and continuously grazed pastures. Improved porosity and pore size distribution and reduced bulk density increase infiltration and water storage and thereby help control NPSP.

Saturated hydraulic conductivity (Ksat) controls water movement in the soils and can affect solute transport through the profile and influence water infiltration and runoff. However, this property depends on pore size distribution and continuity of macropores (diam. >1000μm). Many studies conducted in Missouri with AF have shown greater Ksat values for AF, GB, and biomass crops relative to row crops and grazing management (Seobi et al. 2005; Kumar et al. 2008; Akdemir et al. 2016; Alagele et al. 2019b). At the GAF, Kumar et al. (2012) observed 31 and 46 times greater quasi-steady-state infiltration (qs) and Ksat for AF buffers as compared to pasture treatments. At the RAF site, Seobi et al. (2005) observed 3 and 14 times greater Ksat in grass and AF buffers compared to a corn-soybean rotation. Recent studies on the same watersheds have shown greater improvements in Ksat in AF areas than in crop areas (Akdemir et al. 2016). Improved Ksat at these two long-term study sites has been attributed to improvements in soil physical properties as influenced by the perennial vegetation. Furthermore, soils in these buffer areas are not disturbed or compacted by machinery traffic. Deep tree roots and undisturbed areas in general have more root channels that can help increase water movement through the profile and reduce NPSP. Deep roots and dead root channels provide conduits for deeper soil water infiltration. Even though deep channels were not tested at these two sites, there is evidence that water moves deeper into the soil through structural cracks and root channels and along soil ped surfaces.

Adhikari et al. (2014) evaluated soil heat capacity and thermal conductivity among AF, prairie, and row crop management practices including soil cores from the RAF site. The study showed that thermal conductivity was significantly higher and heat capacity was significantly lower for row crop management compared to AF and prairie systems. The data also showed that soil carbon was positively correlated with heat capacity and negatively correlated with thermal conductivity. Perennial vegetation and denser litter material reduce thermal conductivity and heat transfer into the soil and are inclined to provide higher water content, lower temperatures, and lower soil heat fluxes. Results also imply that reduced heat flow and greater buffer capacity of soils under AF can help climate mitigation, provide favorable conditions for diverse floral and faunal communities, and enhance ecosystem resilience.

## Soil Chemical Properties

Chu et al. (2010) evaluated three soils obtained from RAF, GAF, and Southwest Farms (a third University of Missouri Agricultural Experiment Station farm) to understand antibiotic retention efficiencies influenced by crop, grass, and AF vegetation types (Fig. 5). Both antibiotics, oxytetracycline and sulfamethazine, were more strongly sorbed to AF soils than crop soils. Agroforestry systems with greater biodiversity promote greater degradation and stronger binding of contaminants (Andrews et al. 2004; Chu et al. 2010; Dominati et al. 2010; Lin et al. 2010). Authors have attributed these beneficial effects to biological diversity, exudates by these organisms, and reactions in the soil and rhizosphere.

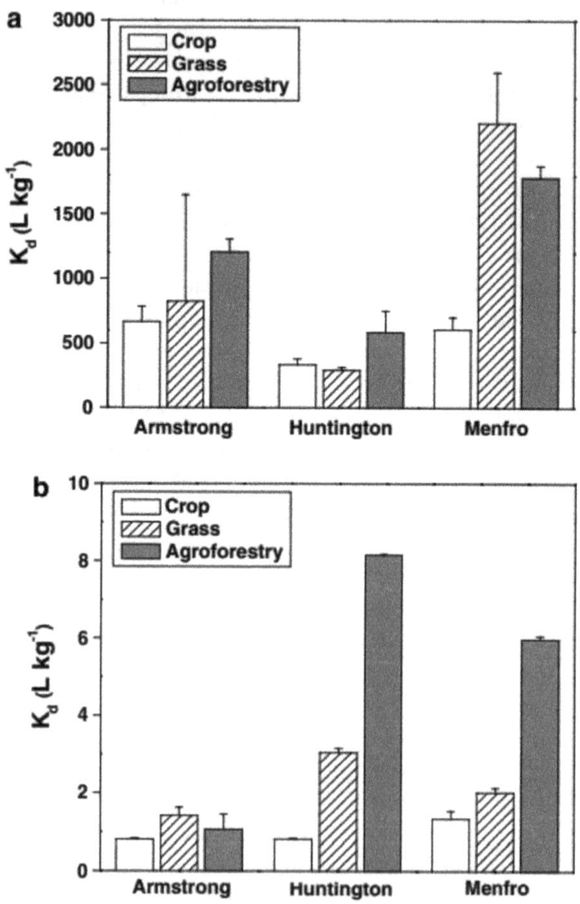


**Fig. 5** Oxytetracycline (**a**) and sulfamethazine (**b**) solid-solution distribution coefficients (Kd) for Armstrong, Huntington, and Menfro soils under crop, grass, and agroforestry management (source: Chu et al. 2010)

Research has shown various phytoremediation procedures mediated by soil organisms and chemicals associated with perennial vegetative buffers. In this study, we did not evaluate all water and soil quality benefits of perennial vegetation and trees. Shorter half-life and quicker degradation of herbicides, antibiotics, and degradation products were found by Lin et al. (2004, 2007, 2010, 2011) in multispecies AF buffers than in non-tree practices.

## Soil Biological Activity

Trees, other perennial vegetation, reduced disturbance, and minimized chemical inputs directly influence the biological activity, soil environment, microclimate, and ecosystem processes in agricultural watersheds (Jose 2009; Banarjee et al. 2016; Udawatta et al. 2019). Inevitable belowground interactions occur within the soil because of the high density of roots of annual and perennial species (Jose et al. 2000, 2004). These favorable conditions promote survival and activity of diverse communities of living organisms in multispecies vegetative systems. Various sizes of living and dead organisms ranging from body width <100μm microfauna and mesofauna (100μm to 2 mm) to >2 mm macrofauna are vital for many soil functions. They play a vital role and contribute to many essential soil functions including decomposition, degradation of chemicals, nutrient cycling, disease suppression, regulation of plant growth, and primary productivity. Therefore, the composition of soil microbiota, their activities, and biomass are used to quantify land management effects on the environment (Anderson and Domsch 1990; Boerner et al. 2000; Schloter et al. 2003).

Several studies have been conducted at these two sites on enzyme activities and soil organisms to understand management effects on their populations and activities (Udawatta et al. 2009; Paudel et al. 2011, 2012; Weerasekara et al. 2016; Alagele et al. 2019a, 2020b). At the RAF site, Udawatta et al. (2009) noticed greater enzyme activities in perennial vegetation areas than in the crop areas 6 years after buffers were established. On the same watersheds, Weerasekara et al. (2016) noticed significantly greater enzyme activities between AF and conventional crop areas 10 years after the establishment of buffers. Supporting these results, at a nearby site Mungai et al. (2005) found greater microbial biomass and activity in perennial buffer soils compared to crop areas of a mature silver maple alley cropping practice. Twenty years after buffers were established, Alagele et al. (2019a) confirmed earlier findings of greater enzyme activities in buffers than crop areas at RAF (Table 2). Furthermore, this study also showed that perennial grasses help improve soil enzymes within 4 years, similar to findings of Udawatta et al. (2009). Soil enzyme activity at this RAF site gradually increased in alleys and tree rows relative with time compared to crop areas (Udawatta et al. 2009; Weerasekara et al. 2016; Alagele et al. 2019a).

At the grazing watersheds, Paudel et al. (2011, 2012) assessed enzyme activities among grazing areas, AF buffers, GB, and a row cropping (corn-soybean rotation) system. All indicators of biological activity and soil carbon were greater in

**Table 2** β-Glucosidase (product = *p*-nitrophenol [PNP]), β-glucosaminidase (product = PNP), fluorescein diacetate hydrolase (FDA; product = fluorescein), and dehydrogenase (product = iodonitrotetrazolium formazan [INTF]) activities (μg product $g^{-1}$ soil $h^{-1}$) for row crop, grass buffer, biomass crop, agroforestry buffer at 50 cm distance (Agroforestry-50), agroforestry buffer at 150 cm distance (Agroforestry-150), and grass waterway for 2018 at the row crop agroforestry watersheds (RAF) at the Greenley Research Center. Different letters within a column are significantly different at $p \leq 0.05$ based on Duncan LSD test (Adapted from Alagele et al. 2019a)

| | Enzyme activity | | | |
|---|---|---|---|---|
| Management | β-Glucosidase | β-Glucosaminidase | FDA | Dehydrogenase |
| Row crop | 118[d] | 71[d] | 62[c] | 34[e] |
| Grass buffer | 127[d] | 75[cd] | 74[b] | 42[d] |
| Biomass crop | 139[c] | 80[c] | 79[b] | 51[c] |
| Agroforestry-50 | 162[b] | 87[b] | 96[a] | 56[b] |
| Agroforestry-150 | 137[c] | 79[c] | 77[b] | 48[c] |
| Grass waterway | 219[a] | 121[a] | 101[a] | 67[a] |

perennial vegetation compared to the row crop, including enzymatic activity in the grazed pastures, demonstrating the benefits of including livestock in the silvopasture component of this AF practice.

Among the microbial parameters, enzyme activity is a promising indicator of rapid response to changes in soil management (Bandick and Dick 1999; Schloter et al. 2003). Soil enzymes play key biochemical functions in the overall process of organic matter decomposition, nutrient mineralization and cycling, nutrient availability, biodegradation of synthetic compounds, and synthesis of plant growth-regulating substances, thereby mediating critical roles in most biochemical and ecological processes in the soil ecosystem (Sinsabaugh et al. 1991; Bardgett and van der Putten 2014). In various AF practices, selected soil enzymes are significantly higher in both grass and grass plus tree strips than in continuously cropped alleys and grazing areas (Meyers et al. 2001; Mungai et al. 2005; Udawatta et al. 2009; Paudel et al. 2011, 2012; Weerasekara et al. 2016; Alagele et al. 2019a). This enhanced enzyme activity in the AF and perennial vegetative areas indicates increased potential to degrade cellulose, hemicellulose, chitin, peptidoglycan, and proteins, which leads to subsequent improved mineralization and nutrient cycling. However, as the system matures these differences between crop and perennial areas gradually diminish. Authors have attributed these changes to the even distribution of organic matter, leaves, litter, and roots with system maturity (Bardhan et al. 2013; Mungai et al. 2005; Weerasekara et al. 2016).

A soil PLFA characterization in 2018 at the RAF site showed significantly greater microbial biomass, bacteria, fungi, and protozoa in AF, GB, and grass waterways than in crop areas (Fig. 6, Alagele et al. 2020b). These counts were larger near the trees and decreased with the distance from trees. The greater PLFA values suggest complex community structure relative to sites with lower total PLFA. The increased microbial biomass (higher total PLFA) in the perennial vegetation soils demonstrates that diverse vegetation favors greater diversity and supplements

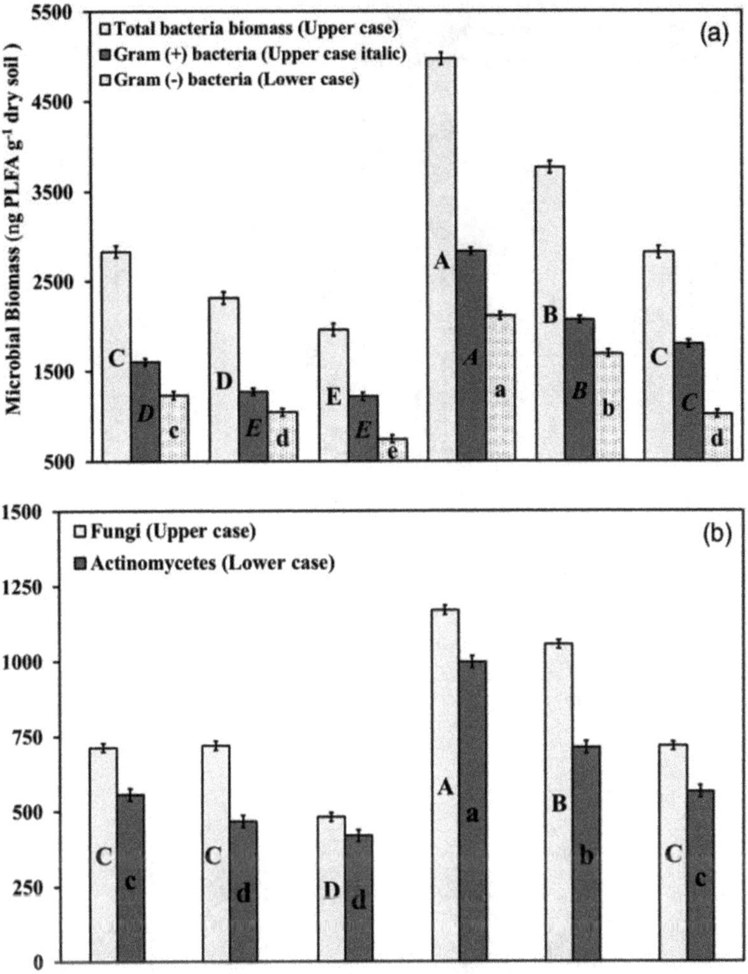


**Fig. 6** Total bacteria, gram-positive (+) bacteria, gram-negative (−) bacteria, fungi actinomycetes, saprophytes, arbuscular mycorrhizae protozoa, and rhizobia biomass for grass buffer (GB), biomass crop (BC), row crop (RC), grass waterway (GWW), agroforestry buffer at 50 cm distance (AF50), and agroforestry buffer at 150 cm distance (AF150) for 2018 at the Greenley paired watersheds. Bars with different uppercase, uppercase italic, and lower case letters represent significant differences among treatments at $p \leq 0.05$ based on Duncan LSD test for specific microbial population biomass (source: Alagele et al. 2020b)

overall AF productivity, similar to that suggested by LaCanne and Lundgren (2018) for regenerative agriculture systems.

PLFA determined greater fungi densities and community structures at the RAF site (Alagele et al. 2020b). Similar to Missouri results, greater abundance and diversity of fungi have been reported in tree-based systems compared to adjacent cropping systems or pastures without trees elsewhere (Chifflot et al. 2009; Lacombe

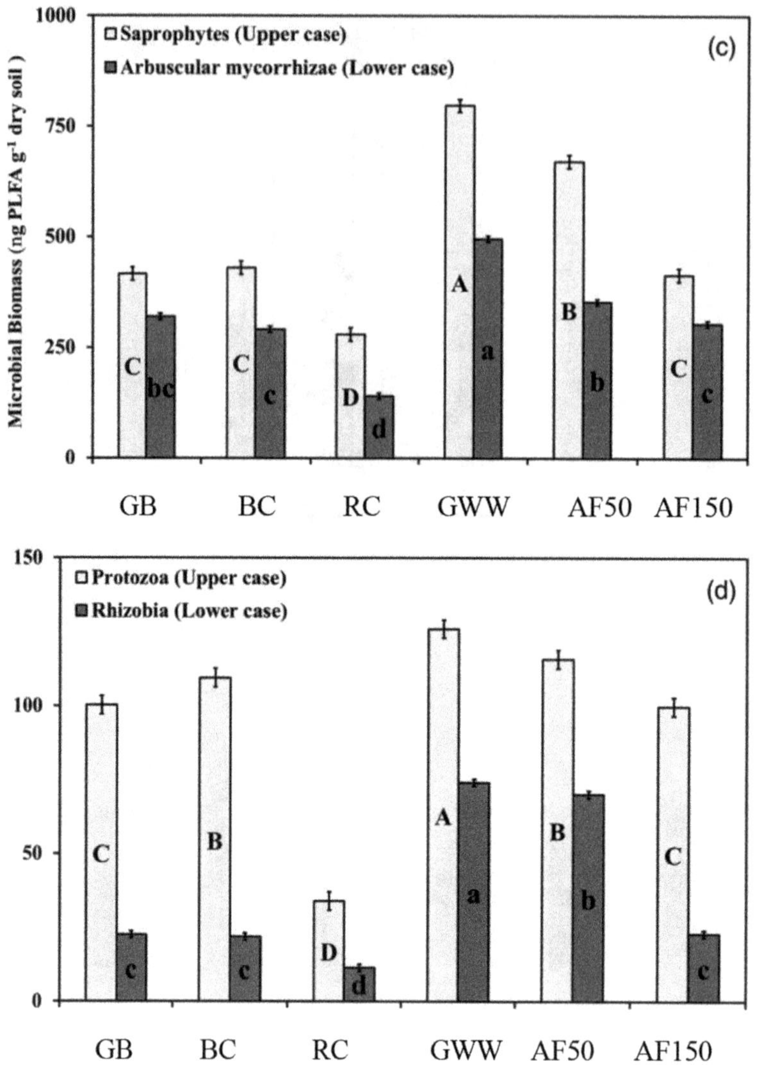


**Fig. 6** (continued)

et al. 2009; Beuschela et al. 2019). In Germany, Beuschela et al. (2019) showed an increasing fungal C:bacterial C ratio which indicates a potential increase of fungal C in tree rows of the study. Fungi in general help improve soil structure, aggregate stability, and nutrient availability, and thereby help improve water quality by reducing losses of sediment and nutrients and also degrading of some chemicals (Bainard et al. 2012, 2013). According to a review by Bainard et al. (2011a), integration of AF increases fungal diversity and abundance more than monocrop systems.

The total soil bacteria count was also greater within the perennial management system. Between the two types of bacteria, the gram-negative bacteria were particularly abundant, which rapidly metabolize readily available C release via rhizodeposition. These bacteria are also involved in nutrient cycling, including N mineralization, plant growth stimulation, and antibiotics production that suppress soil and root pathogens. Kremer et al. (2015) showed that a long-term fruit tree AF practice with perennial native vegetation had a greater number of arbuscular mycorrhizal fungi (AMF) compared to tree fruit and single-species alleys of tall fescue.

The soil biodiversity (SBD) can be defined as the "variation on soil life, from genes to communities, and the variations in soil habitats, from micro aggregates to entire landscapes (Turbé et al. 2010)." Greater SBD found within tree areas of RAF and GAF indicates potential improvements in water quality, nutrient-use efficiencies, soil health, and degradation of chemical compounds (Brussaard et al. 2007). Similar to the two Missouri sites, significantly greater diversity, microbial communities, and functions have been reported at many other AF sites compared to crop, pasture, forest, and tree monoculture management practices (Stamps and Linit 1998; Huang et al. 2002; Bainard et al. 2011b; Unger et al. 2013; Thorup-Kristensen and Rasmussen 2015; Sistla et al. 2016; Torralba et al. 2016; Zhang et al. 2018). Several review papers and a meta-analysis have emphasized the importance of SBD for enhanced ecosystem services (Hooper et al. 2005; Balvanera et al. 2005; Stachowicz et al. 2007; Cardinale et al. 2007; Hillebrand and Matthiessen 2009). Conversion of monocrops to AF systems increases biodiversity as AF harbors greater species richness and diversity (Perfecto et al. 1996; Lawton et al. 1998; Schroth et al. 2004; Jose 2012; Varah et al. 2013).

## *Buffer Vegetation*

The growth of trees at RAF was evaluated from the time of establishment to quantify biomass, carbon, and nutrient accumulation. Trees were planted in November 1997, but they lost height by 1999 due to deer browsing. Wire mesh protection barriers (1 m diam. and 1.2 m tall) were installed around all 333 trees in 1999. After protection was provided, diameter and height growth were significantly greater compared to the two previous years (Udawatta et al. 2005). At the end of 2002, pin oak was taller than the other two species. There was a marked greater growth in height, 10 cm diameter, and woody biomass of pin oak from 2001 to 2002 than for the other two species. The average tree height was only 2.3 m in 2002 after 5 years at the site and ranged from 2.4 m for pin oak to 2.1 m for bur and swamp white oak. The height of pin oak increased by a factor of 2.1 and the diameter by 2.4 in 4 years. Diameter and height of bur and swamp white oak doubled during this time. The total biomass (dry weight) was only 100 kg ha$^{-1}$ at the end of five growing seasons in 2002. At the end of 2002 growing season, these trees had accumulated 45 kg ha$^{-1}$ of C, assuming that 45% of the biomass is C.

The mean height of these trees in 2015 was 7.5 m and ranged from 2 to 11 m. The average DBH was 17 cm (range 6.5–34 cm). These open-grown trees have more branches and greater leaf areas compared to normal forestry conditions. Thus, more nutrients and carbon accumulation can be expected within the main stem, branches, and leaves of these trees. The total aboveground green biomass was 9135 kg ha$^{-1}$ in 2015 using volume estimation equations for oak trees of central hardwoods (Myers et al. 1997). Using the nutrient concentrations of oaks (Pardo et al. 2005), RAF trees of 4567 kg could store 2783, 34, and 3.5 kg per ha C, N, and P in their aboveground biomass. We used the biomass estimation equation for trees under forest conditions and thus these C, N, and P estimations are lower than the true biomass of these open-grown trees at RAF. Open-grown trees of RAF have more leaves and leaf biomass could be between 10 and 15% of the total biomass. Thus, these open-grown tree leaves (>2% N) annually store and return approximately 9–13 kg N ha$^{-1}$. Trees in the agroforestry watershed also provide numerous other ecosystem services including enhanced biodiversity (Udawatta et al. 2019), soil buffering (Adhikari et al. 2014), and microclimate benefits (Svoma et al. 2016). All these beneficial effects collectively help improve water quality on agricultural watersheds.

Perennial buffer vegetation leafs out before the crop is established and utilizes residual nutrients left in the field. Most of the surface runoff on these watersheds occurs during fallow periods and thus perennial vegetative buffers of trees and grasses which leaf out early in the growing season can be effective in reducing nutrients in runoff. The dense fibrous roots of perennial vegetation can remove soluble nutrients from the subsurface horizons and contribute to water quality benefits by reducing nutrients in subsurface flow. Subsurface flow accounts for ~80% of the annual nitrate discharge to streams in Northwestern Missouri (Burwell et al. 1976; Schmitt 1999) and therefore nutrient filtration by trees of agricultural watersheds could be considered as a significant water quality service.

The RAF is located in the claypan region (Major Land Resource Area 113) and these soils consist of a horizon with 30–70% shrink-swell clay. Hydraulic conductivity which is very low restricts downward water movement and root growth below the clay horizon. During dry periods, these clay particles shrink and form cracks while forming an impervious layer during wet periods. This perennial vegetation can help reduce nutrients in runoff by the development of roots through the claypan and improvements in soil water movement through the restrictive layer. Initial root excavation studies at the site showed greater restriction of vertical root distribution due to this restrictive horizon. Increased penetration of roots is anticipated through the claypan as trees grow and develop stronger roots. Tree survival and growth may indicate that roots have penetrated the claypan as no trees died during the 2012 drought despite 45% lower precipitation than the long-term mean. These beneficial effects of the perennial vegetation and associated soil physical and biological improvements, as well as utilization of soil water by the permanent vegetation, were obvious factors in helping reduce NPSP losses in grazing and row crop agriculture in our studies.

The amount of rainwater lost in runoff is determined by vegetation and land management. During the study period, the control watershed lost 35% of the rain in runoff while AF and GB watersheds lost only 25% and 30%. Soil water dynamics at the paired row crop watersheds showed consistent differences among vegetation types from 2003 to 2018 (Fig. 7; Anderson et al. 2009; Udawatta et al. 2011b; Sahin

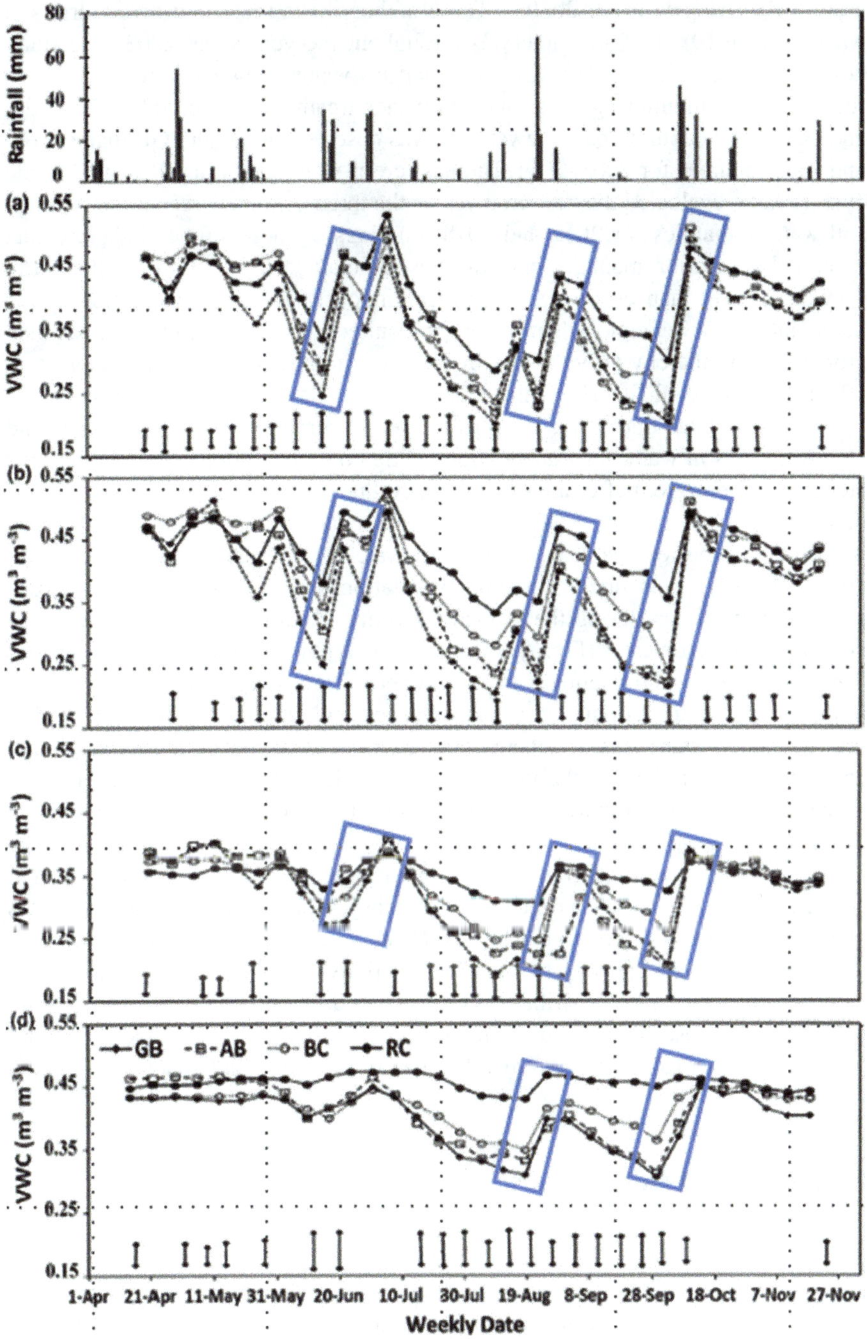


**Fig. 7** Rainfall distribution and effects of agroforestry buffers (AF), grass buffers (GB), biomass crop (BC), and row crop (RC) treatments on weekly volumetric water content (VWC, detected at 12:00 pm each Friday) at A = 5 cm, B = 10 cm, C = 20 cm, and D = 40 cm depths in 2018. Bars indicate the least significant difference (LSD) (0.05) for VWC. Blue rectangles show greater recharge for soils under perennial vegetation (source: Alagele et al. 2020a)

et al. 2016; Alagele et al. 2020a). These authors have measured volumetric soil water content (θ) at 15-min intervals throughout the year since 2003. The study design consisted of AF buffers, GB, corn, and soybean management treatments with four replicated monitoring locations within each treatment to record θ at 5, 10, 20, and 40 cm soil depth. Greater dewatering was observed in AF and GB than in corn and soybean areas for 2003–2011. Biomass crops were added in 2012 and Alagele et al. (2020a) evaluated the new crop versus the three perennial vegetation types on soil water dynamics for 2017 and 2018. Findings of these studies indicated that perennial vegetation management practices maintained lower θ during the entire growing season than corn or soybean. Continuously greater transpiration of the perennial buffer vegetation than the crops resulted in lower θ in buffer areas than crop areas and thereby provided more soil water storage potential (Anderson et al. 2009; Udawatta et al. 2011b; Sahin et al. 2016; Alagele et al. 2020a). Perennial buffer vegetation leafs out in March/April as the temperature becomes favorable and begins to use soil water for transpiration, while no transpiration occurs in the crop area soils. Therefore, buffer areas had lower θ than the crop areas at the beginning of the crop-growing season. This has helped reduce spring runoff on row crop watersheds, as buffered soils can store more water during runoff.

Additionally, soils under perennial vegetation stored more water (recharged soils) by rain events during the growing season and after the cash crop was harvested than did crop areas (Fig. 7, blue boxes). This enhanced soil water storage has been attributed to improvements in soil properties including soil carbon, porosity, infiltration, hydraulic conductivity, and lower soil bulk density. Furthermore, during the November–March main recharge periods, soils in the buffer areas had greater amounts of water than in the crop areas. According to Anderson et al. (2009), 6-year-old grass and AF buffers stored 0.9 and 1.1 cm greater water in a 30 cm soil profile than in crop areas.

Deep roots of trees and grasses of AF extract water from deeper soil layers that shallow-rooted row crops, vegetables, and many other perennials cannot access (Jose et al. 2000). A recent review by Alagele et al. (2021) showed that deep-rooted perennial vegetation can improve land productivity by hydraulically lifting water from deep horizons and redistributing this water to drier upper horizons for shallow-rooted crops to absorb. These soil water dynamics help reduce runoff and store more water in the soil profile and thereby influence water quality and quantity.

## Summary and Conclusions

The two case studies presented in this chapter evaluated the effects of agroforestry and grass buffers on runoff, sediment, nitrogen, and phosphorus loss from row crop and grazing watersheds. Results indicate that AF and GB in these long-term studies have reduced NPSP losses and provided significant water quality benefits on row crop and grazing watersheds. Agroforestry buffers at the grazing watersheds improved groundwater quality and significantly reduced nitrate and total N concentrations in groundwater. Additionally, buffers with deeper roots also helped reduce subsurface drainage losses of nutrients and returned them to the surface soils for other uses. These benefits have improved over time. Accurately calibrated and validated model simulations have further strengthened above findings and predicted enhanced benefits over time.

Water quality and quantity benefits can be attributed to changes in soil properties, soil water dynamics, vegetation characteristics, and soil biology. On the soil surface, resistance from tree roots, fallen branches, live vegetation, and litter material enhance sedimentation and reduce flow velocities, and this reduction of flow rate helps reduce NPSP. In the two studies presented in this chapter, perennial vegetation of AF improved soil physical, chemical, and biological properties, which in turn helped improve water quality. The perennial buffer vegetation was also found to utilize nutrients more efficiently and store them in the biomass for longer periods than annual crops and thereby helped reduce nutrient losses on agricultural watersheds. Trees and perennial vegetation begin dewatering the soil profile before the cash crop is established and continue to use more water even after the cash crop is harvested, thus reducing runoff. Multispecies vegetation of AF encourages biological diversity and thereby enhances physical, chemical, and biological activities on agricultural watersheds. Various exudates and biological reactions can help degrade and bind those chemicals strongly to soil particles, and thus losses from agricultural watersheds are reduced by the perennial vegetative agroforestry buffers. We presume that water quality benefits of these buffers will increase with time as the perennial vegetation improves all of the above mechanisms.

Agroforestry could be a partial solution for many regions of the world for resilient agricultural practices to address future challenges of climate changes, pandemics, and food security. In addition to water and soil quality benefits, AF can provide various ecosystem services including diversified food sources and income potentials. Future studies may evaluate specific water quality benefits of various buffer species by landscape positions for alley cropping and silvopasture practices. Information from these studies can be used to develop soil–site–climate-suitable management practices for enhanced ecosystem services.

At the row crop agroforestry site 3–4 m wide buffers occupy 10–12% of the watershed area and field-edge buffers at the grazing agroforestry site occupy 15% of the area. The upland in-field buffers similar to RAF keep soil and nutrients within the field as compared to edge-of-field buffers while reducing slope length and slope steepness. Buffers help improve water quality and provide many other ecosystem services; however, they take productive land out of production. Landowners could consider other income-generating options like hay, specialty crops, and shade-tolerant crops for buffer areas for watershed diversification and supplemental income. Nut, leguminous species, fruit, and biomass trees can be planted alternately for marketable products until trees are harvested. Buffer design should consider landowner preferences, farm equipment dimensions, easy maintenance, and cost-effectiveness for long-term efficacy and longevity of these practices.

## References

Adhikari P, Udawatta RP, Anderson SH, Gantzer CJ (2014) Soil thermal properties under prairies, conservation buffers, and corn/soybean land use systems. Soil Sci Soc Am J 78:1977–1986

Akdemir E, Anderson SH, Udawatta RP (2016) Influence of agroforestry buffers on soil hydraulic properties relative to row crop management. Soil Sci 181:368–376. https://doi.org/10.1097/SS.0000000000000170

Alagele SM, Anderson SH, Udawatta RP, Veum KS, Rankoth LM (2019a) Effects of conservation practices on soil quality compared to a corn/soybean rotation on a claypan soil. J Environ Qual 48:1694–1702. https://doi.org/10.2134/jeq2019.03.0121

Alagele SM, Anderson SH, Udawatta RP (2019b) Biomass and buffer management practice effects on soil hydraulic properties compared to grain crops for claypan landscapes. Agrofor Syst 93:1609–1625. https://doi.org/10.1007/s10457-018-0255-1

Alagele SM, Anderson SH, Udawatta RP (2020a) Agroforestry, grass, biofuel, and row crop management effects on soil water dynamics for claypan landscapes. Soil Sci Soc Am J 84:203–219

Alagele SM, Anderson SH, Udawatta RP, Veum KS, Rankoth LM (2020b) Long-term perennial management and cropping effects on soil microbial biomass for claypan watersheds. Agron J 112:815–827

Alagele SM, Jose S, Anderson SH, Udawatta RP (2021) Hydraulic lift: processes, evidence, and implications for improving crop production and the environment. Agrofor Syst. https://doi.org/10.1007/s10457-021-00614-w

Al-Kaisi MM, Douelle A, Kwaw-Mensah D (2014) Soil microaggregate and macroaggregate decay over time and soil carbon change as influenced by different tillage systems. J Soil Water Conserv 69:574–580

Anderson TH, Domsch KH (1990) Application of eco-physiological quotients (qCO2 and qD) on microbial biomass from soils of different cropping histories. Soil Biol Biochem 22:251–255

Anderson SH, Udawatta RP, Seobi T, Garrett HE (2009) Soil water content and infiltration in Agroforestry buffer strips. Agrofor Syst 75:5–16. https://doi.org/10.1007/s10457-008-9128-3

Andrews SS, Karlen DL, Cambardella CA (2004) The soil management assessment framework: a quantitative soil quality evaluation method. Soil Sci Soc Am J 68:1945–1962

Bainard KLD, Klironomos JN, Gordon AM (2011a) Arbuscular mycorrhizal fungi in a tree-based intercropping system: a review of their abundance and diversity. Pedobiologia Int J Soil Biol 54:57–61

Bainard LD, Kochb AM, Gordon AM, Newmaster SG, Thevathasan NV, Klironomosb JN (2011b) Influence of trees on the spatial structure of arbuscular mycorrhizal communities in a temperate tree-based intercropping system. Agric Ecosyst Environ 144:13–20

Bainard LD, Kochb AM, Gordon AM, Klironomos JN (2012) Temporal and compositional differences of arbuscular mycorrhizal fungal communities in conventional monocropping and tree-based intercropping systems. Soil Biol Biochem 45:172–180

Bainard LD, Koch AM, Gordon AM, Klironomos JN (2013) Growth response of crops to soil microbial communities from conventional monocropping and tree-based intercropping systems. Plant Soil 363:345–356

Balvanera P, Kremen C, Martinez-Ramos M (2005) Applying community structure analysis to ecosystem function: examples from pollination and carbon storage. Ecol Appl 15:360–375

Banarjee S, Baah-Acheamfour M, Carlyle CM, Bissett A, Richardon AE, Siddique T, Bork EW, Chang SX (2016) Determinants of bacterial communities in Canadian agroforestry systems. Environ Microbiol 18:1805–1816

Bandick AK, Dick RP (1999) Field management effects on soil enzyme activities. Soil Biol Biochem 31:1471–1479

Bardgett RD, van der Putten WH (2014) Belowground biodiversity and ecosystem functions. Nature 515:505–511

Bardhan S, Jose S, Udawatta RP, Fritschi F (2013) Microbial community diversity in a 21-year old temperate alley cropping system. Agrofor Syst 87:1031–1041

Beuschela R, Piephob H, Joergensena RG, Wachendorfa C (2019) Similar spatial patterns of soil quality indicators in three poplar-based silvo-arable alley cropping systems in Germany. Biol Fertil Soils 55:1–14. https://doi.org/10.1007/s00374-018-1324-3

Bharati L, Lee KH, Isenhart TM, Schultz RC (2002) Soil-water infiltration under crops, pasture, and established riparian buffer in Midwest USA. Agrofor Syst 56:249–257

Boerner REJ, Decker KLM, Sutherland EK (2000) Prescribed burning effects on soil enzyme activity in a southern Ohio hardwood forest: a landscape-scale analysis. Soil Biol Biochem 32:899–908

Brussaard L, Ruiter PC, Brown GG (2007) Soil biodiversity for agricultural sustainability. Agric Ecosyst Environ 121:233–244. https://doi.org/10.1016/j.agee.2006.12.013

Burwell RE, Schuman GE, Saxton KE, Heinemann HG (1976) Nitrogen in subsurface- discharge from agricultural watersheds. J Environ Qual 5:325–329

Cardinale BJ, Wright JP, Cadotte MW, Carroll IT, Hector A, Srivastava DS, Loreau M, Weis JJ (2007) Impacts of plant diversity on biomass production increase through time because of species complementarity. PNAS 104:18123–18128

Chifflot V, Rivest D, Olifindvier A, Cogliastro A, Khasa D (2009) Molecular analysis of arbuscular mycorrhizal community structure and spores distribution in tree-based intercropping and forest systems. Agric Ecosyst Environ 131:32–39

Chu B, Goyne KW, Anderson SH, Lin CH, Udawatta RP (2010) Veterinary antibiotic sorption to agroforestry buffer, grass buffer, and cropland soils. Agrofor Syst 79:67–80

Dollinger J, Jose S (2018) Agroforestry for soil health. Agrofor Syst 92(2):213–219

FAO (2019) World's food and agriculture statistical pocketbook. FAO, Rome, p 254

Gantzer CJ, Buyanovsky GA, Alberts EE, Remley PA (1987) Effects of soybean and corn residue decomposition on soil strength and splash detachment. Soil Sci Soc Am J 51:202–206

Garrett HE, Kurtz WB, Lassoie MA, Gold MA, Pearson HA, Hardesty LH, Buck LE Slusher JP (1994) Agroforestry: an integrated land use management system for production and farmland conservation. Final report for the USDA soil conservation service project. USDA SCS 68-3A75-3-134: Resource Conservation Act Appraisal, Soil Conservation Service, Washington, DC

Gold MA, Garrett HE (2009) Agroforestry nomenclature, concepts, and practices. In: Garrett HE (ed) North American agroforestry, an integrated science and practice, 2nd edn. American Society of Agronomy, Inc, Madison, WI, pp 45–55

Hillebrand H, Matthiessen (2009) Biodiversity in a complex world: consolidation and progress in functional biodiversity research. Ecol Lett 12:1405–1419

Hooper DU, Chapin FSI, Ewel JJ, Hector A, Inchausti P, Lavorel S, Lawton JH, Lodge DM, Loreau M, Naeem S, Schmid B, Setälä H, Symstad AJ, Vander-meer J, Wardle DA (2005) Effects of biodiversity on ecosystem functioning: a consensus of current knowledge and needs for future research. Ecol Monogr 75:3–35

Huang W, Luukkanen O, Johanson S, Kaarakka V, Räisänen S, Vihemäki H (2002) Agroforestry for biodiversity conservation of nature reserves: functional group identification and analysis. Agrofor Syst 55:65–72

Jose S (2009) Agroforestry for ecosystem services and environmental benefits: an overview. Agrofor Syst 76:1–10. https://doi.org/10.1007/s10457-009-9229-7

Jose S (2012) Agroforestry for conserving and enhancing biodiversity. Agrofor Syst 85:1–8

Jose S (2019) Environmental impacts and benefits of agroforestry. In: Oxford Encyclopedia of Agriculture and Environment. Oxford University Press, USA. https://doi.org/10.1093/acrefore/9780199389414.013.195

Jose S, Gillespie AR, Seifert JR, Biehle DJ (2000) Defining competition vectors in a temperate alley cropping system in the Midwestern USA. 2. Competition for water. Agrofor Syst 48:41–59

Jose S, Gillespie AR, Pallardy SG (2004) Inter-specific interactions in temperate agroforestry. Agrofor Syst 61:237–255

Kenny JF, Barber NL, Hutson SS, Linsey KS, Lovelace JK, Maupin MA (2009) Estimated use of water in the United States in 2005. Circular 1344

Koltun GF, Landers MN, Nolan KM, Parker RS (1997) Sediment transport and geomorphology issues in the water resources division. Proceedings of the U.S. Geological Survey Sediment Workshop, February 4–7, 1997

Kremer RJ, Kussman RD (2011) Soil quality in a pecan - kura clover alley cropping system in the Midwestern USA. Agrofor Syst 83:213–223

Kremer RJ, Hezel LF, Veum KS (2015) Soil health improvement in an organic orchard production system in northwest Missouri. Proc Organ Agric Res Symp 1-4. LaCrosse, WI

Kumar KC, Gupta S, Chander Y, Singh A (2005) Antibiotic use in agriculture and its impact on the terrestrial environment. Adv Agron 87:1–54

Kumar S, Anderson SH, Bricknell LG, Udawatta RP, Gantzer CJ (2008) Soil hydraulic properties influenced by agroforestry and grass buffers for grazed pasture systems. J Soil Water Conserv 63:224–232

Kumar S, Anderson SH, Udawatta RP (2010) Agroforestry and grass buffer influences on macropores measured by computed tomography under grazed pasture systems. Soil Sci Soc Am J 74:203–212

Kumar S, Anderson SH, Udawatta RP, Kallenbach RL (2012) Water infiltration influenced by agroforestry and grass buffers for a grazed pasture system. Agrofor Syst 84:325–335

LaCanne CE, Lundgren JG (2018) Regenerative agriculture: merging farming and natural resource conservation profitably. PeerJ 6:e4428. https://doi.org/10.7717/peerj.4428

Lacombe S, Bradley RL, Hamel C, Beaulieu C (2009) Do tree-based intercropping systems increase the diversity and stability of soil microbial communities? Agric Ecosyst Environ 131:25–31

Lawton JH, Bignell DE, Bolton B, Bloemers GF, Eggleton P, Hammond PM, Hodda M, Holt RD, Larsen TB, Mawdsley NA, Stork NE, Srivastava DS, Watt AD (1998) Biodiversity inventories, indicator taxa and effects of habitat modification in tropical forest. Nature 391:72–76. https://doi.org/10.1038/34166

Lin CH, Lerch RN, Garrett HE, George MF (2004) Incorporating forage grasses in riparian buffers for bioremediation of atrazine, isoxaflutole and nitrate in Missouri. Agroforest Syst 63:91–99

Lin CH, Lerch RN, Garrett HE, Li YZ, George MF (2007) An Improved HPLC-MS/MS Method for Determination of Isoxaflutole (Balance™) and Its Metabolites in Soils and Forage Plants. J Agric Food Chem 55:3805–3815

Lin C-H, Goyne KW, Kremer RJ, Lerch RN, Garrett HE (2010) Dissipation of sulfamethazine and tetracycline in the root zone of grass and tree species. J Environ Qual 39:1269–1278

Lin C-H, Lerch RN, Goyne KW, Garrett HE (2011) Reducing herbicides and veterinary antibiotics losses from agroecosystems using vegetative buffers. J Environ Qual 40:791–799

Massé D, Saady N, Gilbert Y (2014) Potential of biological processes to eliminate antibiotics in livestock manure: an overview. Animals 4(2):146–163. PMC 4494381. PMID 26480034. https://doi.org/10.3390/ani4020146

Meyers RT, Zak DR, White DC, Peacock A (2001) Landscape-level patterns of microbial community composition and substrate use in upland forest ecosystems. Soil Sci Soc Am J 65:359–367

Missouri Department of Natural Resources (2012) Missouri water quality report (Section 305(b) report). http://www.dnr.mo.gov/env/wpp/waterquality/305b/2012-305b.pdf. Accessed 30 Mar 2013

Missouri Department of Natural Resources (2017) Soil and water conservation districts. Conserving Missouri's soil and water. Missouri Department of natural Resources, Soil and Water Conservation Program. PUB002181. p 2

Mungai NW, Motavalli PP, Kremer RJ, Nelson KA (2005) Spatial variation of soil enzyme activities and microbial functional diversity in temperate alley cropping systems. Biol Fertil Soils 42:129–136

Myers C, Polak D, Stortz L (1997) Full tree weight equations and tables for selected central hardwoods. Department of Forestry, Southern Illinois University, Carbondale, IL

Nearing MA (2001) Potential changes in rainfall erosivity in the US with climate change during the 21st century. J Soil Water Conserv 56:229–232

NRC (National Research Council) (2010) Toward sustainable agricultural systems in the 21st century, Committee on Twenty-First Century Systems Agriculture, Board on Agriculture and Natural Resources. National Academies Press, Washington, DC

Pardo LH, Robin-Abbott M, Duarte N, Miller EK (2005) Tree chemistry database (version 1.0) general technical report NE-324, Forest Service. Northeastern Research Station, Newtown Square, PA, pp 19073–13294

Paudel BR, Udawatta RP, Kremer RJ, Anderson SH (2011) Agroforestry and grass buffer effects on soil quality parameters for grazed pasture and row-crop systems. App Soil Ecol 48:125–132

Paudel BR, Udawatta RP, Kremer RJ, Anderson SH (2012) Soil quality indicator responses to crop, grazed pasture, and agroforestry buffer management. Agrofor Syst 84:311–323

Perfecto I, Rice RA, Greenberg R, Van Der Voort ME (1996) Shade coffee: a disappearing refuge for biodiversity. Bioscience 46:598–608

Sahin H, Anderson SH, Udawatta RP (2016) Water infiltration and soil water content in claypan soils influenced by agroforestry and grass buffers compared to row crop management. Agrofor Syst 90:839–860. https://doi.org/10.1007/s10457-016-9899-x

Sarmah AK, Meyer MT, Boxall ABA (2006) A global perspective on the use, sales, exposure pathways, occurrence, fate, and effects of veterinary antibiotics (VAs) in the environment. Chemosphere 65(5):725–759. PMID 16677683. https://doi.org/10.1016/j.chemosphere.2006.03.026

Sauer TJ, Cambardella CA, Brandle JR (2007) Soil carbon and litter dynamics in a red cedar-scotch pine shelterbelt. Agrofor Syst 71:163–174

Schloter M, Dilly O, Munch JC (2003) Indicators for evaluating soil quality. Agric Ecosyst Environ 98:255–262

Schmitt SJ (1999) Application of a flow source mixing model and remote sensing to the hydrology and water quality of two small watersheds in Northern Missouri. MS Thesis, Univ. of Missouri, Columbia, MO, USA

Schroth G, da Fonseca GAB, Harvey CA, Gascon C, Vasconcelos HL, Izac A-MN (2004) Agroforestry and biodiversity conservation in tropical landscapes. Island, Washington, DC

Schultz RC, Isenhart TM, Colletti JP, Simpkins WW, Udawatta RP, Schultz PL (2009) Riparian and upland buffer practices. In: Garrett HE (ed) North American agroforestry an integrated science and practice. American Society of Agronomy, Madison, WI, pp 163–218

Senaviratne GMMMA, Udawatta RP, Baffaut C, Anderson SH (2013) Agricultural policy environmental extender simulation of three adjacent row-crop watersheds in the claypan region. J Environ Qual 42:726–736. https://doi.org/10.2134/jeq2012.0241

Senaviratne GMMMA, Udawatta RP, Anderson SH, Baffaut C, Thompson A (2014a) Use of fuzzy rainfall-runoff predictions for claypan watersheds with conservation buffers. J Hydrol 507:1008–1018. https://doi.org/10.1016/j.jhydrol.2014.06.023

Senaviratne GMMMA, Udawatta RP, Baffaut C, Anderson SH (2014b) Evaluation of a stepwise, multiobjective, multivariable parameter optimization method for the APEX model. J Environ Qual 43:1381–1391. https://doi.org/10.2134/jeq2013.12.0509

Senaviratne GMMMA, Udawatta RP, Baffaut C, Lory J, Nelson NO, Bhandari A (2018) Evaluation of four parameterization strategies for the APEX model. Am Soc Agric Biol Eng 61:1603–1617. https://doi.org/10.13031/trans.12656

Seobi T, Anderson SH, Udawatta RP, Gantzer CJ (2005) Influences of grass and agroforestry buffer strips on soil hydraulic properties. Soil Sci Soc Am J 69:893–901

Sinsabaugh RL, Antibus RK, Linkins AE (1991) An enzymic approach to the analysis of microbial activity during plant litter decomposition. Agric Ecosyst Environ 34:43–54

Sistla SA, Roddy AB, Williams NE, Kramer DB, Stevens K, Allison SD (2016) Agroforestry practices promote biodiversity and natural resource diversity in Atlantic Nicaragua. PLoS One 11(9):e0162529. https://doi.org/10.1371/journal.pone.0162529

Stachowicz JJ, Bruno JF, Duffy JE (2007) Understanding the effects of marine biodiversity on communities and ecosystems. Annu Rev Ecol Evol Syst 38:739–766

Stamps WT, Linit MJ (1998) Plant diversity and arthropod communities: implications for temperate agroforestry. Agrofor Syst 39:73–89

Svoma BM, Fox NI, Pallardy Q, Udawatta RP (2016) Evapotranspiration differences between agroforestry and grass buffer systems. Agric Water Manag 176:214–221. https://doi.org/10.1016/j.agwat.2016.06.018

Thorup-Kristensen K, Rasmussen CR (2015) Identifying new deep rooted plant species suitable as underground nitrogen catch crops. J. Soil Water Conserv 70:399–409

Torralba M, Fagerholm N, Burgess PJ, Moreno G, Plieninger T (2016) Do European agroforestry systems enhance biodiversity and ecosystem services? A meta-analysis. Agric Ecosyst Environ 230:150–161

Turbé A, DeToni A, Benito P, Lavelle P, Lavelle P, Ruiz N, Van der Putten WH, Labouze E, Mudgal S (2010) Soil biodiversity: functions, threats, and tools for policymakers. Bio Intelligence Service, IRD, and NIOO, Report for European Commission, DG Environment

Udawatta RP, Anderson SH (2008) CT-measured pore characteristics of surface and subsurface soils as influenced by agroforestry and grass buffers. Geoderma 145:381–389. https://doi.org/10.1016/j.geoderma.2008.04.004

Udawatta RP, Krstansky JJ, Henderson GS, Garrett HE (2002) Agroforestry practices, runoff, and nutrient loss: a paired watershed comparison. J Environ Qual 31:1214–1225

Udawatta RP, Motavalli PP, Garrett HE (2004) Phosphorus loss and runoff characteristics in three adjacent agricultural watersheds with claypan soils. J Environ Qual 33:1709–1719

Udawatta RP, Nygren PO, Garrett HE (2005) Growth of three oak species during establishment in an agroforestry practice for watershed protection. Can J For Res 35:602–609

Udawatta RP, Motavalli PP, Garrett HE (2006a) Nitrogen losses in runoff from three adjacent agricultural watersheds with claypan soils. Agric Ecosyst Environ 117:39–48

Udawatta RP, Anderson SH, Gantzer CJ, Garrett HE (2006b) Agroforestry and grass buffer influence on macropore characteristics: a computed tomography analysis. Soil Sci Soc Am J 70:1763–1773

Udawatta RP, Gantzer CJ, Anderson SH, Garrett HE (2008) Agroforestry and grass buffer effects on high resolution X-ray CT-measured pore characteristics. Soil Sci Soc Am J 72:295–304. https://doi.org/10.2136/sssaj2007.0057

Udawatta RP, Kremer RJ, Garrett HE, Anderson SH (2009) Soil enzyme activities and physical properties in a watershed managed under agroforestry and row-crop systems. Agric Ecosyst Environ 131:98–104

Udawatta RP, Garrett HE, Kallenbach RL (2010) Agroforestry and grass buffer effects on water quality in grazed pastures. Agrofor Syst 79(1):81–87. https://doi.org/10.1007/s10457-010-9288-9

Udawatta RP, Garrett HE, Kallenbach RL (2011a) Agroforestry buffers for nonpoint source pollution reductions from agricultural watersheds. J Environ Qual 40(3):800–806

Udawatta RP, Anderson SH, Motavalli PP, Garrett HE (2011b) Calibration of a water content reflectometer and soil water dynamics for an agroforestry practice. Agrofor Syst 82:61–75. https://doi.org/10.1007/s10457-010-9362-3

Udawatta RP, Gantzer CJ, Jose S (2017) Agroforestry practices and soil ecosystem services. In: Al-Kaisi MM, Lowery B (eds) Soil health and intensification of agroecosystems. Elsevier AP, USA, pp 305–334

Udawatta RP, Rankoth LM, Jose S (2019) Agroforestry and biodiversity. Sustainability 11, 02879, 22p. doi:https://doi.org/10.3390/su11102879

Udawatta RP, Anderson SH, Kremer RJ, Garrett HE (2020) Soil benefits of agroforestry. In: Garrett HE, Jose S, Gold M (eds) North American agroforestry: an integrated science and practice, 3rd edn. American Society of Agronomy-Crop Science Society of America-Soil Science Society of America (ASA-CSSA-SSSA), Madison, WI

Unger IM, Goyne KW, Kremer RJ, Kennedy AC (2013) Microbial community diversity in agroforestry and grass vegetative filter strips. Agrofor Syst 87:395–402

Uri ND (2001) The environmental implications of soil erosion in the United States. Environ Monit Assess 66:293–312

USDA-ERS (2012). https://www.ers.usda.gov/amber-waves/2012/march/data-feature-how-is-land-used/. Accessed 23 Feb 2020

USDA-NRCS (2007a) Summary Report 2007 National Resources Inventory. http://www.nrcs.usda.gov/Internet/FSE_DOCUMENTS/stelprdb1041379.pdf. Accessed 30 Mar 2020

USDA-NRCS (2007b) Soil erosion on cropland 2007. https://www.nrcs.usda.gov/wps/portal/nrcs/detail/national/technical/?cid=stelprdb1041887. Accessed 15 Apr 2019

USDA-NRCS (2012) Assessment of the effects of conservation practices on cultivated cropland in the Missouri River Basin. USDA-NRCS. 225

USDA-NRCS (2013) Assessment of the effects of conservation practices on cultivated cropland in the Lower Mississippi River Basin

USEPA (1993) Paired watershed study design 841-F-93-009. Office of Water, United States Environmental Protection Agency, Washington, DC, pp 2–460

USEPA (2017) National water quality inventory: report to congress. U.S Environ Prot Agency: 1–22. https://www.epa.gov/sites/production/files/2017-12/documents/305brtc_finalowow_08302017.pdf

Varah A, Jones H, Smith J, Potts SG (2013) Enhanced biodiversity and pollination in UK agroforestry systems. J Sci Food Agric 93(9):2073–2075

Weerasekara C, Udawatta RP, Jose S, Kremer RJ, Weerasekara C (2016) Soil quality differences in a row-crop watershed with agroforestry and grass buffers. Agrofor Syst 90:829–838. https://doi.org/10.1007/s10457-016-9903-5

Wickramaratne N (2017) Hydrologic regime and nitrogen cycling: understanding the difference between claypan and loess watersheds in Missouri M.S. Thesis, University of Missouri

Zhang Q, Zhang M, Zhou P, Fang Y, Ji Y (2018) Impact of tree species on barley rhizosphere-associated fungi in an agroforestry ecosystem as revealed by 18S rDNA PCRDGGE. Agrofor Syst 92:541–554

# Enhanced Ecosystem Services Provided by Silvopastures

**Gabriel J. Pent and John H. Fike**

## Abbreviations

| | |
|---|---|
| BMP | Best management practices |
| LER | Land equivalency ratio |
| VEB | Vegetative environmental buffers |

## Introduction

Silvopasture is a land management practice that intentionally integrates trees, forages, and livestock into the same system. The interactions between these three components may be numerous and varied and can range from competitive to facilitative. Even with competitive interactions, however, the overall systemic benefits within well-managed silvopastures can outweigh the competitive interactions and provide for successful production and environmental outcomes. Silvopastoral practices, which include integrated forest grazing and silvopasture, can be applied at the pasture or landscape scale to create outcomes that are environmentally sustainable, economically viable, and socially acceptable. While integrated forest grazing practices utilize livestock extensively to manage forest or savanna vegetation and produce specific forest or environmental outcomes, silvopastures are more intensive in nature (Sharrow et al. 2009). The interactions between livestock and trees are

G. J. Pent (✉)
Shenandoah Valley Agricultural Research and Extension Center, Virginia Polytechnic Institute and State University, Raphine, VA, USA
e-mail: gpent@vt.edu

J. H. Fike
School of Plant and Environmental Sciences, Virginia Polytechnic Institute and State University, Blacksburg, VA, USA
e-mail: jfike@vt.edu



R. P. Udawatta, S. Jose (eds.), *Agroforestry and Ecosystem Services*,
https://doi.org/10.1007/978-3-030-80060-4_7

planned and mutually beneficial. The benefits that the trees convey to livestock or forages are not just incidental, but are an intentional consideration in the design process.

Silvopastures can provide numerous benefits, but they have met resistance to implementation from conservation communities, particularly in the context of thinning woodlots and establishing forages. This has occurred in large part from a misunderstanding of the intent and management inputs needed for these systems (Arbuckle 2009). Silvopastures require intensive management, where other practices that combine livestock with trees often do not involve management application and thus may not result in the same beneficial interactions. For example, silvopasture is not just turning cows into the woods with little or no consideration of the forages present or the implications for trees. Such practices have flavored many of the discussions between livestock producers and foresters or other technical service providers. Livestock producers interested in developing silvopasture systems have often faced concern from conservation agencies whose historical rallying cry has (deservedly) been to keep cattle (*Bos* spp.) out of the woods. Without providing adequate forage or feed to livestock or managing intervals of rest between stocking events, many producers have allowed livestock to severely damage forest resources and ecosystems (Patric and Helvey 1986). Soil compaction, browsing, and nutrient buildup can create timber stains and, in severe cases, damage can result in tree death (Lutz 1930; Gill 1992). Often, in these scenarios, the low-light environment under the trees supports limited understory growth, and when livestock are given unmanaged access, this vegetation is destroyed. The lack of understory vegetation can lead to severe erosion and nutrient pollution issues. While allowing livestock unmanaged access to a forest site may be necessary in some isolated instances (e.g., during periods of unexpected, severe environmental stress), it is clear that this practice differs considerably from silvopasture.

In the same manner, silvopasture is not providing livestock with unmanaged access to one or a few trees in a pasture. Although isolated trees can provide shade or a windbreak, they are also sites for manure deposition and excessive disturbance. Such conditions can provide habitat for infectious agents and thus isolated trees may serve as vectors for disease. These trees may also experience early mortality if unable to tolerate the soil compaction, root damage, and nutrient loading caused by the livestock. Even management of pastures with several trees (which may benefit livestock) might not constitute a silvopastoral practice when the manager does not give consideration to the potential value and productivity of the trees. In contrast, silvopastures are managed to create a desirable balance between tree production, forage growth, and suitable animal performance. Even distribution of trees helps ensure that livestock utilize the whole area uniformly, minimizing their impacts across the site.

Silvopasture systems are managed in space as well as in time. Animal products provide short-term income for the system, while the tree component typically is managed for long-term returns. Depending on the species, trees may also offer opportunity for annual income through the provision of non-timber forest products such as fruits, nuts, or pine straw. In either scenario, sufficient tree numbers should be produced to ensure that timber and tree crops will provide a reasonable return on investment over time. Too many trees, however, would suppress forage production and result in an open-canopy forest system.

Ungrazed forests and open pastures are the predominant forms of timber and grassland systems across North America. Appropriately managed, these lands provide numerous goods, services, and societal benefits. We argue, however, that silvopasture practices may improve on these provisions given their enhanced biological efficiency relative to the distinct management of trees or pastures. The body of work surrounding appropriately managed silvopastures indicates that these practices can enhance many of the ecosystem services provided by trees or pastures alone. In the subsequent sections of this chapter we discuss relevant findings and make a case for greater adoption of silvopastoral management. We also consider some factors that should be considered in the integration and strategic placement of silvopastures across a landscape.

## Silvopasture Ecology

The interactions between two or more organisms may be categorized as positive (facilitative or commensal) or negative (competitive, amensal, or parasitic). While facilitative interactions convey benefits to both interacting species, competitive interactions result in harm to (or at least reduce the productivity of) both individuals. In commensalism, one individual benefits while the other individual is unaffected, and amensal relationships result in harm to one individual while the other is unaffected. In parasitic relationships, one species benefits at the expense of another. These interactions may be direct or indirect and thus difficult to define. In a complex system, such as a silvopasture, often the only visible outcomes are the net effects of these processes.

The goal of silvopasture managers is to design and manage systems that support facilitative or commensal interactions which in turn outweigh competitive or amensal interactions among organisms (Jose et al. 2019). Trees and forages in silvopastures compete both directly and indirectly for resources such as light, nutrients, and water. This competition is often cited as a primary concern for the feasibility and management of silvopastures (Workman et al. 2003). Interference, or the negative influence of one plant on another, may also be evident. Examples of this include providing habitat for diseases and pests of another species. Livestock may also have negative influences on plants. Damage to trees, including both reduced health and survival, has long been cited as a sufficient rationale to never allow livestock access to woodlands (Arbuckle 2009). However, livestock may also confer benefits to trees—indirectly through the defoliation of competing understory vegetation (Doescher et al. 1987) or directly via nutrient deposition (Ponder et al. 2005). Until recently, ecological studies have largely focused on competitive interactions (Brooker et al. 2008). Although the presence of facilitative interactions has been broadly recognized in the context of successional communities, this has been less so in communities that are considered stable (Brooker et al. 2008).

Understanding plant-plant and plant-animal interactions may have practical implications for agroecosystem managers (e.g., Doescher et al. 1987; Gomez-Aparicio et al. 2004). The facilitative interactions possible among trees, forages, and livestock may result in substantial overyielding—when the sum of the products managed in an intercrop exceeds the sum of the products managed separately on the same amount of land area. However, overyielding should not be the only metric of productivity of an agroecosystem. Survival and fitness parameters, such as reproductive success, are also important metrics for assessing the performance of an agroecosystem, and plant interactions may often affect either of these metrics prior to or exclusive of any change in yield. Plants may positively affect each other by altering the abiotic environment or through modifications in biotic relationships, and nine different beneficial mechanisms have been described (Hunter and Aarssen 1988). Beneficial abiotic modifications include microclimate adjustment, physical support, changing of soil characteristics, and nutrient production or transfer. Beneficial biotic modifications include protection from herbivory or disease, indirect pest or competitor suppression (an indirect facilitative effect for the benefiting individual), alterations in the composition and behavior of rhizobial communities, pollinator attraction, and attraction of dispersal agents.

Silvopasture systems may be by nature successional as the forest canopy moves towards closure. Facilitative and competitive interactions are not static and may shift in space or time with changing successional conditions. Where stability is valued over succession, it is important to understand and anticipate the potential interactions possible between trees, forages, and livestock. Land managers can manipulate or delay successional processes to achieve desired outcomes. For example, during the tree establishment phase of silvopasture, the land manager often takes steps to minimize herbivore or fire damage to young trees. Additionally as the system matures, trees may be thinned or pruned to prevent canopy closure if a closed-canopy forest is not the ultimate goal.

Facilitative interactions may increase with increasing severity of environmental conditions, but this is not always the case (Brooker et al. 2008). Shrubs may have either neutral or positive impacts on annual species with increasing water availability, and the magnitude of the effects is often species dependent (Tielborger and Kadmon 2000). Whether plant-plant interactions are considered to be positive or negative may depend largely on the defined parameter of plant productivity, which may include survival over time or biomass production among other metrics (Maestre et al. 2005). Nevertheless, facilitation may provide a basis for niche expansion into environments that are typically stressful to individual species (Choler et al. 2001). For example, warm-season grasses (which utilize a different photosynthetic process than cool-season grasses) may be dominant in warmer or dryer reaches of temperate humid zones because they have greater water-use efficiency and are much better adapted to high light and warm temperature regimes. However, in such climatic zones, cool-season forages may be more competitive within silvopastures specifically because of the tree-induced alterations to the understory microclimate (Karki and Goodman 2015). Such changes may expand the potential niche of the more nutritious cool-season forage species, given their greater adaptation to lower light environments and cooler temperatures. In a similar way, shrubs can serve as nurse plants for trees in the reforestation of semiarid environments (Gomez-Aparicio et al. 2004). Rather than competing excessively for resources with young trees as traditionally had been thought, shrubs protect young seedlings from high temperatures and radiation in these moisture-limiting environments and even increase the nutrients and moisture available to the seedlings in the soil. As a result, planting tree seedlings next to undisturbed shrubs greatly improves the survival and vigor of the young trees. The concept of succession served as a model for this practice, thus minimizing the negative impact of such restoration projects on the ecosystems. Indirect facilitation, whereby species A has a direct negative effect on species B, which results in a positive effect for species C, is illustrated in the practice of prescribed grazing. In water-limited situations, where drought may cause high conifer seedling mortality rates, livestock can be utilized to defoliate grasses, thereby reducing the competition for water and improving tree survival (Doescher et al. 1987). The reduction of aboveground herbage largely is mirrored underground as forage plants shed roots following defoliation. If timed properly, this loss of roots reduces the ability of the grazed plant to exploit soil moisture, leaving more soil water available for young conifers. Of course, the timing and intensity of the stocking events are critical for minimizing tree damage, but when done properly, prescription grazing may result in substantial increases in tree growth and survival.

## Supporting Services of Silvopastures

Silvopastures provide the habitat and life-supporting functions necessary for the survival of some wild and domestic animals as well as naturalized and native plants. The unique habitat of silvopastures may convey additional benefits to certain species beyond those provided by open grasslands or closed forests. Silvopastures can also be managed to enhance nutrient cycling through niche partitioning and nitrogen fixation by some plant-microbe associations.

### *Wildlife Habitat*

The multi-strata, savanna-like habitat provided by silvopastures may be favored by a wider range of species than forests or grasslands managed alone. Both woodland and grassland bird species utilize silvopastures in the British Isles (Mcadam et al. 2007), and increases in invertebrate populations in silvopastures (relative to pastures or woodlands) were largely responsible for utilization by this more diverse range of birds. Although silvopastures can provide greater food resources, a substantial value of these systems likely is the provision of alternative and unique habitat. Thus, silvopastures have been promoted as a means of increasing eastern wild turkey (*Meleagris gallopavo*) habitat in the USA (Robinson 2005). The direct impacts of silvopasture on wildlife, however, are largely anecdotal. Few studies have compared the faunal populations or use of silvopastures compared to forest or pasture systems.

Forages and trees within silvopastures can provide both shelter and food for wildlife. While forages provide leaves and seeds, the trees may produce browse, leaf litter, and soft and hard mast. For example, the honey locust tree (*Gleditsia triacanthos* L.), integrated within some silvopasture systems, may produce copious amounts of pods filled with sugars. These pods are preferred by numerous wildlife species, including deer (*Odocoileus virginianus*), squirrels (*Sciurus* spp.), raccoons (*Procyon lotor*), and opossums (*Didelphis virginiana*). Other trees used in some silvopastures, such as chestnut (*Castanea* spp.) and oaks (*Quercus* spp.), are often marketed for their potential for wildlife attraction. The overstory canopy of trees in silvopastures can provide nesting habitat and cover for many birds and small mammal species. The value and utilization of this habitat will vary by species and stage of silvopasture development. In the establishment phases, silvopastures may more closely resemble early successional habitat. In the mature, open-canopy stage, silvopastures resemble savanna ecosystems and as the canopy closes, silvopastures will begin to resemble closed-canopy forests. Silvopastures may be managed to prevent canopy closure, although that may be desirable in some wildlife management scenarios.

Livestock can be used to manipulate ecosystem structure for desired wildlife goals (National Research Council 1970). Timing, as well as the type (grazing or treading) and intensity of disturbance, can be managed for the desired objectives. The degree of intensity may be managed by adjusting stocking density (the livestock stocked per land area) as well as the amount of time the livestock are held in a given location, and the recovery time afforded by the vegetation before restocking. These disturbances can affect floral and faunal populations, similarly to other disturbances such as fire (Derner et al. 2009).

Silvopasture can provide both grassy and woody shrub cover for eastern wild turkey, in addition to an abundance of food sources (Robinson 2005). Habitat quality for ground-nesting birds can be managed by providing patches of brush and shrubs, utilizing native warm-season grasses, and harvesting the forage by grazing (rather than mechanically). In almost all cases, rotational stocking should be used to protect both trees and the health of the ecosystem. By allowing for appropriate timing, intensity, and recovery intervals, rotational stocking can be used to create the patchy growth preferred by ground-nesting birds.

Some ungulates, such as white-tailed deer (*Odocoileus virginianus*), prefer forest edge habitats, which may have similar structure and floral associations as silvopasture systems. As browsers, deer are highly selective. Systems that incorporate nutritious forages (e.g., clovers (*Trifolium* spp.)), digestible leafy browse (e.g., honey locust and black locust (Robinia pseudoacacia L.)), or hard mast and soft mast (e.g., from oaks, chestnut, persimmon (*Diospyros virginiana*), or honey locust trees) will likely attract larger numbers of deer. Landowners looking to diversify their property's income streams may find that attracting wildlife for viewing or hunting enterprises can provide tremendous economic opportunity. These goals should be determined and accounted for during the planning stage for a silvopasture system.

As noted above, both woodland and grassland birds may utilize silvopastures, increasing species diversity on the landscape. Some bird species, such as the endangered red-cockaded woodpecker (*Picoides borealis* (Vieillot)), prefer and subsist only in savanna-like ecosystems (Walters 1991). However, for specialized forest or grassland birds, silvopastures may be less desirable than closed forests and open fields. For example, silvopastures in the Andes Mountains were not as desirable as forests, for both Neotropical migrants and forest specialists, although they were preferred over treeless pastures (Mcdermott and Rodewald 2014). However, the authors noted that the increased tree cover and habitat connectivity with forests afforded by silvopastures improved the quality of bird habitat.

The impact of silvopasture management on invertebrate populations has received far less attention, although some studies have noticed increases in invertebrate populations as a whole within silvopastures (Mcadam et al. 2007). Increased availability of insects and other invertebrates may have significant benefit for bird populations (Benton et al. 2002). The diversity of food resources available within silvopastures may support these increases in invertebrate population size, as may microclimate modifications (cooler temperatures and greater humidity) or structural changes to habitat (Tripathi et al. 2013). Among pollinators, the preferred habitat for bees is open-canopy woodland with an established herbaceous understory (Hanula et al. 2015).

While the benefits of silvopastures for some forest or grassland specialists may be debated (Arbuckle 2009), the greater species and structural diversity provided by silvopastures can improve available habitat across the biome. In addition, there is some evidence that even specialists can utilize and benefit from silvopasture systems (Mcadam et al. 2007).

## *Plant Communities*

Silvopastures, by definition, support the growth and survival of a greater range of functional plant groups than open pastures or forests. These may include coniferous or deciduous trees or both, along with some combination of woody browse, grasses, and forbs. This assemblage of plants may be intentionally introduced or naturally present in the system. Some managers may seek to maximize system productivity by using introduced or naturalized species that are more productive, nutritious, or palatable to livestock, while others may utilize native plants within silvopastures in an effort to mimic natural systems.

All plants within a silvopasture must be tolerant of or protected from herbivory, trampling, and rubbing from livestock. Trees, in general, require protection, but this is size and species dependent. Unpalatable trees (e.g., *Juniper* spp., or yellow pines) may be less subject to browsing, but could still be subject to rubbing and trampling. With some species and managements, a certain minimal level of livestock damage may be tolerated. In some pine systems, high planting density with light grazing has sustained tree establishment and growth, allowing for economical tree establishment without the added costs of protection (e.g., Pearson et al. 1971, 1990; Pent, personal observation).

Unlike trees, forage species both have evolved and are routinely selected for their ability to tolerate herbivory. Forages that persist in a silvopasture must be able to regrow following defoliation in partially shaded environments. They must also be able to compete with trees for water and nutrients in the soil if these resources are limited. However, although shade tolerance of species and cultivars has been studied, we know of no programs actively engaged in breeding and selection for adaptation to silvopasture environments. We note, too, that these capacities are functions of management. For example, greater residual leaf area following grazing events may be needed to maintain productivity in shaded environments.

The selection of plants for silvopasture plantings is not without controversy. Because silvopastures involve both production agriculture and ecological principles for land management, decisions may be driven by a range of goals and objectives. Arbuckle (2009) defines this spectrum of decisions by the ecological-productivist continuum. On the one end are the groups that emphasize agricultural production while the other end of the spectrum includes groups who view ecosystem health (as defined by the presence of natural plant communities) as the ideal.

Where a land manager places himself or herself on this ecological-productivist continuum will likely determine the goals and approach towards silvopasture design and plant selection. For example, in the Southeastern USA, native longleaf pine is sometimes used in efforts to restore (or at least mimic) the historic pine savanna ecosystems which evolved to suit the climatic and edaphic conditions of the region (Stainback and Alavalapati 2004). Native grasses may be desirable in such situations, and some locally or regionally adapted ecotypes have been selected for site suitability and forage production and nutrition (Franzluebbers et al. 2017). In the southeastern USA, once-prevalent pine savannas (Fig. 1) are now scarce, and some plant species that once thrived in these ecosystems are now threatened or endangered (Noss 2012). Plant diversity recorded in these ecosystems was among the highest found in the Western Hemisphere (Peet and Allard 1993). Silvopastures may be designed to structurally resemble these systems, although domestic livestock may replace the large ungulates once present. It remains to be seen whether healthy systems can be developed that support the recovery of rare or endangered savanna plants.

Not only is the structural diversity of silvopasture flora greater than that of forests and pastures, the management practices often utilized in these systems may result in increased species diversity. When fertilized and grazed, young lodgepole pine (*Pinus contorta* Dougl. ex Loud. var. *latifolia* Engelm.) silvopastures in Canada had increased herb and shrub species richness and diversity (Lindgren and Sullivan 2012). Fertilization also promoted the growth and quality of forages. However, on low-fertility sites without fertilizer applications, shrub richness and diversity

**Fig. 1** Managed silvopastures resemble savannas in structure and function, such as this longleaf and loblolly pine savanna in Florida

declined following grazing. This follows an ecological pattern in which on highly productive sites, disturbances such as fire or grazing can result in greater diversity and richness. In contrast, disturbances may produce a negative effect on diversity and richness on poor sites (Proulx and Mazumder 1998). Integrating silvopastures into the landscape can increase the diversity of species and habitat relative to open pastures or closed forests.

## *Nutrient Cycling*

As intensively managed systems, silvopastures are often productive because of nutrients added to the system through external inputs or biological nutrient fixation. Of the macronutrients (nutrients most needed for plant growth), phosphorus and potassium concentrations in the soil are relatively fixed apart from extractions or leaching from the soil without additional inputs of fertilizer, manure, compost, or organic matter. Nitrogen may be added to the soils, not only through fertilizer, manure, urine, or other additives, but also through biological nitrogen fixation and rainfall. Some prokaryote species convert inert, atmospheric nitrogen into plant-available nitrogen. While some of these bacteria are free-living, others form symbiotic relationships or associations with specific plant species. These plant species may be woody (e.g., autumn olive (*Elaeagnus umbellata* Thunb.) and black locust) or herbaceous (e.g., clovers). Nitrogen is taken up by the host plants and utilized for growth. Nitrogen may become available to other plants through leaching, decomposition of the roots, shoots, or leaves of the plant symbiont, or following herbivory of the plant symbiont and subsequent defecation or urination onto the soil. Clovers are often an integral forage species in many silvopastures and at large enough populations may provide enough nitrogen for all of the nitrogen requirements of a pasture.

Most of the nitrogen, phosphorus, and potassium consumed by livestock are excreted or urinated following digestion. In a silvopasture system, these nutrients are often distributed evenly across the site, becoming available to forages and trees. Of particular note to these nutrient cycles is a study of a loblolly pine (*Pinus taeda* L.) stand on poor, Coastal Plain soils deficient in phosphorus and potassium that was thinned and converted to a pine-goat silvopasture in Alabama (Nyakatawa et al. 2012). The addition of fertilizer increased soil nitrogen and phosphorus levels. On areas of the site where no fertilizer was added, phosphorus and potassium remained unchanged, but there were subsequent increases in soil nitrogen levels as a result of manure and urine deposition by the goats.

The structural differences between forages and trees can also have an impact on the nutrients in the soil through nutrient cycling. In general, trees often have roots that explore soil for nutrients at greater depths than forages. This niche partitioning, or differentiation in the utilization of resources, allows for the exploitation of more nutrients than may be achieved in just pastures or forests alone. Tree species that are primarily shallow rooted and compete vigorously with forages are generally not good choices for silvopasture systems. Some forage species may also compete too vigorously with trees, excluding their utilization in silvopastures, or they may require control during the tree establishment phase. As these nutrients are extracted from the soil, they may be utilized by the plants. A portion of these nutrients will return to the soil as organic material, eventually decomposing and becoming available to other plants over time. In the case of plants that are grazed or cut for fodder, the speed of the cycle is amplified within the gut of the herbivore.

## Provisioning Services of Silvopastures

The productivity of silvopastures is a function of the resources (light, water, and nutrients) available to plants and animals in the system, and the interactions, whether facilitative or competitive, between the system components. Productivity may also be largely influenced by stocking management, which can have an impact on the regrowth and yearly productivity of forages and allocation of nutrients. Niche differentiation under heterogeneous conditions often results in greater primary productivity in silvopastures compared to forests or open pastures. Some silvopasture systems have supported increases in primary productivity of more than 63% relative to open pastures or forests and orchard managed alone (Sharrow et al. 1996; Pent et al. 2020a).

**Table 1** Production of trees and forages or livestock in various silvopastures compared to monocultural systems

| | | | | Production relative to monoculture | | | LER |
|---|---|---|---|---|---|---|---|
| Location | Tree | Forage | Livestock | Tree | Forage | Livestock | |
| Australia[a] | *Pinus radiata* | *Trifolium subterraneum* | *Ovis aries* | 0.21 | — | 0.78 | 0.99 |
| Georgia[b] | *Pinus elliottii* | *Cynodon dactylon, Paspalum spp.* | *Bos taurus* | 0.62 | — | 0.59 | 1.21 |
| Missouri[c] | *Pinus taeda* *Juglans nigra* | *Lolium multiflorum, Secale cereale* | *Bos taurus* | 0.43 | — | 1.0 | 1.43 |
| Oregon[d] | *Pseudotsuga menziesii* | *Trifolium subterraneum* | *Ovis aries* | 0.96 | 0.64 | — | 1.60 |
| Virginia[e] | *Juglans nigra* | *Schedonorus arundinaceus, Dactylis glomerata, Trifolium spp.* | *Ovis aries* | 0.47 | — | 1.0 | 1.47 |

[a]Bird et al. (2010). Sheep production was calculated over 25 years beginning with the year that trees were planted. Tree production in the silvopasture (60 trees per hectare) was compared to tree production in a plantation (815 trees per hectare) using estimates of the volume of 6 m butt logs per hectare at year 25. The value of butt logs in this silvopasture was not calculated, but would have been considerably greater than the value of butt logs in the plantation because the silvopasture trees had less defects and branches

[b]Lewis et al. (1983). Livestock productivity was calculated from 5 years after the trees were planted until year 19. Tree productivity in the silvopasture 20 years after establishment was compared to unfertilized range plantation with a tree spacing of 3.7 x 3.7 m

[c]Kallenbach et al. (2006). Trees were 6–7 years old at the time of the study. Tree production estimates were calculated based on the assumption that tree growth within rows would be similar to pine or walnut plantations and that between- and within-row spacing would be 3.0 m in a plantation planting

[d]Sharrow et al. (1996). Calculations presented in manuscript

[e]Pent and Fike (2016). Trees were 21 years old at the time of the study. Tree production estimates were calculated based on the assumption that walnut production per tree would be similar to production per tree in a black walnut orchard and that between- and within-row spacing would be 9.1 m and 7.6 m, respectively, in an orchard

The land equivalency ratio (LER), a mathematical comparison of intercrop versus monocrop productivity, provides a useful means of illustrating the benefits of intercropping systems. Land equivalency ratios are calculated as the sum of components produced within an intercrop relative to the total output of these same crops managed as individual monocrops (Vandermeer 1981). Even if the production of all crops managed in an intercrop is less than when they are managed as monocrops, the total output (i.e., the sum of all yields) per land area may be greater than the total yield of the same crops managed in monoculture (Table 1). An assemblage of multiple, functionally diverse species will more efficiently utilize resources in heterogeneous situations or conditions, although such characteristics may present greater management challenges. This often stands in contrast to the homogenization of many modern agricultural systems built on monocultures whose production is facilitated by the development and application of numerous exogenous inputs.

Silvopasture design, however, is guided by the biological principles of niche differentiation and the stabilization of productivity through biodiversity. As global populations increase and land available for production agriculture and forestry diminishes, the utilization of ecologically sustainable, high LER land management practices such as silvopasture will become increasingly valuable.

## *Forage Production*

During the tree establishment phase of silvopastures, seedling trees typically have little to no effect on the forage "understory"—if it may be called that. Hardwood seedlings planted into newly established cool-season grass and legume pastures in Missouri (3-by-12 m spacing) had no significant effect on forage availability or nutritive value the following 2 years compared with open pastures (Lehmkuhler et al. 2003). However, as trees get older, they will likely have a corresponding impact on forages. About 15% greater forage production was measured underneath moderately spaced trees in an early-stage silvopasture in Appalachia (Buergler et al. 2005). Black walnut and honey locust trees were distributed in varying levels of density across a slope and trees were 7–8 years old at the time of forage sampling. Greatest forage production occurred underneath medium-density trees compared to sites with low-light (high tree density) or high-light (densities similar to open pasture) environments. Microclimate conditions (young trees protected and lowered soil temperatures) and changes in plant portioning in favor of top growth (Belesky 2005) may have favored the growth of cool-season grasses during the growing season.

Competition for light, nutrients, and water increases more vigorously between trees and forages as silvopastures mature. Forages may face this same level of competition in silvopastures developed from thinned forests where the forages are sown underneath established trees. In such cases, the competitive effects may outweigh the benefits that trees can provide to forages through improvements to the microclimate. However, forest manipulations such as thinning or pruning may reduce these competitive effects. For example, in Virginia, 13-year-old stands of black walnut and honey locust trees planted with a 2.5 m (8 ft) in-row spacing and 12.3 m (40 ft) alleys had reduced forage production relative to open pastures (Fannon et al. 2017). After removing 75% of the trees (leaving a 12.3 m × 12.3 m spacing), forage production 7 years later was greater than in the open pastures underneath the honey locust trees but not under the black walnut trees (Pent et al. 2020a). Thus tree age, density, and species may all impact forage production, depending on their relative contributions to competitive effects.

Forage species also display differential adaptation to the growing conditions present in silvopastures. Cool-season grasses and legumes utilize the C3 photosynthetic pathway and have a lower light saturation point than species with the C4 photosystem. Thus, in temperate environments cool-season species are better suited to the shaded conditions of silvopastures than warm-season grasses, although both types have been managed successfully in silvopastoral systems (Lewis et al. 1983; Clason 1995; Kallenbach et al. 2006; Fannon et al. 2017; Franzluebbers et al. 2017). Even within species, adaptation to shaded conditions may vary by cultivar (Lin et al. 1998).

Competitive effects in silvopastoral systems may also be temporal, and system design decisions can be made intentionally to take advantage of these conditions. In temperate Virginia, deciduous hardwood trees (locust and walnut) used for Appalachian silvopasture research were selected in part for their late leaf development and early leaf drop (Jim Burger and Jim McKenna, personal communication). Because these trees primarily grow in summer, their phenology hypothetically complements the production of the cool-season perennial forages, which have production peaks in spring and fall.

In some cases, annual forages may be utilized to take advantage of their complementarity (Sharrow et al. 1996). In California's annual rangelands, grassland species make use of early season rains before going dormant, thus removing the competition for moisture from the trees. In these seminatural savannas, grass species are more productive under the protective canopy of oak trees (Frost and Mcdoucald 1989). In contrast, pine and walnut trees reduced production of annual forages (ryegrass and cereal rye) by about 20% in an early-stage Missouri silvopasture (Kallenbach et al. 2006). However, trees moderated the understory microclimate early in the growing season, allowing for earlier green-up.

## *Animal Production*

Forage availability and nutritive value are not the only determinants of animal production. Health and well-being also play significant roles in the productivity of domestic animals (Fraser et al. 2013). In particular, heat and cold stressors significantly impact livestock production, although the effects of heat stress have perhaps received greater attention (e.g., Silanikove 2000; St-Pierre et al. 2003). Silvopastures can provide significant midday cooling for livestock, even during relatively mild Appalachian summers (Pent et al. 2018). Even when the environment in silvopastures reduces forage production or nutritive value relative to open pastures, this does not always translate to a corresponding decrease in animal performance (Kallenbach et al. 2006; Pent et al. 2020a; Fannon et al. 2017). Improved animal welfare in silvopastures (e.g., lower body temperatures) and corresponding changes in behavior (greater time spent lying down) likely compensate for reductions in forage yield (Pent et al. 2020a, b). Animals with access to shade in silvopastures also demonstrate less agonistic behavior and improved welfare (Karki and Goodman 2010). Because heat stress can compromise animal immune function, under extreme conditions, morbidity and even mortality may be avoided by managing livestock in silvopastures.

## *Tree Production*

Along with their regulating functions, trees in silvopastures may provide timber and non-timber goods. Compared to traditional forestry and timber plantation management, timber production in silvopastures is often lower as a result of thinner tree counts per land area despite greater individual tree growth (Gibson et al. 1994; Ares and Brauer 2005). Pines in silvopastures may be larger than pines in plantations, although it is less clear whether this is a function of reduced competition with other plants or greater nutrient supply as a function of pasture management.

Trees in silvopasture may also be managed for the production of nuts, fruits, sap, fodder, or other non-timber forest products (Fig. 2). The wide, distributed spacing of pecan trees for optimal nut production lends itself well to the production forage and livestock prior to nut harvest in the fall (Ares et al. 2006). In such systems, the economic returns for nut production may be similar or improved in silvopastures as compared to orchards managed without livestock as a result of lower pest pressure or management costs. Similar results may be achieved in other fruit and nut orchards, which are increasing in popularity especially with small-scale landowners and farmers (Orefice et al. 2017).

**Fig. 2** Pods from honey locust trees can be a nutritious fodder source for livestock

## Regulating Services of Silvopastures

Regulating services encompass the benefits of ecosystem processes that help buffer and maintain the environment. Silvopastures, similar to forests and grasslands, provide critical regulating services such as air and water filtration, flood control, pollinator habitat, and erosion prevention. However, as with the enhancement for provisioning services, integrating trees and grasses in silvopastures both diversifies and enhances the value of the regulating services relative to their production as monocrops.

### *Water Quality*

Livestock behavior and management and their potential impact on water quality is of particular concern in animal production systems. Some livestock management practices can severely impair water quality (Agouridis et al. 2005). Overgrazing or unmanaged access of livestock to waterways can cause erosion and stream bank failure (Pietola et al. 2005). The direct and indirect deposition of nutrients, sediment, and pathogens into surface bodies that may be a source of drinking water and recreation or habitat for wildlife can be a serious issue (McDowell et al. 2008). Nutrients can also reach water bodies, even with what presumably are good stock management practices. For example, nitrogen and phosphorus can leach through soils, enter the groundwater supply, and eventually contaminate waterways where it contributes to eutrophication (Meinikmann et al. 2015). Trees in silvopastures have the potential to serve as a "safety net" by capturing nutrients that have escaped the root zones of most forage plants and preventing their entry into groundwater (Jose 2009). For example, silvopastures on sandy loam soils in Florida had greater phosphorus storage capacity and less phosphorus buildup than open pastures, indicating that silvopastures present less of a risk of losing phosphorus to groundwater (Michel et al. 2007).

A similar "safety net" was observed with respect to lower nitrate leaching losses in an Appalachian silvopasture (Boyer and Neel 2010). However, soils in the silvopasture area released more fecal coliform, perhaps due to the greater macroporosity of forest soils that had been converted to silvopastures. The authors noted that even with silvopasture management, keeping livestock near sensitive or valuable waterways or wells may not be advisable—although wildlife may also impact fecal coliform shedding. Evidence suggests that establishing trees into existing grass- and croplands can lower the potential for fecal coliform contamination. For example, creating an alley-cropping system reduced nitrate and fecal coliform levels measured in tile drain effluents compared with those from monocultural cropping systems (Dougherty et al. 2009). Yet soil macropores may increase in numbers as agroforestry systems mature, which could increase both water infiltration and nutrient movement. This was apparent in the soil of buffer strips with trees which had 2.5 times greater numbers of macropores than grass-only buffers (Jose 2009).

## *Air Quality*

In general, plants provide valuable air quality services by consuming carbon dioxide and releasing oxygen. It might be surmised that if the primary productivity of silvopastures was greater than that of forests or grasslands alone, the oxygen production of these systems may be greater than traditional land uses.

Vegetative environmental buffers (VEB) have demonstrated promise in filtering out odors from swine production facilities (Tyndall and Colletti 2007). These buffers work by filtering out the airborne particulates that carry most of the odors from livestock production facilities. Although structurally different from a VEB, the leaf surface area of trees in a silvopasture should still function similarly to reduce dust transport and thereby mitigate odor movement. Thus, trees in silvopastures could reduce the impact of odors from livestock facilities by reducing the passage of particulates through physical impediments and alterations in wind flow. Interestingly, aesthetics can impact how odors are perceived (Tyndall and Colletti 2007). Therefore, silvopastures, with visually appealing parklike aesthetics, might have a real and perceived impact on observed odor.

## *Soil Conservation*

Trees in silvopastures can help conserve soils by intercepting rainfall and improving soil water infiltration rates (Lunka and Patil 2016). In a similar way that trees in pastures may reduce or slow the leaching of nutrients or pathogens from soil, soil may be conserved through the "safety net" effect provided by the trees. To maximize these benefits, trees may be placed strategically on the landscape in areas particularly susceptible to soil erosion. Although not stocked with livestock, the soils of some agroforestry buffers have greater root length densities than stocked pastures without trees (Kumar et al. 2010), which may help retain soil. In addition, these buffers have greater macroporosity; these large pores provide space for water to collect in the soil, thus improving water infiltration rates (Kumar et al. 2010, 2012). When not absorbed by soil or organic matter during a rainfall event, the overland flow of "free" water can contribute significantly to runoff and erosion, even on lightly sloping land. Improvements to water infiltration and holding capacity means not only that more water is stored and available for use by surrounding plants during subsequent dry weather, but also that soil is conserved and contained. Although cattle and small ruminants may compact soils through treading (Drewry et al. 2008; Lunka and Patil 2016), in some silvopastures this has been short-lived and reversed by natural freeze-thaw cycles (Sharrow 2007).

## *Nutrient Distribution and Use Efficiency*

If available, livestock spend significant portions of their day in shade, particularly during summertime. In contrast to creating distributed shade with silvopastures, livestock producers more typically leave individuals or clusters of trees in the middle of pastures or along fence lines in order to provide livestock with shade. Not only may this limited shade be inadequate for larger groups of animals (Fig. 3) and as the sun changes position throughout the day, but also such practices can lead to negative environmental consequences and destruction of the trees. As livestock congregate in these small shaded areas for much of the day, they often denude the vegetation around these trees (Patric and Helvey 1986). Additionally, large amounts of urine and feces are deposited in these small areas, creating health hazards for the livestock, as well as environmental "hot spots." The concentrated nutrients present contaminant risks via runoff or leaching. Runoff is likely the greatest risk, as these congregation points typically are denuded and have heavily compacted, highly disturbed soils with large amounts of manure on the surface. Even in intensively managed (i.e., frequently rotated) pastures without shade, livestock will spend more time near water sources, thereby skewing the distribution of nutrients towards these heavy-use areas (White et al. 2001).

Trees in silvopastures are sufficiently distributed so that all livestock can utilize shade, and the livestock, in turn, are distributed across a pasture (Karki and Goodman 2010). Although we have not found evidence of such research in the literature, it is likely that nutrients are more evenly distributed in silvopastures compared to open pastures with small amounts of shade. In addition, the even distribution and more uniform grazing of forages in silvopastures safeguard against soil compaction and loss of vegetation in sensitive areas. In turn, since they are not trampled or excessively loaded with nutrients, the forages, trees, and soil micro- and mesofauna can

**Fig. 3** Insufficient shade for livestock can result in negative environmental and animal welfare conditions

take up the nutrients available from urine and manure and utilize them for growth, reducing the risk of volatilization of nitrogen or loss of nutrients through runoff or leaching.

Niche partitioning by various forages and trees in silvopastures also increases nutrient-use efficiency. This is achieved through spatial, temporal, and mechanistic differentiation. Trees generally have deeper roots than grasses and forb species (Schenk and Jackson 2002). Not all tree species are deep rooted, however, and even deep-rooted trees may primarily draw nutrients from the same zone of soil as forages (Schroth 1998). Belowground interactions between woody and herbaceous plants also affect root architecture and depth, and competition with grasses can drive the roots of trees to greater depths (Dawson et al. 2001). Nutrient capture by plants may vary temporally, as well (Schroth 1998). In the humid subtropical zone of the southeastern USA, cool-season forages may grow primarily in the spring and fall months, while some "warm-season" trees may grow primarily in the summer. In these integrated systems, nutrient capture is thus distributed across the growing season, reducing the risk of loss associated with periods of low plant growth. Mechanisms of nutrient capture and utilization also vary by tree and forage species (and their associated mycorrhizae), which has implications for nutrient-use efficiency. This mechanistic diversity ensures that more nutrients are utilized in these intercropping systems.

As noted earlier, the greater nutrient efficiency associated with the integration of trees and forages in silvopastures has positive implications for water quality. Both phosphorus leaching and nitrate leaching have been reduced in silvopastures because of enhanced utilization of these nutrients by plants or increased storage capacity of these nutrients in soil (Michel et al. 2007; Bambo et al. 2009; Boyer and Neel 2010). The safety net provided by niche differentiation in silvopastures is evident, but ensuring that the competitive interactions between trees and forages do not lead to exclusion will be a primary goal of the silvopastoralist.

## *Weeds, Diseases, and Pests*

Although our knowledge and utility of the mechanisms are limited, it is possible that direct biophysical interactions between plants can be manipulated to manage or control weeds in pastures. For example, black walnut trees release juglone, a potent allelochemical of species in the Solanaceae family. Carolina horsenettle (*Solanum carolinense* L.), a competitive invader of well-managed pastures, can reduce forage growth and harm livestock through the production of toxins. However, the juglone produced by black walnut trees may control or eliminate certain plants from pastures. Horsenettle has been noted to be entirely absent in mature black walnut and cool-season perennial forage silvopastures in Virginia, while the same weed is prevalent in adjacent open pastures or honey locust silvopasture (Pent et al. 2020a). Black walnut leaf extracts also have reduced bull thistle seed germination (Downs and Cavers 2002).

Pests affect both plants and livestock in pasture systems. The humid microclimates of silvopasture may be conducive to the growth of endoparasites of ruminants (de Mendonca et al. 2014). Although similar infestations of helminths have been noted for cattle in silvopastures and open pastures in a tropical climate (de Mendonca et al. 2014), lower fecal egg counts were noted from sheep (*Ovis aries*) grazing in silvopastures compared to open pastures in a temperate climate (Pent and Fike 2016). In the latter case, residual growth from tree stumps may have provided a source of anthelminthic chemicals, such as condensed tannins, to the sheep. In both cases, it may be that reductions in animal stress resulted in improved animal immune function. Stable flies (*Stomoxys calcitrans* L.) reduce both weight gains and milk production of cattle, resulting in annual losses in excess of $2 billion for the US cattle industry (Taylor et al. 2012). Some bird species consume flies, at times directly off of cattle, and this predator-prey relationship may reduce the impact of flies on livestock. It may be that improving the attractiveness of habitat to birds will result in an increased predation of pests such as flies. In addition, the uniform distribution of manure across silvopastures potentially can reduce the breeding habitat for flies. There is a need for more evidence on how silvopastures may ultimately influence livestock pests.

## *Environmental Modifications*

As noted in our discussion of provisioning services, trees have substantial modulatory effects on ecosystems. Indeed, the protection from weather conditions that trees provide is a major driver of silvopasture adoption in the USA (Orefice et al. 2017). Trees can modulate the local environment and buffer plants and animals from weather extremes both in summer and winter. Cooler conditions have been documented in pine silvopastures compared to open pastures during summer months (Karki and Goodman 2015), while conifers can also protect forages from frosts in winter (Feldhake 2002). The buffering effect of trees can conserve soil moisture and reduce radiation loads and lower the temperatures experienced by plants and animals in summer. These modifications are desirable for both forages and animals in the North-South transition zone. Livestock prefer shade during hot weather, and will actively pursue and utilize the shade from trees in silvopastures (Pent et al. 2020b). Sheep in black walnut silvopastures had 0.5 °C lower vaginal temperatures than sheep in nearby open pasture systems during the afternoon hours; sheep in honey locust silvopastures had somewhat intermediate vaginal temperatures at the same time, indicating that the type and density of the shade provided by the trees can differ in quality (Pent et al. 2018). It should be noted that the sheep in the silvopastures did not cool down at night as much as sheep in the open pastures, likely because tree canopies reduced the radiation of heat from the sheep and soils into the sky. Thus, silvopasture managers should consider designs that include open areas in paddocks where livestock can gather at night. Depending on species and spatial configuration, silvopastures can also reduce wind flow or increase humidity, factors that can also impact livestock heat stress. However, blocking wind can be beneficial at other times in the year. The shape and planting configuration of trees can be designed to maximize windbreak protection for livestock during cold weather. Trees can also have a warming effect on forages by trapping warm air and reducing radiant heat loss. Temperatures in an Appalachian silvopasture were 11 °C warmer in areas where 80% of the field of view of the sky was obstructed by coniferous trees (Feldhake 2002). This can reduce the damaging effect of frosts on forage species and preserve the nutritional quality of winter-stockpiled forages, although such high-density tree stands may reduce seasonal forage production.

## *Carbon Cycling and Sequestration*

Silvopasture practices have been considered effective for climate change mitigation (Montagnini and Nair 2004), although concrete research in this arena is limited. Soils contain large amounts of carbon and are in fact the greatest terrestrial pool of carbon. Carbon in soil may be lost through soil erosion or tillage. Trees and perennial forages can reduce soil erosion rates, thereby reducing the amount of carbon lost from soils compared to processes inherent to cropland. Photosynthesis is the process by which atmospheric carbon is converted into sugars and fibers that are the building blocks of forage for livestock, timber for construction, or biomass for bioenergy production. This biomass may be used for bioenergy production, acting as carbon substitutes for more energy-intensive materials. In addition, the products produced by trees, such as food or building materials, may be produced without the inherent carbon emissions of cutting old growth forests or producing food in intensive monocultural systems. Because of the high productivity of silvopastures compared to pastures or forests managed alone, silvopastures are a promising candidate for carbon sequestration (Sharrow and Ismail 2004).

Plants in silvopastures can reduce atmospheric carbon by depositing carbon into biomass and soil. These resulting carbon stocks in soil are more stable than the harvested or standing aboveground biomass produced by these systems. The carbon in feeds or harvested materials is only effectively sequestered until the material is digested or decomposes and the carbon is then returned to the atmosphere. The roots or unharvested portions of trees or forage plants also decompose over time, releasing carbon back into the atmosphere. However, plants also release stable carbon-rich exudates into the soil, and these exudates and some litter that plants produce may be stable for long periods of time in the soil as organic matter (Rasse et al. 2005).

While soil carbon stocks are difficult to accurately estimate, there is some evidence to indicate that trees may improve the carbon sequestration of grasslands. Tree roots extend to greater depths than forage roots, and in a study of soil carbon by depth, samples from deep in the soil profile had more tree-derived carbon, indicating that the slash pines (*Pinus elliottii*) contributed to carbon storage at greater depths than bahiagrass (*Paspalum notatum*) (Haile et al. 2010). Similarly, in a study of grasslands that contained either birch (*Betula* spp.) or pine trees, soil carbon storage was greater under birch trees because the slowly decomposing needles from the pines inhibited the vegetative growth of understory plants (Howlett et al. 2011). These results indicate that the growth of the forage component of a silvopasture also has an impact on the carbon storage potential of these systems (Howlett et al. 2011). In Canada, agroforestry systems including silvopasture had more soil carbon than corresponding paired agricultural systems (Baah-acheamfour et al. 2014). This work again emphasized the contribution of the herbaceous understory to soil carbon stocks as more soil carbon was measured in silvopastures than in shelterbelts and hedgerows. This may have been due to the greater ratio of deciduous to coniferous tree species present in the silvopastures compared to the hedgerows or to the herbaceous plants' greater contribution to carbon stocks (Baah-acheamfour et al. 2014). Total soil carbon may also be increased through grazing management in thinned woodlands after only a few years of stocking after forage establishment (Nyakatawa et al. 2012).

## Cultural Services of Silvopastures

There is some evidence that silvopastoral management is a historical practice for farmers around the world (Neel 1939; Joffre et al. 1999; Holl and Smith 2007), and silvopastures may resemble historic landscapes. These landscapes may provide distinctive cultural services compared to other agricultural production practices, including improvements to animal welfare and enhanced aesthetic and recreational opportunities.

### *Animal Welfare*

Animal welfare concerns are of increasing importance for livestock producers and recently ranked within the top five priority management issues by beef cattle producers in the USA (USDA 2016). Farm animal welfare has also become an increasingly important concern to the public (AHA 2014), concomitant to agricultural system intensification. In many cases, the more historic or pastoral landscapes are considered preferable habitat for ruminant livestock relative to modern, industrialized confinement operations. Silvopasture may have even greater emotional resonance given the association of tree shade with greater animal comfort and the visual appeal of livestock grazing in a sheltered, bucolic landscape (Fig. 4). Producers of monogastrics such as chickens and pigs may also capitalize on the aesthetic appeal of silvopastures while tree mast, forages, seeds, and insects may also provide supplemental feed for these livestock. As noted previously, the buffering functions of trees are a primary reason that producers have adopted silvopasture in the northeastern USA (Orefice et al. 2017).

**Fig. 4** Silvopastures, such as this oak and cool-season grass silvopasture in Virginia, may confer visual appeal to the agricultural landscape

## *Aesthetics and Recreation*

Aesthetics have also been considered a primary advantage of silvopastures by a majority of landowners surveyed in the southeastern and northwestern USA (Elwood et al. 2003; Workman et al. 2003). The open landscape of silvopastures with sparse tree cover is often described as "parklike." The appeal of the silvopastoral landscape may be inherent to our species. There is some evidence that humans have an evolved disposition to prefer savanna-like landscapes (Balling and Falk 1982). In addition, the silvopasture may be more similar to the natural savanna state of certain regions compared to timber plantations or open fields. For example, there is significant evidence to indicate that the southeastern USA, including the Coastal Plain and the Piedmont regions, was predominately covered in grasslands with sparse numbers of trees (Noss 2012). Some of the processes that sustained these landscapes were non-anthropogenic (climatic disturbances, lightning-induced fire, and grazing ungulates), while in some areas Native Americans also maintained these landscapes through periodic burning. Silvopasture systems may also be steadily maintained through the management of pseudo-natural processes, such as burning or grazing.

The diversified landscape of silvopastures may improve wildlife habitat as described earlier, thereby increasing the quality of hunting, bird-watching, or other agritourism experiences. Landowners do value such features; in an early survey, a majority considered increased biodiversity as a benefit of agroforestry systems

(Lawrence et al. 1992). These advantages have also been recognized and accounted for in some financial analyses of silvopastures (e.g., Grado et al. 2001; Shrestha and Alavalapati 2004).

## Opportunities for Adoption: Placement, Design, and Management Considerations for Silvopastures

Implementation of silvopasture (and agroforestry practices more broadly) has been limited in the USA (USDA 2012). Technological adaptations often take a generation or more to become commonplace, and adoption of silvopastures potentially will be slower. The complexity of the systems and their different production timescales can be daunting both to agricultural producers and tree growers. Couple these factors with few practical examples and limited economic analyses of the possible systems that could be deployed, and producer reluctance to engage is understandable.

That said, several issues could drive a more rapid adoption of silvopasture systems. Changing climate and the need for shade and heat stress abatement, protection from intense weather events, potential restrictions on livestock access to waterways, or need to conserve soil and improve water quality will likely be important at the farm level. In turn, broader societal efforts to manage and support specific wildlife species with suitable vegetation, habitat, and movement (wildlife corridors) may also provide a rationale for future implementation efforts. In the remainder of this chapter, we will focus on these issues, largely from a production perspective.

At the farm scale producers may find several benefits of converting certain portions of land to silvopasture. While not all agricultural land will or should be converted to agroforestry practices, strategic placement and use may enhance the value of the ecosystem services that these practices provide. Although few data and models exist to define the most effective scale and placement of silvopastures, one can infer some generalizations.

For their land protection functions, silvopasture and other agroforestry practices may be best utilized as an alternative to open pastures on sites where erosion is of greatest risk. Plantings along waterways may serve multiple purposes by functioning as buffer strips for crop land, protecting stream banks, and providing grazing for livestock. It should be noted, however, that such installations will not receive cost-share support from state or federal stream protection programs that preclude grazing in these areas. For systems not reliant on governmental support, grazing may be a useful tool for vegetation management in these areas.

Silvopastures may also be deployed on land with substantial slopes. This application appears to have received less attention in North America, likely because steep hillsides often already are forested. However, creating silvopastures on such sites may present an opportunity both to increase grazing lands and simultaneously improve timber stand management. Where pasture systems exist on sloping lands subject to severe erosion and slippage, strategic tree planting has proven an effective means of soil stabilization (e.g., McIvor et al. 2011) and these practices could be expanded into fully developed silvopasture systems.

Silvopastures can also be implemented strategically to improve soil fertility and nutrient cycling. Nitrogen-fixing trees are commonly used in agroforestry systems of the humid tropics and historically were used to improve pastures in temperate North America (e.g., Neel 1939; Smith 1942). Leguminous forbs can also supply nitrogen to other forages in a pasture following defoliation (Ayres et al. 2007). Choosing and placing trees for improved soil quality in a temperate silvopastoral context represent a means for improving land value and use over time, although this has received little research attention to date.

For all sites, and particularly for sloping land, attention must be paid to livestock management and forage resources in order to minimize damage from soil compaction or overgrazing. Application of rotational stocking management is a fundamental tenant of silvopasture implementation and most often discussed in the contexts of maintaining tree, forage, and livestock production and health. However, this management is also critical for achieving the potential environmental benefits of silvopasture. The rainfall interception, increased water infiltration, soil conservation, and nutrient retention and capture that can be accrued with silvopastures all can be negated with poor management, particularly when sensitive areas are overgrazed or overused, particularly during wet conditions.

In a more general context, appropriate grazing management practices can reduce livestock stream bank use and stream occupancy (Haan et al. 2010). Research from the southeastern USA indicates that increased shade availability and distribution (as well as off-stream water access) reduce the amount of time cattle spend in riparian zones (Byers et al. 2005). Although such environmental considerations may drive some producers towards adoption, a greater number will view animal health and welfare (and attendant benefits for production) as the primary motivator for implementing silvopastures. Surprisingly, welfare-based motivations may be a direct consequence of efforts to improve farm environmental outcomes. Specifically, many producers have undertaken efforts to fence cattle from wetlands and surface waters, only to realize that this has taken away their livestock's only sources of shade. This is a particular issue in the North-South transition zone, where environmental heat loads (due to high radiation, temperatures, and humidity) are further compounded by the consumption of toxins produced by an endophyte (*Epichloë coenophiala*) within tall fescue (*Festuca arundinacea* syn *Lolium arundinaceum* syn *Schedonorus phoenix*) (e.g., Aldrich et al. 1993). Interestingly, the most extensive research on BMPs for stream quality has occurred in the Midwest and West (Agouridis et al. 2005). Particularly in the Southeast, better recognition of the interconnections among forages, shade, grazing management, animal welfare, and stream health should elevate silvopastures as part of the BMP tool kit for meeting both conservation and production needs.

Most livestock producers who consider silvopasture adoption for meeting animal welfare needs will likely face the question of how much land is needed to derive these benefits. The answer will depend largely on the duration of time (per day or per season) that stressful environmental conditions are expected to persist. For example, *Bos taurus* cattle experience extreme heat stress at a temperature humidity index of 75 or greater (Silanikove 2000). Thus an estimate of need may be made based on historical weather data. Such a scale would be based on the number of days a silvopasture may be needed to provide shade or windbreaks during expected periods of inclement weather. An acreage for silvopasture could be calculated that would provide sufficient forage to sustain the livestock during those expected periods of heat stress. Estimates of forage production (and hence grazing days) in silvopastures should be based on tree stocking density and incident light. We note here that most comparisons of silvopastures to treeless pastures have been conducted in a purist sense, with animals managed entirely in one system (silvopastures) or the other (treeless pastures). One might also feasibly design silvopastures so that livestock have access to shaded areas in the daytime and open areas during the evening grazing bout or at night while resting. This would reduce the demand for forage production within the silvopasture and allow livestock to cool and recover more fully from daily heat stress since air temperatures at night in open pastures may be cooler. The strategic design of such systems for specific animal welfare goals should be a focus of the future research of these practices.

As with any economic enterprise, one should consider current or future markets when designing a silvopasture system. Beyond the obvious need to match trees, forages, and livestock to the environment and conditions of a silvopasture, care should be taken to assess the potential value that silvopastures can provide throughout the lifetime of the system. Tree species differ in their effects on available resources (with consequence for forage and livestock production) as well as in their productive potential and end use (including timber or non-timber products). The markets for all these products may vary widely, especially for non-timber forest products, such as black walnuts, or specialty timber products, such as red cedar wood. Even livestock may have different valuation in different regions, although these trends have largely begun to diminish given globalization of world markets. Opportunities for marketing and agritourism experiences are also becoming increasingly important facets of farm design considerations. Silvopastures may have special value in such contexts given the aesthetic appeal of comfortable animals grazing in bucolic landscapes.

## Conclusion

The data are clear: silvopasture practices can improve ecosystem services provided by monocultural timber and pasture production systems. Nevertheless silvopasture practices have not been widely adopted in temperate regions. There are no estimates of land managed as silvopasture in the USA, but integrated forest grazing, which to some extent resembles silvopasture practices of an extensive nature, is practiced on about one-fifth of forest land (Bigelow and Borchers 2017). Improving the management of these practices by adopting the principles of silvopasture management may have a significant impact on the health of these lands and the resources and services produced by this land base.

The productivity of silvopasture is certainly not an impediment to their adoption as the land production values often exceed those of timberland or pasture alone. Instead, the barriers to the adoption of silvopasture practices are largely cultural (Mayerfeld et al. 2016). Keeping trees out of pastures has been a dominant legacy handed down by generations of farmers (Raedeke et al. 2003). Certainly the absence of trees simplified crop management during a time when soil tillage was a necessity. From the perspective of foresters, keeping cattle out of the woods is also a dominant legacy of forestry (Arbuckle 2009; Mayerfeld et al. 2016). Livestock access to woodlands without any management of the forage resources or stocking duration or intensity can result in irreversible damage to the timber or the ecosystem. Nevertheless, a growing number of foresters and conservation consultants are recognizing the potential value of silvopastures on the landscape. Together, these professionals and farmers that recognize the productive potential of silvopastures will drive the implementation of these practices.

## References

Agouridis CT, Workman SR, Warner RC, Jennings GD (2005) Livestock grazing management impacts on stream water quality: a review. J Am Water Resour Assoc 41:591–606. https://doi.org/10.1111/j.1752-1688.2005.tb03757.x

Aldrich CG, Paterson JA, Tate JL, Kerley MS (1993) The effects of endophyte-infected tall fescue consumption on diet utilization and thermal regulation in cattle. J Anim Sci 71:164–170

Arbuckle JG (2009) Cattle and trees don't mix !?!: competing agri-environmental paradigms and silvopasture agroforestry in the Missouri Ozarks. In: Franzluebbers AJ (ed) Farming with grass: achieving sustainable mixed agricultural landscapes. Soil and Water Conservation Society, Ankeny, IA, pp 116–133

Ares A, Brauer D (2005) Aboveground biomass partitioning in loblolly pine silvopastoral stands: spatial configuration and pruning effects. For Ecol Manag 219:176–184. https://doi.org/10.1016/j.foreco.2005.08.042

Ares A, Reid W, Brauer D (2006) Production and economics of native pecan silvopastures in Central United States. Agrofor Syst 66:205–215. https://doi.org/10.1007/s10457-005-8302-0

Association AH (2014) Humane heartland farm animal welfare survey. American Humane Association, Washington, DC

Ayres E, Dromph KM, Cook R et al (2007) The influence of below-ground herbivory and defoliation of a legume on nitrogen transfer to neighbouring plants. Funct Ecol 21:256–263. https://doi.org/10.1111/j.1365-2435.2006.01227.x

Baah-acheamfour M, Carlyle CN, Bork EW, Chang SX (2014) Forest Ecology and Management Trees increase soil carbon and its stability in three agroforestry systems in central Alberta, Canada. For Ecol Manag 328:131–139. https://doi.org/10.1016/j.foreco.2014.05.031

Balling JD, Falk JH (1982) Development of visual preference for natural environments. Environ Behav 14:5–28. https://doi.org/10.1177/0013916582141001

Bambo SK, Nowak J, Blount AR et al (2009) Soil nitrate leaching in silvopastures compared with open pasture and pine plantation. J Environ Qual 38:1870–1878. https://doi.org/10.2134/jeq2007.0634

Belesky DP (2005) Growth of Dactylis glomerata along a light gradient in the central Appalachian region of the eastern USA: II. Mechanisms of leaf dry matter production. Agrofor Syst 65:91–98. https://doi.org/10.1007/s10457-004-5726-x

Benton TG, Bryant DM, Cole L, Crick HQP (2002) Linking agricultural practice to insect and bird populations: a historical study over three decades. J Appl Ecol 39:673–687. https://doi.org/10.1046/j.1365-2664.2002.00745.x

Bigelow DP, Borchers A (2017) Major uses of land in the United States, 2012. USDA-Economic Research Service EIB-178

Bird PR, Kellas JD, Jackson TT, Kearney GA (2010) Pinus radiata and sheep production in silvopastoral systems at Carngham, Victoria, Australia. Agrofor Syst 78:203–216. https://doi.org/10.1007/s10457-009-9258-2

Boyer DG, Neel JPS (2010) Nitrate and fecal coliform concentration differences at the soil/bedrock interface in Appalachian silvopasture, pasture, and forest. Agrofor Syst 79:89–96. https://doi.org/10.1007/s10457-009-9272-4

Brooker RW, Maestre FT, Callaway RM et al (2008) Facilitation in plant communities: the past, the present, and the future. J Ecol 96:18–34. https://doi.org/10.1111/j.1365-2745.2007.01295.x

Buergler AL, Fike JH, Burger JA et al (2005) Botanical composition and forage production in an emulated silvopasture. Agron J 97:1141–1147. https://doi.org/10.2134/agronj2004.0308

Byers HL, Cabrera ML, Matthews MK et al (2005) Phosphorus, sediment, and loads in unfenced streams of the Georgia Piedmont, USA. J Environ Qual 34:2293. https://doi.org/10.2134/jeq2004.0335

Choler P, Michalet R, Callaway RM (2001) Facilitation and competition on gradients in alpine plant communities. Ecology 82:3295–3308

Clason TR (1995) Economic implications of silvopastures on southern pine plantations. Agrofor Syst 29:227–238

Dawson LA, Duff EI, Campbell CD, Hirst DJ (2001) Depth distribution of cherry (Prunus avium L.) tree roots as influenced by grass root competition. Plant Soil 231:11–19. https://doi.org/10.1023/A:1010383506446

de Mendonca RMA, Leite RC, Lana AMQ et al (2014) Parasitic helminth infection in young cattle raised on silvopasture and open-pasture in southeastern Brazil. Agrofor Syst 88:53–62. https://doi.org/10.1007/s10457-013-9655-4

Derner JD, Lauenroth WK, Stapp P, Augustine DJ (2009) Livestock as ecosystem engineers for grassland bird habitat in the Western Great Plains of North America. Rangel Ecol Manag 62:111–118

Doescher PS, Tesch SD, Alejandro-Castro M (1987) Livestock grazing: a silvicultural tool for plantation establishment. J For 85:29–37

Dougherty MC, Thevathasan NV, Gordon AM et al (2009) Nitrate and Escherichia coli NAR analysis in tile drain effluent from a mixed tree intercrop and monocrop system. Agric Ecosyst Environ 131:77–84. https://doi.org/10.1016/j.agee.2008.09.011

Downs MP, Cavers PB (2002) Physical and chemical factors associated with the reduction or delay of seed germination and seedling emergence of bull thistle, Cirsium vulgare (Savi) ten. Under leaf litter. Écoscience 9:518–525. https://doi.org/10.1080/11956860.2002.11682739

Drewry JJ, Cameron KC, Buchan GD (2008) Pasture yield and soil physical property responses to soil compaction from treading and grazing - a review. Aust J Soil Res 46:237–256

Elwood NE, Hansen EN, Oester P (2003) Management plans and oregon's NIPF owners: a survey of attitudes and practices. West J Appl For 18:127–132

Fannon AG, Fike JHH, Greiner SP et al (2017) Hair sheep performance in a mid-stage deciduous Appalachian silvopasture. Agrofor Syst 93:81–93. https://doi.org/10.1007/s10457-017-0154-x

Feldhake CM (2002) Forage frost protection potential of conifer silvopastures. Agric For Meteorol 112:123–130

Franzluebbers AJ, Chappell JC, Shi W, Cubbage FW (2017) Greenhouse gas emissions in an agroforestry system of the southeastern USA. Nutr Cycl Agroecosyst 108:85–100. https://doi.org/10.1007/s10705-016-9809-7

Fraser D, Duncan IJH, Edwards SA et al (2013) General principles for the welfare of animals in production systems: the underlying science and its application. Vet J 198:19–27

Frost WE, Mcdoucald NK (1989) Tree canopy effects on herbaceous production of annual rangeland & wing drought. J Range Manag 42:281–283. https://doi.org/10.2307/3899494

Gibson MD, Clason TR, Grozdits GA (1994) Effects of silvopasture management on growth and wood quality of young loblolly pine. In: Edwards MB (ed) Abstracts of the Eighth Biennial Southern Silvicultural Research Conference. U.S. Forest Service Southern Forestry Experiment Station, New Orleans, LA, p 48

Gill RMA (1992) A review of damage by mammals in north temperate forests: 3. Impact on trees and forests. Forestry 65:363–388

Gomez-Aparicio L, Zamora R, Gomez JM et al (2004) Applying plant facilitation to forest restoration: a meta-analysis of the use of shrubs as nurse plants. Ecol Appl 14:1128–1138

Grado SC, Hovermale CH, Louis DGS (2001) A financial analysis of a silvopasture system in southern Mississippi. Agrofor Syst 53:313–322. https://doi.org/10.1023/A:1013375426677

Haan MM, Russell JR, Davis JD, Morrical DG (2010) Grazing management and microclimate effects on cattle distribution relative to a cool season pasture stream. Rangel Ecol Manag 63:572–580. https://doi.org/10.2111/REM-D-09-00045.1

Haile SG, Nair VD, Nair PKR (2010) Contribution of trees to carbon storage in soils of silvopastoral systems in Florida, USA. Glob Chang Biol 16:427–438. https://doi.org/10.1111/j.1365-2486.2009.01981.x

Hanula JL, Horn S, O'Brien JJ (2015) Have changing forests conditions contributed to pollinator decline in the southeastern United States? For Ecol Manag 348:142–152. https://doi.org/10.1016/j.foreco.2015.03.044

Holl K, Smith M (2007) Scottish upland forests: history lessons for the future. For Ecol Manag 249:45–53

Howlett DS, Mosquera-losada MR, Nair PKR et al (2011) Soil carbon storage in silvopastoral systems and a treeless pasture in northwestern Spain. J Environ Qual 40:825–832. https://doi.org/10.2134/jeq2010.0145

Hunter AF, Aarssen LW (1988) Plants helping plants. Bioscience 38:34–40

Joffre R, Rambal S, Ratte JP (1999) The dehesa system of southern Spain and Portugal as a natural ecosystem mimic. Agrofor Syst 45:57–79. https://doi.org/10.1023/A:1006259402496

Jose S (2009) Agroforestry for ecosystem services and environmental benefits: an overview. Agrofor Syst 76:1–10. https://doi.org/10.1007/s10457-009-9229-7

Jose S, Kumar BM, Walter D (2019) Ecological considerations in sustainable silvopasture design and management. Agrofor Syst 93:317–331

Kallenbach RL, Kerley MS, Bishop-Hurley GJ (2006) Cumulative forage production, forage quality and livestock performance from an annual ryegrass and cereal rye mixture in a pine walnut silvopasture. Agrofor Syst 66:43–53. https://doi.org/10.1007/s10457-005-6640-6

Karki U, Goodman MS (2010) Cattle distribution and behavior in southern-pine silvopasture versus open-pasture. Agrofor Syst 78:159–168. https://doi.org/10.1007/s10457-009-9250-x

Karki U, Goodman MS (2015) Microclimatic differences between mature loblolly-pine silvopasture and open-pasture. Agrofor Syst 89:319–325. https://doi.org/10.1007/s10457-014-9768-4

Kumar S, Udawatta RP, Anderson SH (2010) Root length density and carbon content of agroforestry and grass buffers under grazed pasture systems in a Hapludalf. Agrofor Syst 80:85–96. https://doi.org/10.1007/s10457-010-9312-0

Kumar S, Anderson SH, Udawatta RP, Kallenbach RL (2012) Water infiltration influenced by agroforestry and grass buffers for a grazed pasture system. Agrofor Syst 84:325–335. https://doi.org/10.1007/s10457-011-9474-4

Lawrence JH, Hardesty LH, Chapman RC, Gill SJ (1992) Agroforestry practices of non-industrial private forest landowners in Washington State. Agrofor Syst 19:37–55

Lehmkuhler JW, Felton EED, Schmidt DA et al (2003) Tree protection methods during the silvopastoral-system establishment in midwestern USA: cattle performance and tree damage. Agrofor Syst 59:35–42

Lewis CE, Burton GW, Monson WG, McCormick WC (1983) Integration of pines, pastures, and cattle in South Georgia. Agrofor Syst 1:277–297

Lin C, McGraw R, George M, Garrett H (1998) Shade effects on forage crops with potential in temperate agroforestry practices. Agrofor Syst 44:109–119. https://doi.org/10.1023/A:1006205116354

Lindgren PMF, Sullivan TP (2012) Response of plant community abundance and diversity during 10 years of cattle exclusion within silvopasture systems. Can J For Res 42:451–462. https://doi.org/10.1139/X2012-003

Lunka P, Patil SD (2016) Impact of tree planting configuration and grazing restriction on canopy interception and soil hydrological properties: implications for flood mitigation in silvopastoral systems. Hydrol Process 30:945–958. https://doi.org/10.1002/hyp.10630

Lutz HJ (1930) Effect of cattle grazing on vegetation of a virgin forest in northwestern Pennsylvania. J Agric Res 41:561–570

Maestre FT, Valladares F, Reynolds JF (2005) Is the change of plant – plant interactions with abiotic stress predictable? A meta-analysis of field results in arid environments. J Ecol 93:748–757. https://doi.org/10.1111/j.1365-2745.2005.01017.x

Mayerfeld D, Rickenbach M, Rissman A (2016) Overcoming history: attitudes of resource professionals and farmers toward silvopasture in Southwest Wisconsin. Agrofor Syst 90:723–736. https://doi.org/10.1007/s10457-016-9954-7

Mcadam JH, Sibbald AR, Teklehaimanot Z, Eason WR (2007) Developing silvopastoral systems and their effects on diversity of fauna. Agrofor Syst 70:81–89. https://doi.org/10.1007/s10457-007-9047-8

Mcdermott ME, Rodewald AD (2014) Conservation value of silvopastures to Neotropical migrants in Andean forest flocks. Biol Conserv 175:140–147. https://doi.org/10.1016/j.biocon.2014.04.027

McDowell RW, Houlbrooke DJ, Muirhead RW et al (2008) Grazed pastures and surface water quality. Nova Science Publishers, New York

McIvor I, Douglas G, Dymond J et al (2011) Pastoral hill slope erosion in New Zealand and the role of poplar and willow trees in its reduction. In: Godone D (ed) Soil erosion issues in agriculture. InTech Open, London.

Meinikmann K, Hupfer M, Lewandowski J (2015) Phosphorus in groundwater discharge - a potential source for lake eutrophication. J Hydrol 524:214–226. https://doi.org/10.1016/j.jhydrol.2015.02.031

Michel G, Nair VD, Nair PKR (2007) Silvopasture for reducing phosphorus loss from subtropical sandy soils. Plant Soil 297:267–276. https://doi.org/10.1007/s11104-007-9352-z

Montagnini F, Nair PKR (2004) Carbon sequestration: an underexploited environmental benefit of agroforestry systems. Agrofor Syst 61–62:281–295. https://doi.org/10.1023/B:AGFO.0000029005.92691.79

National Research Council (1970) Influence of land management on wildlife. In: Land use and wildlife resources. National Academies Press, Washington, DC

Neel JR (1939) The effect of shade on pasture. University of Tennessee Agricultural Experiment Station, Knoxville, TN

Noss RF (2012) Forgotten grasslands of the south: natural history and conservation. Island Press, Washington, DC

Nyakatawa EZ, Mays DA, Naka K, Bukenya JO (2012) Carbon, nitrogen, and phosphorus dynamics in a loblolly pine-goat silvopasture system in the Southeast USA. Agrofor Syst 86:129–140. https://doi.org/10.1007/s10457-011-9431-2

Orefice J, Carroll J, Conroy D, Ketner L (2017) Silvopasture practices and perspectives in the northeastern United States. Agrofor Syst 91:149–160. https://doi.org/10.1007/s10457-016-9916-0

Patric JH, Helvey JD (1986) Some effects of grazing on soil and water in the eastern forest. USDA Northeastern Forest Experiment Station, Broomall, PA

Pearson HA, Whitaker LB, Duvall VL (1971) Slash pine regeneration under regulated grazing. J For 69:744–746

Pearson HA, Baldwin VC, Barnett JP (1990) Cattle grazing and pine survival and growth in subterranean clover pasture. Agrofor Syst 10:161–168. https://doi.org/10.1007/BF00115364

Peet R, Allard D (1993) Longleaf pine vegetation of the southern Atlantic and eastern Gulf Coast regions: a preliminary classification. Proc Tall Timbers Fire Ecol Conf:45–81. https://doi.org/10.1086/284833

Pent GJ, Fike JH (2016) Sheep performance, grazing behavior, and body temperatures in silvopasture systems. American Forage and Grassland Council, Baton Rouge, LA

Pent GJ, Fike JH, Kim I (2018) Ewe lamb vaginal temperatures in hardwood silvopastures. Agrofor Syst 95(1):21–32. https://doi.org/10.10007/s10457-018-0221-y

Pent GJ, Greiner SP, Munsell JF, Tracy BF, Fike JH (2020a) Lamb performance in hardwood silvopastures, I: animal gains and forage measures in summer. Trans Animal Sci 4:385–389. https://doi.org/10.1093/tas/txz154

Pent GJ, Greiner SP, Munsell JF, Tracy BF, Fike JH (2020b) Lamb performance in hardwood silvopastures, II: animal behavior in summer. Trans Animal Sci 4:363–375. https://doi.org/10.1093/tas/txz177

Pietola L, Horn R, Yli-Halla M (2005) Effects of trampling by cattle on the hydraulic and mechanical properties of soil. Soil Tillage Res 82:99–108

Ponder F, Jones JE, Mueller R (2005) Using poultry litter in black walnut nutrient management. J Plant Nutr 28:1355–1364. https://doi.org/10.1081/PLN-200067454

Proulx M, Mazumder A (1998) Reversal of grazing impact on plant species richness in nutrient-poor vs. nutrient-rich ecosystems. Ecology 79:2581–2592

Raedeke AH, Green JJ, Hodge SS, Valdivia C (2003) Farmers, the practice of farming and the future of agroforestry: an application of Bourdieu's concepts of field and habitus. Rural Sociol 68:64–86

Rasse DP, Rumpel C, Dignac M (2005) Is soil carbon mostly root carbon? Mechanisms for a specific stabilisation. Plant Soil 269:341–356. https://doi.org/10.1007/s11104-004-0907-y

Robinson J (2005) Silvopasture and eastern wild turkey. Agrofor Notes 28

Schenk HJ, Jackson RB (2002) Rooting depths, lateral root spreads and belowground aboveground allometries of plants in water limited ecosystems. J Ecol:480–494. https://doi.org/10.1046/j.1365-2745.2002.00682.x

Schroth G (1998) A review of belowground interactions in agroforestry, focusing on mechanisms and management options. Agrofor Syst 43:5–34. https://doi.org/10.1023/a:1026443018920

Sharrow SH (2007) Soil compaction by grazing livestock in silvopastures as evidenced by changes in soil physical properties. Agrofor Syst 71:215–223. https://doi.org/10.1007/s10457-007-9083-4

Sharrow SH, Ismail S (2004) Carbon and nitrogen storage in agroforests, tree plantations, and pastures in western Oregon, USA. Agrofor Syst 60:123–130. https://doi.org/10.1023/B:AGFO.0000013267.87896.41

Sharrow SH, Carlson DH, Emmingham WH, Lavender D (1996) Productivity of two Douglas fir/subclover/sheep agroforests compared to pasture and forest monocultures. Agrofor Syst 34:305–313

Sharrow SH, Brauer D, Clason TR (2009) Silvopastoral practices. In: North American agroforestry: an integrated science and practice, 2nd edn, American Society of Agronomy, Madison, WI, pp 105–131

Shrestha RK, Alavalapati JRR (2004) Effect of ranchland attributes on recreational hunting in Florida: a hedonic price analysis. J Agric Appl Econ 36:763–772

Silanikove N (2000) Effects of heat stress on the welfare of extensively managed domestic ruminants. Livest Prod Sci 67:1–18. https://doi.org/10.1016/S0301-6226(00)00162-7

Smith RM (1942) Some effects of black locusts and black walnut on southeastern Ohio pastures. Soil Sci 53:385–398

Stainback AG, Alavalapati JRR (2004) Restoring longleaf pine through silvopasture practices: an economic analysis. For Policy Econ 6:371–378. https://doi.org/10.1016/j.forpol.2004.03.012

St-Pierre NR, Cobanov B, Schnitkey G (2003) Economic losses from heat stress by US livestock industries. J Dairy Sci 86:52–77. https://doi.org/10.3168/jds.S0022-0302(03)74040-5

Taylor DB, Moon RD, Mark DR (2012) Economic impact of stable flies (Diptera: Muscidae) on dairy and beef cattle production. J Med Entomol 49:198–209. https://doi.org/10.1603/ME10050
Tielborger K, Kadmon R (2000) Temporal environmental variation tips the balance between facilitation and interference in desert plants. Ecology 81:1544–1553. https://doi.org/10.2307/177305
Tripathi G, Deora R, Singh G (2013) The influence of litter quality and micro-habitat on litter decomposition and soil properties in a silvopasture system. Acta Oecol 50:40–50
Tyndall J, Colletti J (2007) Mitigating swine odor with strategically designed shelterbelt systems: a review. Agrofor Syst 69:45–65. https://doi.org/10.1007/s10457-006-9017-6
USDA (2012) Census of agriculture. USDA, Washington, DC
USDA (2016) Beef 2017: information needs assessment survey results for the upcoming NAHMS beef 2017 study. USDA, Washington, DC
Vandermeer J (1981) The interference production principle: an ecological theory for agriculture. Bioscience 31:361–364
Walters JR (1991) Application of ecological principles to the management of endangered species: the case of the red-cockaded woodpecker. Annu Rev Ecol Syst 22:505–523
White SL, Sheffield RE, Washburn SP et al (2001) Spatial and time distribution of dairy cattle excreta in an intensive pasture system. J Environ Qual 30:2180–2187. https://doi.org/10.2134/jeq2001.2180
Workman SW, Bannister ME, Nair PKR (2003) Agroforestry potential in the southeastern United States: perceptions of landowners and extension professionals. Agrofor Syst 59:73–83

# Silvopasture for Food Security in a Changing Climate

**Ryan Dibala, Shibu Jose, and Ranjith P. Udawatta**

## Abbreviations

| | |
|---|---|
| ADF | Acid detergent fiber |
| ADG | Average daily gain |
| AU | Animal units |
| C | Carbon |
| CAFO | Concentrated agricultural feeding operation |
| CP | Crude protein |
| DM | Dry matter |
| ISPS | Intensive silvopastoral systems |
| LWG | Live weight gain |
| N | Nitrogen |
| NDF | Neutral detergent fiber |
| P | Phosphorus |
| PAR | Photosynthetically active radiation |
| PES | Payment for environmental services |
| THI | Temperature humidity index |
| TNC | Total nonstructural carbohydrates |

R. Dibala (✉) · S. Jose · R. P. Udawatta
College of Agriculture, Food and Natural Resources, University of Missouri, Columbia, Missouri, USA
e-mail: joses@missouri.edu; Udawattar@missouri.edu



R. P. Udawatta, S. Jose (eds.), *Agroforestry and Ecosystem Services*,
https://doi.org/10.1007/978-3-030-80060-4_8

# Introduction

## *The Food-Climate Crisis*

Agroforestry is often praised for the many environmental benefits it provides, such as carbon sequestration, reduction of toxic runoff into waterways, and wildlife enhancement (Udawatta et al. 2011; Udawatta and Jose 2012; McDermott and Rodewald 2014). However, there remains an important and often overlooked value of the ecosystem services provided by agroforestry: food security. In a time when monocultures and chemical inputs of conventional agriculture prevail, there is growing concern about the future of food production, particularly in regard to soil loss and degradation, indiscriminate use of agrochemicals, and environmental and ethical challenges of industrial animal agriculture. Globally, the human population is exploding and is expected to reach 9.1 billion by 2050, urbanization is increasing, and incomes are rising. This has resulted in a rapidly growing demand for animal products and continued natural resource degradation, all of which have profound effects on food security (Delgado et al. 1999).

Although global grain production has more than doubled and global meat production has more than tripled over the last half-century (FAOSTAT 2010), food yield may need to increase by 50% or more in the next half-century to keep up with demands (Godfray et al. 2010). Projected demands for meat and milk production were expected to grow at respective rates of 2.8 and 3.2% annually up to 2020 (Delgado et al. 1999). All the while, food producers are experiencing greater competition for land, water, and energy.

Climate change is exacerbating consequences for animal production through its effects on forage productivity and heat-related stress on the animal. Under climate change scenarios, water will become the main limiting factor to all livestock systems (Steinfeld et al. 2006; de Fraiture et al. 2010) and extended droughts will become the norm. In the face of climate change, producing more food for a growing population while diminishing poverty and hunger is a daunting task, but a challenge that must be heeded. An even greater challenge is not only to increase productivity, but also to do so while treading more lightly on the land (Cribb 2010).

## *Sustainable Livestock Production*

Many decades of research have demonstrated that livestock management is critical for maintaining healthy pastures and optimal productivity (Gerrish 2004; Rayburn 2007). In 1959, farmer and scientist André Voisin coined the term *rational grazing* (Voisin 1988), where he described the basic guidelines necessary for good grazing management: short periods of occupation followed by an ample recovery period. More recently, authors have built on these management guidelines with the introduction of terms such as *prescribed grazing* (USDA-NRCS 2010), *management intensive grazing* (Gerrish 2004), holistic planned grazing (HPG), and mob grazing (Savory and Butterfield 2016). All these terms apply to the same key grazing principles proposed by Voisin, ultimately favoring important pasture species, improving soil health, and increasing forage productivity and nutritional quality (Flack 2016).

These sustainable livestock production methods can be implemented in open pasture or alternatively under dispersed tree cover in a silvopastoral setting. Silvopasture is an agroforestry practice where trees and livestock are combined with improved pasture plants and managed intensively, effectively integrating intensive animal husbandry, silviculture, and forage agronomy practices (Sharrow et al. 2009; Jose and Dollinger 2019). The simultaneous production of timber and livestock can increase the diversity of on-farm products, improve land-use efficiency, and provide better welfare for animals (Murgueitio et al. 2011; Calle et al. 2012b; Broom et al. 2013). Despite numerous accounts of silvopasture's ability to strike an optimal balance between production and conservation (Ibrahim et al. 2010; Galindo et al. 2013; Jose et al. 2019), many producers remain skeptical, arguing that productivity is too greatly reduced under tree cover.

In this chapter, we review a number of studies from various regions of the world that highlight silvopasture's contribution to achieving food security. We focus on the production of forage, meat, and milk in silvopastoral systems as direct indicators of food supply as well as indirect indicators such as thermal stress in livestock, animal health, and habitat provisioning for pollinators. We conclude by addressing some of the problems of modern-day animal agriculture and how silvopasture could play a critical role in the sustainable intensification of livestock production systems.

## Silvopasture: A Contribution to Food Security

### *Forage Production*

It is well established that trees have both competitive (negative) and facilitative (positive) effects on the microenvironments beneath them (Jose et al. 2004; Jose et al. 2019). Canopy solar interception results in lower light transmittance, decreasing the photosynthetic rate of understory vegetation. Trees have been shown to compete vigorously for water and nutrients and can even emit allelopathic chemicals that impede the growth of surrounding vegetation. However, canopy interception can also provide protection from desiccating winds and reduce soil surface temperature and soil evapotranspiration (Belsky et al. 1989; Belesky 2005), which can increase overall soil moisture content (Vetaas 1992). Some trees can fix atmospheric nitrogen (N) and provide up to 650 kg N yr.$^{-1}$, more than enough to fulfill crop N needs for sustained yield (Nygren et al. 2012). Leaf litter under trees has been shown to improve the physical properties of the surface soil and increase chemical properties including soil nutrients and organic matter (Belsky 1994). As a result, the content of carbon (C), phosphorus (P), and N has been shown to gradually decline as a function from the distance of the trunk, resulting in significantly lower levels in the open ground than in sub-canopy soil (Belsky et al. 1989; Tiedemann and Klemmedson 2008).

Elevated nutrient levels can improve the forage quality of sub-canopy grasses, attracting grazers that return nutrients to the soil. This, combined with the trapping of wind and waterborne sediments by trees, can contribute to an "island of fertility" effect (Belsky et al. 1989; Dohn et al. 2013). Tree roots can also decrease the bulk density of the soil, creating the macroporosity favorable to the infiltration of water, increasing water-holding capacity (Malmer et al. 2010). Additionally, integrated perennial systems have better soil thermal properties that can help improve C storage and microenvironment (Adhikari et al. 2014). These benefits, combined with the selection of appropriate tree and forage species, can sometimes result in increased levels of productivity when compared with monocultures.

Tree canopy effects on the growth and nutritive value of understory forages depend on many factors, including forage type, local climate and topography, season, soil fertility and structure, and amount of photosynthetically active radiation (PAR). It is well known that shading has a more detrimental effect on warm-season (C4) grasses than it does on cool-season (C3) grasses (Kephart and Buxton 1993; Lin et al. 1998; Buergler et al. 2005; Pang et al. 2019a). This is because the physiology of C4 grasses allows for greater biomass accumulation per unit of PAR—or radiation-use efficiency—than does the physiology of C3 species. The amount of rainfall appears to be important in determining forage production under shade. In xeric environments where water is the limiting factor, growth and development of many herbaceous species are facilitated by tree canopies through the improvement in moisture regimes (Joffre and Rambal 1993), soil nutrients, and organic matter (Kellman 1979). Several studies have demonstrated that under certain conditions, moderate shading can provide the optimal environment for grass growth and quality (Belsky 1994; Ibrahim et al. 2007; DeBruyne et al. 2011; Orefice et al. 2016b). Hernández and Guenni (2008) concluded that guinea grass [*Megathyrsus maximus* (Jacq.) B.K. Simon & S.W.l. Jacobs] benefited from a compensatory effect from trees that increased soil humidity and improved total forage biomass. Andrade et al. (2004) found that guinea grass var. *Massai* growing under artificial shade reached its highest dry matter (DM) accumulation rate under 30% shade cover in both the rainy and dry seasons. Moustakas et al. (2013) demonstrated that tree effects on grass biomass across a precipitation gradient in a subtropical savanna were facilitative in drier sites, with greater grass biomass observed beneath tree canopies than outside.

Conversely, many studies in temperate environments with more rainfall have shown that canopy coverage either maintains (DeBruyne et al. 2011) or reduces the quantity of understory forage (Feldhake et al. 2010; Orefice et al. 2016b). In a study conducted in the Appalachian Mountains, USA, Neel and Belesky (2015) showed that hardwood silvopasture DM production was 60–70% that of open pasture in the spring and equal to only 40–60% of it in summer. Studying an alder (*Alnus spp.*) and willow (*Salix spp.*) silvopasture in New Zealand, Devkota et al. (2001) concluded

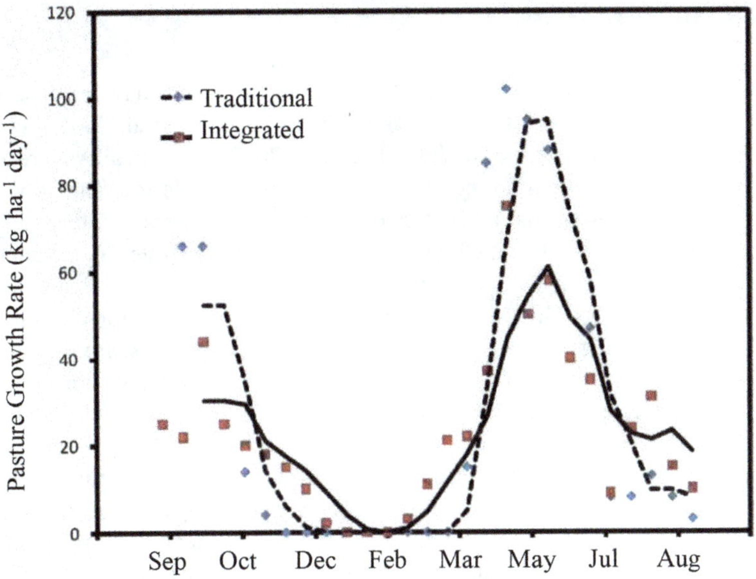


**Fig. 1** Pasture growth rates in traditional (open) and integrated (25% of land area under silvopasture) pasture systems at the Horticulture and Agroforestry Research Center near New Franklin, Missouri, USA (source: Kallenbach 2009)

that a 40–50% canopy closure would maintain pasture production at approximately two-thirds of that in unshaded pasture.

However, research in both temperate and tropical environments has suggested that silvopasture may extend forage longevity and provide more forage than conventional pastures during certain times of the year. Kallenbach (2009) compared the growth of cool season grasses in traditional open pastures to that of integrated pastures where silvopasture was used on only 25% of the total land area. Forage growth on integrated pastures outperformed that of traditional pastures early in the spring, midsummer, and late fall, all times when cool season grasses likely benefit from more moderate microclimates in the understory (Fig. 1).

Similarly, a study examining the growth of guinea grass in the understory of native tree plantations in Panama found that forage DM accumulation early in the dry season was greatest under moderate tree coverage but was greater in open pasture throughout the rainy season (Dibala et al. 2021). In the driest month of February, pooled mean DM grass production was 38% greater under moderate canopy when compared to open pasture. These studies indicate that producers may achieve maximum gains by integrating silvopastures into larger open pasture operations and using them only during periods of relative scarcity.

## *Forage Nutritive Value*

A plethora of research indicates that forage nutritive value may increase when grown under tree canopies (Lin et al. 1998, 2001; Buergler et al. 2006; Feldhake et al. 2010; Neel and Belesky 2017; Orefice et al. 2017). Specifically, increases in crude protein (CP) content are commonly observed. A shade tolerance screening trial in Missouri showed that all 22 tested forages (16 grasses and 6 legumes) had equal or higher percent CP and CP yield (g $pot^{-1}$) under moderate shade than in the control (Pang et al. 2019a, b). Percent CP of forages grown in dense shade was 4.5, 6.1, and 6.11% higher than that of forages grown in full sun for "benchmark" orchardgrass (*Dactylis glomerata* L.), smooth brome (*Bromus inermis* Leyss.), and timothy (*Phleum pretense* L.), respectively. In Veracruz, Mexico, Medinilla-Salinas et al. (2013) found that guinea grass growing under a 12-year-old canopy of [*Gliricidia sepium* (Jacq.) Kunth ex Walp.] trees contained 1.9% greater CP than those growing in the open during the windy season. This is likely due to adaptive mechanisms and changes in plant physiology such as elongation of the cell wall (Kephart and Buxton 1993) and increases in the specific leaf area and shoot:root ratio (Paciullo et al. 2017). The presence of N-fixing trees may also indirectly increase CP content in forages through leaf decomposition, root exudation, and direct nutrient exchange (Sierra and Nygren 2006; Sierra et al. 2007; Jalonen et al. 2009). Xavier et al. (2014) found that N recycled via the litter pathway in a silvopastoral system exceeded that in a monoculture by 34 kg $ha^{-1}$, concluding that the extra N recycled in the system—along with biological N fixation—would confer increases in quality and longevity of forage when compared to grass monocultures.

Typically, the structural carbohydrate metrics acid detergent fiber (ADF) and neutral detergent fiber (NDF) are either increased or unaffected by shade for most forage species (Ladyman et al. 2003; Kallenbach et al. 2006; Sousa et al. 2010; Paciullo and de Castro 2011; Neel and Belesky 2015). However, there are a number of studies that report decreasing values with increased levels of shade (Kephart and Buxton 1993; Obispo et al. 2008; Medinilla-Salinas et al. 2013), indicating lower levels of lignification and overall higher digestibility.

It is well known that the nutritive value of a plant changes throughout its growth stages of maturity, containing greater contents of total nonstructural carbohydrates (TNC) in the early stages of growth and developing larger quantities of lignin and cellulose later in the season (Ball et al. 2001; Pang et al. 2019b). This increase in lignification reduces digestibility and palatability of the plant, resulting in decreased animal intake. Thus, it is important for producers to manage livestock dynamically in response to temporal changes in both the quantity and quality of forages. Silvopasture has been shown to improve the quality of forage at specific times of the year when the quality of open-grown forages declines. Kallenbach et al. (2006) reported that the forage quality of an annual ryegrass (*Lolium perenne* L.) and cereal rye (*Secale cereale* L.) mixture growing in moderate shade frequently outperformed that of open pasture, particularly late in the summer grazing season when ambient temperatures were too high for cool-season grasses.

## *Tree Fodder Production and Nutritive Value*

One way producers can respond to loss of forage productivity and quality is to rely on trees and shrubs to provide alternative and highly nutritious forage sources during critical periods. In the tropics, fodder shrubs can be a strategic resource for farmers during the worst drought periods that often occur during the dry season. For example, in the Yucatán Peninsula of Mexico, mixed stands of the fodder shrubs [*Leucaena leucocephala* (Lam.) de Wit.] and [*Guazuma ulmifolia* Lam.] have been shown to produce up to 5.18 Mg of edible DM $ha^{-1}$, with no statistical differences in yield between dry and wet seasons (Casanova-Lugo et al. 2015). This is a substantial contribution to forage availability, particularly during the dry season, when herbaceous forage yields may be reduced by 5–6 times relative to yields attained during the rainy season (Santiago-Hernández et al. 2016). Fodder shrubs like *Calliandra calothyrsus* Meisn., *G. ulmifolia* Lam., *L. leucocephala* (Lam.) de Wit., and *Tithonia diversifolia* (Hemsl.) A. Gray retain green foliage amidst even the harshest droughts. As the dry season progresses, forage shrubs have been shown to lose nutritive value, digestibility, and palatability at a slower rate than herbaceous forages (Talamuci and Pardini 1999), providing relatively high-quality supplemental forage to both ruminants and nonruminants during times of scarcity.

A widely touted silvopasture model that includes the use of native and non-native trees, shrubs, and herbaceous forages is known as intensive silvopasture (Fig. 2). Intensive silvopastoral systems (ISPS) include the planting of timber trees that are intercropped with high-density (~10,000 plants $ha^{-1}$) plantings of fodder shrubs and highly productive pasture grasses in a system that can be directly grazed by livestock (Murgueitio et al. 2011).

Shrubs are periodically coppiced to encourage low, dense growth of the foliage. Cattle are provided permanent supplies of drinking water and rotated periodically with the use of electric fences to prevent overgrazing and to allow time for pastures to recover. ISPS first began in Australia more than 40 years ago, but it is now becoming the technology of choice in Colombian and regional livestock sectors because they can help reduce the seasonality of production and therefore help to mitigate and adapt to climate change (Cardona et al. 2013).

There is compelling evidence that demonstrates how ISPS can increase overall forage production when compared to open pastures. An ISPS using the shrubs *L. leucocephala* and *G. sepium* combined with guinea grass in the humid tropics of West Africa produced over 20 Mg of DM $ha^{-1}$ of mixed tree-grass fodder (Atta-Krah and Reynolds 1989). Bacab-Pérez and Solorio-Sánchez (2011) compared forage availability and voluntary intake on two ISPS ranches with a conventional ranch in Michoacán, Mexico, and found that the available forage in both ISPS ranches was at least 2.6 times greater than that in the conventional ranch (17,290 and 18,851 versus 6636 kg DM $yr.^{-1}$). Furthermore, only 9% of the available *L. leucocephala* forage was rejected by cattle on both ISPS farms (Table 1). Shelton and Dalzell (2007) reported that *L. leucocephala*-grass pastures are the most productive, profitable, and sustainable beef production systems in northern Australia.

**Fig. 2** Intensive silvopastoral system (ISPS) in Colombia, South America, where it has been widely promoted and implemented (source: Zoraida Calle Diaz/CIPAV)

**Table 1** Forage availability, refusal, and utilization efficiency (kg DM $ha^{-1}$) at three farms in Michoacán, Mexico (Bacab-Peréz and Solorio-Sánchez 2011)

| Farm | *Forage* | Edible forage | Rejection | Use | Use (%) |
|---|---|---|---|---|---|
| Los Huarinches | *L. leucocephala* | 8386 | 826 | 7560 | 91 |
| | *Guinea grass* | 8904 | 4655 | 4249 | 48 |
| | *Total* | 17,290 | 5481 | 11,809 | 68 |
| El Aviador | *L. leucocephala* | 9156 | 826 | 8330 | 91 |
| | *Guinea grass* | 9695 | 3542 | 6153 | 63 |
| | *Total* | 18,851 | 4368 | 14,483 | 77 |
| Conventional | *Cynodon plectostachyus* | 6636 | 2660 | 3976 | 60 |

The use of woody trees and shrubs for livestock fodder in temperate regions has been limited primarily due to a relatively limited plant selection and existing cultural and behavioral norms. Temperate regions lack the diversity of nutritious, N-fixing woody plants capable of coppicing that exists in the tropics. Trees of temperate regions produce palatable fodder during the growing season when highly preferred herbaceous forage is available, unless compromised by extreme weather. Cultural norms such as stockpiling and hay-baling are used instead of the cut-and-carry systems more commonplace in the tropics. However, researchers in temperate regions have explored the production and intake of densely planted forage shrubs and some species have shown particular promise (Papachristou and Papanastasis 1994; Papanastasis et al. 1998, 2008). In North Carolina, black locust (*Robinia pseudoacacia* L.) fodder banks were highly preferred by meat goats with a mean

DM yield of 3213 kg $ha^{-1}$ when planted on a 50 cm spacing and coppiced at 50 cm (Addlestone et al. 1999). In New Zealand, full access to willow (*Salix* spp.) fodder banks was beneficial for ewe reproductive rates (Pitta et al. 2005). Other promising species for temperate ISPS include *Paulownia* (Mueller et al. 2001) and mulberry (*Morus* spp.) (Sánchez 2000). When planted with subterranean clover (*Trifolium subterraneum* L.) in a silvopasture in central Italy, white mulberry (*Morus alba* L.) produced between 4.2 and 5.3 Mg DM $ha^{-1}$ (Talamuci and Pardini 1999). Armand and Meuret (1993) demonstrated that the Japanese white mulberry cultivar Kokuso 21 produced up to 2.2 Mg DM $ha^{-1}$ on good sites in France, but on poorer sites production was much lower at 444 kg DM $ha^{-1}$.

Silvopastoral systems containing forage shrubs are effective at improving animal production because tree foliage is often of higher nutritional quality than grasses (Mueller et al. 2001). Sosa Rubio and others (2004) analyzed the nutritive value of 30 perennial woody species and found that 70% of them contained 12% or more CP. In the case of tropical legumes, even seeds are browsed, which provide nutrients in excess of that required for digestion and metabolism, potentially correcting nutritional deficiencies in mature roughage (Aganga and Tshwenyane 2003).

The overall nutritive value of woody perennial forage can often be hindered by the presence of anti-nutritional compounds that have the ability to severely restrict nutrient utilization (Papanastasis et al. 2008). Secondary compounds such as condensed tannins, alkaloids, saponins, and oxalates are known to occur in many woody perennials and can have detrimental effects to the animal if consumed in high quantities. However, diets containing herbaceous forage with a high level of digestible CP have been shown to counteract the negative effects of tannins (Yiakoulaki 1995). Furthermore, tannins in low to moderate concentrations (20–40 g $kg^{-1}$ DM) can induce beneficial effects, which are associated with suppression of bloat in ruminants (Jones et al. 1973). Research has shown that feeding tannin- and saponin-containing compounds to cattle can increase intake of endophyte-infected tall fescue (*S. arundinacea* L.) and reduce its overall toxicity (Provenza et al. 2009). With the endophyte infecting a large percentage of the estimated 14 million ha of tall fescue in the United States (Ball et al. 2015), the incorporation of woody fodder to animal diets could help mitigate damages and have an enormous economic impact on the beef industry.

## *Tree Fruit Production*

The more obvious food product of perennial trees and shrubs is fruit. In 1929, author J. Russell Smith exposed the masses to the agricultural wealth of trees in his seminal work *Tree Crops: A Permanent Agriculture*. In this masterpiece, Smith expounds on the overlooked abundance of food for both humans and animals produced by woody perennials. He describes the fruiting patterns and yields of common trees like oak (*Quercus* spp.), hickory and pecan (*Carya* spp.), walnut (*Juglans* spp.), chestnut (*Castanea* spp.), persimmon (*Diospyros* spp.), carob (*Ceratonia siliqua* L.), mulberry, and honey locust (*Gleditsia triacanthos* L.). Many anecdotes from producers are found throughout the book, with statements like:

"I never weighed my pigs at the beginning and close of the mulberry season, but I think I can safely say that a pig weighing 100 pounds at the start would weight 200 pounds at the close" and

"I let the cattle pick them (honeylocust pods) up where they can; and where they cannot graze, the beans are gathered and fed to them. My herd of heifers get a great part of their winter pasture from the honeylocust pods."

Since then, accounts like these have been corroborated with empirical evidence. Gold and Hanover (1993) noted that the edible seedpods from honey locust trees can serve as supplemental feed for livestock over several months in autumn and winter when cool-season grass production is limited or negligible. In Virginia, whole-ground honey locust seedpods from the "Millwood" cultivar had a nutritional profile comparable to that of ground whole-ear dent corn (*Zea mays* L.) or oat (*Avena sativa* L.) grain (Johnson et al. 2013). In that same study, mean DM yields of pod-bearing trees were 15.8, 4.8, and 14.7 kg tree$^{-1}$ in 2008, 2009, and 2010, respectively. In good years, a honey locust crop can easily exceed 66 kg of cleaned seed per tree (Gold and Hanover 1993).

In the Mediterranean oak woodland known as the *dehesa*, Iberian pigs are raised extensively on acorns and grass during a 2-month fattening period that coincides with the fruiting period of surrounding holm oak (*Quercus ilex* Lam. spp. *ballota*) and cork oak (*Quercus suber* L.; Fig. 3). In the managed *dehesa*, where mean tree

**Fig. 3** Iberian, acorn-finished pigs under the canopy of holm oak (*Quercus ilex* Lam. spp. *ballota*) in the Mediterranean *dehesa* (source: https://foodism.co.uk/features/long-reads/origins/cinco-jotas-iberico-pork/)

density ranges between 30 and 50 trees $ha^{-1}$, the productivity of acorns is reported to be ten times higher than a dense *Quercus ilex* forest (Pulido 1999; Pulido et al. 2001). Although extremely variable, mean acorn yield was estimated to be 300–700 kg $ha^{-1}$, with yields of 8–14 kg $tree^{-1}$ for *Q. ilex* and 5–10 kg $tree^{-1}$ for *Q. suber* (Rodríguez-Estévez et al. 2007). Individual pigs can consume 7–10 kg of acorns $day^{-1}$, and generally will increase their weight from 100 to 160 kg during the finishing period (Nieto et al. 2002). In Spain, conventional pork finishing operations have resulted in average daily gains (ADG) of 0.66 kg (Agostini et al. 2013), while acorn-finished operations have resulted in ADGs of 0.76 kg (Rodríguez-Estévez et al. 2011).

In Southeast Asia, the presence of livestock has been shown to increase yields of commercially important tree crops like coconut (*Cocos nucifera* L.), palm oil (*Elaeis guineensis* Jacq.), and rubber [*Hevea brasiliensis* (Willd. Ex A. Juss.) Mull. Arg.] (Alexandratos 1995). The establishment of mixed pastures under coconuts in Sri Lanka resulted in increases of 17% and 11% in nut and copra yields, respectively (Liyanage et al. 1993). Moreover, the nutrients from 73 kg of fresh manure and 30 L of urine $palm^{-1}$ $year^{-1}$ reduced the cost of fertilizing the coconuts by 69% (Devendra and Ibrahim 1999). Livestock can also help reduce the cost of weed maintenance, as is the case with Chee and Faiz (1991), who reported a reduction of 20–40% in weeding costs due to regular grazing by cattle.

## Animal Performance

Several important measurements of silvopasture's sustainable contribution to food security are livestock ADG, conception rate, reproductive rate, and stocking rate (animal units (AU) $ha^{-1}$). An increase in any of these metrics can translate into income generation for ranchers. Historically, most studies on silvopastoral systems in temperate regions have demonstrated either decreased or equal animal performance when compared to open pastures (Teklehaimanot et al. 2002; Kallenbach et al. 2006, 2010; Sharrow et al. 2009; Neel and Belesky 2015). More recently, Pent and Fike (2018) compared ADGs of lambs in black walnut (*J. nigra*) and honey locust (*G. triacanthos*) silvopastures with open pasture of stockpiled tall fescue (*S. arundinaceus*) during the winter in Virginia. During the first three weeks of the trial, lambs did not consume honey locust pods due to naivety, but after the fourth week, consumption of pods was so high that lamb ADG was significantly greater than that in plots without honey locust. Future study is needed to determine whether honey locust supports even greater lamb weight gains when there has been previous exposure to pods and higher quality herbaceous forages are available (Pent and Fike 2018). In a study previously described evaluating integrated silvopastures—rotational stocking with a combination of open pasture and silvopasture—Kallenbach (2009) reported that cows in integrated silvopastures lost approximately 10% less weight over winter, reducing the need for supplementation by about 12%.

**Table 2** Performance of cow-calf pairs in a traditional (open) pasture system compared to those in an integrated (a combination of open pasture and silvopasture) system (adapted from Kallenbach 2009)

| Treatment | Winter weight loss (kg) | Calving difficulty (%) | Calf weaning weight (kg) |
|---|---|---|---|
| Traditional | *105* | 15.00 | 270 |
| Integrated | *93* | 3.00 | 295 |
| *P*-value | 0.02 | 0.04 | <0.01 |

**Table 3** Average daily gain (ADG; g animal$^{-1}$) and gain per area (kg ha$^{-1}$), according to rearing systems and experimental year, in the rainy and dry seasons (source: Paciullo and de Castro 2011)

| Experimental year | Rainy season | | Dry season | |
|---|---|---|---|---|
| | Silvopasture | Monoculture | Silvopasture | Monoculture |
| ADG | | | | |
| 2004/2005 | 722Aa | 624Ba | 348ab | 387a |
| 2005/2006 | 647ab | 563ab | 298b | 274b |
| 2006/2007 | 628Ab | 515Bb | 420a | 352ab |
| Gain per area | | | | |
| 2004/2005 | 298Aa | 256Ba | 88 | 97 |
| 2005/2006 | 242ab | 230ab | 75 | 68 |
| 2006/2007 | 258Ab | 211Bb | 105 | 89 |

Means followed by different letters, for each season of the year, capital in the row and lowercase in the column, are different at $p < 0.05$

Additionally, cows that gave birth in integrated silvopastures were 12% less likely to experience calving difficulty (Table 2).

More examples of silvopasture's positive influence on animal performance can be found from the tropics. A silvopastoral system in Brazil including signal grass [*Brachiaria decumbens* (Staph) R.D. Webster] and leguminous shrubs *G. sepium* and [*Mimosa caesalpiniifolia* Benth.] planted at a density of 2500 plants ha$^{-1}$ yielded similar livestock production per unit land area compared with signal grass in monoculture (de M. Costa et al. 2016). Furthermore, additional income and ecosystem services provided by the woody components demonstrate the overall beneficial contributions of this system.

Paciullo and de Castro (2011) evaluated dairy heifer performance in Brazilian silvopastures planted in signal grass with four species of 105 dispersed mature trees ha$^{-1}$ and drew comparisons with performance in similar signal grass open pasture. The authors concluded that a 13% increase in the CP content of signal grass in silvopasture compared with open pasture was sufficient to increase live weight gain of dairy heifers by 17% during the rainy season (Table 3). They posited that this increase in annual average gain could contribute to a reduction in the age at first conception and, consequently, of the first calving event.

A study at the Embrapa Dairy Cattle Center in Brazil concluded that Zebu-Friesian heifers grazing in a silvopasture planted in signal grass accompanied by *Acacia mangium* Willd., *Mimosa artemisiana* Heringer and Paula, and *Eucalyptus*

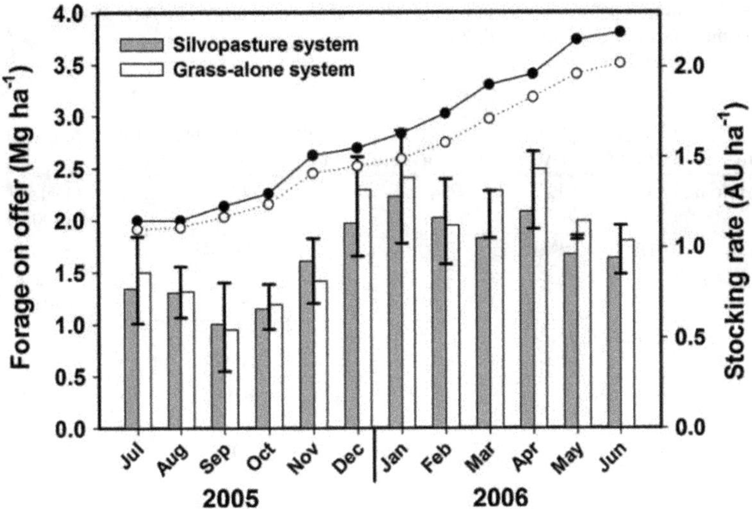


**Fig. 4** Total dry matter yield of forage on offer (Mg ha$^{-1}$; bars) and stocking rate of heifers (AU ha$^{-1}$; lines) from July 2005 to June 2006 in the silvopasture system and the signal grass monoculture. Values are means of 20 replicate samples. Error bars represent least significant differences between means. One AU is equivalent to 450 kg of live weight (source: Xavier et al. 2014)

*grandis* Hill ex Maiden at a density of 198 trees ha$^{-1}$ had significantly greater live weight gain (LWG) five years after system establishment than those grazing signal grass monocultures (Xavier et al. 2014). Silvopasture-raised cattle averaged annual LWGs of 205 kg head$^{-1}$ while those in monocultures averaged 177 kg head$^{-1}$ year$^{-1}$. This equates to a 16% increase in silvopasture-raised heifer annual LWG. The total annual animal intake was estimated to be 4.0 Mg ha$^{-1}$ in the silvopasture compared to 3.5 Mg ha$^{-1}$ in the signal grass monoculture. Forage DM annual mean in the monoculture was marginally greater than that in the silvopasture, but the authors could not determine whether this was due to shading or higher forage intake by heifers in the silvopasture (Fig. 4).

A study on sheep performance in Quintana Roo State, Mexico, analyzed five different feeding rations made up of various percentages of grasses and tree fodders and found that diets consisting of 75% or 100% tree fodder resulted in the greatest weight gains (Sosa Rubio et al. 2004). Similarly, sheep fed *G. sepium* (Chadhocar and Kantharaju 1980) and *Brosimum alicastrum* SW. leaves gained more weight than sheep grazing grass monocultures alone (Pérez et al. 1995). In Bali, Indonesia, the development of a shrub layer creates a three-strata forage system that has resulted in an increase in stocking rates by one animal ha$^{-1}$ and an increase in LWG by 153 kg ha$^{-1}$ year$^{-1}$ (Devendra 2012).

Yamamoto et al. (2007) used data on herd, milk production, and land use from 74 farms in central Nicaragua to quantify the effects of silvopastoral systems on milk production. The data indicated that silvopastoral areas, especially pasturelands with moderate tree density (tree cover approximately 20%), have significant positive

**Table 4** Production parameters of conventional and ISPS farming systems in Australia, Mexico, and Colombia (source: Cardona et al. 2013)

| | | Parameter | | | Reference |
|---|---|---|---|---|---|
| System | Country | Stocking rate (AU ha$^{-1}$) | Live weight gain (g animal$^{-1}$ day$^{-1}$) | Meat production (kg ha$^{-1}$ year$^{-1}$) | |
| Conventional | Australia | 1.5 | 411 | 225 | Dalzell et al. (2006) |
| | Mexico | 1 to 2.5 | 500 | 182,456 | Solorio-Sanchez et al. (2011) |
| | Colombia | 1.2 | 130 | 56.9 | Cordoba et al. (2010) |
| ISPS | Australia | 3 | 822 | 910 | Dalzell et al. (2006) |
| | Mexico | 6 | 900 | 1971 | Solorio-Sanchez et al. (2011) |
| | Colombia | 3.5 to 4.7 | 651–790 | 827–1341 | Cordoba et al. (2010) |
| | | 3.5 | 793–863 | 1013–1103 | Mahecha et al. (2011) |

impacts on annual milk production when overgrazing was avoided. The authors suggested that changing land use from low-density trees with natural pasture to moderate-density trees with conventional pasture using palisade grass [*Brachiaria brizantha* (A.Rich.) Stapf] could result in greatest improvements in yield.

Research has shown that when installed and managed effectively, ISPS can increase carrying capacity by as much as fourfold per hectare (4.3 heads ha$^{-1}$), milk production by as much as 130% to 16,000 liters ha$^{-1}$ year$^{-1}$, and meat production by as much as tenfold (Table 4). These gains, largely due to better distributions of biomass throughout the year, have been shown to increase farm income by at least $440 USD ha$^{-1}$ year$^{-1}$ while sustaining long-term system resiliency (Murgueitio et al. 2011; Calle et al. 2012a; Cardona et al. 2013). Meat quality of ISPS stock rivals the quality of those fed in feedlots, in terms of slaughtering weight and age, fat thickness and color, meat color, and marbling score (Dalzell et al. 2006). Additionally, ISPS has been shown to completely eliminate the use of chemical fertilizers from operations that once relied on inputs of 400 kg urea ha$^{-1}$ year$^{-1}$ (Murgueitio et al. 2011).

The main reason for greater productivity in ISPS is that a diversity of forages is offered to the animal. Evidence indicates that the contribution of legumes to the ruminant diet results in higher performance on mixed forages compared with those grazing grass only (Tudsri and Prasanpanich 2001). This may be due to synergistic effects between grasses and roughage within the animal's gut. Carbohydrates are needed to supply energy for rumen microbial activity to efficiently digest and synthesize proteins. Thus, synchronous availability of TNC and CP has been shown to be critical in the improvement of animal nutrition (Neel and Belesky 2015).

Another way to increase overall system productivity and output of silvopasture is to integrate a variety of livestock, either simultaneously or via the leader-follower grazing system. Manríquez-Mendoza et al. (2011) observed significantly greater annual meat production in a mixed-species silvopasture including both cattle and sheep than for silvopastures grazed by cattle or sheep alone. Leader-follower systems can often outproduce other grazing systems for total animal weight gain because each animal tends to consume its optimal foods first (Shepard 2013).

## Thermal Stress

Thermal stress has been shown to be responsible for reductions in feed intake, ADG, and milk production in dairy cows and can be caused by changes in air temperature, relative humidity, wind speed, and solar radiation (Kendall et al. 2006). Symptoms of heat stress, such as increased respiration rate and body temperature, begin to occur at 30 °C and shade typically becomes beneficial to livestock when the temperature-humidity index (THI) is over 72° Fahrenheit (Blackshaw and Blackshaw 1994). Thermal comfort is especially important for European or mixed European × Zebu cattle breeds, which are more sensitive to the high temperatures of the tropics than pure Zebu breeds (Kendall et al. 2006). A study conducted in Alabama demonstrated that even when artificial shade was made available, cattle preferred the shade provided by trees (Zuo and Goodman 2004).

Several studies have shown that trees modify understory microclimates, creating environments that can mitigate heat stress in animals (Tucker et al. 2008; Karki and Goodman 2015), increasing overall grazing time, ADG, lactation, and reproductive rates (Mitlöhner et al. 2001; Kallenbach 2009; Galindo et al. 2013). Kallenbach (2009) reported that cows using silvopastures experienced less difficulty calving (3% compared to 15%) and weaned heavier calves (295 kg compared to 270 kg) than those using traditional pastures. A study in New Zealand comparing four groups of cattle reported that milk production was significantly higher in cattle that had access to shade (Kendall et al. 2006). In turn, livestock have been shown to modify their behavior in the presence of trees, leading to more consistent and uniform grazing across the landscape (McIlvain and Shoop 1971; Karki and Goodman 2010). It has also been suggested that trees can protect animals against the dangers of extreme cold temperatures (Webster 1970; McArthur 1991).

## Animal Health

Managed intensive rotational grazing and silvopasture can have direct impacts on animal health, helping to prevent the spread of parasites and disease. One of the most economically damaging and widespread ectoparasites affecting livestock production is the horn fly (*Hydrotaea irritans* Fall.), a Eurasian fly that relies on feces or vegetative refuse for reproduction, often causing irritation and transmitting disease in livestock (Giraldo et al. 2011; Broom et al. 2013). The continual animal movement seen in rotational grazing lowers the rate at which livestock return to paddocks where dung patties have yet to fully decompose, reducing host-parasite interactions. Additionally, multispecies leader-follower systems can be used, where free-range poultry follow livestock and actively forage on horn fly larvae developing in dung patties (Greg Judy, personal communication).

Silvopastures provide environments that are conducive to the establishment of beneficial insects, including many that help rapidly degrade cattle manure, further inhibiting the spread of the horn fly. In Colombia, Giraldo et al. (2011) documented significantly greater numbers of dung beetles in ISPS than in conventional pasture. The authors observed an inverse relationship between dung beetle and horn fly abundance in the two cattle-raising systems, which they attributed to both plant cover and contribution of plant litter provided by *L. leucocephala*. Plant litter favors the establishment of not only dung beetles, but also other beneficial fauna that can control pest populations and predatorial beetles (Giraldo et al. 2011). Silvopastures have been shown to support increased numbers of birds (McDermott and Rodewald 2014), ants (Rivera et al. 2013), and other beneficial predators that can lower the populations of ticks and reduce the incidence of diseases such as anaplasmosis, which has been shown to drop from 25 to <5% (Yadav et al. 2019).

ISPS contributes ample amounts of tree foliage to the diet, much of which contains condensed tannins, phenols, saponins, and other anti-nutritive secondary compounds that may have anti-parasitic effects. *T. diversifolia*, a widely planted forage shrub in ISPS throughout the tropics, appears to have promising effects on ruminal microbial ecology, reducing the methanogen and protozoa population and increasing the population of cellulolytic bacteria (Ruíz et al. 2014).

Still, there is some concern that silvopastoral environments could increase the presence of parasitic helminths. In southeastern Brazil, Costa et al. (2013) tested this hypothesis throughout a six-month period and found no significant differences in overall weight, weight gain, or helminth infestation between crossbred Holstein and Gir heifers grazed in silvopasture environments and traditional open pasture environments. In contrast, Francisco et al. (2009) studied two groups of wild horses in Spain and concluded that silvopasture increased the presence of infection by gastrointestinal nematodes.

A relatively new area of research has examined livestock social interactions in silvopastoral systems as a diagnostic for social welfare. Améndola et al. (2015) reported that heifers in an ISPS maintained more stable social hierarchies and expressed more socio-positive behaviors, suggesting that animal welfare was enhanced.

### *Habitat for Pollinators*

Pollinator richness and density have been declining in recent years on a global scale (Thomann et al. 2013). Declines in wild bees and butterflies are linked to historical landscape modification (Burkle et al. 2013) and loss of key nesting and foraging sites (Baude et al. 2016). Pollinator decline not only threatens food security, but could also lead to the extinction of pollinator-dependent plants and ultimately the collapse of modern-day agriculture (Dubeux Junior et al. 2017). A report published by the Intergovernmental Science-Policy Platform for Biodiversity and Ecosystem Services identified agriculture as both a threat to pollinators and a potential solution to support them (Duvic-Paoli 2017). One key way to achieve this is through "ecological intensification," or the process of maintaining or enhancing agricultural productivity through the cultivation and management of beneficial biodiversity—a process achieved with silvopasture.

A study examining pollinator presence on two silvoarable and four silvopastoral systems in the UK found that butterfly diversity was significantly higher on the agroforestry plots when compared to conventional pasture (Brosi et al. 2008). However, hoverfly and bumblebee abundance was higher in the silvoarable treatments, but not for the silvopastoral treatments. The authors attributed this to strips of forbs and grasses retained in tree rows within the silvoarable plots. These so-called pollination reservoirs have been shown to be crucial—even in small strips—to the provisioning of adequate pollinator habitat (Brosi et al. 2008). Moreover, planting insect-pollinated tree species may make silvopastures more attractive to pollinators (Varah et al. 2013).

## Conclusion

Silvopasture has been shown to be an effective strategy to ecologically intensify and increase food supply in livestock production systems, but it should not be promoted in isolation of other important food security considerations. In an eye-opening report, Steinfeld et al. (2006) claimed that the livestock sector emerges as one of the top two or three most significant contributors to the most serious environmental problems. With more than 20 billion domestic farm animals on the planet, they may be even more of a burden for the Earth's biosphere than the current 7.7 billion humans (Hahlbrock 2009). It is time we took a careful look at where and how livestock is being produced and whether or not they hinder or advance our aims to sustain the land in perpetuity (Janzen 2011).

Much of the world's increase in livestock production is occurring through intensive concentrated animal feeding operations (CAFOs), using feed produced on arable lands that could be growing food crops for humans (Pollen 2006). A large portion of food energy in plant biomass is lost when it passes through animals, so that the number of people fed $ha^{-1}$ of cropland declines when grain is diverted through livestock (Godfray et al. 2010). Stresses in which livestock are implicated include land-use change, excretion of pollutants (nutrients, antibiotics, pathogens), overuse of freshwater, inefficient use of energy, diverting food for use as feed, and emission of greenhouse gasses (Janzen 2011). Thus, a worthy and prudent goal would be to decrease livestock product consumption and increase awareness of the origin of livestock products, if they are to be consumed.

With that said, many authors make cogent arguments for the role of animal agriculture (Janzen 2011; Hahn Niman 2014; Savory and Butterfield 2016). Livestock may compete with humans for food, but they also create protein from resources we cannot use directly—namely cellulose, from vast grasslands that cannot, or at least should not, be cultivated (Garnett 2009). Most grasslands have coevolved with large ungulates and have even been shown to thrive best under periodic animal impact, restorative disturbances that naturally aerate and return nutrients to soils. Unlike arable cropland, perennial grasses are not regularly tilled, reducing erosion and sequestering large amounts of carbon to help mitigate climate change (Janzen 2004; Mbow et al. 2014). Carbon sequestration can be enhanced even further when combined with trees in silvopastoral systems (Udawatta and Jose 2012). One study found that long-term storage of soil carbon in silvopasture was up to five times greater than traditionally grazed systems, and that did not take into account the carbon sequestered by trees (Toensmeier 2017).

Animal agriculture is now widely engrained in the fabric of many cultures and societies. In fact, meat, milk, and other animal products account for about a third of the protein consumed by humans globally and account for 40% of the global agricultural gross domestic product (Steinfeld et al. 2006). This, combined with the growing stigma of affluence surrounding the consumption of meat, is reason to believe that animal agriculture is here to stay. Silvopasture is an age-old practice that could augment the benefits and minimize the stresses of livestock production if adopted more widely.

Establishing agroforestry on land that currently has low tree cover has been identified as one of the most promising strategies to raise food production without additional deforestation (Garrity et al. 2010). Creating silvopastures from existing monocultures is the low-hanging fruit for the sustainable intensification of livestock production systems. Some have also proposed the thinning, seeding, and management of private woodlots currently under no form of management (Orefice et al. 2016a). In the Central Hardwood Region of the United States, there is an estimated 2.3 million ha of forest being pastured without the benefit of intensive management (Garrett et al. 2004). Managing this acreage under silvopasture would help prevent damages caused by extensive grazing and increase the overall pasture area available.

The establishment of silvopasture is often easier said than done. In many developing countries, a lack of land tenure makes farmers reluctant to invest in the long-term endeavor of establishing trees that may ultimately benefit others than themselves. Where landholdings are small, farmers are often unwilling or unable to spare land for agroforestry establishment, even if it promises higher long-term returns (Mbow et al. 2014). In the case of ISPS, start-up costs can be relatively expensive—with return on investment taking as long as 3–4 years—and may be entirely prohibitive without the availability of subsidies (Murgueitio et al. 2011; Calle 2013) or incentive programs like payment for environmental services (PES) (Pagiola et al. 2005). ISPS is also inherently complex, often requiring extensive capacity building, training, and deployment of new technologies through outreach and extension programs (Calle 2013).

National and regional policy makers across the globe would be wise to support and promote the multiple benefits of silvopasture: sound pasture management, simultaneous timber and livestock production, seasonal increases in meat and milk production, increased biodiversity and forage diversity, better welfare for animals, and carbon sequestration are all advantages of this land-use practice. There is a strong need for programs connecting producers who have successfully implemented silvopasture with others who have not. Policy makers should also address the obstacles faced by landholding producers and create programs to incentivize the adoption and utilization of silvopasture. Prohibitive start-up costs, lack of access to technical information, and poor understanding of existing government-subsidized programs are all issues that need to be addressed. As climate change continues to intensify and jeopardize global food security, silvopasture should no longer be treated as an anomaly, practiced by the few; it should be widely recognized, supported, and promulgated for the effective food provisioning tool that it is, expanding and facilitating green ranching opportunities to farmers around the world.

## References

Addlestone B, Mueller J, Luginbuhl J-M (1999) The establishment and early growth of three leguminous tree species for use in silvopastoral systems in the southeastern USA. Agrofor Syst 44:253–265. https://doi.org/10.1023/A:1006254812236

Adhikari P, Udawatta RP, Anderson SH (2014) Soil thermal properties under prairies, conservation buffers, and corn-soybean land use systems. Soil Sci Soc Am J 78:1977–1986. https://doi.org/10.2136/sssaj2014.02.0074

Aganga A, Tshwenyane S (2003) Feeding values and anti-nutritive factors of forage tree legumes. Pak J Nutr 2:170–177

Agostini PS, Gasa J, Manzanilla EG et al (2013) Descriptive study of production factors affecting performance traits in growing-finishing pigs in Spain. Span J Agric Res 11:371–381. https://doi.org/10.5424/sjar/2013112-3011

Alexandratos N (1995) World agriculture towards 2010. Rome, Italy

Améndola L, Solorio FJ, Ku-Vera JC et al (2015) Social behaviour of cattle in tropical silvopastoral and monoculture systems. Animal 10:863–867. https://doi.org/10.1017/s1751731115002475

Andrade CMS, Valentim JF, Carneiro JD, Vaz FA (2004) Crescimento de gramíneas e leguminosas forrageiras tropicais sob sombreamento. Pesq Agrop Brasileira 39:263–270. https://doi.org/10.1590/S0100-204X2004000300009

Armand D, Meuret M (1993) Du mûrier fourrager dans les systèmes d'élevage ovin préalpin. In: Fodder trees and shrubs in the Mediterranean production systems: objectives and expected results of the €C research contract, Agricultur. pp 53–59

Atta-Krah A, Reynolds L (1989) Utilization of pasture and fodder shrubs in the nutrition of sheep and goats in the humid tropics of West Africa. Sheep and goat meat production in the humid tropics of West Africa. Rome, Italy

Bacab HM, Solorio FJ, Solorio SB (2012) Efecto de la altura de poda en Leucaena leucocephala y su influencia en el rebrote y rendimiento de Panicum maximum. Av Investig Agropecu 16:65–77

Ball D, Collins M, Lacefield G, et al (2001) Understanding forage quality. 1–16

Ball DM, Lacefield GD, Schmidt SP, et al (2015) Understanding the tall fescue endophyte. Salem, OR

Baude M, Kunin WE, Boatman ND et al (2016) Historical nectar assessment reveals the fall and rise of Britain in bloom Europe PMC Funders Group. Nature 530:85–88. https://doi.org/10.5285/69402002-1676-4de9-a04e-d17e827db93c

Belesky DP (2005) Growth of Dactylis glomerata along a light gradient in the central Appalachian region of the eastern USA: I. dry matter production and partitioning. Agrofor Syst 65:81–90. https://doi.org/10.1007/s10457-004-5725-y

Belsky A (1994) Influences of trees on savanna productivity: tests of shade, nutrients, and tree-grass competition. Ecology 75:922–932

Belsky AJ, Amundson RG, Duxbury JM et al (1989) The effects of trees on their physical, chemical and biological environments in a semi-arid savanna in Kenya. J Appl Ecol 26:1005–1024. https://doi.org/10.1098/rstb.1981.0005

Blackshaw JK, Blackshaw AW (1994) Heat stress in cattle and the effect of shade on production and behaviour: a review. Aust J Exp Agric 34:285–295. https://doi.org/10.1071/EA9940285

Broom DM, Galindo FA, Murgueitio E (2013) Biodiversity and good welfare for animals. Sustainable, efficient livestock production with high biodiversity and good welfare for animals. Proc R Soc B 280:1–9

Brosi BJ, Armsworth PR, Daily GC (2008) Optimal design of agricultural landscapes for pollination services. Conserv Lett 1:27–36. https://doi.org/10.1111/j.1755-263x.2008.00004.x

Buergler AL, Feldhake CR, Teutsch CD et al (2005) Botanical composition and forage production in an emulated silvopasture. Agron J 97:1141. https://doi.org/10.2134/agronj2004.0308

Buergler AL, Fike JH, Burger JA et al (2006) Forage nutritive value in an emulated silvopasture. Agron J 98:1265–1273. https://doi.org/10.2134/agronj2005.0199

Burkle LA, Marlin JC, Knight TM (2013) Plant-pollinator interactions over 120 years: loss of species, co-occurrence, and function. Science (80-) 340:1611–1615. https://doi.org/10.1126/science.1232728

Calle Z (2013) A strategy for scaling-up intensive silvopastoral systems in Colombia. J Sustain For 32:677–693

Calle Z, Murgueitio E, Chara J (2012a) Integrating forestry, sustainable cattle-ranching and landscape restoration. Unasylva 63:31–40

Calle Z, Murgueitio E, Chará J (2012b) Integrating forestry, sustainable cattle-ranching and landscape restoration. Unasylva 63:31–40

Cardona CAC, Ramírez JFN, Morales AMT et al (2013) Contribution of intensive silvopastoral systems to animal performance and to adaptation and mitigation of climate change. Rev Colomb Ciencias Pecu 27:76–94

Casanova-Lugo F, Ramírez-Avilés L, Ku-Vera JC et al (2015) Forage yield and quality of Leucaena leucocephala and Guazuma ulmifolia in tropical silvopastoral systems. Trop Grasslands Forrajes Trop 2:24. https://doi.org/10.17138/tgft(2)24-26

Chadhocar P, Kantharaju H (1980) Effect of Gliricidia maculata on growth and breeding of Bannur ewes. Trop Grasslands 14:78–82

Chee Y, Faiz A (1991) Forage resources in Malaysian Rubber Estates. In: Shelton H, Stur W (eds) Forages for plantation crops, Bali, Indonesia, p 168

Costa JO, Leite RC, Toth G et al (2013) Parasitic helminth infection in young cattle raised on silvopasture and open-pasture in southeastern Brazil. Agrofor Syst 88:53–62. https://doi.org/10.1007/s10457-013-9655-4

Cribb J (2010) The coming famine: the global food crisis and what we can do to avoid it. University of California Press, Berkeley, CA

Dalzell S, Shelton H, Mullen B et al (2006) Leucaena: a guide to establishment and management. Meat and Livestock Australia Ltd., Sydney
de Fraiture C, Molden D, Wichelns D (2010) Investing in water for food, ecosystems, and livelihoods: an overview of the comprehensive assessment of water management in agriculture. Agric Water Manag 97:495–501. https://doi.org/10.1016/j.agwat.2009.08.015
de M. Costa SB, de Mello ACL, Dubeux JC Jr et al (2016) Livestock performance in warm-climate silvopastures using tree legumes. Agron J 108:2026. https://doi.org/10.2134/agronj2016.03.0180
DeBruyne SA, Feldhake CM, Burger JA, Fike JH (2011) Tree effects on forage growth and soil water in an Appalachian silvopasture. Agrofor Syst 83:189–200. https://doi.org/10.1007/s10457-011-9376-5
Delgado C, Rosegrant M, Steinfeld H et al (1999) Livestock to 2020: The Next Food Revolution. IFPRI Food, Agriculture, and the Environment Discussion Paper 28. Washington, D.C. (USA)
Devendra C (2012) Intensification of integrated natural resources use and agricultural systems in the developing world. Agrotechnology 01:1–4. https://doi.org/10.4172/2168-9881.1000e101
Devendra C, Ibrahim M (1999) Silvopastoral systems as a strategy for diversification and productivity enhancement from livestock in the tropics. 2020:8–22
Devkota NR, Wall AJ, Kemp PD, et al (2001) Relationship between canopy closure and pasture production in deciduous tree based temperate silvopastoral systems
Dohn J, Dembélé F, Karembé M et al (2013) Tree effects on grass growth in savannas: competition, facilitation and the stress-gradient hypothesis. J Ecol 101:202–209. https://doi.org/10.1111/1365-2745.12010
Dubeux Junior JCB, Muir JP, Apolinário VX et al (2017) Tree legumes: an underexploited resource in warm-climate silvopastures. Rev Bras Zootec 46:689–703. https://doi.org/10.1590/S1806-92902017000800010
Duvic-Paoli LA (2017) The intergovernmental science-policy platform for biodiversity and ecosystem services or the framing of scientific knowledge within the law of sustainable development. Int Community Law Rev 19:231–269. https://doi.org/10.1163/18719732-12341355
Dibala R, Jose S, Gold M et al (2021) Tree density effects on soil, herbage mass and nutritive value of understory Megathyrsus maximus in a seasonally dry tropical silvopasture in Panama. Agroforest Syst 95:741–753. https://doi.org/10.1007/s10457-021-00628-4
FAOSTAT (2010) Food and agricultural commodities production, Rome
Feldhake CM, Neel JPS, Belesky DP (2010) Establishment and production from thinned mature deciduous-forest silvopastures in Appalachia. Agrofor Syst 79:31–37. https://doi.org/10.1007/s10457-010-9289-8
Flack S (2016) The art and science of grazing: how grass farmers can create sustainable systems for healthy animals and farm ecosystems. Chelsea Green Publishing, White River Juntion, VT
Francisco I, Arias M, Cortiñas FJ et al (2009) Silvopastoralism and autochthonous equine livestock: analysis of the infection by endoparasites. Vet Parasitol 164:357–362. https://doi.org/10.1016/j.vetpar.2009.06.018
Galindo F, Olea R, Suzán G (2013) Animal welfare and sustainability. In: International Workshop on Farm Animal Welfare, Sao Paolo, SP
Garnett T (2009) Livestock-related greenhouse gas emissions: impacts and options for policy makers. Environ Sci Pol 12:491–503. https://doi.org/10.1016/j.envsci.2009.01.006
Garrett H, Kerley M, Ladyman K et al (2004) Hardwood silvopasture management in NA: new vistas in agroforestry. Agrofor Syst 61:21–33. https://doi.org/10.1007/978-94-017-2424-1_2
Garrity DP, Akinnifesi FK, Ajayi OC et al (2010) Evergreen agriculture: a robust approach to sustainable food security in Africa. Food Security 2:197–214. https://doi.org/10.1007/s12571-010-0070-7
Gerrish J (2004) Management-intensive grazing: the grassroots of grass farming. Green Park Press, Ridgeland, MO

Giraldo C, Escobar F, Chará JD, Calle Z (2011) The adoption of silvopastoral systems promotes the recovery of ecological processes regulated by dung beetles in the Colombian Andes. Insect Conserv Divers 4:115–122. https://doi.org/10.1111/j.1752-4598.2010.00112.x

Godfray HCJ, Beddington JR, Crute IR et al (2010) Food security: the challenge of feeding 9 billion people. Science (80- ) 327:812–818. https://doi.org/10.1126/science.1185383

Gold MA, Hanover JW (1993) Honey locust (Gleditsia triacanthos), a multipurpose tree for the temperate zone. Int Tree Crop J 7:189–207. https://doi.org/10.1080/01435698.1993.9752919

Hahlbrock K (2009) In: Skogley D (ed) Feeding the planet: environmental protection through sustainable agriculture. Haus Publishing, London

Hahn Niman N (2014) Defending beef: the case for sustainable meat production. Chelsea Green Publishing, White River Junction, VT

Hernández M, Guenni O (2008) Producción de biomasa y calidad nutricional del estrato graminoide en un sistema silvopastoril dominado por samán (Samanea saman Jacq Merr). Zootec Trop 26:439–453

Ibrahim M, Villanueva C, Casasola F, Rojas J (2007) Sistemas silvopastoriles como una herramienta para el mejoramiento de la productividad y rehabilitación ecológica de paisajes ganaderos en Centro América. Arch Latinoam Prod Anim 15:14. https://doi.org/10.1002/mc.21974

Ibrahim M, Guerra L, Casasola F, Neely C (2010) Importance of silvopastoral systems for mitigation of climate change and harnessing of environmental benefits. 189–196

Jalonen R, Nygren P, Sierra J (2009) Transfer of nitrogen from a tropical legume tree to an associated fodder grass via root exudation and common mycelial networks. Plant Cell Environ 32:1366–1376. https://doi.org/10.1111/j.1365-3040.2009.02004.x

Janzen HH (2004) Carbon cycling in earth systems - a soil science perspective. Agric Ecosyst Environ 104:399–417. https://doi.org/10.1016/j.agee.2004.01.040

Janzen HH (2011) What place for livestock on a re-greening earth? Anim Feed Sci Technol 166–167:783–796. https://doi.org/10.1016/j.anifeedsci.2011.04.055

Joffre R, Rambal S (1993) How tree cover influences the water balance of Mediterranean rangelands. Ecology 74:570–582

Johnson J, Fike J, Fike W et al (2013) Millwood honey locust trees: seedpod nutritive value and yield characteristics. Agrofor Syst 87:849–856. https://doi.org/10.1007/s10457-013-9601-5

Jones W, Anderson L, Ross M (1973) Bloat in cattle. New Zeal J Agric Res 16:441–446. https://doi.org/10.1080/00288233.1973.10421128

Jose S, Dollinger J (2019) Silvopasture: a sustainable livestock production system. Agrofor Syst 93:1–9. https://doi.org/10.1007/s10457-019-00366-8

Jose S, Gillespie AR, Pallardy SG (2004) Interspecific interactions in temperate agroforestry. Agrofor Syst 61–62:237–255. https://doi.org/10.1023/B:AGFO.0000029002.85273.9b

Jose S, Walter D, Kumar B (2019) Ecological considerations in sustainable silvopasture design and management. Agrofor Syst. https://doi.org/10.1007/s10457-016-0065-2

Kallenbach R (2009) Integrating silvopastures into current forage-livestock systems. In: Agroforestry comes of age: putting science into practice, pp 455–461

Kallenbach RL, Kerley MS, Bishop-Hurley GJ (2006) Cumulative forage production, forage quality and livestock performance from an annual ryegrass and cereal rye mixture in a pine walnut silvopasture. Agrofor Syst 66:43–53. https://doi.org/10.1007/s10457-005-6640-6

Kallenbach RL, Venable EB, Kerley MS, Bailey NJ (2010) Stockpiled tall fescue and livestock performance in an early stage midwest silvopasture system. Agrofor Syst 80:379–384. https://doi.org/10.1007/s10457-010-9322-y

Karki U, Goodman MS (2010) Cattle distribution and behavior in southern-pine silvopasture versus open-pasture. Agrofor Syst 78:159–168. https://doi.org/10.1007/s10457-009-9250-x

Karki U, Goodman MS (2015) Microclimatic differences between mature loblolly-pine silvopasture and open-pasture. Agrofor Syst 89:319–325. https://doi.org/10.1007/s10457-014-9768-4

Kellman M (1979) Soil enrichment by neotropical savanna trees. J Ecol 67:565–577

Kendall PE, Nielsen PP, Webster JR et al (2006) The effects of providing shade to lactating dairy cows in a temperate climate. Livest Sci 103:148–157. https://doi.org/10.1016/j.livsci.2006.02.004

Kephart K, Buxton D (1993) Forage quality responses of C3 and C4 perennial grasses to shade. Crop Sci 33:831–837

Ladyman KP, Kerley MS, Kallenbach RL, et al (2003) Quality and quantity evaluations of shade grown forages. Notes

Lin C, McGraw R, George M, Garrett H (1998) Shade effects on forage crops with potential in temperate agroforestry practices. Agrofor Syst 44:109–119. https://doi.org/10.1023/A:1006205116354

Lin CH, Mcgraw ML, George MF, Garrett HE (2001) Nutritive quality and morphological development under partial shade of some forage species with agroforestry potential. Agrofor Syst 53:269–281. https://doi.org/10.1023/A:1013323409839

Liyanage ADES, Jayasundara HPS, Fernando DNS, Fernando MTN (1993) Integration of legume-based pasture and cattle into coconut farming systems in Sri Lanka. J Asian Farming Syst 1:579–588

Malmer A, Murdiyarso D, (Sampurno) Bruijnzeel L, Ilstedt U (2010) Carbon sequestration in tropical forests and water: a critical look at the basis for commonly used generalizations. Glob Chang Biol 16:599–604. https://doi.org/10.1111/j.1365-2486.2009.01984.x

Manríquez-Mendoza L, López-Ortiz S, Olguín-Palacios C et al (2011) Productivity of a silvopastoral system under intensive mixed species grazing by cattle and sheep. Trop Subtrop Agroecosyst 13:573–584

Mbow C, Van Noordwijk M, Luedeling E et al (2014) Agroforestry solutions to address food security and climate change challenges in Africa. Curr Opin Environ Sustain 6:61–67. https://doi.org/10.1016/j.cosust.2013.10.014

McArthur AJ (1991) Forestry and shelter for livestock. Forest Ecol Manag 45(1-4):93–107

McDermott ME, Rodewald AD (2014) Conservation value of silvopastures to neotropical migrants in Andean forest flocks. Biol Conserv 175:140–147. https://doi.org/10.1016/j.biocon.2014.04.027

McIlvain EH, Shoop MC (1971) Shade for improving cattle gains and rangeland use. J Range Manag 24:181–184

Medinilla-Salinas L, de la Cruz Vargas-Mendoza M, López-Ortiz S et al (2013) Growth, productivity and quality of Megathyrsus maximus under cover from Gliricidia sepium. Agrofor Syst 87:891–899. https://doi.org/10.1007/s10457-013-9605-1

Mitlöhner FM, Morrow JL, Dailey JW et al (2001) Shade and water misting effects on behavior, physiology, performance, and carcass traits of heat-stressed feedlot cattle. J Anim Sci 79:2327–2335. https://doi.org/10.2527/2001.7992327x

Moustakas A, Kunin WE, Cameron TC, Sankaran M (2013) Facilitation or competition? Tree effects on grass biomass across a precipitation gradient. PLoS One 8. https://doi.org/10.1371/journal.pone.0057025

Mueller J, Luginbuhl J-M, Bergmann B (2001) Establishment and early growth characteristics of six paulownia genotypes for goat browse in Raleigh, NC, USA. Agrofor Syst 52:63–72. https://doi.org/10.1023/A:1010641602384

Murgueitio E, Calle Z, Uribe F et al (2011) Native trees and shrubs for the productive rehabilitation of tropical cattle ranching lands. For Ecol Manag 261:1654–1663. https://doi.org/10.1016/j.foreco.2010.09.027

Neel JPS, Belesky DP (2015) Herbage production, nutritive value and animal productivity within hardwood silvopasture, open and mixed pasture systems in Appalachia, United States. Grass Forage Sci 72:137–153. https://doi.org/10.1111/gfs.12211

Neel JPS, Belesky DP (2017) Herbage production, nutritive value and animal productivity within hardwood silvopasture, open and mixed pasture systems in Appalachia, United States. Grass Forage Sci 72:137–153. https://doi.org/10.1111/gfs.12211

Nieto R, Ribera M, García M, Aguilera J (2002) Amino acid availability and energy value of acorn in the Iberian pig. Livest Prod Sci 77:227–239. Livest Prod Sci 77:227–239
Nygren P, Fernandez MP, Harmand JM, Leblanc HA (2012) Symbiotic dinitrogen fixation by trees: an underestimated resource in agroforestry systems? Nutr Cycl Agroecosyst 94:123–160. https://doi.org/10.1007/s10705-012-9542-9
Obispo NE, Espinoza Y, Gil JL et al (2008) Efecto del sombreado sobre la producción y calidad del pasto guinea (Panicum maximum) en un sistema silvopastoril. Zootec Trop 26:285–288
Orefice J, Carroll J, Conroy D, Ketner L (2016a) Silvopasture practices and perspectives in the Northeastern United States. Agrofor Syst 91:1–12. https://doi.org/10.1007/s10457-016-9916-0
Orefice J, Smith RG, Carroll J et al (2016b) Forage productivity and profitability in newly-established open pasture, silvopasture, and thinned forest production systems. Agrofor Syst:1–15. https://doi.org/10.1007/s10457-016-0052-7
Orefice J, Smith RG, Carroll J et al (2017) Soil and understory plant dynamics during conversion of forest to silvopasture, open pasture, and woodlot. Agrofor Syst 91:729–739. https://doi.org/10.1007/s10457-016-0040-y
Paciullo DSC, de Castro CRT, Gomide CA de M, et al (2011) Performance of dairy heifers in a silvopastoral system. Livest Sci 141:166–172. doi: https://doi.org/10.1016/j.livsci.2011.05.012
Paciullo DSC, Gomide CAM, Castro CRT et al (2017) Morphogenesis, biomass and nutritive value of Panicum maximum under different shade levels and fertilizer nitrogen rates. Grass Forage Sci 72:590–600. https://doi.org/10.1111/gfs.12264
Pagiola S, Agostini P, Gobbi J et al (2005) Paying for biodiversity conservation services. Mt Res Dev 25:206–211. https://doi.org/10.1659/0276-4741(2005)025[0206:PFBCS]2.0.CO;2
Pang K, Van Sambeek JW, Navarrete-Tindall NE, Lin C-H, Jose S, Garrett HE (2019a) Responses of legumes and grasses to non-, moderate, and dense shade in Missouri, USA. II. Forage quality and its species-level plasticity. Agrofor Syst 93:25–38. https://doi.org/10.1007/s10457-017-0068-7
Pang K, Van Sambeek JW, Navarrete-Tindall NE et al (2019b) Responses of legumes and grasses to non-, moderate, and dense shade in Missouri, USA. I. Forage yield and its species-level plasticity. Agrofor Syst 93:11–24. https://doi.org/10.1007/s10457-017-0067-8
Papachristou TG, Papanastasis VP (1994) Forage value of Mediterranean deciduous woody fodder species and its implication to management of silvo-pastoral systems for goats. Agrofor Syst 27:269–282. https://doi.org/10.1007/BF00705061
Papanastasis VP, Platis PD, Dini-papanastasi O (1998) Effects of age and frequency of cutting on productivity of Mediterranean deciduous fodder tree and shrub plantations. Forest Ecol Manag 110:283–292
Papanastasis VP, Yiakoulaki MD, Decandia M, Dini-papanastasi O (2008) Integrating woody species into livestock feeding in the Mediterranean areas of Europe. Anim Feed Sci Technol 140:1–17. https://doi.org/10.1016/j.anifeedsci.2007.03.012
Pent GJ, Fike JH (2018) Lamb productivity on stockpiled fescue in honey locust and black walnut silvopastures. Agrofor Syst 93:1–9. https://doi.org/10.1007/s10457-018-0264-0
Pérez J, Zapata B, Sosa R (1995) Brosimum alicastrum Swartz. como forraje en la alimentación de ovinos en crecimiento. Agroforestería las Am 2:17–21
Pitta DW, Barry TN, Lopez-villalobos N, Kemp PD (2005) Effects on ewe reproduction of grazing willow fodder blocks during drought. Anim Feed Sci Technol 120:217–234. https://doi.org/10.1016/j.anifeedsci.2005.02.030
Pollen M (2006) The omnivore's dilemma: the search for a perfect meal in a fast-food world. Bloomsburg Publishing, Plc, London
Provenza FD, Villalba JJ, Wiedmeier RW et al (2009) Value of plant diversity for diet mixing and sequencing in herbivores. Rangelands 31:45–49. https://doi.org/10.2111/1551-501X-31.1.45
Pulido F (1999) Herbivorismo y regeneración de la encina (Quercus ilex L.) en bosques y dehesas. Universidad de Extremadura, Cáceres

Pulido F, Díaz M, Hidalgo S (2001) Size-structure and regeneration of Spanish holm oak Quercus ilex forest and dehesas: effects of agroforestry use on their long-term sustainability. For Ecol Manag 146:1–13

Rayburn EB (2007) Forage production for pasture-based livestock production, 2007. Natural Resource, Agriculture, and Engineering Service (NRAES), Ithaca, NY

Rivera LF, Armbrecht I, Calle Z (2013) Silvopastoral systems and ant diversity conservation in a cattle-dominated landscape of the Colombian Andes. Agric Ecosyst Environ 181:188–194. https://doi.org/10.1016/j.agee.2013.09.011

Rodríguez-Estévez V, García Martínez A, Perea Muñoz J et al (2007) Producción de bellota en la dehesa: factores influyentes. Arch Zootec 56:25–43

Rodríguez-Estévez V, Sánchez-Rodríguez M, García AR, Gómez-Castro AG (2011) Average daily weight gain of Iberian fattening pigs when grazing natural resources. Livest Sci 137:292–295. https://doi.org/10.1016/j.livsci.2010.11.015

Ruíz T, Febles G, Galindo J et al (2014) Tithonia diversifolia, its possibilities in cattle rearing systems. Cuba J Agric Sci 48:79–82

Sánchez MD (2000) Mulberry: an exceptional forage available almost worldwide. World Rev Anim Prod 93:1–21

Santiago-Hernández F, López-Ortiz S, Ávila-Reséndiz C et al (2016) Physiological and production responses of four grasses from the genera Urochloa and Megathyrsus to shade from Melia azedarach L. Agrofor Syst 90:339–349. https://doi.org/10.1007/s10457-015-9858-y

Savory A, Butterfield J (2016) Holistic management: a commonsense revolution to restore our environment, 3rd edn. Island Press, Washington, DC

Sharrow S, Brauer D, Clason T (2009). Silvopastoral Practices. In: North American Agroforestry: An Integrated Science and Practice, 2nd edition, H.E. Garrett (ed.). American Society of Agronomy, 677 S. Segoe Rd., Madison, WI 53711, USA

Shelton HM, Dalzell SA (2007) Production, economic and environmental benefits of Leucaena pastures. Trop Grasslands 41:174–190

Shepard M (2013) Restoration agriculture: real-world permaculture for farmers. Acres USA, Austin, TX

Sierra J, Nygren P (2006) Transfer of N fixed by a legume tree to the associated grass in a tropical silvopastoral system. Soil Biol Biochem 38:1893–1903. https://doi.org/10.1016/j.soilbio.2005.12.012

Sierra J, Daudin D, Domenach AM et al (2007) Nitrogen transfer from a legume tree to the associated grass estimated by the isotopic signature of tree root exudates: a comparison of the 15N leaf feeding and natural 15N abundance methods. Eur J Agron 27:178–186. https://doi.org/10.1016/j.eja.2007.03.003

Sosa Rubio E, Pérez Rodríguez D, Ortega Reyes L, Zapata Buenfil G (2004) Evaluacion del Potencial Forrajero de arboles y arbustos tropicales para la alimentacion. Técnica Pecu en México 42:129–144

Sousa LF, Maurício RM, Moreira GR et al (2010) Nutritional evaluation of "Braquiarão" grass in association with "Aroeira" trees in a silvopastoral system. Agrofor Syst 79:189–199. https://doi.org/10.1007/s10457-010-9297-8

Steinfeld H, Gerber P, Wassenaar T et al (2006) Livestock's long shadow: environmental issues and options. FAO, Rome

Talamuci P, Pardini A (1999) Pastoral systems dominated by fodder crops harvesting and grazing. In: Etienne M. (ed.) Dynamics and sustainability of Mediterranean pastoral systems. Zaragoza: CIHEAM. Cahiers Options Mediterranean; no. 39. Pages 29–44

Teklehaimanot Z, Jones M, Sinclair FL (2002) Tree and livestock productivity in relation to tree planting configuration in a silvopastoral system in North Wales, UK. Agrofor Syst 56:47–55. https://doi.org/10.1023/A:1021131026092

Thomann M, Imbert E, Devaux C, Cheptou PO (2013) Flowering plants under global pollinator decline. Trends Plant Sci 18:353–359. https://doi.org/10.1016/j.tplants.2013.04.002

Tiedemann AR, Klemmedson JO (2008) Effect of Mesquite on physical and chemical properties of the soil. J Rangel Manag 26:27–29. https://doi.org/10.2307/3896877
Toensmeier E (2017) Perennial staple crops and agroforestry for climate change mitigation. In: Montagnini F (ed) Integrating landscapes: agroforestry for biodiversity conservation and food sovereignty. Springer, Cham
Tucker CB, Rogers AR, Schütz KE (2008) Effect of solar radiation on dairy cattle behaviour, use of shade and body temperature in a pasture-based system. Appl Anim Behav Sci 109:141–154. https://doi.org/10.1016/j.applanim.2007.03.015
Tudsri S, Prasanpanich S (2001) Effect of pasture production systems on milk production in the central plains of Thailand. Trop Grasslands 35:246–253
Udawatta RP, Jose S (2012) Agroforestry strategies to sequester carbon in temperate North America. Agrofor Syst 86:225–242. https://doi.org/10.1007/s10457-012-9561-1
Udawatta RP, Garrett HE, Kallenbach R (2011) Agroforestry buffers for nonpoint source pollution reductions from agricultural watersheds. J Environ Qual 40:800–806. https://doi.org/10.2134/jeq2010.0168
USDA - NRCS (2010) Prescribed grazing - practice introduction - practice code 528
Varah A, Jones H, Smith J, Potts SG (2013) Enhanced biodiversity and pollination in UK agroforestry systems. J Sci Food Agric 93:2073–2075. https://doi.org/10.1002/jsfa.6148
Vetaas OR (1992) Micro-site effects of trees and shrubs in dry savannas. J Veg Sci 3:337–344. https://doi.org/10.2307/3235758
Voisin A (1988) Grass productivity. Island Press, Washington, DC
Webster A (1970) Direct effects of cold weather on the energetic efficiency of beef production in different regions of Canada. Can J Anim Sci 50:563–573
Xavier DF, da Silva Lédo FJ, de Campos Paciullo DS et al (2014) Nitrogen cycling in a Brachiaria-based silvopastoral system in the Atlantic forest region of Minas Gerais, Brazil. Nutr Cycl Agroecosyst 99:45–62. https://doi.org/10.1007/s10705-014-9617-x
Yadav A, Gendley MK, Sahu J, Patel PK (2019) Silvopastoral system: a prototype of livestock agroforestry. Pharma Innov J 8:76–82
Yamamoto W, Dewi IA, Ibrahim M (2007) Effects of silvopastoral areas on milk production at dual-purpose cattle farms at the semi-humid old agricultural frontier in Central Nicaragua. Agric Syst 94:368–375. https://doi.org/10.1016/j.agsy.2006.10.011
Yiakoulaki MD (1995) Intake by goats grazing kermes oak shrublands with varying cover in northern Greece. Small Rumin Res 17:223–228
Zuo H, Goodman MSM (2004) Landscape use by cattle affected by pasture developments and season. J Range Manag 57:426–434

# Potential Indicators to Assess the Sustainability of South American Forests Under Silvopastoral Management: Case Study of an Old Roble Forest in Chile's Andes

**María Gabriela Cristina Medina Rivero and Francis Dube**

## Introduction

According to the UN Forest and Agriculture Organization (FAO 2018), the world's forest area, including planted forests, encompasses around four billion hectares, covering 31% of the Earth's surface. South America hosts 23% of the world's forests. These ecosystems include tropical and temperate forests covering of 831.5 million hectares, corresponding to 46% of the land area. Most of these forests are in the Amazon, Mesoamerica, the Southern Cone, and the Caribbean. However, the FAO report notes that between 1990 and 2015, the global forest area decreased from 31.6% to 30.6%. The biggest reduction occurred in tropical regions, especially in Latin America, sub-Saharan Africa, and South and Southeast Asia, mainly due to deforestation for agricultural and/or livestock production.

This situation is worrying, since in these regions native forests provide essential life services such as food, water regulation, soil conservation, and economic resources; they alleviate effects of climate change, provide spaces for recreation and tourism, and host social, cultural, and spiritual values (Guzmán et al. 2012; Franco 2015). Additionally, these forests are important as livelihoods for rural populations play a big role in the conservation of biodiversity and maintenance of carbon reserves.

In this sense, forests that are managed with timber production, non-timber forest products, agriculture and/or livestock, and environmental services in mind have increased in recent years. Simultaneously, recognition and concern for the social,

M. G. C. M. Rivero
Department of Agricultural Sciences, Universidad de los Andes,
Trujillo, Trujillo State, Venezuela

F. Dube (✉)
Faculty of Forestry Sciences, University of Concepción, Concepción, Biobío Region, Chile
e-mail: fdube@udec.cl



R. P. Udawatta, S. Jose (eds.), *Agroforestry and Ecosystem Services*,
https://doi.org/10.1007/978-3-030-80060-4_9

cultural, and spiritual values associated with forests have grown. Payment mechanisms for environmental services and other financial means have contributed to value the multi-functionality of forests, and protected areas play an important role in maintaining the goods and services provided through forest ecosystems. As a result, the rate of forest loss has slowed in recent years, which is an encouraging development, as it is expected that the global population will rise from 7.6 billion to 10 billion by 2050, meaning food demand could rise by 50% and with it the pressure on natural resources, especially forests.

Current positive statistics suggest that many countries have improved forest management and progressed toward meeting the UN's Sustainable Development Goals. Thus, sustainable forest management has contributed to a reduction in net forest loss from 0.18% in the 1990s to 0.08% over the past 5 years and the maintenance of biomass stocks. Forests in protected areas established by law significantly increased in the tropics. They make up 17% of the total forest area, while the area subject to long-term management plans increased to 2.1 billion hectares in 2010. Globally, the area of certified forests also increased from 285 million hectares to 440 million hectares between 2010 and 2014 (FAO 2018).

South America faces the challenge of having to boost agricultural production to meet the demands of a growing population without reducing the forest area. Trends in the region show that agricultural, livestock, and especially forestry production are shifting toward more sustainable systems by implementing models that contemplate diversification and harmonious social, environmental, and economic development.

One option for diversified production, promoted since 1990 as a practice for sustainable production systems, is agroforestry. It entails the use and management of trees or shrubs on the surface to sustain crops and/or animals, either simultaneously or sequentially (Krishnamurth 1999; Nair 2004; Murgueitio et al. 2012; Montagnini et al. 2015). Around 20% of the world's population (1.5 billion people) in developing countries depend directly on products and services from agroforestry systems (AFSs), most common in the tropics, which help improve the life quality of socially vulnerable communities and the conservation of natural resources (Alonso 2011; Medina et al. 2013; FAO 2014b; Kássio et al. 2018).

Agroforestry practices include silvopastoral systems. These involve the combination or association of a woody component with livestock and/or pastures or crops on the same land, facilitating ecological, economic, and social interactions among these components. These associations are also referred to as silvopastures. That practice has been successfully implemented in many regions, and especially in Latin America has proved its enormous potential in, for example, Costa Rica, Cuba, Mexico, Colombia, Brazil, Venezuela, and Argentina. The latter is among the South American countries with temperate areas that have the greatest experience with silvopasture in native forests.

This system has been applied for a long time, incorporating livestock activity in the mountains with satisfactory results. Research has consistently concluded that this practice can constitute a sustainable productive alternative, including the possibility of recovering degraded ecosystems (Peri et al. 2008; Peri et al. 2009; Gargaglione 2011; Peri 2011; Soler Esteban 2012; Bahamonde et al. 2012). In

Chile, silvopasture and agroforestry have been underdeveloped as studies focused only on some areas and species (Fernández et al. 2002; Dube 2017). Historically a forest country, the concept spread in Chile from the 1960s. Before, forest plantations were managed under traditional systems, which excluded animals from the forest.

However, over the past 12 years this perception has changed even more as studies on agroforestry as an alternative model for forest establishment and industrial purposes have deepened, emphasizing tree planting on fields of small agricultural producers, considering the cultural identity, environment, and productive systems of farmers (Sotomayor 2015). In Chile, silvopastoral systems—within the agroforestry category—are well known and mainly used by small landowners of dry soils, without irrigation. For bigger producers, business vision has prevented silvopasture at forest sites because economic interests prevail over environmental and social criteria. Yet, during more than 10 years 1114 hectares on 1600 properties have been established in which forestry systems occupy 44.4% and windbreaks 43.7% (Sotomayor 2011, 2015).

According to Sotomayor, various studies support this preference. They verified decreases in erosive processes after introducing trees together with agricultural crops. Soil losses fell over 1700%, wind reduced by 200% after planting trees in meadows, while meadow productivity increased 41%. Silvopastoral systems have also led to less pollutants in channels through the use of biofilters and helped mitigate climate change, among other social benefits.

Comparatively few experiences of its use in native forest exist, and the situation of these forests has followed the global trend. They have historically been subjected to destructive and unsustainable use, principally overexploitation, changed land use, and a belief that trees represent a competitor or productivity hindrance. Under this belief, forest species reduce agricultural production, so trees and shrubs were extracted, cut, or burned, leading to a changed rural landscape and often scarcity of trees (Sotomayor 2010).

Damage then results from an impediment to the forest's regeneration ability, which leads to falling biodiversity and soil fertility, and increasing compaction and erosion (INFOR 2015). Wood availability for energy production in a domestic context also falls.

However, Dube (2017) confirmed that silvopastoral systems represent a good management alternative for forests on livestock farms. Using long-lived native trees in silvopastoral systems could allow carbon sequestration in biomass and soil, preserve the quality of the soil's organic carbon, and help balance socio-environmental and economic concerns of land use. The University of Concepción develops at its Ranchillo Alto site in the Andean Precordillera, Ñuble Region, one experiment on large-scale native forest silvopastoralism. Since 2015 researchers have been seeking ways to rejuvenate an overripe forest with native species, valuing the quantity and quality of grass planted under different degrees of tree cover and its effect on animal production. This way, they research methods to encourage and advise on implementing silvopastoral modules on the properties of the surrounding communities to combine peasant experience and scientific innovation. The effort has led to beneficial effects in the recovery of eroded and/or degraded areas, improved pasture productivity and livestock productive performance, and ability to sequester large amounts of carbon (Dube et al. 2015; Dube et al. 2016; Dube 2017).

Research on plantations and in native forests across the country showed the potential of silvopastoral systems as enablers of productive and social systems to greatly contribute to solving management problems of the Chilean forestry sector in rural areas (Sotomayor et al. 2002; Sotomayor and Cabrera 2008; Sotomayor et al. 2009; Sotomayor and Teuber 2011; Sotomayor and Soto 2011; Squella and Squella 2011; Dube et al. 2011; Sotomayor et al. 2012; Schmidt et al. 2013).

Such research documents the beneficial effects and advantages resulting from integrating tree and animal species into a silvopastoral system on livestock farms, agricultural dryland, and/or traditional Chilean forestry. Their contribution is reflected in earning income in the short and medium term, increased production of meadows or interleaved crops, production of high-value timber (at the end of rotation), higher value of land because of tree planting, recovery of degraded land, erosion control and water protection, mitigation of greenhouse gas effects and carbon capture, production of food with non-wood forest products, protection of livestock during winter (biological sheds), an aesthetically more pleasing landscape, and more wildlife.

The Forestry Institute (INFOR), through its National Agroforest Program, supported by the National Institute of Agricultural Research (INIA) and the Institute for Agricultural Development (INDAP), starting in 2002, has also achieved meaningful results. Studies among producers taking part in agroforestry projects highlighted, among others, 81.8% higher household incomes, 57.7% higher life quality and higher productivity, and 50.5% better erosion control. Energy supply (heating and feeding) became 44.13% more secure. Aspects such as incorporation of working sources and conservation of natural resources also appear as relevant factors (Sotomayor 2015).

The importance of native forests managed under silvopastoral systems relates to the productive capacity of grassland, livestock, and possibility of obtaining timber and non-wood products such as poles, rods and firewood, fungi, and fruits, among others, from silvicultural interventions. However, despite the great impact these systems have on the region, their adoption remains marginal. This can be attributed in part to the lack of technical knowledge regarding management plans to ensure a contribution to native forest—mainly regarding managing interactions between the system's components (Braun 2016)—insufficient public policies (Laclau 2015), financing and/or credit plans, and low economic incentives (Miranda et al. 2011). The latter represents a great barrier for the dissemination, adoption, and appropriation of silvopastoral technology, especially for small and medium-sized producers (Miranda et al. 2018).

The Recovery of Native Forest and Forest Development and Regulations Act (Law No. 20.283), enacted in 2008, remains the legal cornerstone. It has contributed to Chile standing out among developing countries for having drastically reduced deforestation through the plantation system and measures to counter the deterioration of native forest (FAO 2014a; FAO 2018).

Sustainability and sustainable development (SD) have been defined as "that development that meets the needs of the present without compromising the capacity of future generations to meet their own needs" (Brundtland 1987). But after publication of the Brundtland Report, debate on the meaning of sustainability and difficulties in the practical application of this concept emerged. Its ambiguous nature has led to different positions on how to make sustainability work and thus on the methods and instruments for evaluating it (Masera et al. 2000; Castro 2010; Arnés et al. 2013; Olmos and González 2013; Sarandón and Flores 2014). Yet sustainability is impossible to measure because of the term's ambiguity, its complexity stemming from its multifunctional character, timeframe, and lack of universal evaluation methodology (Sarandón et al. 2006a; Nahed 2008; Galván et al. 2008; Martín 2008; Sarandón and Flores 2009; Fernandez et al. 2010; Leccardi and Feixa 2011; González 2011; Sarandón et al. 2014; Barrezueta 2015).

Although no universal process to compare numerous sustainability measurement methods exists (Perez et al. 2016), initiatives to quantitatively evaluate sustainability at spatial and temporal scales plus its various environmental, social, and economic characteristics and conditions can be grouped into four approaches (Gómez-Limón 2010): (1) sustainability indicator analysis, (2) temporary productivity trend studies, (3) agricultural system resilience and sensitivity analysis, and (4) simulation techniques.

Each of these approaches has advantages and drawbacks. However, the scientific literature has weighed the peculiarities of each approach and has highlighted the first, involving the construction, calculation, and analysis of sustainability indicators, and its usefulness for translating complex sustainability variables into easy-to-interpret values (Dumanski et al. 1998; Masera et al. 2000; Galván et al. 2008; Kumaraswamy 2012; Arnés et al. 2013; Olmos and González 2013; Sarandón and Flores 2014; Martínez-Castro et al. 2015; Bustamante et al. 2017).

Although widely used, views on the best methodology to construct and select sustainability indicators diverge. Most investigations thus confront serious questions, since many exercises have been designed to measure global or regional functioning, making implementation at local geographical scales difficult, and often ignore cultural specificities and needs of the population or the context in which they develop. Therefore, they do not deliver rich and practical results.

In addition, lack of consensus on the weighting, categorization, and type of indicators to be integrated into the models for sustainability analysis creates confusion and facilitates bias. Researchers thus use methods they individually deem most pertinent, based on either literature, international standards, participatory forums, interviews, work with experts, or self-design (Glave and Escobal 2000; Delgado et al. 2010; Nahed 2008; Galván et al. 2008; Astier et al. 2008; Arocena 2009; Bolívar 2011; Ibáñez 2012; Acevedo-Osorio and Angarita 2013; Kú et al. 2013; Sarandón et al. 2014; Sarandón 2014; Barrezueta 2015; Bustamante et al. 2017; Silva-Santamaría and Ramírez-Hernández 2017). This means few empirical studies based on local knowledge and participation of farmers that have a stake in the production processes exist. All of this creates greater controversy and calls into question the relevance, coherence, and veracity of the proposed indicator models.

Although sustainability—and hence indicators to measure it—is difficult to define, attempting to do so is necessary, given the major challenges modern society faces to balance food production, socioeconomic growth, and conservation of natural resources and the environment. This requires the application of new holistic, systemic analytical approaches and validating them through case studies.

Along these considerations, research was carried out to build a preliminary set of economic, environmental, and social indicators in a participatory manner through methodological triangulation. This is a viable tool to simplify data, ensure the validity of results, mitigate bias and prejudice in the methodological framework, and obtain a preliminary base group of indicators with potential to help assessing the sustainability of a forest of old oak groves managed under silvopastoral conditions in Yungay commune in the Andean Precordillera.

## Materials and Methods

The study was carried out at the fiscal property called "Ranchillo Alto," on an estimated area of 673 hectares, extending into the El Avellano and Calabozo sectors of Yungay, Ñuble Region, Chile. The property is 33 km from the town of Yungay, 111 kilometers from Chillán, and 120 km from the regional capital Concepción. The research was holistic, systemic, and multidisciplinary and included a case study (Contreras 2002). It followed the principles of participatory action research (PAR) (Park 2005).

Indicators were built according to the "EVA" methodology of participatory sustainability assessment of peasant production systems as part of a research project called "Silvopasture in old oak groves with varying degrees of coverage as a sustainable management option at a native forest site in the Andean pre-cordillera, Ñuble Region, Chile." The research took place under the responsibility of the Faculty of Forestry Sciences at the University of Concepción (UdeC), with financial support from the National Forestry Corporation (CONAF). It included three major phases: Phase I Study of the Analysis Unit (E), Phase II Participatory Valuation of Sustainability (V), and Phase III Analysis and Feedback of Results (A). Each phase contained several stages. The present work will focus only on stage one (evaluation contextualization) and stage two (participative sustainability indicator generation) of Phase II, and for stage two only on the construction, identification, and selection of potential indicators.

*Participative Sustainability Indicator Generation*: This step refers to the development of a limited group of context-appropriate indicators that also work under similar conditions to assess progress toward sustainability and identify trends that favor sustainability and related risks to facilitate decision-making. Generating indicators is fundamental and determining for sustainability assessment, as only adequate indicator selection will ensure success (Loaiza et al. 2014). For this phase, the guidelines, recommendations, and suggestions proposed by Masera et al. (2000), Sarandón (2002), Astier and González (2008), Gallopin (2006), Nahed (2008),

Galván et al. (2008), Astier et al. (2008), Sarandón et al. (2006b), Bolívar (2011), Acevedo-Osorio and Angarita (2013), Sarandón et al. (2014), Sarandón et al. (2014), and Barrezueta (2015) proved useful.

**Stage 1: Evaluation Contextualization**

This comprised of the participatory definition of key elements in the evaluation process and was carried out through focused discussion groups (Aliaga et al. 2012). This approach allowed the community to build the conceptual framework of sustainability (Sarandón et al. 2014) to be used for the research. The community defined sustainable forest management as "that management and care of the forest ecosystem that guarantees all the products, goods and services it **provides permanently for the well-being of people and the current and future development of the community**" (*protection and care, union, food, health, future, justice, obedience, patience, stay in time, equality, forest care, use what the forest gives us, protect the soil, safety, food, always living from the forest, manage animals well, earn money, benefit for people*).

According to this framework, forest management under silvopastoral conditions must meet three fundamental aspects as objectives for the community: **sufficiently productive and economically viable** (*money*), **ecologically appropriate** (*care for the environment*), ***and*** being **socially and culturally acceptable** (*help improve local life quality*). Subsequently, the spatial and temporal framework of the evaluation was established (Gómez-Limón 2010; Toro-Mújica et al. 2011; Sarandón et al. 2014; Zarazúa et al. 2015). The community decided that the spatial scale will be at site level and the timescale range from 2015 to 2019. Residents also defined the sustainability assessment criterion to be used—weak, strong, or superstrong. After a debate, they opted for the superstrong sustainability approach (*people are most important, life quality of everyone improves*). This kind of sustainability coined the concept of "natural heritage" which considers that nature has other than just economic values, recognizing sociocultural, ecological, and mystical factors as equally important. In this sense, the development focus is not economic growth but life quality, positioning citizens as political subjects, where participation is a strategy of co-responsibility in development models (Arias 2017).

For the definition of evaluation levels, a hierarchical evaluation structure was built in a participatory manner, and ranged from general to particular or specific, considering Sarandón et al.'s (2014) assertion that sustainability is a multidimensional concept and must be "simplified" to understand it. Community members decided to form the hierarchical system with evaluation categories including (1) dimension (economic, environmental, social) and (2) indicators (economic, environmental, social). Under that participatory construct it was selected how to evaluate sustainability in the future (Sarandón and Flores 2014). Residents agreed to use longitudinal (over time), retrospective (What happened?), and prospective (What will happen?) comparative assessment. They reasoned that these aspects can be recognized in a cyclical evaluation-action-evaluation process to establish momentum for strengthening sustainability through constant feedback between the generation of alternatives and their evaluation (Astier et al. 2008).

In addition, the indicators were typified, defining the characteristics of those that should guide sustainability assessment (Galván et al. 2008). The population stipulated that these should be simple, mainly qualitative, and constructed through participatory work (Sotelo et al. 2011).

**Stage 2: Construction, Identification, and Selection of Potential Sustainability Indicators**

Refers to obtaining and selecting "potential" or "possible" indicators to determine the best candidates, once their ability to reflect qualitative and/or quantitative characteristics of the object under study is ascertained. They were selected from a menu built on previous empirical or scientific experiences in combination with an intuitive (subjective) on-site approach based on local knowledge of producers in a specific context. These strategies were identified by Glave and Escobal (2000) and belong to the group recommended for the construction of a preliminary or partial list and the incorporation of local knowledge systems as part of processes of technology appropriation and strengthening of rural communities.

This group of potential indicators was created with the help of methodological triangulation as an enriching tool to add rigor, depth, and complexity. It also enabled bias reduction and increased the understanding of the phenomenon (Okuda and Gómez-Restrepo 2005). One priority of methodological triangulation relates to enhancing validity so the potential local indicators have the required characteristics to use them in the case study and for sustainability assessments of production systems. Methodological triangulation contemplates consultation or bibliographic review of the area in which a systematic documentary review of data from 200 case studies (Table 1) took place. They appeared in scientific journals, conference reports, symposia, theses, and books from 2000 to 2019 and related to sustainability assessments, mostly in Latin America, via indicators in different systems (agricultural, livestock, forestry, agroforestry, aquaculture, etc.). Following Martínez-Castro et al. (2016), they must include in their analysis the assessment of some agroecosystem, the final list of case studies should be based on quality and quantity of available information, and they should have focused mostly on Latin America (the latter was modified because the author emphasized Mexico).

The second method was based on a consultation with producers and community members, in which participatory research (focused discussion groups) was carried out with residents of Ranchillo Alto and Los Avellanos (Aliaga et al. 2012), including brainstorming (Geilfus 2009). This was based on studies by Delgado et al. (2007), Fawaz and Vallejos-Carte (2011), Silva-Santamaría and Ramírez-Hernández (2017), and Bustamante et al. (2017). These authors emphasized that sustainability assessment must include the active participation of actors and rural communities so nothing is imposed on them.

In a workshop, all indicators resulting from bibliographic review were analyzed and discussed to familiarize producers with the subject and to present the sustainability assessment variables. From then on, participants were identifying on a flip chart the indicators that related most to their reality and perceived economic, environmental, and social needs. By brainstorming, the group decided vocally if the

**Table 1** Case studies consulted on the evaluation of sustainability in various types of agroecosystems in Latin America and other countries of the world

| N° | Country | System | Reference | Source |
|---|---|---|---|---|
| 1 | Mexico | Commercial agricultural | Aguilar-Jiménez et al. (2011) | Rev. FCA UNCUYO |
| 2 | Spain | Agroecological | Alonso and Guzmán (2004) | Agroecología |
| 3 | Mexico | Commercial agricultural | Sánchez-Morales et al. (2014) | Agroecología |
| 4 | Colombia | Commercial livestock | Arias-Giraldo and Camargo (2007) | Livestock Research for Rural Development |
| 5 | Nicaragua | Family farming | Arnes (2013) | Revista española de estudios agroecológicos y pesqueros, |
| 6 | Mexico | Family farming | Astier and González (2008) | International J. Sustainable Development & World Ecology |
| 7 | Mexico | Family farming | Astier et al. (2012) | Ecology and Society |
| 8 | Venezuela | Commercial livestock | Bechara-Dikdan et al. (2014) | Rev. Fac. Agron. (LUZ) |
| 9 | Chile | Family farming | Blanco et al. (2001) | Revista EURE |
| 10 | Argentina | Commercial agricultural | Blandi (2016) | Thesis |
| 11 | Argentina | Commercial agricultural | Blandi et al. (2015) | Revista de la Facultad de Agronomía |
| 12 | Venezuela | Commercial agricultural | Bolívar (2011) | CICAG |
| 13 | Mexico | Family farming | Bustamante et al. (2017) | Book chapter |
| 14 | Mexico | Commercial agricultural | Candelaria-Martínez et al. (2014) | Cuad. Desarro. Rural |
| 15 | Colombia | Commercial agricultural | Cardona and Granobles (2015) | Book chapter |
| 16 | Mexico | Commercial agricultural | Casas-Cázares et al. (2009) | Agrociencia |
| 17 | Mexico | Commercial agricultural | Castelán et al. (2014) | Ecosistemas y Recursos Agropecuarios |
| 18 | Mexico | Family livestock | Castillo et al. (2012) | Revista Científica UDO Agrícola |
| 19 | Italy | Agroecological | Certomà and Migliorini (2005) | Series book: Environmental Earth Science |
| 20 | Mexico | Livestock | Cruz-Mendoza et al. (2016) | Revista Mexicana de Agroecosistemas |
| 21 | Nicaragua | Commercial agricultural | De Miguel et al. (2009) | Book chapter |
| 22 | Venezuela | Farming | Delgado et al. (2010) | Agroalimentaria |

(continued)

**Table 1** (continued)

| | | | | |
|---|---|---|---|---|
| 23 | Colombia | Farming | Díaz and Valencia (2010) | Revista de Investigación Agraria y Ambienta |
| 24 | Philippines | Forest | Dolom (2003) | Unasylva |
| 25 | Uruguay | Commercial agricultural | Dieste (2011) | Thesis |
| 26 | Honduras | Agroforestry | Duarte (2005) | Thesis |
| 27 | Argentina | Agroforestry | Escribano et al. (2014) | ITEA 110 |
| 28 | Chile | Family farming | Fawaz and Vallejos (2011) | Cuad. Desarro. Rural |
| 29 | Argentina | Agroecological | Flores and Sarandón (2015) | Rev. Fac. Agron. La Plata |
| 30 | Mexico | Livestock | Espinosa et al. (2004) | Téc. Pecu. Méx. |
| 31 | Salvador | Agroforestry | Estrada (2014) | Thesis |
| 32 | Brazil | Agricultural | Fernández et al. (2010) | Administração e Sociologia Rural |
| 33 | Argentina | Commercial agricultural | Flores et al. (2007) | Rev. bras. Agroecologia |
| 34 | Colombia | Family farming | Fonseca-Carreño et al. (2015) | Revista Ciencia y Agricultura |
| 35 | Argentina | Commercial livestock | Gaeta and Muñoz (2014) | Ciencias Agronómicas |
| 36 | Argentina | Livestock | García (2009) | ITEA |
| 37 | Mexico | Commercial agricultural | Gerritsen and González (2008) | Reports/bulletin |
| 38 | Colombia | Family farming | Giraldo-Díaz et al. (2015) | Libre Empresa |
| 39 | Mexico | Commercial agricultural | Gutiérrez (2006) | Espacio y Desarrollo |
| 40 | Argentina | Agricultural/ livestock | Loewy (2008) | Revista Iberoamericana de Economía Ecológica |
| 41 | Ecuador | Commercial agricultural | Luna (2016) | Thesis |
| 42 | Argentina | Commercial agricultural | Manzoni (2015) | Memorias congreso |
| 43 | Mexico | Commercial agricultural | Mazabel-Domínguez (2010) | Revista de Sociedad, Cultura y Desarrollo Sustentable |
| 44 | Uruguay | Livestock | Molina (2008) | Ganadería |
| 45 | Chile | Forest | Glaría (2013) | Polis Revista Latinoamericana |
| 46 | Peru | Agroecological | Gomero and Velásquez (2003) | LEISA Revista de Agroecología |
| 47 | Uruguay | Family farming | Molina (2016) | Familias y campo. Rescatando estrategias de adaptación: |
| 48 | Nicaragua | Agroforestry | Morán (2014) | Medio ambiente, Tecnología y Desarrollo Humano, |

(continued)

**Table 1** (continued)

| | | | | |
|---|---|---|---|---|
| 49 | Brazil | Agroecological | Moura (2002) | Thesis |
| 50 | Mexico | Agricultural | López (2015) | Revista Científica Ecociencia |
| 51 | Mexico | Family agriculture | Moya et al. (2003) | LEISA Revista de Agroecología: |
| 52 | Argentina | Livestock | Nasca et al. (2006) | Zootecnia tropical |
| 53 | Brazil | Agricultural | Oliveira et al. (2009) | Rev. bras. De Agroecología |
| 54 | Spain | Agroecological | Rodríguez (2015) | Thesis |
| 55 | Chile | Agroecological | Pino et al. (2011) | Producción hortofrutícola orgánica |
| 56 | Mexico | Commercial agricultural | Priego-Castillo et al. (2009) | Universidad y Ciencia |
| 57 | Costa Rica | Commercial agricultural | Ramírez et al. (2008) | Agronomía Costarricense |
| 58 | Mexico | Family farming | Romero et al. (2011) | Revista de Geografía Agrícola |
| 59 | Spain | Commercial agricultural | Sánchez (2009) | Thesis |
| 60 | Mexico | Commercial agricultural | Sánchez (2012) | Thesis |
| 61 | Mexico | Natural resources | Rodríguez (2015) | Studia politicae |
| 62 | Argentina | Commercial agricultural | Sarandón et al. (2006) | Revista Brasilera de Agroecología, |
| 63 | Argentina | Agricultural | Sarandón et al. (2006) | Revista Agroecología |
| 64 | Cuba | Commercial agricultural | Silva (2014) | Thesis |
| 65 | Mexico | Aquifer | Neri-Ramírez et al. (2013) | RCHSCFA |
| 66 | Mexico | Farming | Brunett et al. (2005) | Livestock Research for Rural Development |
| 67 | Colombia | Agroecological | Varela (2010) | Book chapter |
| 68 | Italy | Agricultural/farm | Tellarini and Caporali (2000) | Agriculture, ecosystems and environment: |
| 69 | Peru | Agricultural | Glave and Escobal (2009) | Newsletter/technical reports |
| 70 | Mexico | Diversified | López-Ridaura (2001) | Newsletter/technical reports |
| 71 | Uruguay | Forest | Crosara (2001) | Thesis |
| 72 | Mexico | Diversified | López-Ridaura (2002) | Ecological indicators |
| 73 | Mexico | Agricultural | Astier et al. (2003) | Revista de Agroecología |
| 74 | Mexico | Diversified | Speelman et al. (2007) | International J. Sustainable Development & World Ecology |
| 75 | Mexico | Agricultural | González (2006) | Convergencia [online] |
| 76 | Costa Rica | Farming | Rodríguez (2006) | Revista Pensamiento Actual |
| 77 | Bolivia | Natural resources | Delgadillo and Delgado (2003) | LEISA Revista de Agroecología |

(continued)

**Table 1** (continued)

| | | | | |
|---|---|---|---|---|
| 78 | Argentina | Agricultural | Strassera et al. (2009) | Rev. Bras. de Agroecologia |
| 79 | Bolivia | Family farming | Frías and Delgado (2003) | LEISA Revista de Agroecología |
| 80 | Chile | Natural resources | Páez (2003) | Revista MAD |
| 81 | Mexico | Agroforestry | Alemán et al. (2003) | LEISA Revista de Agroecología |
| 82 | Panama | Agricultural/farms | Castillo (2004) | Investig. Pens. Crit |
| 83 | Argentina | Farming/farms | Vega et al. (2015) | RIA |
| 84 | Argentina | Agricultural | Wehbe et al. (2009) | Newsletter/technical reports |
| 85 | Argentina | Agricultural/ regional | Flores and Sarandón (2006) | Revista Brasilera de Agroecología |
| 86 | Mexico | Family farming | Neri et al. (2008) | Ra Ximhai Rev. Sociedad, Cultura y Desarrollo Sustentable |
| 87 | Argentina | Agroecological | Dellepiane and Sarandón (2008) | Revista Brasileira de Agroecologia |
| 88 | Mexico | Agroforestry | Nahed (2008) | Avances en investigación agropecuaria (AIA) |
| 89 | Panama | Forest | Chifarelli (2008) | Congress book of abstracts |
| 90 | Chile | Polycultures | Vega (2009) | Thesis |
| 91 | Uruguay | Agricultural/ livestock | Albicette (2009) | Agrociencia |
| 92 | Argentina | Family farming | Cáceres (2009) | Agrociencia |
| 93 | Peru | Agricultural | Paiva (2009) | Rev. Bras. De Agroecologia |
| 94 | Mexico | Natural resources | Uribe (2009) | Thesis |
| 95 | Bolivia | Agroforestry | Gruberg and Azero (2009) | Acta Nova |
| 96 | Costa Rica | Agroecological | Fallas et al. (2009) | Cuadernos de Investigación UNED |
| 96 | Spain | Family farming | Pilarte (2010) | Revista Científica-FAREM Estelí/ Ciencias Ambientales |
| 97 | Spain | Agricultural | Gómez-Limón (2010) | CUIDES |
| 98 | Paraguay | Agricultural | Vargas Insfrán (2010) | Investigación Agraria |
| 99 | Mexico | Natural resources | García (2010) | Book chapter |
| 100 | Chile | Agricultural | Loyola and Rivas (2010) | Tiempo y Espacio |
| 101 | Venezuela | Pastures | Lok (2010) | RET. Revista de Estudios Transdisciplinarios |
| 102 | Nicaragua | Family farming | Arnés (2011) | TFM |
| 103 | Venezuela | Livestock | Delgado et al. (2007) | Gaceta de Ciencias Veterinarias |
| 104 | Spain | Agroecological | Mestre (2011) | Thesis |
| 105 | Colombia | Commercial agricultural | Ramírez et al. (2014) | Rev. Fac. Nal. Agr. Medellín |
| 106 | Spain | Livestock | Toro (2011) | Thesis |

(continued)

**Table 1** (continued)

| | | | | |
|---|---|---|---|---|
| 107 | Mexico | Agroecological | Gutiérrez et al. (2012) | Tropical and Subtropical Agroecosystems |
| 108 | Argentina | Farming | Lageyre (2012) | Thesis |
| 109 | Cuba | Agricultural | Silva-Santamaría et al. (2017) | Luna Azul |
| 110 | Ecuador | Ecological farm | Cruz et al. (2016) | Livestock Research for Rural Development |
| 111 | Argentina | Livestock | Otta and Quiroz (2016) | Rev. FCA UNCUYO |
| 112 | Ecuador | Agricultural | Guerra (2016) | Thesis |
| 113 | Mexico | Agricultural | Pastor et al., (2016) | Entreciencias |
| 114 | Ecuador | Agricultural | Armijos (2016) | Thesis |
| 115 | Argentina | Agricultural | Viani et al. (2015) | Memorias congreso |
| 116 | Colombia | Agricultural | Machado et al. (2014) | IDESIA (Chile) |
| 117 | Ecuador | Agricultural | Barrezueta (2016) | Thesis |
| 118 | Ecuador | Agricultural | Armijos (2016) | Thesis |
| 119 | Mexico | Agricultural | Montejo (2015) | Dzemocut, Yucatán Economía |
| 120 | Peru | Agricultural | Meza and Julca (2015) | Ecología Aplicada |
| 121 | Peru | Agricultural | Collantes and Rodríguez (2015) | Tecnología & Desarrollo |
| 122 | Argentina | Agricultural | Vásquez and Vignolles (2015) | Soc. & Nat. Uberlândia |
| 123 | Mexico | Agricultural | Martínez-Castro et al. (2015) | International journal |
| 124 | Peru | Agricultural | Márquez and Julca (2015) | Revista de la Facultad de Ingeniería de la USIL |
| 125 | Brazil | Agroecological | De Ataíde et al. (2015) | Ambiente & Sociedad São Paulo, XVIII |
| 126 | Argentina | Agricultural | Sarandón et al. (2014) | Book chapter |
| 127 | Colombia | Agricultural | Loaiza et al. (2014) | Colombia Forestal |
| 128 | Ecuador | Agricultural subsistence | Villavicencio (2014) | Thesis |
| 129 | Colombia | Farming | Quiroz et al. (2014) | Thesis |
| 130 | Ecuador | Agroecological | Chango (2014) | Thesis |
| 131 | The Savior | Agroecological | Escobar (2014) | Thesis |
| 132 | Peru | Agroecological | Alvarado (2013) | Natura@economía |
| 134 | Mexico | Livestock | Domínguez (2014) | Revista Científica Biológico-Agropecuaria Tuxpan, |
| 135 | Argentina | Agricultural | Abraham et al. (2014) | Rev. FCA UNCUYO |
| 136 | Bolivia | Agroforestry | Wilkes (2013) | Thesis |
| 137 | Colombia | Agroecological | Acevedo-Osorio, Angarita (2013) | Book chapter |

(continued)

**Table 1** (continued)

| | | | | |
|---|---|---|---|---|
| 138 | Mexico | Agricultural | Kú et al. (2013) | Avances en Investigación Agropecuaria, |
| 139 | Mexico | Livestock | Domínguez 2013 | Revista Iber. para la Inv. y el Desarrollo Educativo |
| 140 | Peru | Agricultural | Merma and Julca (2012) | Ecología aplicada |
| 141 | Venezuela | Agricultural | Gravina and Leiva (2012) | Cultivos tropicales |
| 142 | Argentina | Agricultural | Castro (2010) | Quebracho |
| 143 | Peru | Agricultural subsistence | Paiva and Greta (2009) | Rev. bras. De Agroecología |
| 144 | Venezuela | Agricultural | Bolivar (2010) | Revista vinculando |
| 145 | Mexico | Agroecological | Gutierrez et al. (2011) | Tropical and Subtropical Agroecosystems |
| 146 | Colombia | Livestock | Ríos (2010) | Thesis |
| 147 | Argentina | Agricultural | Abraham et al. (2014) | Rev. FCA UNCUYO |
| 148 | Mexico | Agricultural subsistence | Bustamante et al. (2012) | Cuadernos de trabajo de la UACJ |
| 149 | Colombia | Agroecological | Varela (2010) | Thesis |
| 151 | Colombia | Agricultural subsistence | Guzmán (2016) | Tesis |
| 152 | Peru | Agricultural | Ilasaca et al. (2018) | Revista de Investigaciones Altoandinas |
| 153 | Mexico | Agricultural | Torres et al. (2008) | Revista Región y Sociedad |
| 154 | Chile | Agricultural | Peredo et al. (2016) | IDESIA (Chile) |
| 155 | España | Livestock | García-Diez et al. (2011) | Revista científica de la Sociedad Española de Acuicultura |
| 156 | Ecuador | Agroecological | Alvarado W. (2013) | Tesis |
| 157 | Chile | Agricultural subsistence | Mora M. A (2015) | Tesis |
| 158 | Mexico | Livestock | Montes- Perez et al. (2016) | Revista Abanico Veterinario |
| 159 | Mexico | Agricultural | Primo et al. (2014) | Revista Agroecología 9 |
| 160 | Mexico | Agricultural | Garcés et al. (2018) | Revista Latinoamericana el Ambiente y las Ciencias |
| 161 | Ecuador | Agricultural | Clavijo y Cuvi (2017) | Letras Verdes. Revista Latinoamericana de Estudios Socioambientales |
| 162 | Mexico | Agroecological | Gastón et al. (2011) | Tropical and Subtropical Agroecosystem |
| 163 | Colombia | Agricultural | Figueroa (2016) | Revista Tendencia |
| 164 | Peru | Agricultural | Espinola et al. (2017) | REVISTA INVESTIGACION OPERACIONAL |
| 165 | Uruguay | Livestock | Oyhantçaba (2010) | Tesis |
| 166 | Mexico | Agricultural | González et al. (2014) | Revista Mexicana de Agronegocios |

(continued)

**Table 1** (continued)

| | | | | |
|---|---|---|---|---|
| 167 | Mexico | Agricultural | Ayala y Guerrero (2009) | Revista Iberoamericana de Economía Ecológic |
| 168 | Peru | Family farming | IICA (2017 | Libro |
| 169 | Colombia | Agricultura organica | Ortiz (2017) | Tesis |
| 170 | Mexico | Agricultural | Neri et al. (2008) | Ra Ximhai Revista de la Sociedad y cultura Sustentable |
| 171 | Ecuador | Agricultural | Viteri (2013) | Tesis |
| 172 | Mexico | Agroecologicos | Álvarez (2015) | Tesis |
| 173 | Colombia | Agricultural | Garzón y López (2017) | Tesis |
| 174 | Peru | Agricultural | Pinedo-Taco et al. (2018) | Revista Ecosist. Recur. Agropec. |
| 175 | Venezuela | Agricultural | Bolívar (2010) | Revista VINCULANDO |
| 176 | Argentina | Agricultural | Suarez J. C. (2003) | Tesis |
| 177 | Colombia | Agricultural | Herney (2018) | Tesis |
| 178 | Venezuela | Agricultural | Hernández y Leyva (2012). | Cultivos tropicales |
| 179 | Ecuador | Agricultural | *Gaibor* et al. *(2017)* | Revista DELOS: Desarrollo Local Sostenible |
| 180 | Nicaragua | Agricultural | Prieto et al. (2013) | Revista Española de Estudios Agrosociales y Pesqueros |
| 181 | Uruguay | Livestock | García (2009) | Rev. bras. De Agroecologia |
| 182 | Brazil | Family farming | Nicoloso et al. (2015) | Revsita AIDA |
| 183 | Uruguay | Family farming | Chiappe et al. (2008) | Memorias congresos |
| 184 | Ecuador | Agricultural | Bravo Medina et al. (2017) | Revista bioagro |
| 185 | Peru | Family farming | Chavez (2016 | Tesis |
| 186 | Mexico | Forest | Cano et al. (2007) | Inifap Capitulo de libro |
| 187 | Colombia | Livestock | Ruiz et al. (2017) | Livestock Research for Rural Development |
| 188 | Peru | Livestock | Culquimboz (2017) | Tesis |
| 189 | Mexico | Agricultural | Velazquez (2017) | Revista Espacio I + D (Innovación más Desarrollo) |
| 190 | España | Agricultural | Fernadez et al. (2010) | Memorias congresos |
| 191 | Colombia | Agricultural | Cerón et al. (2014) | Revista Colombia Forestal |
| 192 | Peru | Agricultural | Calle (2018) | Tesis |
| 193 | Argentina | Agricultural | Fontana (2010) | Tesis |
| 194 | Mexico | Agricultural | Morales y Romero (2018) | Revista de El Colegio de san Lui |
| 195 | Argentina | Livestock | Bonnefon et al. (2016) | Tesis |
| 196 | Ecuador | Agricultural | Palomeque (2015) | Tesis |

(continued)

**Table 1** (continued)

| | | | | |
|---|---|---|---|---|
| 197 | Peru | Agricultural | Barreto (2017) | Tesis |
| 198 | Cuba | Agroturismo | Pérez et al. (2009) | Revista Investigación Operacional |
| 199 | Argentina | Livestock | Nazca et al (2006) | Capítulo de libro |
| 200 | Argentina | Commercial agricultural | Canelada et al. (2015) | Rev. Agron. Noroeste Argent |

indicators identified were of high, medium, or low importance, and which ones should be discarded. They could also propose new, context-specific indicators.

Expert consultation or judgment, involving specialist views on a certain aspect, represented the final validation method (Cabero and Llorente 2013). The consultation developed through the interview technique or structured dialog (survey), which resorted to individual, anonymous, and confidential aggregation. Under this technique, 30 academics—mostly from Biobío Region or Latin American countries other than Chile—with scientific backgrounds in economics, environmental sciences, and agroforestry, among others, judged all indicators coming from the bibliographic review.

Criteria suggested by Rodríguez (2006) and Escobar-Pérez and Cuervo-Martínez (2008) guided this process. They included characteristics for expert selection, optimal number of judges, and steps to follow in the consultation. A simple model was designed, in which the experts determined the indicators most important for sustainability assessment. They listed indicators according to a previously assigned value, with (0) equaling rejection/irrelevance, (1) low importance/relevance, (2) medium importance/relevance, and (3) high importance/relevance.

## *Statistical Analysis*

Several statistical strategies were employed for the processing and analysis of data for obtaining, selecting, and prioritizing indicators through methodological triangulation. For bibliographic consultation, data were processed in an Excel spreadsheet, analyzing the indicators by frequency of use or occurrence within the sample studied, using descriptive statistics. Data from the consultation with producers was analyzed qualitatively. Expert judgment analysis involved the degree of agreement among experts, using Kendall's external W concordance coefficient. This is a nonparametric test, commonly used in the social sciences. In case of low concordance, the item was adjusted or removed until the desired measurement objective was achieved (Escobar-Pérez and Cuervo-Martínez 2008). Assessment led to values between 0 (total disagreement) and 1 (total agreement). The higher the Kendall value, the stronger the association that validates the instrument.

Automatic classification analysis (conglomerates) helped to strengthen the data derived in this step. This means indicators were grouped trying to achieve maximum homogeneity in each group to emphasize the greatest difference between the groups as a selection strategy (Van der Kloot et al. 2005). Finally, the results of each consultation (bibliography-experts-producers) went into a database and were submitted to automatic classification analysis (conglomerates), using Euclidean distance as a differentiation criterion. This made the most relevant indicators and their relationships visible. Statistical package SPSS v. 23 (IBMCorp. Released 2017) was used for data analysis and processing.

## Results and Discussion

### *Results of Bibliographic Consultation*

From the bibliographic review, 244 indicators were obtained that are used mainly—but not exclusively—in Latin America to evaluate the sustainability of production systems through case studies (Table 2). Results showed that indicators are heterogeneous and particular to the processes of which they form part, since local climate, topography, economic and cultural relations, and local history are never universal. In this step the claim of Astier and González (2008) and Sarandón et al. (2014) was corroborated, who posited that these factors make finding a unique and universal concept of sustainability—and even more so benchmarks for its evaluation—difficult.

For the economic dimension 74 indicators were found. Among high, medium, infrequent, and very infrequent use in the case studies, the most common ones—considered also the most representative—were determined. This group comprised 20 items ranging from 14 to 54% of frequency (Table 2). The most frequently cited (>20%) were *income/expenses, productivity/productive efficiency, commercialization and marketing, dependence on economic activity, product diversity for sale, benefit/cost relationship, and economic return.*

A multitude of parameters that measure economic viability in the short, medium, and long term exist already. Most of the economic indicators obtained are commonly used to determine the conditions that an activity, project, etc. must meet to be economically profitable while others are linked to specific situations of each system and problematic. However, most frequently used indicators were recommended in the topic-specific literature for assessing the benefits and impacts of any economic activity.

For the environmental dimension 79 indicators were found. When dividing indicators into groups characterized by high or low frequency of use (Table 2), 22 indicators emerged with 52–14% of frequency, 13 with values above 20%. They included *use of recursos naturales conservation practices, biodiversity, soil chemistry, % of plant cover of the soil, agrochemical and fertilizer usage, erosion and soil loss risks, water quality, (8) soil management, number of agrodiversity species, energy efficiency, agricultural/animal/forest management practices, soil fauna conservation, growing health status, and animals/woods.*

**Table 2** Results of the bibliographic consultation of the indicators according to their frequency of use in the evaluation of small farmers (campesinos) sustainability in Latin America

| Dimensions | High frequency of use | % of use | Medium frequency of use | % of use | Low frequency of use | % of use | Very low frequency of use | % of use |
|---|---|---|---|---|---|---|---|---|
| **Economic** | Income/expenses, productivity/productive efficiency, marketing and commercialization, dependence on economic activity, diversity of products for sale, (6) profit/cost relationship, economic profitability, quantity of agrodiversity species in production/for production, income diversification, income, profit/profit/net income, access to credits/other types of external financing, product quality, economic risk, self-financing capacity, level of use/dependence on inputs and external resources, net present value, relationship supply/demand, period/time of recovery of investment, production costs/other types of costs | Between 54 and 14% | Economic value of land, annual distribution of income/expenses, profitability over time, gross margin, investment, economic benefits, economic well-being, availability of pasture, value added to products, savings capacity, improvement of the family economy, evaluation financial, prices/fluctuations, economic value of sales/production/products, internal rate of return | Between 13.9 and 7% | Economic value of risk, gross domestic product, contribution to GDP, income/opportunity cost ratio, postharvest practices, planting planning and operation/production records, economic stability, family financial contribution, type of financing, technical efficiency, improvement crop/animal/tree genetics, source of income, amount of labor used in production, integration into production chains, current and potential markets, economic efficiency | Between 6.9 and 3% | Economic value of risk, gross domestic product, contribution to GDP, income/opportunity cost ratio, postharvest practices, planting planning and operation/production records, economic stability, family financial contribution, type of financing, technical efficiency, improvement crop/animal/tree genetics, source of income, amount of labor used in production, integration into production chains, current and potential markets, economic efficiency | Less than 3% |
| **Total** | **20** | | **15** | | **16** | | **23** | |

| | | | | | | | | |
|---|---|---|---|---|---|---|---|---|
| **Envi-ronmen-tal** | Use of practices for conservation of natural resources, biodiversity, soil chemistry, % of plant soil cover, level of use of agrochemicals and fertilizer, erosion and risks of soil loss, water quality, soil management, amount of agrodiversity species present, energy efficiency, agricultural/animal/forestry management practices, conservation of edaphic fauna, sanitary status of crop, animals/forests, degree or level of soil compaction, rate of sown land/area, alternative/agroecological/sustainable technologies in use, evaluation of the behavior of the components of the system, diversification of the land, soil fertility, deforestation/reforestation activities, soil physics, water management | Between 52 and 14% | % of areas intervened/degraded by anthropic action, irrigation water quality, area under AFS/diversification, evaluation of climate behavior, risk of contamination, domestic waste and production management, areas/area under conservation/areas of biological support, agroforestry management practices, air quality, ability to adapt to ecological/environmental changes, consumption and efficiency in the use of fossil energy, use of local inputs or resources, current status of the natural resources, use of alternative energy/alternative energy sources, recycling of waste, energy balance, % of natural regeneration of vegetation, botanical composition, diversity of habitats, % of slopes/landslides/landslides | Between 13.9 and 7% | Support activities for agroecology/organic/sustainable production, exploitation level of the natural resources (NR), degree of alteration or degradation of the natural resources/ecosystem, ecotoxicity, diversity and visual quality of the landscape/conserved landscape, % of natural or nonintervention areas, awareness activities for conservation and preservation of environment and NR, interaction/interrelation between the components of the system, greenhouse effect/carbon footprint, % of reduction of degraded or recovered areas, risk of salinization, environmental impact, soil productivity, capacity to use or load the land, forest management, environmental education, fragility of the ecosystem, payment for environmental services | Between 6,9–3% | Animal welfare, physiology of the crop/animal, type of exploitation, consumption and efficiency in the use of water/ecological footprint, life cycle analysis (LCA), carbon sequestration/fixation, global warming, compliance with environmental regulations/legislation, level of optimization in the use of NR, evaluation of the strata of the system, forest certification, level of eutrophication, heterogeneity of the landscape, forest restoration, % of forest cover, biological soil activity/soil biology, acidification potential, ratio of use of renewable/nonrenewable energy, % of slopes/landslides, photochemical contamination | Less than 3% |
| **Total** | **22** | | **20** | | **18** | | **19** | |

(continued)

**Table 2** (continued)

| Dimensions | High frequency of use | % of use | Medium frequency of use | % of use | Low frequency of use | % of use | Very low frequency of use | % of use |
|---|---|---|---|---|---|---|---|---|
| **Social** | Social participation/integration, capacity building/training/training activities, innovation capacity/ technology adoption/resistance to change, degree or level of social/ community organization, level of external dependence, generation of jobs, self-sufficiency and food security, satisfaction for innovation/technology/ desire for technological adoption, level or degree of satisfaction of basic needs, level of empowerment, level of environmental awareness/vision/ relationship with nature, use of local knowledge or knowledge, level of self-management, government/ institutional support, equitable distribution of system benefits, degree of family integration, participation of women, quality of life, educational level, access and quality of basic services, land tenure regime, equity of gender, area destined to production for self-consumption/subsistence, capacity to adapt to intra- and extra-farm changes, need for extra-farm employment, use and quality of technical assistance, generational transfer/inheritance of land/knowledge, unemployment rate/level | Between 50 and 13% | Demand/supply of internal and external labor, capacity for self-sufficiency, health status/risks, desires to remain in place, effective availability of inputs/resources, satisfaction with productive activity/risks of abandonment, migration, extension activities and technological transfer, equitable distribution of land, socialization of knowledge, autonomy in the purchase and use of productive inputs/ resources, presence of conflicts, use of family labor, free time dedicated to leisure and distraction, existence of claims and resilience, sense of belonging and identity, cultural diversity, labor law/ legal support for workers/ labor justice, autonomy and income management, capacity for teamwork, sense/level of cooperativism and solidarity, ties of neighborhood, desires to stay in the place | Between 12,9–6% | Level of technification of the producer, access and equitable use of the natural resources of the sector, occupational health and safety/risks, nutrition/nutritional quality of food, labor organization/unions, conflicts over the use of land, control over the system and decision-making, efficiency in the organization, working conditions, satisfaction with the role/ performance of the leader, level of self-esteem, compliance with the rules of coexistence and good customs, cultural erosion/loss of local knowledge, personological profile of the producers/elements, psychosocial, citizen security | Between 5,9–3% | Recovery of knowledge/culture/ traditions, level of importance of social capital, level of leadership, social function of the company, promotion of peasant research, accessibility to the property or community/state of the access roads, satisfaction with the quality of the environment and natural resources, work remuneration capacity, economic level, seasonality of the workforce, cultural practices/ cultural support and promotion activities, ritual practices, level of intensification of productive activity, number of ethnic groups involved, strategies to overcome problems, expropriation and concentration of land, life expectancy, democracy in management and decision-making, level of paternalism, continuity/seasonality of work, need to hire specialized personnel, number of wages/ salaries, need for intermediaries, relationships/exchange with regional institutions, human development index | Less than 3% |
| **Total** | **28** | | **23** | | **15** | | 25 | |

Thus, it was possible to identify numerous highly complex indicators for the evaluation process, because many aspects were incorporated that appear as determining factors of environmental sustainability in the scientific literature. The ecological value of sustainability relates to ensuring the permanent availability of ecosystem functions. This is because the indicators are related to various edaphic factors, water, energy, ecological, biological, productive, technological, management, legislative, political, and other aspects. Such relations generate multiple interactions and dynamics between functions, values, and processes so cause-and-effect relationships between the production of the agroecosystem and ecological processes appear, evidencing the possible impacts of production models on ecosystems and sustainability trends of ecosystems (Altieri 1999; Altieri et al. 2000).

For the social dimension 91 indicators were counted. Of these, 28 were derived with frequencies of 50–13% (Table 2). The group highlights 20 indicators with at least 20% frequency: *participation/social integration, capacity building/training activities, innovation capacity/adoption of technology/resistance to change, degree of social/community organization, external dependency level, job creation, self-reliance and food security, innovation satisfaction/technology/technological adoption wishes, degree of satisfaction of basic needs, level of empowerment, environmental awareness/vision/relationship with nature, use of local knowledge, self-management level, government/institutional support, equitable system benefit sharing, degree of family integration, female participation, quality of life, educational level, and access to and quality of basic services.*

In this dimension, most indicators were deemed more important than economic and environmental ones. This confirms that a large number of works deal not just with the ecological approach. The importance of the social approach over the economic one highlights a trend to break with an economistic paradigm. In theory, a consensus exists on the inability of conventional indicators such as gross domestic product (GDP) and per capita income to assess development in terms of "human satisfaction." In practice, these indicators are still used almost exclusively not just to measure human development and life quality, but also to design, plan, and implement development policies.

Such importance of the social approach underscores the role of communities, local institutions, and family farming, which are essential to food security and the economic, commercial, and social dynamics of local territories (Fawaz and Vallejos-Carte 2011). The social approach also involves the need to keep encouraging the development of more holistic and systemic indicators that stem from local knowledge, experiences, and most urgent needs of residents.

## *Expert Consultation or Judgment Results*

Only the most relevant results, determined through Kendall's W coefficient (Table 3), are presented. For the economic dimension, 74 indicators were evaluated, of which the experts considered 62, while 12 had a significant W value of at least

**Table 3** Coefficient of agreement among experts (W Kendall) in the selection of indicators of the economic, environmental, and social dimensions

| Dimension | Indicator | W Kendall[a] (*Chi squared*) | Significance (*P*) |
|---|---|---|---|
| Economic | Farm planning/records | 0.338 (13.200) | 0.004 |
| | Dependence on external inputs | 0.573 (14.556) | 0.002 |
| | Income diversification | 0.436 (17.000) | 0.001 |
| | Access to credit/other funding sources | 0.510 (19.889) | 0.000 |
| | Income/expenses | 0.585 (22.800) | 0.000 |
| | Productivity | 0.510 (19.889) | 0.000 |
| | Commercialization/marketing | 0.769 (30.000) | 0.000 |
| | Cost-effectiveness | 0.585 (22.800) | 0.000 |
| | Economic feasibility | 0.363 (14.143) | 0.003 |
| | Dependence on economic activity | 0.510 (19.889) | 0.000 |
| | Diversity of products for sale | 0.338 (13.200) | 0.004 |
| | Technical efficiency | 0.510 (19.889) | 0.000 |
| | Carbon sequestration/fixation | 0.372 (14.556) | 0.002 |
| Environmental | Biodiversity | 0.436 (17.000) | 0.001 |
| | Soil management | 0.462 (18.000) | 0.000 |
| | Agricultural management practices | 0.510 (19.909) | 0.000 |
| | Amount of agrodiversity | 0.338 (13.200) | 0.004 |
| | Deforestation/reforestation activities | 0.363 (14.143) | 0.003 |
| | Erosion/risks of soil loss | 0.583 (21.000) | 0.000 |
| | Conservation practices of natural resources | 0.385 (15.000) | 0.002 |
| | Chemistry and soil fertility/quality | 0.583 (21.000) | 0.000 |
| | Degree of satisfaction of needs | 0.325 (12.667) | 0.005 |
| Social | Use of local knowledge | 0.350 (13.667) | 0.003 |
| | Ability to adapt to changes | 0.338 (13.200) | 0.004 |
| | Job creation | 0.441 (17.200) | 0.001 |
| | Self-management | 0.357 (15.000) | 0.002 |
| | Satisfaction for technology | 0.510 (19.889) | 0.000 |
| | Migration | 0.692 (27.000) | 0.000 |
| | Community organization | 0.538 (21.000) | 0.000 |
| | Food self-sufficiency | 0.385 (15.000) | 0.002 |
| | Government support | 0.462 (18.000) | 0.000 |
| | Capacity building | 0.585 (22.800) | 0.000 |
| | Quality of life | 0.692 (27.000) | 0.000 |

$P \leq 0.001$ highly significant; $P \leq 0.01$ fairly significant

[a]Under the context of this research, the values of W $\geq$ 0.400 are considered to be of relevant agreement

0.4. These were *income/expenses, commercialization and marketing, level of dependence on economic activity, diversity of products for sale, economic profitability, level of dependence on inputs and external resources, productivity, technical efficiency, access to credits, income diversification, estate planning, and economic/*

*financial feasibility*. Experts agreed most frequently on *income/expenses, commercialization and marketing, economic profitability, productivity, level of dependence on economic activity, technical efficiency, access to credits, or other sources of external financing*.

A similar situation emerged in the environmental dimension. Overall, of the 79 indicators evaluated, 64 were taken into account. Experts agreed most frequently (W coefficient >0.4) on *practices of natural resource conservation, erosion and risk of soil loss, soil management, soil chemistry and fertility, carbon sequestration/fixation, reforestation/afforestation, agricultural/animal/forest management practices, number of agrodiversity species present, and biodiversity*. As remarkably consistent stood out: *level of erosion and risk of soil loss, soil chemistry and quality, soil management, and agricultural/animal/forest management practices*.

In the social dimension 91 indicators were suggested, 64 were evaluated, and 12 were most frequently agreed on: *quality of life, job creation, capacity building, community organization, satisfaction with technology and innovation, migration, use of local knowledge, government support, adaptability and response to changes, self-management, level or degree of satisfaction of basic needs, self-sufficiency, and food security*. High concordance was found for *quality of life, capacity building, community organization, satisfaction with technology, migration*, and *government support*.

None of the indicators had an absolute level of concordance. In general, a wide range of low values and certain medium cases of evaluation emerged, suggesting little overall agreement among experts. This low level of concordance can be attributed to a weak holistic view detected in the survey results, as several specialists did not judge outside their field of expertise. In such cases, the evaluation criteria could be biased (Robles and Rojas 2015).

Further, many specialists have diverse views on the approach of indicators that are appropriate for assessing sustainability. That is why the literature proposes a wide variety of approaches that according to the views researchers hold allow SD to be operationalized. These approaches are the product of the ambiguity and complexity of the term that have been discussed for years and were mentioned above.

Given the low level of concordance among experts, an alternative method, automatic classification analysis (conglomerates), was carried out to determine the most relevant indicators. Results revealed the presence of multiple groups, depending on the dimension analyzed. For the economic dimension (Fig. 1), 7 subgroups are recognized, of which 4 include 12 indicators: *commercialization and marketing, quantity of agrodiversity species in production/targeted for production, technical efficiency, income diversification, feasibility/economic feasibility, product diversity for sale, income/expenses, planting planning and operation records/production, productivity/productive efficiency, access to credits/other sources of external financing, level of use/dependence on external inputs and resources, and level of dependence on economic activity*.

In the environmental dimension (Fig. 2) six subgroups were defined, highlighting three subgroups that include eight indicators of high importance: *soil chemistry*

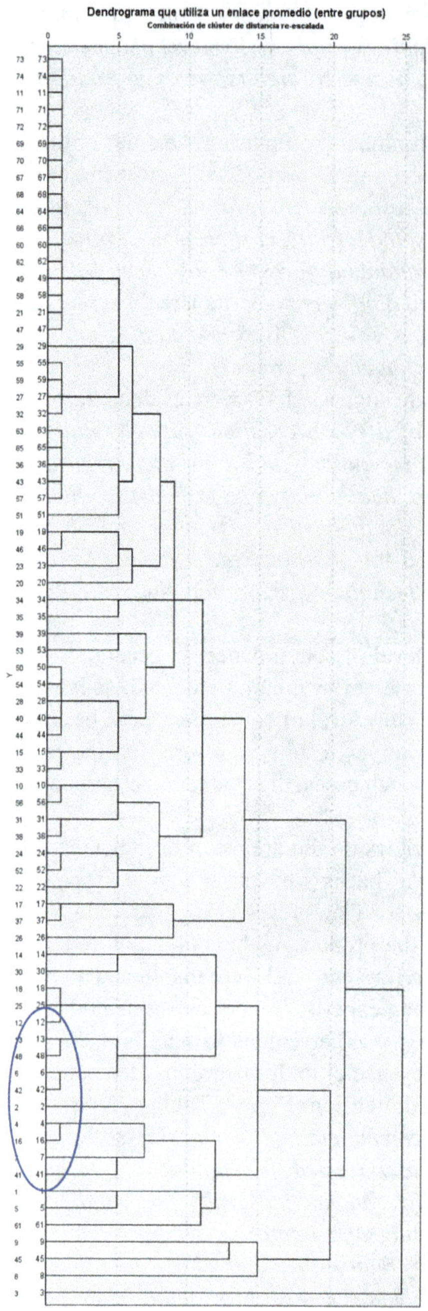


**RELEVANT INDICADTORS**

(1) Income / expenses,
(2) Productivity / Productive efficiency,
(3) Commercialization and marketing,
(4) Dependence on economic activity,
(5) Diversity of products for sale,
(6) Benefit / Cost Ratio,
(7) Access to credits / other types of external funding
(8) Quantity of agrodiversity species in production / for production,
(9) Income diversification,
(10) Improvement or increase of yields / production / products,
(11) Net income / gain / income,
(12) Economic stability,
(13) Product quality,
(14) Economic risk,
(15) Self-financing capacity,
(16) Level of use / dependency of external inputs and resources,
(17) Net Present Value,
(18) Supply / demand relationship,
(19) Period / time of recovery of the investment,
(20) Production costs / other types of costs,
(21) Economic value of the land,
(22) Annual distribution of income / expenses,
(23) Profitability over time,
(24) Gross margin,
(25) Investment,
(26) Economic benefits,
(27) Economic welfare,
(28) Grass availability,
(29) Value added to products,
(30) Savings capacity,
(31) Improvement of the family economy,
(32) Financial evaluation,
(33) Prices / fluctuations,
(34) Economic value of sales / production / products,
(35) Internal rate of return,
(36) Economic value of the risk,
(37) Gross Domestic Product,
(38) Contribution to GDP,
(39) Income / opportunity costs ratio,
(40) Postharvest practices,
(41) Planting planning and operation / production records,
(42) Economic profitability,
(43) Economic contribution of the family,
(44) Type of financing,
(45) Technical / productive efficiency,
(46) Genetic improvement of crops / animals / trees,
(47) Origin of income,
(48) Amount of labor used in production,
(49) Integration into productive chains,
(50) Current and potential markets,
(51) Economic efficiency,
(52) Cash flow,
(53) Financial strategies,
(54) Reproductive index,
(55) Harvest index,
(56) Supply / demand relationship,
(57) Reproductive efficiency,
(58) Reinvestment of income,
(59) Productive strategies,
(60) Number of people linked to the management of the system,
(61) Feasibility / economic viability,
(62) Product lifetime,
(63) Business plan,
(64) Certification of product quality / organic certification,
(65) Economic loss of crops / livestock / timber,
(66) Seasonality of income,
(67) Consumption *per capita* / *per capita* income,
(68) Economic analysis,
(69) Capital assets,
(70) Payment of incentives for production / technology / innovation,
(71) Economic sensitivity,
(72) Level of indebtedness,
(73) Economic value of the land,
(74) Economic balance.

**Fig. 1** Dendrogram of the automatic classification analysis (cluster analysis) that groups the indicators in the economic dimension

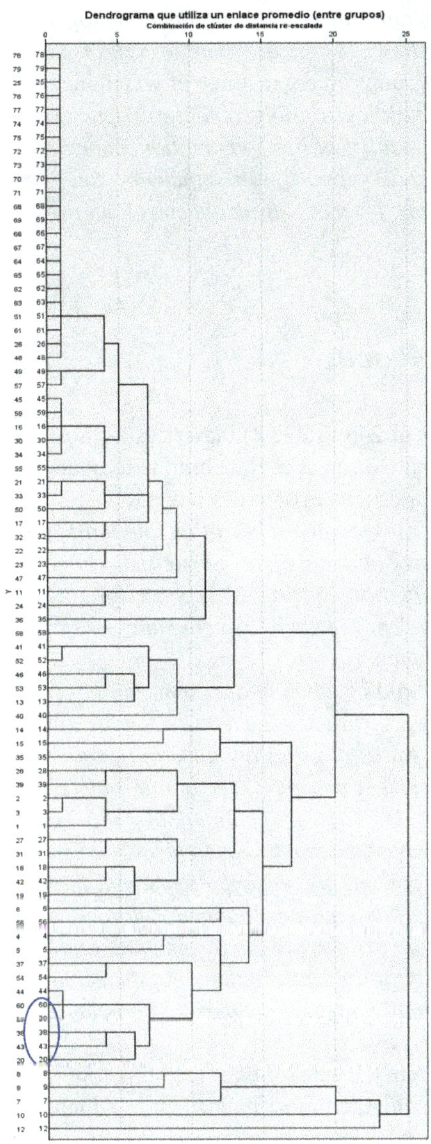


**RELEVANT INDICADTORS**

1) Use of natural resource conservation practices,
(2) Amount of agrodiversity species present,
(3) Water quality,
(4) % vegetation cover of the soil,
(5) Level of use of agrochemicals and fertilizer,
(6) Conservation of edaphic fauna,
(7) Chemistry and soil quality,
(8) Soil management,
(9) Biodiversity,
(10) Agricultural / animal / forestry management practices,
(11) Energy efficiency,
(12) Erosion and risks of soil loss,
(13) Health status of crops, animals / forests,
(14) Degree or level of soil compaction,
(15) Rate of land / area sown,
(16) Alternative / agroecological / sustainable technologies in use,
(17) Evaluation of the behavior of the components of the system,
(18) Diversification of the property,
(19) Natural soil fertility,
(20) Deforestation / reforestation activities,
(21) Physics and soil quality,
(22) Water management,
(23) % areas intervened / degraded by anthropic action,
(24) Irrigation water quality,
(25) Area under AFS / diversification,
(26) Evaluation of climate behavior,
(27) Risk of contamination,
(28) Domestic waste and production management
(29) Areas / area under conservation / areas of biological support,
(30) Agroforestry management practices,
(31) Air quality,
(32) Ability to adapt to ecological / environmental changes,
(33) Consumption and efficiency in the use of fossil energy,
(34) Use of local inputs or resources,
(35) Current status of the natural resources,
(36) Use of alternative energy / alternative energy sources,
(37) Recycling of waste,
(38) Energy balance,
(39) % natural regeneration of vegetation,
(40) Botanical composition,
(41) Diversity of habitats,
(42) Support activities for agroecology / organic / sustainable production,
(43) Exploitation level of the natural resources (NR),
(44) Degree of alteration or degradation of the natural resources / ecosystem,
(45) Ecotoxicity,
(46) Diversity and visual quality of the landscape / conserved landscape,
(47) % natural or non-intervention areas,
(48) Awareness activities for conservation and preservation of environment and NR,
(49) Interaction / interrelation between the components of the system,
(50), Greenhouse effect / carbon footprint,
(51) % reduction of degraded or recovered areas,
(52) Risk of salinization,
(53) Environmental impact,
(54) Soil productivity,
(55) Capacity to use or load the land,
(56) Forest management,
(57) Environmental education,
(58) Fragility of the ecosystem,
(59) Payment for environmental services,
(60) Animal welfare,
(61) Physiology of the crop / animal,
(62) Type of exploitation,
(63) Consumption and efficiency in the use of water / ecological footprint,
(64) Life Cycle Analysis (LCA),
(65) Carbon sequestration / fixation,
(66) Global warming,
(67) Compliance with environmental regulations / legislation,
(68) Level of optimization in the use of NR,
(69) Evaluation of the strata of the system,
(70) Forest certification,
(71) Level of eutrophication,
(72) Heterogeneity of the landscape,
(73) Forest restoration,
(74) % of forest cover,
(75) Biological soil activity / soil biology,
(76) Acidification potential,
(77) Ratio of use of renewable / non-renewable energy,
(78) % of slopes / landslides / landslides,
(79) Photochemical contamination.

**Fig. 2** Dendrogram of the automatic classification analysis (cluster analysis) that groups the indicators in the environmental dimension *and quality, soil management, biodiversity, agricultural/forest/animal management practices, deforestation/reforestation activities, erosion, and land loss risks.*

In the social dimension (Fig. 3), ten subgroups emerged, three of which included ten most important indicators: *capacity building/training activities/capacitation, satisfaction with technology or innovation, quality of life, job creation, community/social organization, government/institutional support, self-sufficiency and food security, degree of satisfaction of basic needs, level of self-management, migration, and social participation/integration.*

## *Community Member Consultation Results*

Members of the communities agreed on 37 of 244 (Table 4) indicators as most relevant for sustainability assessment of the native forest at Ranchillo Alto under silvopastoral conditions. The 12 most important economic indicators for the community were *productivity, income/expenses, commercialization and marketing, access to credits/alternative sources of external financing, economic risk, economic well-being, yields, planting planning/production or predial records, production improvements or increases, technical efficiency, feasibility/economic feasibility, income diversification,* and *economic efficiency.*

For the social dimension 13 variables should be taken into account: *practices for recursos naturales conservation, natural soil fertility, soil chemistry/quality, water quality, erosion and risks/soil loss, health status of crops/animals/trees, soil management, agricultural/animal/forest management practices, use of alternative/ecological/sustainable production technologies, diversification of the farm, reforestation/deforestation activities, and area under AFS/diversification.* Finally, they chose 13 most important parameters: *community participation/social integration, capacity building/training activities, job creation, capacity/willingness to work as a team, government/institutional support, women's participation, ability to adapt to intra/extra-predial changes, quality of life, satisfaction with the economic activity carried out/risk of abandonment, family integration, desire to remain in the place, and access to and quality of basic services.*

The remaining indicators were grouped into the categories of medium, low, or no relevance/importance. The latter category contained most indicators, in which 105 were discarded. The results show that while agreement or concordance between the communities varied, members of both agreed on many indicators that relate to various aspects the literature points out as relevant or decisive for rural development. High concordance also emerged for indicators derived from bibliographic review and expert judgment.

This proves the extensive and important local knowledge residents possess, representing a key finding of the research. With that finding we could confirm the claims of Suset et al. (2002), Machado et al. (2009), and Miranda et al. (2007) that participation constitutes an effective means through which the local population

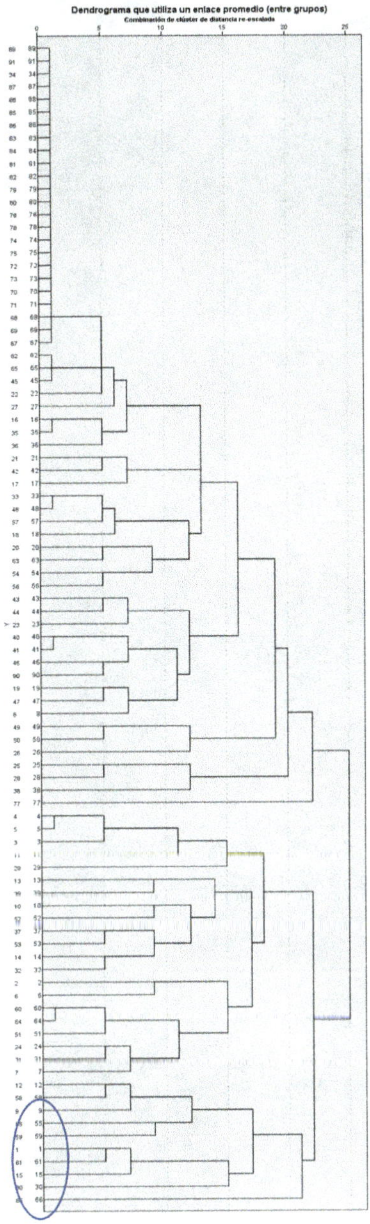


**RELEVANT INDICADTORS**

(1) Social participation / integration,
(2) Equitable distribution of system benefits,
(3) Ability to innovate / adopt technology / resistance to change,
(4) State of health / risks,
(5) Level of external dependence,
(6) Efficiency in the organization,
(7) Self-sufficiency and food security,
(8) Satisfaction for innovation / technology / technological adoption wishes, Citizen security,
(9) Level or degree of satisfaction of basic needs,
(10) Level of empowerment,
(11) Level of environmental awareness / vision / relationship with nature,
(12) Migration, (13) Self-supply capacity,
(14) Satisfaction with the leader's role / performance,
(15) Capacity building / training / training activities,
(16) Degree of family integration,
(17) Participation of women,
(18) Nutrition / nutritional quality of food
(19) Educational level,
(20) Access and quality of basic services,
(21) Land tenure regime,
(22) Gender equity,
(23) Area allocated to production for self-consumption / subsistence,
(24) Adaptation capacity to intra and extra property changes,
(25) Need for extra farm employment,
(26) Use and quality of technical assistance,
(27) Generational transfer / inheritance of land / knowledge,
(28) Unemployment rate / level,
(29) Demand / supply of internal and external labor,
(30) Degree or level of social / community organization,
(31) Efficiency in the organization,
(32) Desires to remain in place,
(33) Effective availability of inputs / resources,
(34) Satisfaction for productive activity / abandonment risks,
(35) Use of local knowledge or knowledge,
(36) Outreach and technology transfer activities,
(37) Equitable distribution of land,
(38) Socialization of knowledge,
(39) Autonomy in the purchase and use of inputs / productive resources,
(40) Presence of conflicts,
(41) Use of family labor,
(42) Free time dedicated to leisure and entertainment,
(43) Existence of claims and resilience,
(44) Sense of belonging and identity,
(45) Cultural diversity,
(46) Labor law / legal support to workers / labor justice,
(47) Autonomy and management in the management of income,
(48) Ability to work in a team,
(49) Sense / level of cooperativism and solidarity,
(50) Neighborhood ties,
(51) Desires to remain in place,
(52) Level of technification of the producer,
(53) Access and equitable use of the sector's natural resources,
(54) Occupational health and safety / risks,
(55) Quality of life,
(56) Labor organization / unions,
(57) Conflicts over the use of land,
(58) Level of self-management
(59) Job creation, (60) Working conditions,
(61) [illegible] / institutional support
(62) Self-esteem level,
(63) Compliance with the rules of coexistence and good customs,
(64) Cultural erosion / loss of local knowledge,
(65) Personological profile of psychosocial producers / elements,
(66) Satisfaction for innovation / technology / technological adoption wishes
(67) Rescue of knowledge / culture / traditions,
(68) Level of importance of social capital,
(69) Leadership level,
(70) Social function of the company,
(71) Promotion of peasant research,
(72) Accessibility to the property or community / state of the access roads,
(73) Satisfaction for the quality of the environment and natural resources,
(74) Work remuneration capacity,
(75) Economic level, (76) Seasonality of the workforce,
(77) Cultural practices / cultural support and promotion activities,
(78) Ritual practices, (79) Level of intensification of productive activity,
(80) Number of ethnic groups involved,
(81) Strategies to overcome problems,
(82) Expropriation and concentration of land,
(83) Life expectancy, (84) Democracy in management and decision-making,
(85) Level of paternalism,
(86) Continuity / labor seasonality,
(87) Need to hire specialized personnel,
(88) Amount of wages / salaries,
(89) Need for intermediaries,
(90) Relations / exchange with regional institutions,
(91) Human Development Index.

**Fig. 3** Dendrogram of the automatic classification analysis (cluster analysis) that groups the indicators in the social dimension

**Table 4** Economic indicators selected by community members according to their level of importance

| Dimension | Level of importance | Indicators where both communities were in full agreement | Indicators where only community 1 agreed | Indicators where only community 2 agreed |
|---|---|---|---|---|
| Economic | Highly relevant | Productivity, income/expenses, marketing and commercialization, access to credits/other alternative sources of external financing, economic risk, economic well-being, yields, sow planning/registers of farm or production performance, improvements or increase of production, efficiency technical, feasibility/economic viability, income diversification, and economic efficiency | Number of species of agrodiversity in production/for production, number of people linked to the management of the system, reproductive efficiency, economic analysis, level of use/dependence on external inputs and resources, economic stability, economic loss of crops/animals/wood, self-financing capacity | Family financial contribution, product quality, quality/organic certification, economic profitability |
| | Moderately relevant | Genetic improvement of crops/animals | Business plan, reinvestment of income, supply/demand ratio, availability of fodder | Dependence on economic activity, economic value of risk |
| | Little bit relevant | Diversity of products for sale | Investment, net profit/profit/income, dependence on economic activity, economic profitability, level of use/dependence on external inputs and resources, payment of incentives or incentives for production/technology/innovation | Added value to products, product quality, quality/organic certification, supply/demand relationship |

| Dimension | Level of importance | Indicators where both communities were in full agreement | Indicators where only community 1 agreed | Indicators where only community 2 agreed |
|---|---|---|---|---|
| | Not relevant | Level of indebtedness, product lifetime, annual distribution of income/expenses, gross margin, payment of incentive or stimulus to production/technology/innovation, type of financing, cash flow, financial strategies, seasonality of income, consumption/per capita income, net present value, harvest index, productive strategies, gross domestic product, contribution to GDP, period/time to recover the investment, production costs/other types of costs, economic value of land, net present value, profitability over time, economic benefit, savings capacity, financial evaluation, prices/fluctuations, economic value of sales, income/opportunity cost ratio* postharvest practices, reproductive index, harvest index, capital goods, economic sensitivity, improvement of the family economy, financial evaluation, current and potential markets, source of income, integration to chains productive, internal rate of return, profitability over time, economic balance | Economic losses of cultivation/livestock/wood, economic contribution of the family, value added to products, net profit, economic value of risk | Amount of agrodiversity in production/destined for production, number of people linked to system management, level of use/dependence on external inputs, reproductive efficiency, economic analysis, economic stability, loss of crops/animals/livestock, capacity for self-financing, availability forage, dependence on economic activity, reinvestment of income, business plan, availability of pasture |

exercises influence and control over decisions affecting it, and disadvantaged groups are mobilized to meet their own demands. Thus, farmers' contribution to the definition, prioritization, and resolution of their problems is crucial in adopting strategies adequate to local reality, committing to local development and assessing the impact of development policies to achieve rural SD. Farmers face great challenges regarding information management, innovation, market positioning, management of environmental variables, and sustainability of agroproductive activity. Their participation can contribute to effective responses and a view toward development from the endogenous strengths of the territory.

It was also possible to confirm through the exercise that the community members have clear aspirations and expectations regarding the native forest management system proposed. Accuracy and speed in the selection of indicators were evident. This highlights the locals' identification with territory, their roots, knowledge of the sector, level of awareness of their reality, and ability to be active participants in developing silvopastoral systems.

## Methodological Triangulation Results

To obtain and preliminarily select potential local indicators for sustainability assessment of the system under study, the data from the bibliography-experts-producers consultation were subjected to automatic classification analysis. For the economic dimension, data triangulation revealed the formation of two major groups, containing 12 and 62 indicators (Fig. 4). The group of greatest concordance included 12 indicators, which fell into two subgroups of three and nine indicators, respectively.

The indicators of subgroup 1 were ***income/expenses, productivity,*** **and** ***commercialization/marketing***. These must figure in any sustainability assessment, regardless of evaluation objectives and place of assessment.

Analysis in the environmental dimension (Fig. 5) led to two groups with a significant difference in importance. The largest group contained 60 indicators and was of lower comparative relevance. Another group contained 19 indicators with the highest overall connotation. This group was subdivided into subgroups consisting of nine and ten important indicators, of which the former was, integrally, the most relevant (Table 5).

In this sense, *practices for the conservation of natural resources, biodiversity, soil chemistry/quality, level of use of agrochemicals and fertilizers, erosion/risk of soil loss, and water quality* and *practices of crop/animals/trees management, energy efficiency, and soil management* made up the first group. They represent the best balance and are most representative in the methodological triangulation. Generally, in this dimension the greatest affinities were found between the three types of queries relating to the most appropriate indicators for evaluation. Socially, two groups

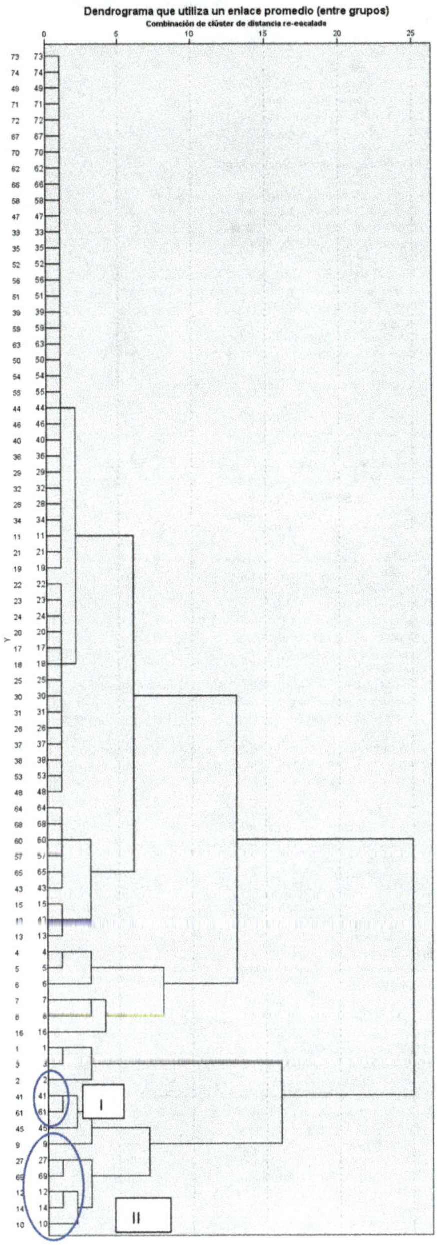


**RELEVANT INDICATORS**

(1) Income / expenses,
(2) Productivity / productive efficiency
(3) Commercialization and marketing,
(4) Dependence on economic activity,
(5) Diversity of products for sale,
(6) Benefit / Cost Ratio,
(7) Economic profitability,
(8) Quantity of agrodiversity species in production / for production,
(9) Income diversification,
(10) Improvement or increase of Returns / production / products,
(11) Net income / gain / income,
(12) Access to credits / other types of external financing,
(13) Product quality,
(14) Economic risk,
(15) Self-financing capacity,
(16) Level of use / dependency of external inputs and resources,
(17) Net Present Value,
(18) Supply / demand relationship,
(19) Period / time of Recovery of the investment,
(20) Production costs / other types of costs,
(21) Economic value of the land,
(22) Annual distribution of income / expenses,
(23) Profitability over time,
(24) Gross Margin,
(25) Investment,
(26) Economic benefits,
(27) Economic welfare,
(28) Grass availability,
(29) Value added to products,
(30) Savings capacity,
(31) Improvement of the family economy,
(32) Financial evaluation,
(33) Prices / fluctuations,
(34) Economic value of sales / production / products,
(35) Internal rate of return,
(36) Economic value of the risk,
(37) Gross Domestic Product,
(38) Contribution to GDP,
(39) Income / opportunity costs ratio,
(40) Postharvest practices,
(41) Planting planning and operation / production records,
(42) Economic stability,
(43) Economic contribution of the family,
(44) Type of financing,
(45) Technical efficiency,
(46) Genetic improvement of crops / animals / trees,
(47) Origin of income,
(48) Amount of labor used in production,
(49) Integration into productive chains,
(50) Current and potential markets,
(51) Economic efficiency,
(52) Cash flow,
(53) Financial strategies,
(54) Reproductive index,
(55) Harvest index,
(56) Supply / demand relationship,
(57) Reproductive efficiency,
(58) Reinvestment of income,
(59) Productive strategies,
(60) Number of people linked to the management of the system,
(61) Feasibility / economic viability,
(62) Product lifetime,
(63) Business plan,
(64) Certification of product quality / organic certification,
(65) Economic loss of crops / livestock / timber,
(66) Seasonality of income,
(67) Consumption per capita / per capita income,
(68) Economic analysis,
(69) Capital assets,
(70) Payment of incentives for production / technology / innovation,
(71) Economic sensitivity,
(72) Level of indebtedness,
(73) Economic value of the land,
(74) Economic balance.

**Fig. 4** Dendrogram of the automatic classification analysis (cluster analysis) that groups the indicators in the environmental dimension

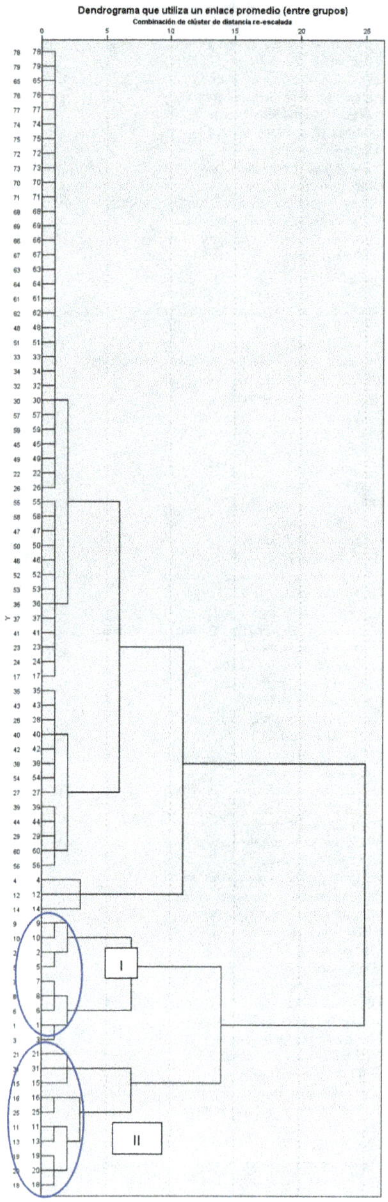


**RELEVANT INDICATORS**

(1) Use of natural resource conservation practices,
(2) Biodiversity,
(3) Chemistry and soil quality,
(4) % vegetation cover of the soil,
(5) Level of use of agrochemicals and fertilizer,
(6) Erosion and risks of soil loss,
(7) Water quality,
(8) Soil management,
(9) Amount of agrodiversity species present,
(10) Energy efficiency,
(11) Agricultural / animal / forestry management practices,
(12) Conservation of the edaphic fauna,
(13) Health status of crops, animals / forests,
(14) Degree or level of soil compaction,
(15) Rate of land / area sown,
(16) Alternative / agroecological / sustainable technologies in use,
(17) Evaluation of the behavior of the components of the system,
(18) Diversification of the property,
(19) Natural soil fertility,
(20) Deforestation / reforestation activities,
(21) Physics and soil quality,
(22) Water management,
(23) % areas intervened / degraded by anthropic action,
(24) Irrigation water quality,
(25) Area under AFS / diversification,
(26) Evaluation of climate behavior,
(27) Risk of contamination,
(28) Domestic waste and production management
(29) Areas / areas / area under conservation / areas of biological support,
(30) Agroforestry management practices,
(31) Air quality,
(32) Ability to adapt to ecological / environmental changes,
(33) Consumption and efficiency in the use of fossil energy,
(34) Use / use of local inputs or resources,
(35) Current status of the natural resources,
(36) Use of alternative energy / alternative energy sources,
(37) Recycling of waste,
(38) Energy balance,
(39) % natural regeneration of vegetation,
(40) Botanical composition,
(41) Diversity of habitats,
(42) Support activities for agroecology / organic / sustainable production,
(43) Exploitation level of the natural resources (NR),
(44) Degree of alteration or degradation of the natural resources / ecosystem,
(45) Ecotoxicity,
(46) Diversity and visual quality of the landscape / conserved landscape,
(47) % of natural or non-intervention areas,
(48) Awareness activities for the conservation and preservation of the environment and NR,
(49) Interaction / interrelation between the components of the system,
(50), Greenhouse effect / carbon footprint,
(51) % reduction of degraded or recovered areas,
(52) Risk of salinization,
(53) Environmental impact,
(54) Soil productivity,
(55) Capacity to use or load the land,
(56) Forest management,
(57) Environmental education,
(58) Fragility of the ecosystem,
(59) Payment for environmental services,
(60) Animal welfare,
(61) Physiology of the crop / animal,
(62) Type of exploitation,
(63) Consumption and efficiency in the use of water / ecological footprint,
(64) Life Cycle Analysis (LCA),
(65) Carbon sequestration / fixation,
(66) Global warming,
(67) Compliance with environmental regulations / legislation,
(68) Level of optimization in the use of the natural resources,
(69) Evaluation of the strata of the system,
(70) Forest certification,
(71) Level of eutrophication,
(72) Heterogeneity of the landscape,
(73) Forest restoration,
(74) % forest cover,
(75) Biological soil activity / soil biology,
(76) Acidification potential,
(77) Ratio of use of renewable / non-renewable energy,
(78) % slopes / landslides / landslides,
(79) Photochemical contamination.

**Fig. 5** Dendrogram of the automatic classification analysis (cluster analysis) that groups the indicators in the economic dimension

**Table 5** Environmental indicators selected by community members according to their level of importance

| Dimension | Level of importance | Indicators where both communities were in full agreement | Indicators where only community 1 agreed | Indicators where only community 2 agreed |
|---|---|---|---|---|
| Environmental | Highly relevant | Practices for the conservation of natural resources, natural soil fertility, soil chemistry/quality, water quality, erosion and risks/soil loss, crop/animal/tree health status, soil management, agricultural/animal management practices/forestry, use of alternative technology/agroecological/sustainable production, diversification of the farm, reforestation/deforestation activities, surface under agroforestry systems/diversification | Agrodiversity, use of agrochemicals and fertilizers, air quality, soil physics/quality | Land/planted area rate, biodiversity, energy efficiency |
| | Moderately relevant | % of natural or non-intervened areas, awareness activities for the conservation and preservation of natural resources and the environment, behavior of the climate, diversity of the landscape and quality of the landscape, use/exploitation of inputs or local resources | Risk of contamination, natural resource exploitation level, % of slopes/landslides, soil productivity, soil compaction, carbon capture/fixation, life cycle analysis | % of vegetal cover of the soil, quality of the water for irrigation, % of reduction of degraded areas/reclaimed areas, water management |
| | Little bit relevant | Animal welfare, % of areas degraded/intervened by anthropogenic action, capacity of use/load of the land | Conservation of edaphic fauna, evaluation of the behavior of system components, biodiversity, support for agroecological/organic/sustainable production, greenhouse effect/water footprint, waste management | Consumption and efficiency in the use of water/water footprint, land rate/planting area |

(continued)

**Table 5** (continued)

| Dimension | Level of importance | Indicators where both communities were in full agreement | Indicators where only community 1 agreed | Indicators where only community 2 agreed |
|---|---|---|---|---|
| | Not relevant | Compliance with environmental regulations/legislation, fragility of the ecosystem, physiology of the crop/animal, ecotoxicity, interaction/intercalation between the components, consumption and efficiency in the use of fossil energy, risk of salinization, type of exploitation, conservation zones/areas of biological support, agroforestry management practices, degree of alteration/degradation of natural resources/ecosystem, current state of natural resources, use of alternative energy/alternative energy sources present, recycling of waste, energy balance, % of natural regeneration of vegetation, botanical composition, habitat diversity, environmental impact, forest management, environmental education, payment for environmental services, type of exploitation, relation of renewable/nonrenewable energy use, level of optimization in the use of natural resources, quantity and evaluation of systems strata, forest certification, eutrophication level, heterogeneity of the landscape, forest restoration, forest cover, soil biological activity/soil biology, acidification potential, global warming, capacity to adapt to environmental/ecological changes, photochemical contamination | Water management, % of reduction of degraded areas/reclaimed areas, consumption and efficiency in the use of water/water footprint | Use of agrochemicals and fertilizers, % of ground cover, % of slope/landslide, level of exploitation of natural resources, energy efficiency, life cycle analysis, soil physics/quality, air quality, risks of contamination, carbon sequestration/fixation, agrodiversity, greenhouse effect/carbon footprint, support for agroecological production/organic agriculture, conservation of edaphic fauna, evaluation of the behavior of system components, support for agroecological/organic/sustainable production, waste management, water quality for irrigation, rate of land/planted area, biodiversity, productivity of the soil, degree or level of soil compaction |

emerged with, respectively, 72 and 19 indicators. The group with the fewest indicators was the most important comparatively (Fig. 6). It was divided into two subgroups of ten and nine indicators, and the former contained the most important ones, including *participation/social integration, capacity building/training/education, self-sufficiency and food security, level or degree of satisfaction of basic needs, level of external dependency, quality of life, job creation, migration, degree or level of social/community organization, and capacity for innovation/adoption of technology/resistance to change.*

In this first approach toward defining fourth-generation indicators, a group of partial or potential indicators to evaluate production systems' sustainability emerged. These indicators are highly consistent with key aspects outlined in the literature regarding each sustainability dimension. Overall, 22 of 244 indicators were considered most relevant for assessing sustainability while 28 were complementary, and 50 were validated through methodological triangulation. Hence, the degree of triangulation was high and adequate, in line with the criteria described by, among others, Ato et al. (2006), Cabero and Llorente (2013), and Robles and Rojas (2015).

In this regard, social scientists consider that a great variety of methods, data, and researchers in the analysis of a specific problem contribute to the reliability of final results (Beltrán et al. 2013; Alzás et al. 2016; Dorantes-Nova et al. 2016; Quintriqueo et al. 2017). The multiple methods in this research suggest a valid and consistent interpretation of the research context, although they remain subject to progressive improvement. This confirms that the main objective of the triangulation process was achieved, including increasing the validity of results by debugging the intrinsic deficiencies of a single data collection method and control of methodological, data, and researcher biases (Oppermann 2000) (Table 6).

For that, authors such as Salgado (2007) maintain that, for the analysis of an increasingly polyhedral reality, a need to combine research techniques to achieve complementary findings and develop knowledge arises. Also, Blaikie (1991) and Paul (1996) note that in the analysis of complex organizational systems diversity of data collection methods is needed. Data must be "triangulated" to reflect the complexity they are trying to describe; otherwise data would be obtained "under suspicion."

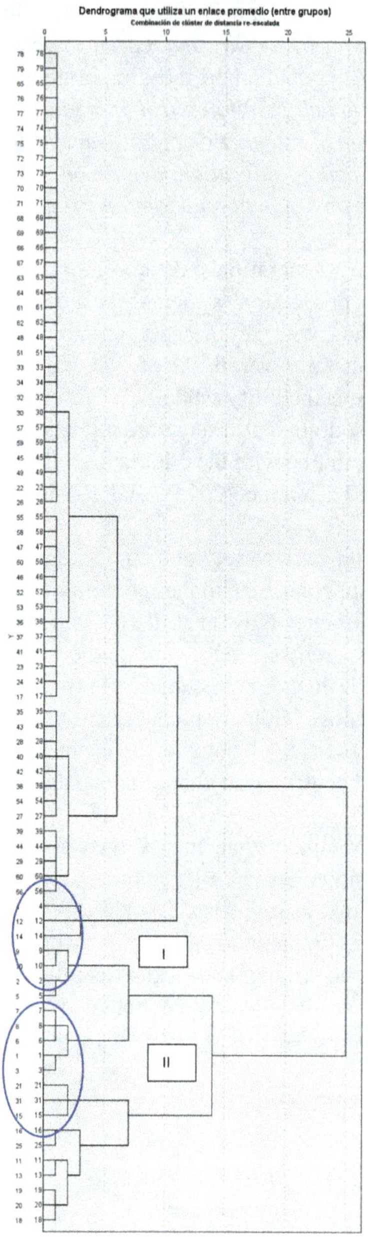


**RELEVANT INDICATORS**

(1) Social participation / integration,
(2) Capacity building / training / training activities,
(3) Ability to innovate / adopt technology / resistance to change,
(4) Degree of family integration, (5) Level of external dependence,
(6) Generation of jobs, (7) Self-sufficiency and food security,
(8) Migration, (9) Level or degree of satisfaction of basic needs,
(10) Quality of life,
(11) Level of environmental awareness / vision / relationship with nature,
(12) Need for extra property employment,
(13) Level of self-management,
(14) Degree or level of social / community organization,
(15) Equitable distribution of system benefits,
(16) Participation of women,
(17) Level of empowerment
(18) Government / institutional support,
(19) Educational level,
(20) Access and quality of basic services,
(21) Satisfaction for innovation / technology / technological adoption wishes,
(22) Gender equity,
(23) Area allocated to production for self-consumption / subsistence,
(24) State of health / risks,
(25) Use of local knowledge or knowledge,
(26) Use and quality of technical assistance,
(27) Generational transfer / inheritance of land / knowledge,
(28) Unemployment rate / level,
(29) Demand / supply of internal and external labor,
(30) Self-supply capacity,
(31) Adaptation capacity to intra and extra property changes,
(32) Desires to remain in place,
(33) Effective availability of inputs / resources,
(34) Satisfaction for productive activity / abandonment risks,
(35) Land tenure regime,
(36) Outreach and technology transfer activities,
(37) Equitable distribution of land,
(38) Socialization of knowledge,
(39) Autonomy in the purchase and use of inputs / productive resources,
(40) Presence of conflicts,
(41) Use of family labor,
(42) Free time dedicated to leisure and entertainment,
(43) Existence of claims and resilience,
(44) Sense of belonging and identity,
(45) Cultural diversity,
(46) Labor law / legal support to workers / labor justice,
(47) Autonomy and management in the management of income,
(48) Ability to work in a team,
(49) Sense / level of cooperativism and solidarity,
(50) Neighborhood ties,
(51) Desires to remain in place,
(52) Level of technification of the producer,
(53) Access and equitable use of the sector's natural resources,
(54) Occupational health and safety / risks,
(55) Nutrition / nutritional quality of foods,
(56) Labor organization / unions,
(57) Conflicts over the use of land,
(58) Control over the system and decision making,
(59) Efficiency in the organization,
(60) Working conditions,
(61) Satisfaction with the leader's role / performance,
(62) Self-esteem level,
(63) Compliance with the rules of coexistence and good customs,
(64) Cultural erosion / loss of local knowledge,
(65) Personological profile of psychosocial producers / elements,
(66) Citizen security, (67) Rescue of knowledge / culture / traditions,
(68) Level of importance of social capital,
(69) Leadership level, (70) Social function of the company,
(71) Promotion of peasant research,
(72) Accessibility to the property or community / state of the access roads,
(73) Satisfaction for the quality of the environment and natural resources,
(74) Work remuneration capacity,
(75) Economic level, (76) Seasonality of the workforce,
(77) Cultural practices / cultural support and promotion activities,
(78) Ritual practices, (79) Level of intensification of productive activity,
(80) Number of ethnic groups involved,
(81) Strategies to overcome problems,
(82) Expropriation and concentration of land,
(83) Life expectancy, (84) Democracy in management and decision-making,
(85) Level of paternalism, (86) Continuity / labor seasonality,
(87) Need to hire specialized personnel, (88) Amount of wages / salaries,
(89) Need for intermediaries,
(90) Relations / exchange with regional institutions,
(91) Human Development Index.

**Fig. 6** Dendrogram of the automatic classification systems (cluster analysis) that groups the indicators in the social dimension

**Table 6** Social indicators selected by community members according to their level of importance

| Dimension | Level of importance | Indicators where both communities were in full agreement | Indicators where only community 1 agreed | Indicators where only community 2 agreed |
|---|---|---|---|---|
| Social | Highly relevant | Community participation/social integration, capacity building/training/training activities, job creation, capacity/willingness to work as a team, government/institutional support, participation of women, capacity to adapt to intra/extra farm changes, quality of life, satisfaction with the economic activity carried out/abandonment risk, family integration, desire to remain in place, access and quality of basic services | Demand/supply of labor, promotion/encouragement of peasant research, equity/gender equality, environmental awareness/vision/relationship with nature, socialization of knowledge, neighborhood ties, access and quality of basic services, labor law/legal support to the worker/social justice, rate/level of unemployment | Self-sufficiency and food security, organization/social/community, capacity for innovation/adoption of innovation/technology/resistance to change, citizen security, level/degree/capacity for self-management, degree of cooperativism/solidarity, migration, educational level, need for off-campus employment, level of satisfaction of basic needs, level of external dependency, access and quality of technical assistance |
| | Moderately relevant | Use of local knowledge or knowledge, degree or level of leadership, land tenure regime, cultural practices/cultural support and promotion activities, accessibility to the property and quality of access roads | Generation transfer/inheritance of land/knowledge, community organization, migration, degree of satisfaction of basic needs, compliance with the rules of coexistence and good customs, loss of culture and local knowledge, health status/risks, strategies or actions to overcome problems, degree of solidarity/cooperativism | Labor law/legal support for workers/social justice, environmental awareness/vision/relationship with nature, effective availability of inputs and resources, use of family labor, demand/supply of internal/external labor, level of external dependency, level of satisfaction with innovation/technology/wishes for adoption, existence of claims and resilience, capacity for self-sufficiency, health and safety at work/risk, equity/gender equality |

(continued)

**Table 6** (continued)

| Dimension | Level of importance | Indicators where both communities were in full agreement | Indicators where only community 1 agreed | Indicators where only community 2 agreed |
|---|---|---|---|---|
| | Little bit relevant | Sense of belonging, free time for leisure and distraction | Use of family labor, satisfaction with the environment/quality of the environment, effective availability of inputs and resources, capacity for innovation/adoption of technology/resistance to change, capacity for self-sufficiency, level/degree/capacity for self-management, autonomy in the purchase and use of inputs and resources, self-management in the management of income, satisfaction with innovation/technology/desires for adoption, educational level, capacity to remunerate the workforce, level of self-esteem | Producer's level of technification, generation transfer/land inheritance/knowledge, promotion/stimulus to peasant research, degree of satisfaction with economic activity/risk of abandonment, labor law/legal support for workers/social justice, erosion or cultural loss/local knowledge, equitable distribution of benefits, socialization of knowledge, need to hire specialized personnel |

| | | | | |
|---|---|---|---|---|
| | Not relevant | Expropriation and concentration of land, ritual practices, number of ethnic groups involved, presence of conflicts, level of empowerment, social function of the company, area destined for the production of self-consumption/subsistence, activities of extension and technological transfer, equitable distribution of land, cultural diversity, equitable access and use of natural resources of the sector, nutrition/nutritional quality of food, social organization/unions, conflicts over the use and management of land, control over the system and decision-making, satisfaction with role/performance of the leader, level of importance of social capital, economic level, seasonality of the workforce, level of intensification of productive activity, democracy in management and decision-making, level of paternalism, number of wages/salaries, exchange relations universities/other institutions, human development index, rescue of knowledge/culture/traditions, personological profile/other psychosocial aspects, control and power in the management of the system, capacity to remunerate the workforce, life expectancy | Self-sufficiency and food security, level/degree of external dependency, autonomy in the purchase and use of inputs and resources, existence of claims and recovery capacity, citizen security, labor law/legal support/social justice, occupational health and safety/risks, level of technification of the producer, need for off-farm employment, capacity for innovation/adoption of technology/resistance to change, need to hire specialized personnel, erosion/cultural loss/local knowledge, promotion/stimulation of peasant research, distribution equitable benefits, degree of satisfaction with the productive activity/risk of abandonment | Access and quality of basic services, strategies, or actions to overcome problems, neighborhood ties, rules of coexistence and good customs, autonomy in the purchase and use of inputs and resources, level of self-esteem, autonomy in managing income, inhabitants' health status/risks, use of family labor, presence of accidents and resilience |

## Conclusions

The continued dissemination of existing experiences in many countries, by agencies and research centers, regarding sustainability assessment through indicator analysis suggests a growing interest in the subject. Such research in a range of scenarios, conditions, populations, scales (temporary and spatial), characteristics, problems, and complexities proves that since the 1990s society has been becoming more concerned about environmental and social problems arising from the current model of development. Much work is being done to shift the path toward more favorable solutions for rising challenges, which is an encouraging development.

The study found many indicators that can be classified according to specific characteristics. It also found that previous studies do not explicitly mention the shortcomings and problems that prevent progress toward measurable, stronger, and more universal indicators on the sustainability of production systems. This is crucial, as the proposals for sustainability indicators must improve, because the idea that sustainability is an ongoing process is shared, so its definition will keep evolving, as will the methodologies used for its evaluation.

From the creation of indicators, it was found that their construction must be based on the needs and problems of the social actors whose systems are evaluated. Evaluation must be done with local characteristics of the agroecosystems in mind. This phase is decisive and complex and requires rigorousness, detail, and care. The participation of producers in determining potential indicators was an enriching experience. It yielded interesting and decisive results that allowed to corroborate the importance of local knowledge and the exchange of indigenous knowledge, which mirrored expert knowledge and scientific literature.

Thus, the methodology was specifically applied but based on a holistic approach. This resulted in its first stage in the construction and selection of a set of potential indicators that will allow to evaluate trends, establish differences between systems or farms, and detect critical points that could compromise the sustainability of local production systems. This group comprised of 50 preliminary indicators considered best to evaluate sustainability. Of these, 22 were listed as the most decisive, with groups of three, nine, and ten indicators for economic, environmental, and social dimensions. The remaining ones were considered to be optional, depending on the needs of the context.

It was confirmed that the selection process is as important as the indicators, because if they are poorly chosen or redundant, they could distort the assessment of the system under study and the trend toward sustainability. It is also necessary to analyze the potential indicators through a last detailed review and discussion with the community to prioritize, define, and therefore synthesize even more data. This is important to finally define in a participative manner the most viable and appropriate indicators. Undoubtedly, it is a crucial and challenging stage that should not be taken lightly, for the aim is to create a limited group of indicators that enable a precise picture of progress toward sustainability of the system under study and predict associated risks.

The methodology for indicator construction allowed to meet the objectives of simplifying and validating the data. Methodological triangulation generated potential indicators that, although susceptible to progressive improvements, can reflect certain characteristics of the object of study and thereby showed that change occurred in the system. The combination of research methods provided the possibility of corroborating or disputing existing sustainability data. Diversity of criteria, debates, positions, and opinions made the obtained data reliable as they minimized biases that have characterized sustainability assessments. This is an important contribution to scientific work to build an adequate, objective, practical, flexible, and universal methodology for the evaluation of sustainable development.

**Acknowledgments** The authors would like to thank all academics and researchers of Forestry Sciences, Agronomy, Economic, Social and Environmental Faculties of Universidad de Concepción, Universidad del Bío-Bío, Universidad Católica de la Santísima Concepción, Unidad de Desarrollo Tecnológico de la Universidad de Concepción (UDT), Instituto Forestal (INFOR), Bioforest S.A., Universidad Rómulo Gallegos (Venezuela), Instituto Nacional de Investigaciones Agrícolas de Venezuela (INIA), Instituto Venezolano de Investigaciones Científicas (IVIC), Universidad de los Andes (Venezuela), Universidad Central de Venezuela, Universidad de la Plata (Argentina), Universidad de São Paulo (Brazil), Estación Experimental de Pastos y Forrajes "Indio Hatuey" (Cuba), Centro Agronómico Tropical de Investigación y Enseñanza (Costa Rica), and Universidad de Chapingo (Mexico). Their part in the experts' consultation was invaluable. Thanks go also to all members of the Ranchillo Alto and Los Avellanos communities for their participation in the development of the research in its first and second phases. Also, María Gabriela Medina extends a special sense of recognition to Dr. Danny Eugenio García Marrero whose collaboration was invaluable and who contributed with his knowledge and scientific experience. This made the research more solid and rigorous. He always urged and motivated me to work toward quality and excellence which I will transmit to our son Fabio Ernesto García Medina. This study was partially funded by the Native Forest Research Fund of the National Forestry Corporation, Chile (FIBN-CONAF Project N° 001-2014).

## References

Acevedo-Osorio Á (2009) ¿Cómo evaluar el nivel de sostenibilidad de un programa agroecológico? Un procedimiento metodológico para diseñar, monitorear y evaluar programas rurales con enfoque de desarrollo sostenible. Bogotá, Colombia, 80

Aliaga FA, Basulto OF, Cabrera J (2012) El grupo de discusión: elementos para la investigación en torno a los imaginarios sociales. Prisma Soc 9:136–175

Alonso J (2011) Los sistemas silvopastoriles y su contribución al medio ambiente. Rev Cuba Cien Agrícola 45(2):107–115

Alzás T, Casa LM, Luengo R, Carvalho JL, Verissino S (2016) Revisión Metodológica de la Triangulación como estrategia de Investigación. Invest Qualitat Cien Soc 3:639–648

Arias J (2017) La sostenibilidad justa como paradigma sistémico ambiental. Gestión Ambiente 20(2):232–243

Arnés E, Marín O, Merino A, Hernández C (2013) Evaluación de la sustentabilidad de la agricultura de subsistencia en San José de Cusmapa, Nicaragua. Rev Esp Estud AgroecolPesq 236:171–197

Arocena F, Porzecanski R (2010) El desarrollo sustentable: Reflexiones sobre su alcance conceptual y medición. Rev Cien Soc Dep Sociol 26:16–29

Astier M, González C (2008) Formulación de indicadores ambientales para evaluaciones de sustentabilidad de sistemas de manejo complejos. Int J Sustain Dev World Ecol 14:345–361

Astier M, Galván MY, Masera O (2008) Retos para los análisis de sustentabilidad de los SMRN. In: Astier M, Masera O, Galván MY (eds) Evaluación de sustentabilidad. Un enfoque dinámico y multidimensional. Mundiprensa-SEAE-CIGA-CIEco-GIRA, Valencia, España, pp 1993–2000

Ato M, Benavente A, López J (2006) Análisis comparativo de tres enfoques para evaluar el acuerdo entre observadores. Psicothema 18(3):638–645

Bahamonde HA, Peri PL, Alvarez R, Barneix A, Moretto A, Martínez G (2012) Litter decomposition and nutrients dynamics in Nothofagus antarctica forests under silvopastoral use in Southern Patagonia. Agroforest Syst 84:345–360

Barrezueta S (2015) Introducción a la sostenibilidad agraria: Con enfoque de sistemas e indicadores. Universidad Técnica de Machala, Ecuador, p 77

Beltrán E, Galitó N, García N, Jové G, Macarulla M (2013) La triangulación múltiple como estrategia metodológica REICE. Rev Iberoam Sobre Calidad Eficacia Cambio Educ 11(4):5–24

Blaikie NWH (1991) A critique of the use of triangulation in social research. Qual Quant 25:115–136

Bolívar H (2011) Metodologías e indicadores de evaluación de sistemas agrícolas hacia el desarrollo sostenible. Centro Invest Cien Adm Gerenciales (CICAG):1–18

Braun A (2016) Incremento de los sistemas silvopastoriles en América del Sur. In Andrea Braun, Suzanne Van Dijk, Markus Grulke, Katalin Solymosi Monografía del BID; 461

Brundtland GH (1987) Report of the World Commission on Environment and Development: Our Common Future. United Nations. 300

Bustamante TI, Carrera B, Rita Schwentesius R (2017) Sostenibilidad de pequeños productores en Tlaxcala, Puebla y Oaxaca, México. Estudios Regionales en economía, población y desarrollo. Cuadernos de trabajo de la UACJ N° 37

Cabero J, Llorente M (2013) La aplicación del juicio de experto como técnica de evaluación de las tecnologías de la información y comunicación (TIC). Rev Tecnol Infor Comun Educ 7(2)

Castro DC (2010) La sustentabilidad del "Monte - Guasapampa": aspectos ecológicos, sociales y económicos. Quebracho 18(1–2):90–100

Contreras R (2002) La investigación acción participativa: revisando sus metodologías y sus potencialidades. In: Durton J, Miranda F (eds) Experiencias y metodología de la investigación participativa, vol 58. División de Desarrollo Social FAO- CEPAL, Santiago, Chile, pp 9–17

Delgado A, Armas W, Daubeterre R, Araque A (2007) Evaluación de la sostenibilidad de un sistema de producción caprino, utilizando indicadores. Gaceta Cien Vet 13(1):45–52

Delgado A, Armas W, D'Aubeterre R, Hernández C, Araque C (2010) Sostenibilidad del sistema de producción cabra-Hibiscus-Aloe vera en el semiárido de Cauderales (estado Lara, Venezuela). Agroalimentaria 16(31):49–63

Dorantes-Nova JA, Hernández-Mosqueda JS, Tobón-Tobón S (2016) Juicio de expertos para la validación de un instrumento de medición del. Síndrome Burnout Docencia Ra Ximhai 12(6):327–346

Dube F (2017) Sistemas agroforestales. Estudio cmprativo de productividad entre sistemas silvopastoriles y bosques plantados. Espinoza M, Acuña E, García J.; Rodriguez R.; Rubilar, R Silvicultura de bosques plantados con fines productivos. Sello Editorial Universidad de Concepción, Primera Edición

Dube F, Espinosa M, Zagal E, Stolpe N, Thevathasan N, Gordon A (2011) Optimización del secuestro del carbono en sistemas silvopastorales con Pinus ponderosa y Trifolium spp. sobre suelos volcánicos degradados en la Patagonia Chilena. In: Actas III Seminario Investigación y Desarrollo en la pequeña y mediana propiedad: avances de la agroforestería en Chile. Concepción. Instituto Forestal, Concepción, Chile, pp 9–14

Dube F, Müller-Using, Stolpe N, Zagal E., Doussoulin M (2015) Silvopastoreo en robledales viejos con distintos grados de cobertura como una opción de manejo sustentable en la Región del Bíobío, Chile. 3° Congreso Nacional de Sistemas Silvopastoriles: VII Congreso Internacional Sistemas Agroforestales. Compilado por Pablo L. Peri. Primera edición. Ediciones INTA, Santa Cruz, Argentina. p 716

Dube F, Sotomayor A, Loewe V, Müller-Using B, Stolpe N, Zagal E, Doussoulin M (2016) Silvopastoril Systems in temperate zones of Chile. Silvopastoril systems in Southern South Aemerica. Springer, pp P183–P211
Dumanski J, Gameda S, Pieri C (1998) Indicators of land quality and sustainable land management. The World Bank, Washington DC, pp 13–22
Escobar-Pérez J, Cuervo-Martínez Á (2008) Validez de contenido y juicio de expertos: una aproximación a su utilización. Av Medición 6:27–36
FAO (2014a) El estado de los bosques del mundo. Potenciar los beneficios socioeconómicos de los bosques. Organización de las Naciones Unidas para la Alimentación y Agricultura, Roma, p 132
FAO (2014b) El estado de la inseguridad alimentaria en el mundo. Fortalecimiento de un entorno favorable para la seguridad alimentaria y la nutrición. Organización de las Naciones Unidas para la Alimentación y la Agricultura, Roma, p 62
FAO (2018) El estado de los bosques del mundo - Las vías forestales hacia el desarrollo sostenible, Roma, p 132
Fawaz J, Vallejos-Carte R (2011) Calidad de vida, ocupación, participación y roles de género: un sistema de indicadores sociales de sostenibilidad rural (Chile). En Cuad Desarro Rural 8(67):45–68
Fernández F, Ovalle C, Manabe T (2002) Sistemas agroforestales. Producción Moderna de madera y praderas en el secano interior. Instituto Forestal, Chile, pp 89–102
Franco MH (2015) El Desarrollo Rural desde la Agroforestería Agroecológica. Rev Agroforest Neotrop 5:83–92
Gallopin G (2006) Los indicadores de desarrollo sustentable. aspectos conceptuales y metodológicos. FODEPAL, Santiago, Chile, p 36
Galván MY, Masera O, López-Ridaura S (2008) Las evaluaciones de sustentabilidad. In: Astier M, Masera O, Galván MY (eds) Evaluación de sustentabilidad. Un enfoque dinámico y multidimensional. Editorial Mundiprensa-SEAE-CIGA-CIEco-GIRA, Valencia, España, pp 41–57
Gargaglione V (2011) Dinámica de macronutrientes en bosques de Nothofagus antarctica de Patagonia Sur. Tesis Doctorado en Ciencias Agrarias. Buenos Aires, Argentina. Escuela para graduados de la Facultad de Agronomía, Universidad de Buenos Aires (UBA). p 132
Geilfus F (2009) 80 herramientas para el desarrollo participativo: diagnóstico, planificación, monitoreo, evaluación. Instituto Interamericano de Cooperación Para la Agricultura, San José, Costa Rica, p 217
Glave M, Escobal J (2000) Indicadores de sostenibilidad para la agricultura andina. Bol Agroecol 67:32
Gómez-Limón JA (2010) Evolución de la sostenibilidad del olivar en Andalucía: Una propuesta metodológica. CUIDES 5:65–140
González R (2011) La incorporación de la Generación y al mercado laboral. El caso de una Entidad Financiera de la ciudad de Resistencia. Palermo Bus Rev 5:67–93
Guzmán A, Abt M, Brassiolo M (2012) Tipificación de las estrategias de uso del bosque por pequeños productores campesinos en Santiago del Estero. Quebracho 20(1–2):39–48
Ibáñez R (2012) Indicadores de sustentabilidad: Utilidades y limitaciones. Teoría Praxis 11:102–126
IBM Corp (2017) IBM SPSS Statistics for Windows, Version 23.0. IBM Corp, Armonk, NY
INFOR (2015) Anuario forestal 2014. Boletín Estadístico 144. Instituto Forestal, Santiago, Chile, p 159
Kássio J, Benítez V, Santa Cruz R, Posse JP, Santiago R, Jorge Hernández J, Carlos Mantero C, Morales V, Silveira ED, Viñoles C (2018) Oportunidades y desafíos para los sistemas silvopastoriles en Uruguay. Veterinaria (Montevideo) 54(209):20–30
Krishnamurth LM (1999) Agroforestería básica. Programa de las naciones unidas para el medio ambiente. Oficina regional para América Latina y el Caribe. Serie N° 3, México. p 340
Kú VM, Pool L, Mendoza J, Aguirre E (2013) Propuesta metodológica para evaluar proyectos productivos con criterios locales de sustentabilidad en Calakmul, México. Av Invest Agropecuar 17(1):9–34

Kumaraswamy S (2012) Sustainability issues in agro-ecology: Socio-ecological perspective. Agric Sci 3(2):153–169

Laclau P (2015) Instrumentos de Política Pública y Sistemas Silvopastoriles. 3° Congreso Nacional de Sistemas Silvopastoriles: VII Congreso Internacional Sistemas Agroforestales. Compilado por Pablo L. Peri. Primera edición. Ediciones INTA, Santa Cruz, Argentina. p 716

Leccardi C, Feixa C (2011) El concepto de generación en las teorías sobre la juventud, Ultima década N° 34. CIDPA Valparaíso, Chile, pp 11–32

Ley sobre Recuperación del Bosque Nativo y Fomento Forestal y Reglamentos. Núm. 20.283. 2008. Ley Corporación Nacional Forestal, Santiago, Chile, 89

Loaiza W, Carvajal Y, Ávila ÁJ (2014) Evaluación agroecológica de los sistemas productivos agrícolas en la microcuenca centella (Dagua, Colombia). Colomb Forest 17(2):161–179

Machado H, Suset A, Martín GJ, Funes-Monzote FR (2009) Del enfoque reduccionista al enfoque de sistema sostenible en la agricultura cubana: un necesario cambio de visión. Pastos Forrajes 32(3):1–10

Martín MA (2008) La teoría de las generaciones de Ortega y Gasset: Una lectura del siglo XXI. Tiempo Espacio Año 17(20):98–110

Martínez-Castro CJ, Ríos-Castillo M, Castillo-Leal M, Jiménez-Castañeda JC, Cotera-Rivera J (2015) Sustentabilidad de agroecosistemas en regiones tropicales de México. Trop Subtrop Agroecosyst 18:113–120

Martínez-Castro CJ, Ríos-Castillo M, Castillo-Leal M, Cruz-Cabrera BC, Ruiz-Martínez A (2016) Indicadores socioeconómicos y ambientales de sustentabilidad en agroecosistemas de México. Una revisión de literatura. Naturaleza Desarrollo 12(2):12–32

Masera O, Astier M, López S (2000) Sustentabilidad y sistemas campesinos: cinco experiencias de evaluación en el México rural. Mundi Prensa. Instituto de Ecología, México, DF, p 346

Medina MG, García DE, Moratinos P, Iglesias JM, Clavero T (2013) Evaluación del potencial agronómico de Morus alba para su inclusión en sistemas de pastoreo en TrujilloVenezuela. Variabilidad y relación entre variables descriptoras. Rev Fac Agron 30:73–89

Miranda T, Suset A, Cruz A, Machado H, Campos M (2007) El Desarrollo sostenible. Perspectivas y enfoques en una nueva época. Pastos Forrajes 30(2):191–197

Miranda T, Machado H, Suárez J, Sánchez T, Lamela L, Iglesias J, Suset A, Pérez A, Milera M, Martín GJ, Campo M, López O, Simón L (2011) La Innovación y la transferencia de tecnologías en la Estación Experimental "Indio Hatuey": 50 años propiciando el desarrollo del sector rural cubano (Parte I). Pastos Forrajes 34(4):393–412

Miranda T, Machado H, Lezcano J, Suset A, Oropesa K, Tirado F, Luis Lamela L, Montejo I (2018) Contribución de la innovación a la gestión local del desarrollo en un municipio de Matanzas. Pastos Forrajes 41(4):292–299

Montagnini F, Somarriba E, Murgueitio E, Fassola H, Eibl B (2015) Sistemas Agroforestales. Funciones Productivas, Socioeconómicas y Ambientales, Serie técnica. Informe técnico 402. CATIE, Turrialba, Costa Rica. Editorial CIPAV, Cali, Colombia, p 454

Murgueitio E, Chará D, Solarte A, Uribe F, Zapata C, Rivera JE (2012) Agroforestería Pecuaria y Sistemas Silvopastoriles Intensivos (SSPi) para la adaptación ganadera al cambio climático con sostenibilidad. Universidad de Antioquia, Colombia. Rev Colomb Cien Pecuar 26:313–316

Nahed TJ (2008) Aspectos metodológicos en la evaluación de la sustentabilidad de sistemas agrosilvopastoriles. Av Invest Agropecuar 12(3):3–19

Nair, P.K.R. 2004. Agroforestry: Trees in support of sustainable agriculture. In: Hillel, H. Rosenzweig, C., Powlson, D., Scow, K., Singer, M. & Sparks, D. (eds). Encyclopedia of soils in the environment. Elsevier, London pp. 35–44

Okuda M, Gómez-Restrepo C (2005) Métodos en investigación cualitativa: triangulación. Rev Colomb Psiquiatr 34(1):118–124

Olmos MA, González W (2013) Valor de sustentabilidad. Cien Agric 10(1):91–100

Oppermann M (2000) Triangulation. A methodological discussion. Int J Tour Res 2:141–146

Park P (2005) Que es la investigación acción participativa. Perspectivas teóricas y metodológicas. In: Salazar M (ed) La investigación acción participativa. Inicios y desarrollos. Capítulo VI. Cooperativa Laboratorio Educativo, Caracas, Venezuela, pp 135–172

Paul J (1996) Between method triangulation. Int J Organ Anal 4(2):135–153

Perez MP, Ramos AE, Santa María A (2016) Evaluación de la sustentabilidad: una reflexión a partir del caso de la Red Nacional de Desarrollo Rural Sustentable (México). Entreciencias 4(9):61–72

Peri PL (2011) Sistemas silvopastoriles en bosques de Nothofagus antárctica. Revisión del conocimiento en Patagonia Sur, Argentina. Cien Invest Forest 17(2):217–242

Peri PL, Bahamonde HA, Monelosn LH, Martínez-Pastur G (2008) Producción de Hojarasca en Bosques Primarios y bajo Manejo Silvopastoril de Nothofagus antarctica en la Provincia de Santa Cruz, Argentina. In: Segunda reunión sobre Nothofagus en la Patagonia, EcoNothofagus 2008. Esquel, Argentina, pp 149–155

Peri PL, Hansen N, Rusch V, Tejera L, Monelos L, Fertig M, Bahamonde H, Sarasola M (2009) Pautas de manejo de sistemas silvopastoriles en bosques nativos de Nothofagus antarctica (ñire) en Patagonia. In: Actas del Primer Congreso Nacional de Sistemas Silvopastoriles. Posadas, Misiones, Argentina, pp 151–155

Quintriqueo S, Sanhuesa S, Friz M (2017) Triangulación de Metodos como propuesta para el estudio de la competencia comunicativa intercultural en contextos de inmigración e interculturalidad. Andamio 14(34):283–303

Robles P, Rojas M (2015) La validación por juicio de expertos: dos investigaciones cualitativas en Lingüística aplicada. Revista Nebrija. No. 18. Universidad Nebrija, Madrid, España

Rodríguez F (2006) El diseño de indicadores e índices para evaluar el aporte de las fincas agropecuarias a la sostenibilidad ambiental. Análisis de caso en la Microregión Platanar-la Vieja, Cuenca del río San Carlos, Costa Rica. Rev Pensamiento Act Univ Costa Rica 6(7):23–39

Salgado AC (2007) Investigación cualitativa: diseños, evaluación del rigor metodológico y retos. Liberabit 13:71–78

Sarandón SJ (2002) El desarrollo y uso de indicadores para evaluar la sustentabilidad de los agroecosistemas. In: Sarandón SJ (ed) Agroecología. El camino hacia una agricultura sustentable, vol 20. Ediciones Científicas Americanas Capítulo, pp 393–414

Sarandón SJ (2014) El desarrollo y uso de indicadores para evaluar la sustentabilidad de los agroecosistemas. In: Agroecología: El cambio hacia una agricultura sustentable, pp 393–414

Sarandón SJ, Flores CC (2009) Evaluación de la sustentabilidad en agroecosistemas: una propuesta metodológica. Agroecología 4:19–28

Sarandón SJ, Flores CC (2014) La insustentabilidad del modelo de agricultura actual. In: Santiago S, Flores C (eds) Agroecología: bases teóricas para el diseno y manejo de Agroecosistemas sustentables. Ediciones Edulp. Universidad Nacional de La Plata, Argentina, pp 13–41

Sarandón SJ, Marasas MF, Dipietro F, Belaus A, Muiño W, Oscares E (2006a) Evaluación de la sustentabilidad del manejo de suelos en agroecosistemas de la provincia de La Pampa, Argentina, mediante el uso de indicadores. Rev Bras Agroecol 1(1):497–500

Sarandón SJ, Zuluaga MS, Cieza R, Gómez C, Janjetic L, Negrete E (2006b) Evaluación de la sustentabilidad de sistemas agrícolas de fincas en Misiones, Argentina, mediante el uso de indicadores. Rev Agroecol 1:19–28

Sarandón SJ, Flores CC, Gargoloff A, Blandi ML (2014) Análisis y evaluación de agroecosistemas: construcción y aplicación de indicadores. Bases teóricas para el diseño y manejo de Agroecosistemas sustentables. In: Sarandón SJ, Flores CC (eds) Facultad de ciencias Forestales. Universidad de la Plata, Argentina., Editorial Edup, pp 375–410

Schmidt H, Schmidt A, Alonso M (2013) Manejo silvopastoril en el renoval de Lenga. In: Actas del 2° Congreso Agroforestal Patagónico, 16–17 mayo, El Calafate, Argentina. 106

Silva-Santamaría L, Ramírez-Hernández O (2017) Evaluación de agroecosistemas mediante indicadores de sostenibilidad en san josé de las lajas, provincia de mayabeque, cuba. Luna Azul 44:120–152

Soler Esteban R (2012) Regeneración natural de Nothofagus antarctica en bosques primarios, secundarios y bajo manejo silvopastoril. Tesis Doctorado en Ciencias Biológicas. Córdoba, Argentina. Universidad Nacional de Córdoba. 126

Sotelo J, Tolon A, Lastra X (2011) Indicadores por y para el Desarrollo Sostenible. Un Estudio de Caso. Estudios Geográficos LXXII(271):611–654

Sotomayor A (2010) Sistemas silvopastorales, alternativa de producción integrada para un desarrollo sustentable de la agricultura en Chile. Ciencia e Investigación Forest 16(1):19–51

Sotomayor A (2011) El Rol de los árboles en una producción agropecuaria sustentable. In: Actas III Seminario Investigación y Desarrollo en la pequeña y mediana propiedad: avances de la Agroforestería en Chile. Concepción. Instituto Forestal, Concepción, Chile, pp 31–60

Sotomayor A (2015) Sistemas agroforestales y su contribución a un desarrollo silvoagropecuario sustentable en Chile. 3° Congreso Nacional de Sistemas Silvopastoriles: VII Congreso Internacional Sistemas Agroforestales. Compilado por Pablo L. Peri. Primera edición. Ediciones INTA, Santa Cruz, Argentina, p 716

Sotomayor A, Cabrera C (2008) Análisis de un sistema silvopastoral con Pinus radiata D. Don, asociado con ganado ovino en la zona mediterránea costera central de Chile. Cien e Invest Forest 14(2):269–286

Sotomayor A, Soto H (2011) Análisis de la productividad de espinales (Acacia caven Moll) bajo manejo silvopastoral en la comuna de Cauquenes, Chile. En III Seminario Investigación y Desarrollo en la pequeña y mediana propiedad. Concepción. INFOR, Concepción, Chile, p 20

Sotomayor A, Teuber O (2011) Evaluación del efecto de los árboles manejados bajo ordenación silvopastoral en los parámetros climáticos del sitio, con relación a un manejo ganadero sin árboles. Ciencia e Investigación 17(1):23–40

Sotomayor A, Helmke E, García E (2002) Manejo y mantención de plantaciones forestales. INFOR, CORFO, Concepción, Chile, p 51

Sotomayor A, Teuber O, Moya I (2009) Resultados y experiencia sobre manejo de sistemas silvopastorales en la Región de Aysén. In: Teuber O (ed) Sistemas Agroforestales para la Región de Aysén: Cortinas Cortaviento y Silvopastoreo. Instituto de Investigaciones Agropecuarias e Instituto Forestal, Coyhaique, Chile, pp 165–199

Sotomayor A, Teuber O, Moya I, Almonacid P (2012) Productividad de una pradera natural mejorada bajo la influencia de sistemas silvopastorales con Pinus contorta en relación a una situación de manejo ganadero sin árboles. Cien Forest 18(1):7–30

Squella F, Squella F (2011) Estudio técnico – productivo de un sistema silvopastoral de producción ovina en un bosque de pino (Pinus radiata) localizado en el secano Mediterráneo subhúmedo de Chile. In: Actas III Seminario Investigación y Desarrollo en la pequeña y mediana propiedad: avances de la agroforestería en Chile. Concepción. Instituto Forestal, Concepción, Chile, pp 61–79

Suset A, Gonzales A, Machado L, Miranda T (2002) Diagnóstico socio psicológico: herramienta para la viabilidad de cambios tecnológicos. Estudio de caso en el municipio Martí zona Gumende de la provincia de Matanzas, Cuba. Pastos Forrajes 25(2):34–39

Toro-Mújica P, García A, Gómez-Castro AG, Acero R, Perea J, Rodríguez-Estévez V (2011) Sustentabilidad de agroecosistemas. Arch Zootec 60(R):15–39

Van der Kloot WA, Spaans AMJ, Heinser WJ (2005) Instability of hierarchical cluster analysis due to input order of the data: the Permu CLUSTER solution. Psychol Methods 10(4):468–476

Zarazúa JA, Mazabel D, Moncayo R, Rendón-Medel R (2015) Evaluación de la sustentabilidad en espacios rurales relacionados con el ecoturismo. Estudio de caso en la Selva Lacandona, Chiapas, México. Revista Luna Azul 40:47–68

# Agroforestry for Biodiversity Conservation

**Ranjith P. Udawatta, Lalith M. Rankoth, and Shibu Jose**

## Abbreviations

| | |
|---|---|
| AF | Agroforestry |
| AMF | Arbuscular mycorrhizae fungi |
| BD | Biodiversity |
| C | Carbon |
| CS | Carbon sequestration |
| ES | Ecosystem services |
| N | Nitrogen |
| P | Phosphorus |
| SOC | Soil organic carbon |

## Introduction

Habitat loss and environmental degradation caused by population growth, agricultural intensification, and deforestation are major contributors of loss of biodiversity (BD) and associated ecosystem functions (Steffan-Dewenter et al. 2007; Culman et al. 2010). Even though modern agriculture is largely blamed for declining BD, agricultural practices can conserve BD when sustainable management practices are implemented to protect BD (Kleijn and Sutherland 2003; Opermmann et al. 2012). For instance, ~50% of plant and animal species depend on agricultural habitats in

R. P. Udawatta (✉)
The Center for Agroforestry, School of Natural Resources, University of Missouri, Columbia, MO, USA
e-mail: UdawattaR@missouri.edu

L. M. Rankoth
Department of Crop Science, Faculty of Agriculture, University of Peradeniya, Peradeniya, Sri Lanka

S. Jose
College of Agriculture Food and Natural Resources, University of Missouri, Columbia, MO, USA

 
R. P. Udawatta, S. Jose (eds.), *Agroforestry and Ecosystem Services*,
https://doi.org/10.1007/978-3-030-80060-4_10

Europe (Kristensen 2003). Therefore, agricultural practices that provide habitat, better environmental conditions, food, and protection can conserve and improve BD.

More advanced landscape-scale agroforestry (AF) approaches create stronger links between AF and BD conservation (Mcneely and Schroth 2006; Jose 2012; Jose 2009). Agroforestry intentionally integrates trees and/or livestock for increased benefits arising from interactions among components in the system (Gold and Garrett 2009). Trees, grasses, shrubs, and forbs within AF provide canopy, understory, shrub layer, herbaceous layer, and floor for structural and spatial diversity for various animals. Additionally, careful selection of trees to meet landowner objectives and soil-site conditions as well as strategic placement of each component and advanced management of the system can yield numerous production, environmental, and economic benefits. During the last two decades AF's integration of trees and/or livestock with cropping systems as a measure for BD conservation has been receiving increased attention (Sanchez 1995; Dobson et al. 1997; Huang et al. 1997; Huang 1998; Leakey 1999; Boffa 1999; Huang et al. 2002; Buck et al. 2004; Mcneely and Schroth 2006).

As AF plays five major roles in BD conservation (Jose 2009; Jose 2012) AF has been identified as a tool to preserve rich species diversity around the world (Méndez et al. 2001; Hemp 2006; Borkhataria et al. 2012). Agroforestry-induced BD improvements have been reported in both temperate and tropical regions (Noble and Dirzo 1997; Huang et al. 2002; Thevathasan and Gordon 2004; Dollinger and Jose 2018). Some studies have indicated significantly greater diversity in AF compared to forests and tree monoculture management (Huang et al. 2002; Steffan-Dewenter et al. 2007; Sistla et al. 2016). In a meta-analysis Bhagwat et al. (2008) reported 60% greater mean richness of taxa in AF than forests. In Europe, Torralba et al. (2016) demonstrated an overall positive effect of AF on BD using a meta-analysis. Conversion of AF to monocultures has reduced BD (Perfecto et al. 1996; Lawton et al. 1998; Schroth et al. 2004). In spite of greater BD in AF as compared to adjacent forests, AF usually has lower number of endemic species due to intensive management (Noble and Dirzo 1997; Bhagwat et al. 2008).

Despite numerous studies confirming the role of AF in enhancing BD, comprehensive review and synthesis on this topic are limited (Swallow and Boffa 2006; Udawatta et al. 2019). This is an updated version of a previous review on this topic published by Udawatta et al. (2019).

## Agroforestry and Diversity

### *Agroforestry and Floral Diversity*

Modern farming methods have increased crop production by implementing new technologies, modern machinery, planting improved varieties, and applying synthetic chemicals (Montgomery 2007; Nielson 2012). According to Varah et al.

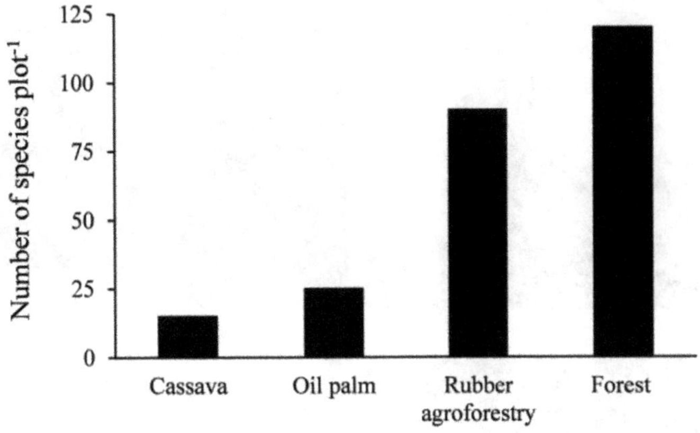


**Fig. 1** Number of plant species in monocrop cassava, oil palm, rubber agroforestry, and natural forest plots (40x5 $m^2$) in Jambi area of central Sumatra, Indonesia. Adapted from Murdiyarso et al. (2002)

(2013) the focus of modern farming is just productivity, and it reduces the environmental complexity and grows large areas of monocultures for better economies of scale. This reduced species diversity has decreased the functional diversity which eventually results in reduced ecosystem functions (Fig. 1; Murdiyarso et al. 2002; Mace et al. 2012). For example, Murdiyarso et al. (2002) observed 15 plant species per plot (200 $m^2$) in continuously cultivated cassava (*Manihot esculenta*), 25 species per plot in oil palm (*Elaeis guineensis*) plantations, 90 species per plot in rubber (*Hevea brasiliensis*) AF, and 120 species per plot in primary forests in Jambi area, central Sumatra (Fig. 1). Conversely, AF favors greater species richness and diversity and therefore integration of AF into farming systems increases BD (Jose 2012; Varah et al. 2013).

Tropical, temperate, and drier regions of the world have shown greater diversity with AF as compared to monocrops (Fig. 2). Dhakal et al. (2012) reported increased species diversity in cardamom plantations in montane forest ecosystems in Sri Lanka compared to the natural montane forests and the cardamom-based traditional AF system in Himalayas, India, and it was proven to be a sustainable land-use system with a higher plant diversity (Sharma et al. 2009). Some AF practices are less complex while Asian homegarden AF practices are more complex with multiple canopy layers and multiple species. For example, the species richness of tropical homegardens varied from 27 (Sri Lanka) to 602 (West Java) (Kumar and Nair 2007). In another study, Kabir and Webb (2008) reported 419 species, of which 59% were native, in a survey of floristic and structural diversity of 402 homegardens from six regions across southwestern Bangladesh. A study conducted to identify plant diversity and multi-use evaluated randomly selected 106 suburban homegardens in Sri Lanka and reported a total of 289 species of which 51% were ornamental plants, 36% were food plants, 12% were medicinal plants, and 6% generated income whereas the rest was used for domestic usage (Kumari et al. 2009).

**Fig. 2** Advancing plant diversity with advancing tree:crop combinations in different agroforestry practices, a single tree in a crop land (**a**); alley cropping in France (**b**); windbreaks in the USA (**c**); riparian forest buffers in the USA (**d**); coffee (*Coffea arabica*) in agroforestry under tropical climate (**e**); black pepper (*Piper nigrum*), areca nut (*Areca catechu*), and coffee intercropping in agroforestry under tropical climate (**f**); mixed cropping in Kandyan homegardens in Sri Lanka (**g**); and medicinal plant production as forest farming in Vietnam (**h**)

Homegarden AF functions as an "intermediary" for conserving tree species diversity in Bangladesh (Bardhan et al. 2012). The species richness in homegardens increased as the size of homegardens increased and 30% of the species were shared by homegardens and natural forests. According to ecologists these homegardens mimic the natural forests to the closest level, both structurally and functionally (Ewel 1999), and neotropical and old-world tropical homegardens are considered as AF systems with high floristic diversity (Jose 2009).

Agroforestry systems in West Africa, Central America, and seasonally dry regions also have exhibited greater diversity than monocrops and some forests (Donald 2004; Gillison et al. 2004; Khan and Arunachalam 2003; Michon and de Foresta 1995). Recent studies have shown greater diversity in AF than previously reported for Cameron, Kenya, and Uganda (Kindt 2002).

In Cameroon and Nigeria, cocoa (*Theobroma cacao*) AF plays a major role in their agriculture than monocrop cocoa. In the south-eastern Rift Valley escarpment of Ethiopia, Negash et al. (2012) investigated native woody species diversity in AF and reported a total of 58 woody species, belonging to 49 genera and 30 families from three different AF systems. The high diversity was attributed to the indigenous, multistrata AF systems. In Ghana, Appiah (2012) investigated understory species richness and family counts 8 years after the establishment of indigenous mixed-species plantation AF on degraded land. In their study species richness increased by 24% and the number of families represented increased by 48%. The author concluded that planted indigenous species enhanced habitats for other forest tree species on degraded sites and thereby helped to recover BD within an agricultural landscape.

Dawson et al. (2014) estimated 244 tree and shrub species in five East African countries (Eritrea, Ethiopia, Kenya, Tanzania, and Uganda) with some value as fodder (Fig. 3). Additionally, 54% and 18% of these species serve as possible alternative or additional uses such as shade and as ethnoveterinary medicines, providing supporting ecosystem service roles for livestock production. The Agroforestree Database (AFTD; www.worldagroforestry.org/resources/databases/agroforestree) contains a list of 295 trees (161 indigenous, 134 exotic) for fodder in Africa, with most species also having additional functions. For dry tropical Africa, Dicko and Sikena (1992) compiled a list of 124 fodder tree and shrub species while Smith (1992) listed 56 trees and shrubs of known fodder value in humid parts of the continent. The AFTD listed 650 tree species that provide a wide range of products and services useful for tropical smallholders (Dawson et al. 2014). Similarly, researchers have compiled lists of trees commonly found on cacao farms in Nigeria and Ghana (Fanaye et al. 2003; Osei-Bonsu et al. 2003; Asare 2005).

During the last three decades, new approaches have evolved; field boundaries were planted with mostly exotic tree fodders like Latin American calliandra (*Calliandra calothyrsus*, a leguminous species). In Kenya, Uganda, Tanzania, and Rwanda more than 200,000 dairy farmers grow calliandra as animal feed, primarily for cattle and goat (Wambugu et al. 2011). Calliandra has increased milk production and thus farm income. Farmers use calliandra as a substitute for dairy meal or as a supplement and 1 kg of dried calliandra leaf is estimated to contain the same amount

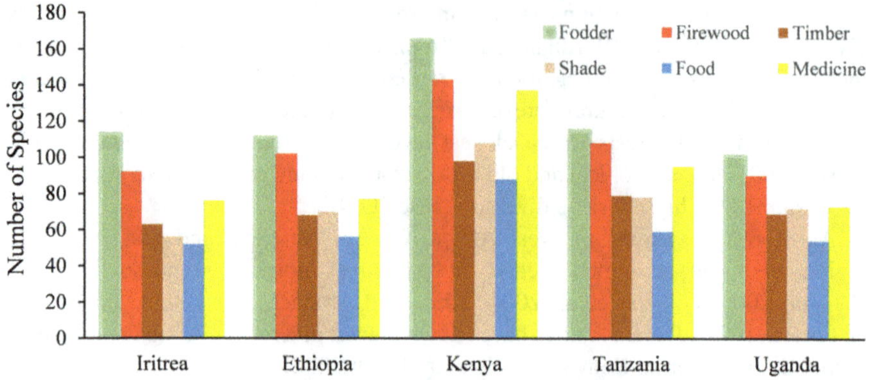


**Fig. 3** Number of tree and shrub species (indigenous and exotic) useful for fodder, firewood, timber, shade, food, and medicine in Eritrea, Ethiopia, Kenya, Tanzania, and Uganda. Adapted from Dawson et al. (2014)

of digestible protein as the same weight of commercial dairy meal (Roothaert et al. 2003).

In North America, alley cropping, forest farming, riparian buffers, silvopasture, and windbreaks are the five main AF practices. According to Sharrow et al. (2009) silvopasture is the most prevalent AF practice found in the USA and Canada. Cold and warm season grass species and hardwood and evergreens as well as N-fixing trees are integrated in these systems. Some landowners plant common forest species in pastures to provide shade for livestock which helps increase floral diversity in silvopasture.

Pecan (*Carya illinoinensis* (Wanganh.) K. Koch), walnut (*Juglans nigra*), chestnut (*Castanea* spp.), oaks (*Quercus* spp.), ashes (*Fraxinus* spp.), and basswood (*Tilia* spp.) can be used in alley cropping while walnut and poplar are the most commonly used trees (Garrett et al. 2009). Forage crops, cereals, vegetables, specialty crops, and biomass crops can be grown in alleys until light, water, and nutrients become limiting or continuously with management practices like tree thinning, branch pruning, and root pruning (Jose et al. 2004; Udawatta et al. 2014, 2016).

In forest farming, landowners draw significant additional income by planting several species of high-value marketable products under the forest canopy. These include edible products such as fruits, nuts, berries, greens, mushrooms, and wild vegetables; medicines and herbal supplements such as ginseng (*Panax ginseng*), echinacea (*Echinacea purpurea*), goldenseal (*Hydrastis canadensis*), black cohosh (*Actaea racemosa*), and witch hazel (*Hamamelis virginiana*); decorative products such as flowers, Spanish moss (*Tillandsia usneoides*), vines, stems, seedheads, leaves, and fruiting structures used in floral arrangements; and handicrafts and specialty woods such as grape vines and branches that are commonly used (Rao et al. 2004; Chamberlain et al. 2009). Riparian forest buffers along water bodies and between crop or pasture lands and water bodies also contribute to floral diversity

(Naiman et al. 2005). In a 4-year study, Elliot and Vose (2016) reported herbaceous and woody species density between 4 and 17 and 3 and 8 $m^{-2}$, respectively.

Over the years, several legislation in the USA have encouraged planting trees and creating some form of AF on the landscape. The title of the land was offered if a homestead would plant trees on 10–40 acres through the Timber Culture Act of 1873, until the offer was repealed in 1891. The Prairie States Forestry Project was implemented to establish windbreaks in 1930s to protect soils from wind erosion after the Dust Bowl. The first tree was planted in Oklahoma and approximately 217 million trees were planted in a 1150-by-100-mile zone from the Canada border to Texas in six states. Windbreaks established in the USA for farmstead use, field use, livestock shelter, living snow fence, wildlife habitat, and screening usually consist of several tree species including a combination of evergreens and deciduous as well as shrubs. Commonly used species include Arizona cypress (*Cupressus arizonica*), Austrian pine (*Pinus nigra*), bur oak (*Quercus macrocarpa*), Eastern redcedar (*Juniperus virginiana*), green ash (*Fraxinus pennsylvanica*), ponderosa pine (*Pinus ponderosa*), poplar (*Populus spp.*), redbud (*Cercis canadensis*), sycamore (*Platanus occidentalis*), walnut, white ash (*Fraxinus Americana*), willow spp. (*Salix spp.*), and white spruce (*Picea glauca*) (Kort and Turnock 1999). Windbreak AF practices were also established in Canada and former USSR to combat drought and improve soil properties to address crop failures, famine, mass exodus, and human losses (Udawatta et al. 2017).

Agroforestry has increased the floral diversity in Central and South America. For example, Sistla et al. (2016) reported that AF and secondary forests shared 38 species, secondary forest and pasture shared 15 species, and AF and pasture shared 28 species in Nicaragua. In their study, species richness per $m^2$ was 8.14 ± 0.15, 7.97 ± 0.15, and 6.18 ± 0.6 for secondary forest, AF, and pasture. Rendón-Sandoval et al. (2020) sampled perennial plant species from nine traditional AF systems and nine tropical forest sites in Tehuacán-Cuicatlán Valley, Mexico, to determine the alpha, beta, and gamma diversity. Their results showed that AF sites maintained, on average, 68% of the species (95% of them native to the region) and 53% of the abundance of individuals occurring in the adjacent forests. Agroforestry sites harbor 30% (39 species) of plants endemic to Mexico. Another study conducted in the same valley by Vallejo et al. (2016) found that on average, AF had 70% of the species found in forests of temperate and semiarid areas. Santos et al. (2019) used 72 studies for a meta-analysis with 143 study sites and 1700 quantitative comparisons to quantify various AF effects on BD in Brazilian Atlantic Forest. They used old-growth forests as a reference ecosystem. Biodiversity and ES provided by AF were 45% and 65% greater than conventional production systems. Highly diverse AF systems possessed higher BD and ES values compared to simple AF systems and conventional production systems.

Studying plant diversity in Southeast Asia, Congo Basin, and Amazon basin Tomich et al. (1998) and Tomich et al. (2001) stated that the diversity of multistrata AF systems was between primary forests and monocrop perennials or field crops. In a review using 89 published data sets, Alkemade et al. (2009) showed that AF had diversity similar to lightly used forests, secondary forests, and forest plantations.

The same review showed that primary forest contained greater diversity and agriculture contained lower diversity than AF. De Beenhouwer et al. (2013) used 74 studies across Africa, Latin America, and Asia and showed that species richness was 46% lower in plantations than coffee-AF and cocoa-AF. Additionally, recent manuscripts, reviews, meta-analysis, and other documents from studies across the world suggest that AF has greater diversity as compared to monocropping practices and perhaps greater than forests in some regions (Bhagwat et al. 2008; De Beenhouwer et al. 2013; Torralba et al. 2016; Niether et al. 2020). Tree species guides, various data sets, AFTD, and literature provide lists of suitable tree and shrub species for greater floral diversity of AF.

## *Agroforestry and Faunal Diversity*

The faunal diversity is closely linked with floral diversity and the higher floristic and structural diversity of AF has been shown to support greater faunal diversity compared to monoculture systems (Jose 2012). Because of this connection, AF farms in temperate, tropical, and subhumid regions have shown numerous direct and indirect benefits including sustainability, land productivity, and environmental services. For instance, global pollination service represents US$195–387 billion annual benefit for domesticated and wild plants (Costanza et al. 1997; Porto et al. 2020). Approximately 90% of flowering plants are pollinated by insects and over 75% of world's most important crops and 35% of food production depend on animal pollination (Kearns and Inouye 1997; Klein et al. 2007). AF practices increase pollinator diversity, which is essential for food production as well as maintenance of population levels of wild plants (Gallai et al. 2009; Varah et al. 2013).

Peters and Carroll (2012) observed similar bee species richness for four out of five flowering periods, but nearly tripled during one high-density flowering period. The study evaluated plant–pollinator interaction (how phenology, particularly flower density) and their influence on bean production in *Coffea arabica* AF in Costa Rica. Coffee flowering phenology, in turn, was proximately controlled by precipitation, and the differences in coffee flowering phenology interacted with bee species richness to influence initial fruit set rates.

Pollinator count is affected by tree density. The pollinator decrease was 40% under monocrops and 15% under mixed-tree AF compared to forests (Barrios et al. 2017). The authors also reported a no increase in the monocrop and 15% increase of less important pollinators in the mixed-tree AF. The mixed-tree AF had 93% of the pollinators found in forests and maintained 85% of the pollinators found in forests. Multiple tree species were more beneficial than a single tree species for greater pollinator BD (Barrios et al. 2017). Increased vegetation cover provided by perennial vegetation, microclimate, flowers, and nesting sites, associated with the diversity of plants, has been identified as a major contributor for increased pollinator BD (Donaldson et al. 2002; Potts et al. 2010; Klein et al. 2006, 2007).

Auad et al. (2012) observed that pastures managed in a silvopastoral environment harbored high numbers of natural enemies and beneficial insects. The objective of the study was to estimate the abundance, diversity, and constancy of families from the order Hymenoptera in a silvopastoral system in Brazil. Their 5841 specimens included 549 species in 11 families. Authors attributed the abundance and diversity of insects to microhabitats, greater protection from predators, and increased availability and diversity of food resources and nesting substrates of the complex structure of the silvopastural practice.

Neita and Escobar (2012) examined the changes in species richness, abundance, biomass, composition, and functional group structure of the dung beetle (Scarabaeinae) communities in Pacific lowlands of Colombia. The understory tree canopy cover and density varied among the three studied AF systems. Beetle diversity was lower in management systems with less tree cover and was very similar to that of abandoned agricultural fields. These authors showed that the structure of the dung beetle assemblage of *B. patinoi* growing below a diversified and permanent tree cover was similar to that of the primary and secondary forest.

Rahman et al. (2012) sampled 15 land-use practices to quantify land-use intensification on the distribution and abundance of soil invertebrate communities in a human-dominated biosphere reserve of international importance in India. The study design consisted of simple and intensively managed annual crop fields through less intensively managed AF and pristine forest ecosystems. Results showed the highest taxonomic richness in forests and the lowest in annual crops and coconut monoculture plantations. Agroforestry had the highest diversity of ants (21) and it was greater than the forests (12). Agroforestry systems, plantations, and forests also had significantly greater abundance of earthworms and millipedes than annual crops. Reviewing faunal diversity, Barrios et al. (2012) reported greater diversity of earthworms, beetles, centipedes, millipedes, termites, and ants in AF compared to continuous cropping.

Tree-crop combinations and spatial arrangements of AF systems create structural and functional diversity and thereby influence insect population density and species diversity (Jose 2009). Many studies have reported that favorable microclimate, safety, food sources, protection, and diversity provided by AF have contributed to insect diversity and richness in AF (Potts et al. 2010; Peters and Carroll 2012; Adhikari et al. 2014). For example, structures like windbreaks had greater density and diversity of insect populations than monocrop areas (Brandle et al. 2004). Insects in AF also provide indirect benefits such as pest and disease control (Pumariño et al. 2015). In West Africa, coffee berry borer infestation was 69% lower in AF coffee practices due to natural enemies such as ants and parasitoids (Perfecto et al. 1996) and birds (Karp et al. 2013) as compared to monocrops (Barrios et al. 2017).

Agroforestry systems of shade coffee and multistrata cacao provided habitat for avian, mammalian, and other species and thereby enhanced faunal diversity (Jose 2012). According to Buck et al. (2004) cocoa and coffee AF had greater bird diversity in Southeast Asia and Central America. Griffith (2000) evaluated bird biodiversity during the fires of 1998 in two AF farms in the buffer zone of the Maya

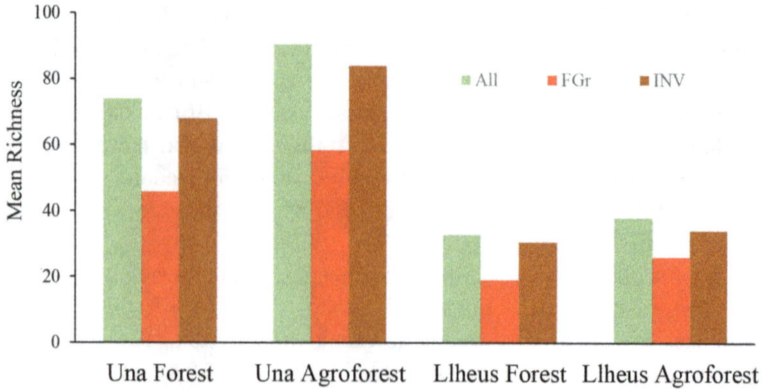


**Fig. 4** Mean richness of all birds (All), species that contribute to seed dispersal (frugivores/granivores; FGr), and species that contribute to invertebrate removal (insectivores; INV) for mature forest and agroforestry sites in Brazil. Adapted from Rocha et al. (2019)

Biosphere Reserve in Guatemala. The study noticed greater number of bird species in AF areas, including forest specialists and forest generalists. This greater diversity was attributed to fruit, nectar, nesting sites, protection from predators, and refuge both in AF and forestry patches. In the Brazilian Atlantic Forests, small cacao AF in the forested landscape sustains functional diversity as diverse as nearby forests when considering the entire community, forest specialist, and habitat generalists (Rocha et al. 2019). They evaluated two mature forest sites (49% Una and 4.8% Ilhéus) and with cacao AF cover (6% and 82%, respectively). Agroforestry sites had greater mean richness of all, seed dispersal, and insect removal birds compared to respective forest sites (Fig. 4).

Evaluating bird diversity on a 10-km$^2$ landscape of mixed rainforest, pasture, AF, and a monoculture in the Caribbean lowlands of northeastern Costa Rica, Greenler and Ebersole (2015) observed greater diversity of birds on AF (Fig. 5). The study recorded the highest number of species in riparian forests, followed by 83 in live fences. Rainforest and cacao plantation had fewer species, 61 and 60, respectively. The rainforest had the most unique species (30), 23 were unique in both riparian forests and live fences, and only 9 in the cacao plantation.

They observed 167 bird species from 36 families under AF in northeastern Costa Rica. In another meta-analysis with 52 peer-reviewed articles on cocoa production, Niether et al. (2020) observed greater BD in cocoa-AF than monocropping.

Birds and bats were significantly greater in AF than monocultures of indigenous reserves of Talamanca, Costa Rica (Fig. 6; Harvey and Villalobos 2007). They studied birds and bats of four mainland uses within the reserves: forests, cacao AF systems, banana AF systems, and plantain monocultures. They have captured a total of 2678 bats of 45 species and 3056 birds of 224 species.

Agroforestry systems recorded bat and bird assemblages that were abundant, diverse, and species rich as forests. The plantain monocultures had highly modified and depauperate assemblages of both birds and bats. The study concluded that

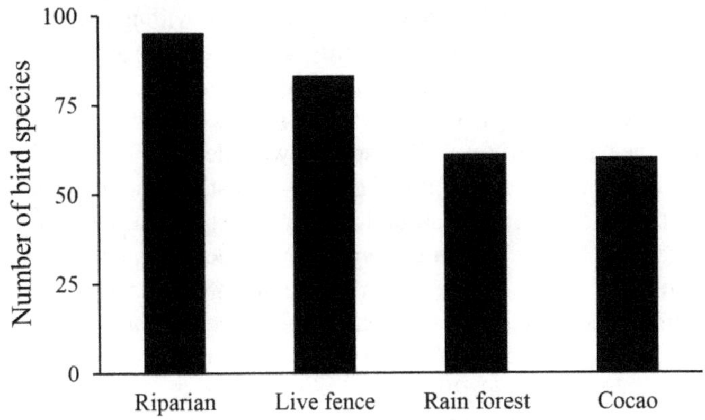


**Fig. 5** Bird species counts for riparian buffer, live fence, rain forest, and cocoa monoculture in Costa Rica. Adapted from Greenler and Ebersole (2015)

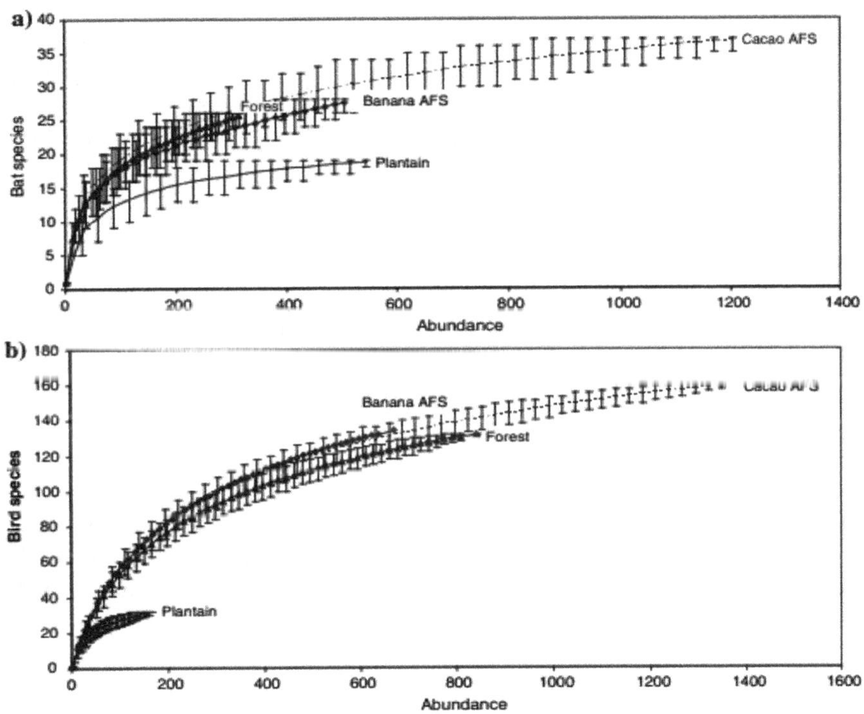


**Fig. 6** Bat (**a**) and bird (**b**) species in forests, cacao agroforestry systems, banana agroforestry systems, and plantains in Talamanca, Costa Rica. Bars represent 95% confidence intervals, based on 500 iterations. Source: Harvey and Villalobos 2007

diverse cacao and banana AF helped conserve BD by providing habitat for bats and birds. The increased species richness of birds in Sweden was attributed to the percent of shrub and tree areas within pastures (Söderström et al. 2001). These diverse systems had greater diversity and abundance of insects and other invertebrates.

In Uganda, the number of forest bird species were determined by the tree density and the distance to intact forest in intact primary forest, regenerated secondary forest, and agricultural fields in and around Mabira forest (Naidoo 2004). The size of the land parcel and human pressure were the main determinants of diversity and richness of birds in Chagga homegardens of Kilimanjaro, Tanzania (Soini 2005). In Ethiopia, coffee under shade tree AF system had twice as many bird species compared to forests (Buechley et al. 2015). Those birds benefit the farmers by controlling insect pests (Johnson et al. 2010).

Studies in Indonesia, the third largest cacao producer in the world, have shown that greater species richness of nectivore and frugivore bird species was associated with higher species richness of shade trees in cacao-based AF systems (Clough et al. 2009). Another study in Thailand showed that a mixed fruit orchard in Thailand had 75% of the bird species found in adjacent forests and was dominated by frugivores, nectarivores, and widespread generalists (Round et al. 2006).

In Australia, Fischer et al. (2010) compared birds and bats with a number of trees. They reported doubled bird richness with the presence of one tree as compared to treeless sites. At the same site bat richness tripled when 3–5 trees were present than no-tree sites, and bat activity increased by a factor of 100. Bat species richness reaches a nearly asymptotic value at roughly 5 trees $ha^{-1}$, while bird species richness keeps increasing slightly even above 100 trees $ha^{-1}$. According to Burgess (1999) bird counts were greater due to introduction of silvopasture in the UK. These farms also indicate greater numbers of airborne arthropods and small mammals.

In a review of BD in tropical AF, Bhagwat et al. (2008) found 112–139% richness of bats and lower plants (bryophytes and ferns) than the neighboring forests. The AF showed 91% of mammals, 98% of insects, and 100% of trees found in the forest. Trees provide connectivity, nesting sites, protection against predators, low-risk areas, breeding areas, food sources, landscape complexity, and heterogeneity, in conventional farming systems, and thereby enhance pollinators, birds, aquatic systems, and beneficial species into the landscape (Harvey and Villalobos 2007; Buechley et al. 2015; Greenler and Ebersole 2015; Barrios et al. 2017). Current literature supports that integration of AF helps improve faunal diversity although literature is limited for each bird species.

Among the many medium-size animals, earthworms receive greater attention as they improve soil physical, chemical, and biological properties (Spurgeon et al. 2013). Reviewing faunal diversity Barrios et al. (2012) reported greater diversity of earthworms, beetles, centipedes, millipedes, termites, and ants in AF compared to the continuous cropping control. Price and Gordon (1999) observed greater earthworm populations in AF as compared to conventional cropping systems in southern Ontario, Canada, and the density was greater near the trees. Earthworm counts ranged from the lowest 71 counts $m^{-2}$ in silver maple to 90 counts $m^{-2}$ for white ash to the highest 182 counts $m^{-2}$ for poplar. In a soil fauna and BD review, Marsden et al. (2020) reported that earthworm categories were more abundant in the tree rows than in the crop alley for 13 French silvoarable systems. Trees provide favorable conditions for earthworms and other fauna and findings of Pauli et al. (2010) indicated that the spatial distribution of earthworm casts was closely related to the distribution of trees. Findings from Portugal by Moreno et al. (2016) support results of Price and Gordon (1999) and Pauli et al. (2010). Earthworm species, numbers, and richness were greater for wooded pastures than open pastures in the extensive silvopastural systems of Iberian dehesas. A survey of 13 AF practices in France showed that soil tillage and inorganic fertilization significantly increased total earthworm abundance and biomass in tree rows due to increased SOC and lack of disturbance relative to alleys and treeless control plots (Cardinael et al. 2018).

Studies in Canada and the USA showed greater population numbers of arthropods and detritivores in AF than conventional cropping (Middleton 2001; Stamps et al. 2002; Thevathasan and Gordon 2004). Barrios et al. (2012) reviewed seven articles to evaluate AF and soil fauna and functions for tropics. The study findings showed 1–6.1 times more micro-, meso-, and macrofauna in AF than continuous cropping. Marsden et al. (2020) used 67 articles mostly from tropics to evaluate AF effects on soil fauna diversity. The study concluded that the effect of AF was mainly positive (70% of the datasets) on fauna abundance and diversity when compared with cropland, and neutral or negative when compared to forests. For their study the response ratio was between 1.2 and 2 and some reaching values close to 10.

Silvopasture with cattle, goat, sheep, swine, chicken, and bison is popular in the temperate zone (Fig. 7). These animals are mainly raised for milk, meat, and high-value products. These large-scale management practices usually raise only one type of animals or rarely two or three species on the same land. In the USA, 54 million ha of forest land is grazed or 20% of the total forest land or 13% of the total land grazed (Sharrow et al. 2009). Forest land areas are 250 and 436 million ha in the USA and Canada and these can be potentially converted to silvopasture for enhanced land productivity and BD. In Spain and Portugal 6.3 million ha of land characterized by a savannah-like system is managed under dehesa raising pigs (Joffre et al. 1999). Similarly, silvopasture is a common practice in South America. Globally, 3315 million ha pasture lands account for 30% of the global ice-free lands for livestock production and contributes 40% of the global agricultural production (Herrero et al. 2013).

Due to animals, these AF areas can have greater diversity of insects, birds, soil microorganisms, and other small and large animals (see also chapter "Enhanced Ecosystem Services Provided by Silvopastures").

In the mid-1980s, goats and sheep under rubber plantations were raised in East Aceh, North Sumatra, and West Java (Ciburuy near Bogor) as income from rubber fluctuated rapidly (Wibawa et al. 2006; Joshi et al. 2006). According to Kartamulia et al. (1993) sheep rearing under rubber plantations has increased farm profit by 22% compared to conventional monoculture plantations and reduced the weeding

**Fig. 7** Silvopastural practices in the world with different animal breeds: cattle (**a**), pigs (**b**), goats (**c**), sheep (**d**), chicken (**e**), turkey (**f**), and bison (**g**)

cost. Amir et al. (1986) concluded that small ruminants are profitable under rubber plantations with a potential increase of farm income by 11%. Additionally, over 60% of the nutritional requirements of small ruminants can be met from undergrowth of rubber plantations. Several studies have shown that sheep grazing under rubber trees was good for rubber trees and latex production due to effective weed control (Kartamulia et al. 1993; Karokaro 1996). These animals provide manure and control weeds. Control of understory also reduces the fire risk (Rigueiro-rodríguez et al. 2004).

In tropical and temperate regions, raising free-range chickens under trees and various plants is receiving increased attention (Fig. 7). In these systems chickens are raised in managed habitats where they have regular access to a diverse diet rich in plants and insects. The chickens consume forage and insects and thus control the understory growth, pest, and insects and return the nutrients back to the system through their manure. The system provides shelter and water.

Only a few studies have reported relationships between BD for medium to large animals. For example, in Makalu Barun National Park and Conservation Area of Nepal, AF with *Alnus nepalensis* and cardamom has contributed to the integrity of riparian corridors for wildlife conservation (Zomer et al. 2001). The smaller parcel size and lack of research studies may have contributed to lack of data on BD and AF for medium to large fauna. Farmers and landowners select desired species that contribute to either farm productivity or farm income. Therefore, AF has very limited potential to conserve BD of harmful and noncontributing large or small fauna. However, only larger structures like windbreaks, riparian buffers, and forest-AF connections offer woody habitat for medium and larger body wildlife in many agriculture-dominated and monocropping landscapes, thus leading to improved diversity of medium and large wildlife (Johnson and Beck 1988), habitat, and species richness (Söderström et al. 2001). Millspaugh et al. (2009) have described changes and management options for AF to improve diversity of medium and large wildlife.

Even though traditionally there are several types of AF systems both in tropical and temperate environments, new combinations of trees and crops and animals were also being practiced for a better productivity which ultimately leads to a greater BD. One such advance is AF integration of fish into the cropping system. Riparian buffer AF improves the stream environment, thus indirectly improving the fish production (Gordon et al. 1992). Agrosilvofishery which is unique to China integrates fish in the AF system where the trees provide shade and litter as food for fish and receive water and nutrients back from the ponds. Crops such as rice, wheat, and melon are also used in these systems and time-to-time sediment removal from the ponds enriches the soil environment (Zou and Sanford 1990). Mangrove AF systems in Indonesia are also known to be facilitating and improving fish, shrimp, and shellfish production and breeding (Weinstock 1994).

## *Agroforestry and Soil Microbial Diversity*

One square meter of soil may contain ~20,000 species of fungi, bacteria, virus, protozoa, nematodes, and many other species (Keesstra et al. 2016). Soil microbial communities are important for most biogeochemical processes including biogeochemical cycles, nutrient cycling, mineralization, nutrient supply, degradation of chemicals, aboveground BD, soil formation, and soil health. Soil microbes can be found on surfaces and inside of soil aggregates, roots, organic matter, and dead and live plants and animals. Soil functions like moisture, temperature, food sources, chemicals, and stress conditions can affect the functions, community structure, and diversity of soil communities. Perennial vegetation and plant diversity change these soil conditions and therefore affect microbial communities, functions, structure, richness, and diversity. Among various soil microbial groups, fungi, bacteria, actinomycetes, nematodes, and virus are considered vital for many soil functions. This section describes AF effects on diversity, richness, and community structure of various soil microbial groups.

Research from both temperate and tropical climates has shown greater fungal densities and diversity in AF than monocropping. For example, two major soil fungal groups, Ascomycota and Basidiomycota, were 12–96 times more abundant in the tree row and at 1 m crop row than at 7 m and 24 m crop rows and the monoculture cropland in Germany (Beule et al. 2020). Lacombe et al. (2009) used PLFA profiling to estimate fungal abundance in Canada, and showed that the abundance was greater in a tree-based system compared to adjacent cropping systems. In another study, Chifflot et al. (2009) observed significantly greater spatial differences and diversity of fungi pores on tree-based intercropping sites compared to a monoculture. Bainard et al. (2012) observed greater arbuscular mycorrhizal fungi (AMF) richness in a tree-based cropping system and several taxa that were not found in monocropped lands at the University of Guelph Research Station, Canada. In their study, AMF composition was significantly different between the monocrop and tree-based cropping systems. The study also showed that tree species influenced fungal communities. In their study AMF communities growing on corn (*Zea mays*) roots were significantly different for Norway spruce (*Picea abies*) and silver maple (*Acer saccharinum*).

Spatial differences and compositional changes occur within the systems. For instance, Beuschel et al. (2019) in Germany found greater abundance of saprotrophic fungi and ectomycorrhizal fungi, and a greater fungal C:bacterial C ratio in tree rows of the silvopasture. As the system matures, more saprotrophic and ectomycorrhizal fungi were noticed and this shift was attributed to more complex and diverse organic material, reduced soil disturbance, and soil pore geometry (Beuschel et al. 2019). Using molecular techniques with three AF systems, Zhang et al. (2018) have noticed greater fungal diversity in the rhizosphere compared to bulk soil.

System age and species composition influence fungal diversity and studies all over the world have reported similar observations. According to Kremer and Kussman (2011) AMF diversity and richness were greater in fruit tree AF practice with perennial native vegetation compared to single-species alleys of tall fescue in their long-term study in Missouri, USA. Significantly greater richness in AMF communities with the tree-based cropping systems was reported in Southeast Asia, Congo Basin, and Amazon Basin (Tomich et al. 1998; Tomich et al. 2001). Supporting these observations Chifflot et al. (2009), Bainard et al. (2011a), and Kremer and Kussman (2011) found significantly greater fungal diversity and spores near trees of tree-based alley cropping than monocrop and forests. The literature also describes higher level of root colonization and greater spore densities in the rhizosphere of crops growing in close proximity to trees than away from trees (Mutabaruka et al. 2002; Pande and Tarafdar 2004; Prasad and Mertia 2005). A review by Bainard et al. (2011a) showed that integration of AF increases fungal diversity and abundance than monocrop systems.

AMF richness and diversity translate into numerous benefits including better soil structure, water dynamics, nutrient status, microbial community structure, and weed suppression (Bainard et al. 2012). Increased AMF in turn favors greater soil bacteria and protozoa within AF systems. These benefits can be attributed to high density of roots of multiple vegetation including trees, shrubs, grass, and crops as well as contributions from livestock and inevitable belowground interactions within the soil (Jose et al. 2000, 2004).

Agroforestry has been shown to have greater bacterial abundance and species richness than monocropping. Banerjee et al. (2015) noticed greater bacterial diversity and richness in AF systems of Canada as compared to monocrop. The difference was attributed to soil C and pH. In Missouri, Kremer and Hezel (2013) investigated microbial communities between alley cropping with fruit trees and alleys with native grasses and forbs. They observed higher total soil microbial biomass and more robust microbial community compositions especially more gram-negative bacteria in AF than alleys of tall fescue, adjacent unmanaged pasture, and row-cropped fields.

Beule et al. (2020) assessed 13 taxonomic groups of microorganisms in temperate poplar-based alley cropping AF in Germany with real-time PCR molecular tools. The abundance of acidobacteria, actinobacteria, alpha- and gamma-proteobacteria, Firmicutes, and Verrucomicrobia was greater in the poplar tree rows than crop rows and monoculture croplands. Differences in poplar root density, input of tree litter, soil moisture, and absence of tillage in the tree rows have contributed to higher abundance in AF. These counts were 2.0–2.9 times greater in the tree row than in the middle of the crop row and the monoculture croplands. Banerjee et al. (2015) observed the abundance of certain bacterial taxa in a large-scale Canadian AF study. Their amplicon sequencing study concluded that AF supported bacterial growth but bacterial diversity was not increased.

Increased bacterial and microbial abundance has been reported for AF systems with the increasing number of plants for many regions of the world (Zak et al. 2003; Bardhan et al. 2013; Banerjee et al. 2015; Beule et al. 2020). For example, Banerjee et al. (2015) reported significantly greater bacterial abundance and species richness in hedgerows and woodlands compared to agricultural lands using 16S rRNA gene copies across a 270 km soil climate gradient in Alberta, Canada. Compared to conventional agriculture higher plant diversity, favorable conditions, higher microbial biomass, and N mineralization rates are reported in temperate windbreaks (Kaur et al. 2000; Wojewoda and Russel 2003). Perennial vegetation and diverse plant communities change soil physical, chemical, and biological properties in addition to reduced disturbance and chemicals. Quantity and quality of litter, complex organic compounds, rhizodeposition products, soil physiochemical properties, aggregate stability, thermal parameters, and microclimate favor diverse soil communities (Bainard et al. 2011a, 2011b; Amador et al. 1997; Boerner et al. 2000; Mungai et al. 2006; Dornbush 2007; Udawatta and Anderson 2008; Udawatta et al. 2008; Helgason et al. 2010; Adhikari et al. 2014).

Field and laboratory studies show that these diverse groups are more resilient due to diversity of the litter and rhizodeposition products of trees (Keith et al. 2008; Rivest et al. 2013). Increased fungi:bacteria ratios were reported in tree rows compared to the crop rows of alley cropping AF systems (Beuschel et al. 2019). Additionally, the integration of trees into agricultural fields decreased the metabolic quotient indicating a greater substrate-use efficiency of soil microorganisms (Rivest et al. 2013; Beuschel et al. 2019). Comparing temperate AF cropland and grassland Beule et al. (2020) showed increased fungi:bacteria ratio under trees. The increased ratio was attributed to alterations of ammonium-oxidizing populations.

Agroforestry increases functional diversity of enzyme activities as well (Seiter et al. 1995; Mungai et al. 2006; Udawatta et al. 2008; Weerasekara et al. 2016; Beule et al. 2020). The arrangement of crops, pasture, shrubs, and trees in AF causes spatially multifaceted landscapes that are different from a monocrop management and these differences influence soil, site, and microclimate and thus affect biological functions of enzymes (Jose 2009; Banerjee et al. 2015). Laboratory and field research has shown that AF has diverse and rich enzyme activities than crop and pasture lands (Mungai et al. 2006; Udawatta et al. 2008; Paudel et al. 2011, 2012; Bainard et al. 2012; Banerjee et al. 2015). A meta-analysis by Bhagwat et al. (2008) reported greater mean richness of taxa in AF as compared to forests. The study also showed 60% greater richness in AF than cropping systems.

Soil enzymes were significantly greater in grass only and grass plus tree buffers than in continuously cropped areas of alley cropping AF systems in Missouri (Meyers et al. 2001; Mungai et al. 2006; Udawatta et al. 2008). On the same watersheds, 10 years after the buffers were established Weerasekara et al. (2016) noticed significantly greater enzyme activities in buffer soils than crop areas. In another study on deep loess soils with buffers and grazing Paudel et al. (2011, 2012) investigated enzyme activities. In their study enzyme activities were greater in eastern cottonwood trees (*Populus deltoides*) plus tall fescue grass [*Festuca arundinacea*] buffers, tall fescue grass buffers, and permanent pasture alleys of tall fescue plus forage legumes, compared with a row-cropping (corn-soybean (*Glycine max*) rotation) system. Greater enzyme activities have been reported in AF than monocropping and pastures for temperate and tropical regions (Chander et al. 1998; Rodrigues et al. 2015).

Enzyme activity has been shown to decrease with increasing distance from tree areas. For example, Seiter et al. (1995) noticed declining bacterial and fungal biomass in an alder (*Alnus rubra*)-sweet corn alley cropping system with increasing distance from tree rows. In Germany, Beule et al. (2020) observed greater fungal abundance in tree areas than 7 m and 24 m from tree rows. Enzymes degrade and convert complex molecules to simple molecules and also synthesize compounds and molecules for their needs and thus greater enzyme activities indicate enhanced potential to degrade cellulose, hemicellulose, chitin, peptidoglycan, and proteins which leads to improved mineralization and nutrient cycling.

Montagne et al. (2017) assessed the effects of intercropped and neighboring trees on the soil of three AF vineyards on indigenous soil microbial communities in south-western France. They used a metagenomic approach that consisted of calculating fungi:bacteria by qPCR and characterizing the microbial diversity by Illumina sequencing of 16S and 18S regions. Microbial abundance and diversity were significantly different between the soils of AF vineyards and neighboring forest. Kaur et al. (2000) evaluated the role of AF in the microbial biomass of a rice–berseem clover (*Trifolium alexandrinum*) crop rotation and agrisilvicultural systems of Acacia, Eucalyptus, and Populus along with rice–berseem and single-species tree plantations. Microbial biomass carbon was low in rice–berseem crops (96.14 $\mu g\ g^{-1}$ soil) and increased in soils under tree plantations (109.12–143.40 $\mu g\ g^{-1}$ soil) and agrisilvicultural systems (133.80–153.40 $\mu g\ g^{-1}$ soil). Microbial biomass C (42%) and N (13%) were significantly greater in tree-based systems as compared to monocropping.

Phospholipid fatty acid analysis investigation for the detection of multiple microbial communities or functional genes in soils of AF systems is limited in the literature. In a recent study, Alagele et al. (2020) found greater bacterial, fungal, actinomycetes, and total microbial biomass in AF buffers and grass buffers than crop areas (Fig. 6, in chapter "Water Quality and Quantity Benefits of Agroforestry and Processes: Long-Term Case Studies from Missouri, USA"). In their study the highest concentrations were found in the grass waterways of the lowest landscape position. Grass waterways were never disturbed, the grass cover on the waterway was established in 1990, and buffers on the watersheds were established in 1997 (Fig. 2, in chapter "Water Quality and Quantity Benefits of Agroforestry and Processes: Long-Term Case Studies from Missouri, USA"). The study also showed decreasing microbial biomass with increasing distance from trees (Fig. 6, in chapter "Water Quality and Quantity Benefits of Agroforestry and Processes: Long-Term Case Studies from Missouri, USA").

## Practical Implication of Agroforestry on Biodiversity

Agroforestry has a greater potential for BD conservation because 43% of the global agricultural land already has at least 10% tree cover (Zomer et al. 2016). This indicates the significant presence of AF, in one form or another, at the global scale although many of these practices may not be formally recognized as AF. Despite this large figure, the current level of intentional adoption of AF is very low. Several conventional production practices currently fail to meet the challenges of sustainable production, food security, and other related economic and environmental goals. Therefore, there is a great opportunity for federal, state, and local governments as well as private agencies to contribute and participate at various levels to enhance AF adoption.

Agroforestry mimics natural forests (Steffan-Dewenter et al. 2007) and therefore AF can be promoted for BD conservation and improvement. Agroforestry causes spatially heterogeneous and/or concentrated communities due to spatial arrangement of trees, shrubs, and grasses (Pauli et al. 2010; Bainard et al. 2012). However, trees, shrubs, or animals in AF can favor certain species over others. For instance, Sileshi and Mafongoya (2007) reported an abundance of earthworms and beetles with legumes producing high-quality biomass and increases of millipedes and beetles with legumes producing low-quality biomass. The spatial heterogeneity and species favoritism emphasize the importance of landscape-level planning and incorporation of many species. There is another reason for incorporation of multiple species as AF often supports fewer species compared to its natural analogues, i.e., adjacent forests or savannahs (Noble and Dirzo 1997; Bhagwat et al. 2008). Barrios et al. (2012) cautioned about generalizing BD improvement of AF. It is imperative to select soil-site-climate-appropriate components, follow a scientifically guided strategic planning, and implement intensive management practices for desired production, economical, BD conservation, and environmental benefits. While AF may not completely conserve BD as in natural habitats, it often serves as an intermediary for species conservation compared to other intensive production systems (Bardhan et al. 2012).

Even though AF mimics complexity of natural forests and is better for wildlife than conventional monocrop agriculture, it is doubtful that AF farmers and landowners will integrate every species of fauna and flora in their farming system (Krause 2019). Each plant or animal is expected to contribute to the production or service aspect of the system and plants or animals that do not contribute may not be integrated into the AF farming practice. These could include small, large, harmful, dangerous, and unwanted animals as well as plants with no production or economic value. Even endangered and threatened species will have very little potential to include in AF practices if they do not contribute to production or economics. Although deliberate species selection for human consumptive use is evident in AF practices (Jose 2011), AF can still harbor a much higher species richness and diversity compared to monoculture cropping systems (Jose 2012). Although larger animals need larger land areas AF can serve as wildlife corridors and food forests and thereby help conserve BD between protected areas or as buffer between protected areas and production-dominated landscapes.

Fertilizers, pesticides, herbicides, and other chemicals exert a strong impact on BD (Kleijn et al. 2008) and agricultural systems with less input promote higher BD (Thiele-Bruhn et al. 2012). A well-designed AF practice that provides required nutrients for the entire farm can reduce synthetic chemical use. This also helps reduce machinery use for chemical applications and thereby helps climate mitigation. This approach of small and diversified AF, particularly in the tropics, is a viable land-use strategy for BD conservation and can contribute to sustainable agricultural intensification and provision of ecosystem services (Swallow and Boffa 2006; Sistla et al. 2016).

## Summary and Future Directions

The recent Intergovernmental Science-Policy Platform on Biodiversity and Ecosystem Services (IPBES 2019) report emphasizes the importance of BD conservation for food security, climate mitigation, reduced desertification, and environmental stability. Biodiversity is declining at an alarming rate at every corner of the Earth, significantly impacting ecosystem services and threatening all inhabitants of the Earth. The declining BD has been attributed to a number of factors, including agricultural intensification, population growth, dietary patterns, industrialization, loss of habitat, and climate change. Agroforestry mimics the complexity of natural ecosystems and provides some level of refuge and habitat for certain flora and fauna, thereby conserving and promoting BD. Adoption of AF could help partially reverse the damage associated with agricultural intensification and deforestation. The above- and belowground structural complexity and multilayer canopy of AF integration assist in many ways to satisfy suitable conditions for BD. Studies have also shown that more complex and older AF systems were more diverse than simple or younger systems. Also, it has been shown that the larger the size of an AF practice, the greater its BD conservation value. Agroforestry-induced BD conservation has been attributed to food, shelter, habitat, protection, refuge, favorable microclimate, improved soil-plant-water relationships, and other resources provided by multispecies vegetation of AF.

The review also emphasizes the importance of multispecies integration for greater heterogeneity of the landscape. The selection of site-climate-suitable combinations can be used to further enhance BD and thereby the services provided by the enhanced BD. However, proper planning must be conducted before the practices are adopted. These include selection of soil-site-climate-suitable species (species combinations) and consideration of social and local needs. Since trees and animals could create spatial heterogeneity in soil and plant properties, species and/or species combinations and spacing should be considered for optimum results. Newer molecular techniques including DNA analysis, PLFA, metagenomics, and metabolomics can be used to understand relationships among various components of the AF practices and select the most suitable combinations and implement best management plans. Simulation of appropriate models may enhance our understanding of interactions and benefits and facilitate predictions for short- and long-term benefits of adopted practices. Long-term impacts of benefits should be evaluated for proposing new management strategies for enhancement of BD and ecosystem services including environmental, production, and economic benefits. Future studies could evaluate potential integration of endangered, threatened, but less commercially important species into AF farming practices for the protection of these species and for overall

BD conservation. Several other main research topics including development of lists of plants and animals for different regions, management strategies by regions, and support services like financial, input, extension, and marketing could help strengthen AF adoption further and thereby increase its BD conservation value on a global scale.

**Acknowledgments** This is an updated and expanded version of a previous paper published by the authors (Udawatta et al. 2019) for which the authors retain the copyright.

## References

Adhikari P, Udawatta RP, Anderson SH, Gantzer CJ (2014) Soil thermal properties under prairies, conservation buffers, and corn/soybean land use systems. Soil Sci Soc Am J 78:1977–1986

Alagele SM, Anderson SH, Udawatta RP, Veum KS, Rankoth LM (2020) Long-term perennial management and cropping effects on soil microbial biomass for claypan soils. Agron J 112:815–827. https://doi.org/10.1002/agj2.20116

Alkemade R, van Oorschot M, Miles L, Nellemann C, Bakkenes L, ten Brink B (2009) GLOBIO3: a framework to investigate options for reducing global terrestrial biodiversity loss. Ecosystems 12:374–390

Amador JA, Glucksman AM, Lyons JB, Gorres JH (1997) Spatial distribution of soil phosphatase activity within a riparian forest. Soil Sci 162:808–825

Amir P, Knipscheer HC, Karokaro S, . Reese A, De Boer AJ, (1986) The potential of small ruminant livestock on rubber plantation schemes: a linear programming analysis. Working Paper No. 86. Winrock International, Morrilton, AR

Appiah M (2012) Changes in plant species composition within a planted forest in a deciduous agroecosystem in Ghana. Agrofor Syst 85:57–74. https://doi.org/10.1007/s10457-011-9459-3

Asare R (2005) Cocoa agroforests in West Africa: a look at activities on preferred trees in the farming systems, Danish Centre for Forest, Landscape and Planning (KVL):77

Auad AM, Resende TT, da Silva DM, Fonseca MDA (2012) Hymenoptera (Insecta: hymenoptera) associated with silvopastoral systems. Agroforest Syst 85:113–119. https://doi.org/10.1007/s10457-011-9449-5

Bainard LD, Klironomos JN, Gordon AM (2011a) Arbuscular mycorrhizal fungi in tree-based intercropping systems: a review of their abundance and diversity. Pedobiologia 54:57–61

Bainard LD, Kochb AM, Gordon AM, Newmaster SG, Thevathasan NV, Klironomosb JN (2011b) Influence of trees on the spatial structure of arbuscular mycorrhizal communities in a temperate tree-based intercropping system. Agric Ecosyst Environ 144:13–20

Bainard LD, Koch AM, Gordon AM, Klironomos JN (2012) Temporal and compositional differences of arbuscular mycorrhizal fungal communities in conventional monocropping and tree-based intercropping systems. Soil Biol Biochem 45:172–180

Banerjee S, Baah-Acheamfour M, Carlyle CN, Bissett A, Richardson AE, Siddique T, Bork EW, Chang SX (2015) Determinants of bacterial communities in Canadian agroforestry systems. Environ Microbiol 18:1805–1816

Bardhan S, Jose S, Biswas S, Kabir K, Rogers W (2012) Homegarden agroforestry systems: an intermediary for biodiversity conservation in Bangladesh. Agrofor Syst 85(1):29–34

Bardhan S, Jose S, Udawatta RP, Fritschi F (2013) Microbial community diversity in a 21-year-old temperate alley cropping system. Agrofor Syst 87:1031–1041

Barrios E, Sileshi GW, Shepherd K, Ferguson S (2012) Agroforestry and soil health: linking trees, soil biota, and ecosystem services. In: Wall DH et al (eds) Soil ecology and ecosystem services, 1st edn. Oxford University Press, pp 315–330

Barrios E, Valencia V, Jonsson M, Brauman A, Hairiah K, Mortimer PE, Okubo S (2017) Contribution of trees to the conservation of biodiversity and ecosystem services in agricultural landscapes. Int J Biodivers Sci Ecosyst Serv Manag 14(1):1–16. https://doi.org/10.1080/21513732.2017.1399167

Beule L, Lehtsaar E, Corre MD, Schmidt M, Veldkamp E, Karlovsky P (2020) Poplar rows in temperate agroforestry croplands promote bacteria, fungi, and denitrification genes in soils. Front Microbiol 10:3108. https://doi.org/10.3389/fmicb.2019.03108

Beuschel R, Piepho HP, Joergensen RG, Wachendorf C (2019) Similar spatial patterns of soil quality indicators in three poplar-based silvo-arable alley cropping systems in Germany. Biol Fertil Soils 55:1–14

Bhagwat SA, Willis KJ, Birks HJB, Whittaker RJ (2008) Agroforestry: a refuge for tropical biodiversity? Trends Ecol Evol 23:261–267. https://doi.org/10.1016/j.tree.2008.01.005

Boerner RE, Decker KL, Sutherland EK (2000) Prescribed burning effects on soil enzyme activity in a southern Ohio hardwood forest: a landscape-scale analysis. Soil Biol Biochem 32(7):899–908

Boffa JM (1999) Agroforestry parklands in sub-Saharan Africa. FAO Conservation Guide 34. Food and Agriculture Organization of the United Nations, Rome. http://www.fao.org/DOCREP/005/X3940E/X3940E00.html

Bongaarts J (2019) IPBES, 2019. Summary for policymakers of the global assessment report on biodiversity and ecosystem services of the intergovernmental science policy platform on biodiversity and ecosystem services. Popul Dev Rev 45(3):680–681

Borkhataria RR, Collazo JA, Groom MJ (2012) Species abundance and potential biological control services in shade vs. sun coffee in Puerto Rico. Agric Ecosyst Environ 151:1–5

Brandle JR, Hodges L, Zhou X (2004) Windbreaks in sustainable agriculture. Agrofor Syst 61:65–78

Buck LE, Gavin TA, Lee DR, Uphoff NT, Behr DC, Drinkwater LE, Hively WD, Werner FR (2004) Ecoagriculture: a review and assessment of its scientific foundations. Cornell University, Ithaca, NY

Buechley ER, Şekercioğlu CH, Atickem A, Gebremichael G, Ndungu JK, Mahamued BA, Beyene T, Mekonnen T, Lens L (2015) Importance of Ethiopian shade coffee farms for forest bird conservation. Biol Conserv 188:50–60

Burgess PJ (1999) Effects of agroforestry on farm biodiversity in the UK. Scott For 53:24–27

Cardinael R, Guenet B, Chevallier T, Dupraz C, Cozzi T, Chenu C (2018) High organic inputs explain shallow and deep SOC storage in a long-term agroforestry system–combining experimental and modeling approaches. Biogeosciences 15:297–317

Chamberlain JL, Mitchell D, Bringham T, Hobby T, Zabek L, Davis J (2009) Forest framing practices. In: Garrett HE (ed) North American agroforestry: an integrated science and practice, 2nd edn. American Society of Agronomy, Madison, WI, pp 219–256

Chander K, Goyal S, Nandal DP, Kapoor KK (1998) Soil organic matter, microbial biomass and enzyme activities in a tropical agroforestry system. Biol Fertil Soils 27:168–172

Chifflot V, Rivest D, Olivier A, Cogliastro A, Khasa D (2009) Molecular analysis of arbuscular mycorrhizal community structure and spores distribution in tree-based intercropping and forest systems. Agric Ecosyst Environ 131:32–39

Clough Y, Putra DD, Pitopang R, Tscharntke T (2009) Local and landscape factors determine functional bird diversity in Indonesian cacao agroforestry. Biol Conserv 142(5):1032–1041

Costanza R, d'Arge R, de Groot R, Farber S, Grasso M, Hannon B, Limburg K, Naeem S, O'Neill RV, Paruelo J, Raskin RG, Sutton P, van den Belt M (1997) The value of the world's ecosystem services and natural capital. Nature 387:253–260

Culman SW, Young-Mathews A, Hollander AD, Ferris H, Sanchez-Moreno S, Green ATO, Jackson LE (2010) Biodiversity is associated with indicators of soil ecosystem functions over a landscape gradient of agricultural intensification. Landsc Ecol 25:1333–1348

Dawson IK, Carsan S, Franzel S, Kindt R, van Breugel P, Graudal L, Lillesø J-PB, Orwa C, Jamnadass R (2014) Agroforestry, livestock, fodder production and climate change adaptation

and mitigation in East Africa: issues and options. ICRAF Working Paper No. 178. Nairobi, World Agroforestry Centre. https://doi.org/10.5716/WP14050.PDF
De Beenhouwer MM, Aerts R, Honnay O (2013) A global meta-analysis of the biodiversity and ecosystem service benefits of coffee and cacao agroforestry. Agric Ecosys Environ 175:1–7. https://doi.org/10.1016/j.agee.2013.05.003
Dhakal B, Pinard MA, Gunatilleke IN, Gunatilleke CS, Weerasinghe HM, Dharmaparakrama ALS, Burslem DF (2012) Impacts of cardamom cultivation on montane forest ecosystems in Sri Lanka. For Ecol Manag 274:151–160
Dicko MS, Sikena LK (1992) Fodder trees and shrubs in range and farming systems in dry tropical Africa. In: Speedy A, Pugliese P-L (eds) Legume trees and other fodder trees as protein sources for livestock, Proceedings of the FAO Expert Consultation Held at the Malaysian Agricultural Research and Development Institute (MARDI), Kuala Lumpur, Malaysia, 14–18 October 1991. Food and Agriculture Organization of the United Nations, Rome, pp 27–41
Dobson AP, Bradshaw AD, Baker AJM (1997) Hopes for the future: restoration ecology and conservation biology. Science 277:515–522
Dollinger J, Jose S (2018) Agroforestry for soil health. Agrofor Syst 92:213–219
Donald PF (2004) Biodiversity impacts of some agricultural commodity production systems. Conserv Biol 18(1):17–37
Donaldson J, Nänni I, Zachariades C, Kemper J (2002) Effects of habitat fragmentation on pollinator diversity and plant reproductive success in Renosterveld shrub-lands of South Africa. Conserv Biol 5:1267–1276
Dornbush ME (2007) Grasses, litter and their interaction affects microbial biomass and soil enzyme activity. Soil Biol Biochem 39:2241–2249
Elliot K, Vose JM (2016) Riparian area harvesting impacts on vegetation composition and diversity: headwaters to estuaries. In: Young T, Parece TE (eds) Advances in watershed science and management. Springer, Berlin/Heidelberg, p 104
Ewel JJ (1999) Natural systems as models for the design of sustainable systems of land use. Agrofor Syst 45(1–3):1–21
Fanaye AO, Adeyemi EA, Olaiya AO (2003) Spacing experiments in cocoa/kola/citrus intercrop (poster). 14th Int. Cocoa Res. Conf. Oct 13-18, 2003. Accra, Ghana.
Fischer J, Stott J, Law BS (2010) The disproportionate value of paddock trees. Biol Conserv 143:1564–1567
Gallai N, Salles JM, Settele J, Vaissière BE (2009) Economic valuation of the vulnerability of world agriculture confronted with pollinator decline. Ecol Econ 68(3):810–821
Garrett HE, McGraw RL, Walter WD (2009) Alley cropping practices. In: Garrett HE (ed) North American agroforestry: an integrated science and practice, 2nd edn. American Society of Agronomy, Madison, WI, pp 133–162
Gillison AN, Liswanti N, Budidarsono S, van Noordwijk M, Tomich TP (2004) Impact of cropping methods on biodiversity in coffee agroecosystems in Sumatra, Indonesia. Ecol Soc 9(2):7
Gold MA, Garrett HE (2009) Agroforestry nomenclature, concepts, and practices. In: Garrett HE (ed) North American agroforestry: an integrated science and practice, 2nd edn. American Society of Agronomy, Madison, WI, pp 45–56
Gordon AM, Williams PA, Kaushik NK (1992) Advances in agroforestry: crops, livestock and fish have it made in the shade. Highlights Agric Food Res Ont 15:2–7
Greenler SM, Ebersole JJ (2015) Bird communities in tropical agroforestry ecosystems: an underappreciated conservation resource. Agrofor Syst 89(4):691–704
Griffith DM (2000) Agroforestry: a refuge for tropical biodiversity after fire. Conserv Biol 14(1):325–326
Harvey CA, Villalobos JAG (2007) Agroforestry systems conserve species-rich but modified assemblages of tropical birds and bats. Biodivers Conserv 16:2257–2292. https://doi.org/10.1007/s10531-007-9194-2
Helgason BL, Walley FL, Germida JJ (2010) No-till and soil management increases microbial biomass and alters community profiles in soil aggregates. App Soil Ecol 46:390–397

Hemp A (2006) The banana forests of Kilimanjaro: biodiversity and conservation of the Chagga homegardens. In: Hawksworth DL, Bull AT (eds) Forest diversity and management. Topics in Biodiversity and Conservation. Springer, Netherlands, pp 133–155

Herrero M, Havlik P, Valin H et al (2013) Biomass use, production, feed efficiencies, and greenhouse gas emissions from global livestock systems. Proc Nat Acad Sci U S A 110:20888–20893

Huang W (1998) Productive coexistence and gain in agroforestry systems. Acta For Fenn 260:1–72

Huang W, Kanninen M, Xu Q, Huang B (1997) Agroforestry in China: present state and future potential. Ambio 26:393–397

Huang W, Luukkanen O, Johanson S, Kaarakka V, Räisänen S, Vihemäki H (2002) Agroforestry for biodiversity conservation of nature reserves: functional group identification and analysis. Agrofor Syst 55(1):65–72

Joffre R, Rambal S, Ratte JP (1999) The dehesa system of southern Spain and Portugal as a natural ecosystem mimic. Agrofor Syst 45(1):57–79

Johnson RJ, Beck MM (1988) Influences of shelterbelts on wildlife management and biology. Agric Ecosyst Environ 22(23):301–335

Johnson MD, Kellermann JL, Stercho AM (2010) Pest reduction services by birds in shade and sun coffee in Jamaica. Anim Conserv 13(2):140–147

Jose S (2009) Agroforestry for ecosystem services and environmental benefits: an overview. Agrofor Syst 76(1):1–10

Jose S (2011) Managing native and non-native plant species in agroforestry. Agrofor Syst 83:101–105

Jose S (2012) Agroforestry for conserving and enhancing biodiversity. Agrofor Syst 85(1):1–8

Jose S, Gillespie AR, Pallardy SG (2004) Interspecific interactions in temperate agroforestry. Agrofor Syst 61:237–255

Joshi L, Wibawa G, Lahang I, Akiefnawati R, Mulyoutami E, Wulandari D, Penot E (2006) Diversified rubber agroforestry for smallholder farmers—a better alternative to monoculture. Paper presented at workshop "Rubber Development in Lao PDR: exploring Improved Systems for Smallholder Rubber Production", Vientiane, Lao PDR, 9–11 May, 2006

Kabir ME, Webb EL (2008) Can homegardens conserve biodiversity in Bangladesh? Biotropica 40:95-103. http://onlinelibrary.wiley.com/doi/10.1111/j.1744-7429.2007.00346.x/full. Verified 6 May 2020

Karokaro S (1996) The economic value of sheep grazing to manage weeds in rubber plantations. Working Paper No. 171. Winrock International, Morrilton, AR, p 7

Karp DS, Mendenhall CD, Sandi RF, Chaumont N, Ehrlich PR, Hadly EA, Daily GC (2013) Forest bolsters bird abundance, pest control and coffee yield. Ecol Lett 16:1339–1347

Kartamulia I, Karokaro S, De Boer J (1993) Economic analysis of sheep grazing in rubber plantation: a case study of OPMM Membang Muda. Working Paper No. 145. Winrock International, Morrilton, AR, p 12

Kaur B, Gupta SR, Singh G (2000) Soil carbon, microbial activity and nitrogen availability in agroforestry systems on moderately alkaline soils in northern India. App Soil Ecol 15:283–294. https://doi.org/10.1016/S0929-1393(00)00079-2Get

Kearns CA, Inouye DW (1997) Pollinators, flowering plants, and conservation biology. Bioscience 47:297–307

Keesstra SD, Bouma J, Wallinga J, Tittonell P et al (2016) The significance of soils and soil science towards realization of the United Nations sustainable development goals. Soil 2:111–128

Keith AM, Van der Wal R, Brooker RW, Osler GHR, Chapman SJ, Burslem DFRP, Elson DA (2008) Increasing litter species richness reduces variability in a terrestrial decomposer system. Ecology 89:2657–2664

Khan ML, Arunachalam A (2003) Traditional agroforestry as a viable choice to conserve agrobiodiversity in the Northeast India. In: Pathak PS, Newaj R (eds) Agroforestry: potentials and opportunities. Agrobios (India) and Indian Society of Agroforestry, Jodhpur, pp 95–105

Kindt R (2002) Methodology for tree species diversification planning for African agroecosystems. PhD thesis, University of Ghent, Ghent, Belgium

Kleijn D, Sutherland WJ (2003) How effective are European agri-environment schemes in conserving and promoting biodiversity? J Appl Ecol 40:947–969

Kleijn D, Kohler F, Báldi A, Batáry P, Concepción ED, Clough Y, Díaz M, Gabriel D, Holzschuh A, Knop E, Kovács A (2008) On the relationship between farmland biodiversity and land-use intensity in Europe. Proc R Soc Lond Biol Sci 276:903–909

Klein AM, Steffan-Dewenter I, Tscharntke T (2006) Rain forest promotes trophic interactions and diversity of trap-nesting hymenoptera in adjacent agroforestry. J Anim Ecol 75:315–323

Klein AM, Vaissiere BE, Cane JH, Steffan-Dewenter I, Cunningham SA, Kremen C, Tscharntke T (2007) Importance of pollinators in changing landscapes for world crops. Proc R Soc B Biol Sci 274(1608):303–313

Kort J, Turnock R (1999) Carbon reservoir and biomass in Canadian prairie shelterbelts. Agrofor Syst 44:175–186

Krause T (2019) Agroforestry needs to take another look at biodiversity, and not just the plants. Swedish International Agricultural Network Initiative. https://www.siani.se/news-story/agroforestry-needs-to-take-another-look-at-biodiversity-and-not-just-the-plants/. Accessed 15 Sep 2020

Kremer RJ, Hezel LF (2013) Soil quality improvement under an ecologically based farming system in Northwest Missouri. Renew Agric Food Syst 28:245–254

Kremer RJ, Kussman RD (2011) Soil quality in a pecan–kura clover alley cropping system in the Midwestern USA. Agrofor Syst 83:213–223

Kristensen P (2003) EEA core set of indicators. Revised version April, 2003. European Environment Agency, Copenhagen

Kumar BM, Nair PKR (2007) Tropical homegardens: a time-tested example of sustainable agroforestry. Springer, Dordrecht

Kumari MAS, Kansuntisukmongkol K, Brockelman WY (2009) Plant diversity in home gardens and its contribution to household economy in suburban areas in Sri Lanka. Environ Nat Resour J 7:12–30

Lacombe S, Bradley RL, Hamel C, Beaulieu C (2009) Do tree-based intercropping systems increase the diversity and stability of soil microbial communities? Agric Ecosyst Environ 131:25–31

Lawton JH, Bignell DE, Bolton B, Bloemers GF, Eggleton P, Hammond PM, Hodda M, Holt RD, Larsen TB, Mawdsley NA et al (1998) Biodiversity inventories, indicator taxa and effects of habitat modification in tropical forest. Nature 391:72–76

Leakey RRB (1999) Agroforestry for biodiversity in farming systems. In: Collins WW, Qualset CO (eds) Biodiversity in agroecosystems. CRC Press, New York, NY, pp 127–145

Mace GM, Norris K, Fitter AH (2012) Biodiversity and ecosystem services: a multilayered relationship. Trends Ecol Evol 27(1):19–26

Marsden C, Martin-Chave A, Cortet J, Hedde M, Capowiez Y (2020) How agroforestry systems influence soil fauna and their functions - a review. Plant Soil 453:29–44

McNeely JA, Schroth G (2006) Agroforestry and biodiversity conservation – traditional practices, present dynamics, and lessons for the future. Biodivers Conserv 15:549–554. https://doi.org/10.1007/s10531-005-2087-3

Méndez VE, Lok R, Somarriba E (2001) Interdisciplinary analysis of homegardens in Nicaragua: micro- zonation, plant use and socioeconomic importance. Agrofor Syst 51(2):85–96

Meyers RT, Zak DR, White DC, Peacock A (2001) Landscape-level patterns of microbial community composition and substrate use in upland forest ecosystems. Soil Sci Soc Am J 65:359–367

Michon G, de Foresta H (1995) The Indonesian agroforest model. Forest resource management and biodiversity conservation. In: Halladay P, Gilmour DA (eds) Conserving biodiversity outside protected areas. The Role of Traditional Agro-ecosystems. World Conservation Union, Forest Conservation Programme, Andalucía, pp 90–106

Middleton H (2001) Agroforestry and its effects on ecological guilds and arthropod diversity. M. Sc. (Forestry) Thesis. Faculty of Forestry, University of Toronto, Toronto, Ontario, Canada

Millspaugh JJ, Schulz JH, Mong TW, Burhans D, Walter WD, Bredesen R, Pritchert RD Jr, Dey DC (2009) Agroforestry wildlife benefits. In: Garrett HE (ed) North American agroforestry: an integrated science and practice, 2nd edn. American Society of Agronomy Inc., Madison, WI, pp 257–285

Montagne V, Nowak V, Guilland C, Gontier L, Dufourcq T, Guenser J, Grimaldi J, Bourgade E, Ranjard L (2017) Agroforestry management in vineyards: effects on soil microbial communities. Proc. 19th EGU General Assembly conference. 23–28 April, 2017, Vienna, Austria p 7385

Montgomery DR (2007) Soil erosion and agricultural sustainability. PNAS 104:13268–13272

Moreno G, Gonzalez-Bornay G, Pulido F, Lopez-Diaz ML, Bertomeu M, Jua'rez E, Diaz M (2016) Exploring the causes of high biodiversity of Iberian dehesas: the importance of wood pastures and marginal habitats. Agrofor Syst 90:87–105. https://doi.org/10.1007/s10457-015-9817-7

Mungai NW, Motavalli PP, Kremer RJ, Nelson KA (2006) Spatial variation of soil enzyme activities and microbial functional diversity in temperate alley cropping practices. Biol Fertil Soils 42:129–136

Murdiyarso D, van Noordwijk M, Wasrin UR, Tomich TP, Gillison AN (2002) Environmental benefits and sustainable land-use options in the Jambi transect, Sumatra, Indonesia. J Veg Sci 13:429–438

Mutabaruka R, Mutabaruka C, Fernandez I (2002) Diversity of arbuscular mycorrhizal fungi associated to tree species in semiarid areas of Machakos, Kenya. Arid Land Res Manag 16:385–390

Naidoo R (2004) Species richness and community composition of songbirds in a tropical forest-agriculture landscape. Anim Conserv 7:93–105

Naiman RJ, Decamps H, McClain ME (2005) Structural patterns. In: Naiman RJ, Decamps H, McClain ME (eds) Riparia: ecology, conservation and management of stream communities. Elsevier, Boston, CA, pp 79–122

Negash M, Yirdaw E, Luukkanen O (2012) Potential of indigenous multistrata agroforests for maintaining native floristic diversity in the south-eastern Rift Valley escarpment, Ethiopia. Agrofor Syst 85(1):9–28

Neita JC, Escobar F (2012) The potential value of agroforestry to dung beetle diversity in the wet tropical forests of the Pacific lowlands of Colombia. Agrofor Syst 85:121–131. https://doi.org/10.1007/s10457-011-9445-9

Nielson RL (2012) Historical corn grain yields for Indiana and the U.S. https://www.agry.purdue.edu/ext/corn/news/timeless/yieldtrends.html. Accessed 1 Oct 2020

Niether W, Jacobi J, Blaser WJ, Andres C, Armengot L (2020) Cocoa agroforestry systems versus monocultures: a multi-dimensional meta-analysis. Environmental Research Letters Published by IOP Publishing Ltd.

Noble IR, Dirzo R (1997) Forests as human-dominated ecosystems. Science 277:522–525

Oppermann R, Beaufoy G, Jones G (2012) High nature value farming in Europe. 35 European countries—experiences and perspectives. Verlag regionanlkultur, Germany PAGES

Osei-Bonsu K, Ameyaw-Oduro C, Tetteh JP (2003) Traditional cocoa agroforestry: 1. Species encountered in the cocoa ecosystem of a typical cocoa-growing district in Ghana. Paper presented at the 14th International Cocoa Research Conference, Accra Ghana. 13–18 October 2003

Pande M, Tarafdar JC (2004) Arbuscular mycorrhizal fungal diversity in neem-based agroforestry systems in Rajasthan. Appl Soil Ecol 26:233–241

Paudel B, Udawatta RP, Kremer RJ, Anderson SH (2011) Agroforestry and grass buffer effects on soil quality parameters for grazed pasture and row-crop systems. Appl Soil Ecol 48:125–132

Paudel B, Udawatta RP, Kremer RJ, Anderson SH (2012) Soil quality indicator responses to row crop, grazed pasture, and agroforestry buffer management. Agrofor Syst 84:311–323

Pauli N, Oberthur T, Barrios E, Conacher AJ (2010) Fine-scale spatial and temporal variation in earthworm surface casting activity in agroforestry fields, western Honduras. Pedobiologia 53:127–139

Perfecto I, Rice RA, Greenberg R, van der Voort ME (1996) Shade coffee: a disappearing refuge for biodiversity. Bioscience 46:598–608

Peters VE, Carroll CR (2012) Temporal variation in coffee flowering may influence the effects of bee species richness and abundance on coffee production. Agrofor Syst 85(1):95–103

Porto RG, de Almeida RF, Cruz-Neto O (2020) Pollination ecosystem services: a comprehensive review of economic values, research funding and policy actions. Food Sec 12:1425–1442. https://doi.org/10.1007/s12571-020-01043-w

Potts SG, Biesmeijer JC, Kremen C, Neumann P, Schweiger O, Kunin WE (2010) Global pollinator declines: trends, impacts and drivers. Trends Ecol Evol 25:345–353

Prasad R, Mertia RS (2005) Dehydrogenase activity and VAM fungi in tree-rhizosphere of agroforestry systems in Indian arid zone. Agrofor Syst 63(3):219

Price GW, Gordon AM (1999) Spatial and temporal distribution of earthworms in a temperate intercropping system in southern Ontario, Canada. Agrofor Syst 44:141–149

Pumariño L, Sileshi GW, Gripenberg S, Kaartinen R, Barrios E, Muchane MN, Midega C, Josson M (2015) Effects of agroforestry on pest, disease, and weed control: a meta-analysis. Basic App Ecol 16:573–582. https://doi.org/10.1016/j.base.2015.08.006

Rahman PM, Varma RV, Sileshi GW (2012) Abundance and diversity of soil invertebrates in annual crops, agroforestry and forest ecosystems in the Nilgiri biosphere reserve of Western Ghats, India. Agrofor Syst 85(1):165–177

Rao MR, Palada MC, Becker BN (2004) Medicinal and aromatic plants in agroforestry systems. Agrofor Syst 61:107–122

Rendón-Sandoval FJ, Alejandro Casas A, Moreno-Calles AI, Torres-García I, García-Frapolli E (2020) Traditional agroforestry systems and conservation of native plant diversity of seasonally dry tropical forests. Sustainability 12:4600. https://doi.org/10.3390/su12114600

Rigueiro-rodríguez A, Losada MRM, Franco RR, Hernández MPG, Urtiaga JJV (2004) Silvopastoral systems as a forest fire prevention technique. Proc. of an international congress on silvopastoralism and sustainable management. Lugo, Spain. pp. 380–387

Rivest D, Lorente M, Olivier A, Messier C (2013) Soil biochemical properties and microbial resilience in agroforestry systems: effects on wheat growth under controlled drought and flooding conditions. Sci Total Environ 463:51–60

Rocha J, Laps RR, Machado CC, Campiolo S (2019) The conservation value of cacao agroforestry for bird functional diversity in tropical agricultural landscapes. Ecol Evol 9:7903–7913. https://doi.org/10.1002/ece3.5021

Rodrigues RC, Araújo RA, Costa CS, Lima AJT, Oliveira ME, Cutrim JAA Jr, Santos FNS, Araújo JS, Santos VM, Araújo ASF (2015) Soil microbial biomass in an agroforestry system of Northeast Brazil. Trop Grasslands 3:41–48. https://doi.org/10.17138/tgft(3)41-48

Roothaert R, Franzel S, Kiura M (2003) On-farm evaluation of fodder trees and shrubs preferred by farmers in Central Kenya. Exp Agric 39(4):423

Round PD, Gale GA, Brockelman WY (2006) A comparison of bird communities in mixed fruit orchards and natural forest at Khao Luang, southern Thailand. Biodivers Conserv 15:2873–2891

Sanchez PA (1995) Science in agroforestry. Agrofor Syst 30:5–55

Santos PZF, Crouzeille R, Sansevero JBB (2019) Can agroforestry systems enhance biodiversity and ecosystem service provision in agricultural landscapes? A meta-analysis for the Brazilian Atlantic Forest. For Ecol Manag 433:140–145

Schroth G, da Fonseca GAB, Harvey CA, Gascon C, Vasconcelos HL, Izac A-MN (2004) Agroforestry and biodiversity conservation in tropical landscapes. Island Press, Washington, DC, p 523

Seiter S, Ingham ER, William RD, Hibbs DE (1995) Increase in soil microbial biomass and transfer of nitrogen from alder to sweet corn in an alley cropping system. In: Ehrenreich JH, Ehrenreich DL, Lee HW (eds) Growing a sustainable future. University of Idaho, Boise, ID, pp 56–158

Sharma G, Sharma R, Sharma E (2009) Traditional knowledge systems in large cardamom farming: biophysical and management diversity in Indian mountainous regions. Indian J Tradit Knowl 8(1):17–22

Sharrow SH, Brauer D, Clason TR (2009) Silvopasture practices. In: Garrett HE (ed) North American agroforestry: an integrated science and practice, 2nd edn. American Society of Agronomy, Madison, WI, pp 105–131

Sileshi G, Mafongoya PL (2007) Quantity and quality of organic inputs from coppicing leguminous trees influence abundance of soil macrofauna in maize crops in eastern Zambia. Biol Fertil Soils 43:333–340

Sistla SA, Roddy AB, Williams NE, Kramer DB, Stevens K, Allison SD (2016) Agroforestry practices promote biodiversity and natural resource diversity in Atlantic Nicaragua. PLoS One 11(9):e0162529. https://doi.org/10.1371/journal.pone.0162529

Smith OB (1992) Fodder trees and shrubs in range and farming systems in tropical humid Africa. In: Speedy A, Pugliese P-L (eds) Legume trees and other fodder trees as protein sources for livestock, Proceedings of the FAO Expert Consultation Held at the Malaysian Agricultural Research and Development Institute (MARDI), Kuala Lumpur, Malaysia, 14–18 October, vol 1991. Food and Agriculture Organization of the United Nations, Rome, pp 43–59

Söderström B, Svensson B, Vessby K, Glimskär A (2001) Plants, insects and birds in semi-natural pastures in relation to local habitat and landscape factors. Biodivers Conserv 10(11):1839–1863

Soini E (2005) Dynamics of livelihood strategies in the changing landscapes of the southern slopes of Mt. Kilimanjaro, Tanzania. Agric Syst 85(3):306–323

Spurgeon DJ, Keith AM, Schmidt O, Lammertsma DR, Faber JH (2013) Land-use and land-management change: relationships with earthworm and fungi communities and soil structural properties. BMC Ecol 13:46

Stamps WT, Woods TW, Linit MJ, Garrett HE (2002) Arthropod diversity in alley cropped black walnut (Juglans nigra L.) stands in eastern Missouri, USA. Agrofor Syst 56:167–175

Steffan-Dewenter I, Kessler M, Barkmann J, Bos MM, Buchori D, Erasmi S, Faust H, Gerold G, Glenk K et al (2007) Tradeoffs between income, biodiversity, and ecosystem functioning during tropical rainforest conversion and agroforestry intensification. PNAS 104:4973–4978

Swallow B, Boffa J (2006) The potential for agroforestry to contribute to the conservation and enhancement of landscape biodiversity. In: Garrity D, Okono A, Grayson M, Parrott S (eds) World agroforestry into the future. World Agroforestry Centre, Nairobi, pp 95–101

Thevathasan NV, Gordon AM (2004) Ecology of tree intercropping system in the north temperate region: experience from southern Ontario, Canada. Agrofor Syst 61:257–268

Thiele-Bruhn S, Bloem J, De Vries FT, Kalbitz K, Wagg C (2012) Linking soil biodiversity and agricultural soil management. Curr Opin Environ Sustain 4:523–528

Tomich TP, van Noordwijk M, Vosti SA, Witcover J (1998) Agricultural development with rainforest conservation: methods for seeking best bet alternatives to slash-and-burn, with applications to Brazil and Indonesia. Agric Econ 19(1–2):159–174

Tomich TP, van Noordwijk M, Budidarsono S, Gillison A, Kusumanto T, Murdiyarso M, Stolle F, Fagi AM (2001) Agricultural intensification, deforestation, and the environment: assessing tradeoffs in Sumatra, Indonesia. In: Lee DR, Barrett CB (eds) Tradeoffs or synergies? Agricultural intensification, economic development and the environment. CAB International, Wallingford, pp 221–244

Torralba M, Fagerholm N, Burgess PJ, Moreno G, Plieninger T (2016) Do European agroforestry systems enhance biodiversity and ecosystem services? A meta-analysis. Agric Ecosyst Environ 230:150–161

Udawatta RP, Anderson SH (2008) CT-measured pore characteristics of surface and subsurface soils as influenced by agroforestry and grass buffers. Geoderma 145:381–389

Udawatta RP, Kremer RJ, Adamson BW, Anderson SH (2008) Variations in soil aggregate stability and enzyme activities in a temperate agroforestry practice. Appl Soil Ecol 39:153–160

Udawatta RP, Nelson KA, Jose S, Motavalli PP (2014) Temporal and spatial differences in crop yields of a mature silver maple alley cropping system. Agron J 106:407–415. https://doi.org/10.2134/agronj2013.0429

Udawatta RP, Gantzer CJ, Reinbott TM, Wright RL, Pierce RA II (2016) Yield differences influenced by distance from riparian buffers and CRP. Agron J 108:647–655. https://doi.org/10.2134/agronj2015.0273

Udawatta RP, Gantzer CJ, Jose S (2017) Agroforestry practices and soil ecosystem services. In: Al-Kaisi MM, Lowery B (eds) Soil health and intensification of agroecosystems. Elsevier AP, USA, pp 305–334
Udawatta RP, Rankoth LM, Jose S (2019) Agroforestry and biodiversity. Sustainability 11:2879. https://doi.org/10.3390/su11102879
Vallejo M, Moreno-Calles AI, Casas A (2016) TEK and biodiversity management in agroforestry systems of different sociological contexts of the Tehuacán Valley. J Ethnobiol Ethnomed 12:31
Varah A, Jones H, Smith J, Potts SG (2013) Enhanced biodiversity and pollination in UK agroforestry systems. J Sci Food Agric 93(9):2073–2075
Wambugu C, Place F, Franzel S (2011) Research, development and scaling-up the adoption of fodder shrub innovations in East Africa. Int J Agric Sustain 9(1):100–109
Weerasekara C, Udawatta RP, Jose S, Kremer RJ, Weerasekara C (2016) Soil quality differences in a row-crop watershed with agroforestry and grass buffers. Agrofor Syst 90:829–838
Weinstock JA (1994) Rhizophora mangrove agroforestry. Econ Bot 48(2):210–213
Wibawa G, Joshi L, Van Noordwijk M, Penot EA (2006) Rubber-based agroforestry systems (RAS) as alternatives for rubber monoculture system. IRRDB Conference, Ho Chi Minh City, Viet Nam, 13–17 November. Kuala Lumpur: International Rubber Research and Development Board
Wojewoda D, Russel S (2003) The impact of a shelterbelt on soil properties and microbial activity in an adjacent crop field. Pol J Ecol 51:291–307
Zak DR, Holmes WE, White DC, Peacock AD, Tilman D (2003) Plant diversity, soil microbial communities, and ecosystem function: are there any links? Ecology 84:2042–2050
Zhang Q, Zhang M, Zhou P, Fang Y, Ji Y (2018) Impact of tree species on barley rhizosphere-associated fungi in an agroforestry ecosystem as revealed by 18S rDNA PCRDGGE. Agrofor Syst 92:541–554
Zomer RJ, Ustin SL, Carpenter CC (2001) Land cover change along tropical and sub-tropical riparian corridors within the Makalu Barun National Park and conservation area, Nepal. Mt Res Dev 21(2):175–183
Zomer RJ, Neufeldt H, Xu J, Ahrends A, Bossio D, Trabucco A, Van Noordwijk M, Wang M (2016) Global tree cover and biomass carbon on agricultural land: the contribution of agroforestry to global and national carbon budgets. Sci Rep 6:29987
Zou X, Sanford RL (1990) Agroforestry systems in China: a survey and classification. Agrofor Syst 11(1):85–94

# The Role of Temperate Agroforestry Practices in Supporting Pollinators

**Gary Bentrup, Jennifer Hopwood, Nancy Lee Adamson, Rae Powers, and Mace Vaughan**

## Introduction

Agroforestry is the intentional integration of trees and/or shrubs with herbaceous crops and/or livestock in an agricultural production system. In temperate regions, agroforestry systems include many different practices such as windbreaks, riparian buffers, alley cropping, hedgerows, shelterbelts, silvopasture, and forest farming. Agroforestry practices can deliver a suite of ecosystem services from provisioning, regulating, cultural, and supporting services (Smith et al. 2013). With some exceptions (e.g., pollinator hedgerows), ecosystem services provided by insect pollinators

The original version of this chapter was revised. The correction to this chapter is available at https://doi.org/10.1007/978-3-030-80060-4_19

G. Bentrup (✉)
U.S. Department of Agriculture, U.S. Forest Service, National Agroforestry Center, Lincoln, NE, USA
e-mail: gary.bentrup@usda.gov

J. Hopwood
Xerces Society for Invertebrate Conservation, Omaha, NE, USA
e-mail: jennifer.hopwood@xerces.org

N. L. Adamson
International Center for Agricultural Research in the Dry Areas, Rabat, Morocco
e-mail: nancy.adamson@xerces.org

R. Powers
Xerces Society for Invertebrate Conservation, Lincoln, NE, USA
e-mail: raeann.powers@xerces.org

M. Vaughan
Xerces Society for Invertebrate Conservation, Portland, OR, USA
e-mail: mace.vaughan@xerces.org



R. P. Udawatta, S. Jose (eds.), *Agroforestry and Ecosystem Services*, https://doi.org/10.1007/978-3-030-80060-4_11

are often not specifically considered in the design and management of agroforestry practices (Udawatta et al. 2019). However, whether using alley cropping or a windbreak, managing a riparian buffer, or forest farming, agroforestry practices can increase the overall diversity of plants and physical structure in landscapes and, as a result, provide habitat for pollinators and other insects beneficial for agriculture such as predators and parasitoids of crop pests and decomposers. Agroforestry plantings can also have indirect benefits for pollinators including habitat connectivity and protection from pesticide exposure. This chapter provides an overview of the current scientific knowledge regarding how agroforestry practices can support pollinators and pollination services.

## Importance of Pollinators

Plant pollination by animals is one of the most well-known and important ecosystem services and is essential in both natural and agricultural landscapes (IPBES 2016). An estimated 85% of the world's flowering plants depend on animals—mostly insects—for pollination (Ollerton et al. 2011). Pollination is a mutually beneficial interaction between plants and pollinators. Animals, particularly insects, visit flowers seeking sustenance, and in the process transfer pollen grains from one flower or plant to another, allowing flowering plants to reproduce. Sugary nectar and/or protein-packed pollen grains are food resources for pollinators.

Insect pollination is critical to agricultural production. Eighty-seven of the world's 124 most commonly cultivated crops (70%) are reliant on or benefit from animal pollination, including crops that produce fruits, vegetables, spices, nuts, and seeds (Klein et al. 2007). Additionally, insect-pollinated plants such as alfalfa and clover provide feed for livestock. Roughly 35% of global crop production is dependent on pollination by animals (Klein et al. 2007). The majority of minerals, vitamins, and nutrients needed to maintain human health (such as vitamin C, calcium, and folic acid) come from crop plants that depend partially or fully on animal pollinators (Eilers et al. 2011). The value of crops directly dependent on pollination by insects (e.g., apples, squash) was estimated in 2009 at $15.1 billion in the United States, and the value of crops indirectly dependent on pollinators (e.g., alfalfa hay, onions) was estimated in 2004 at $12 billion (Calderone 2012).

Pollinators are a keystone group in most terrestrial ecosystems, necessary for plant reproduction and important for wildlife food webs (Kearns et al. 1998). They sustain wildland plant communities that provide food and shelter for myriad wildlife. Fruits, seeds, and nuts, that result from animal pollination, are food for many insects, birds, and mammals. Pollinators can also be direct prey for wildlife. For example, pollinator larvae are an important part of the diet of many young birds (Buehler et al. 2002). Healthy habitat that supports pollinators often confers other ecosystem services such as reduced soil erosion, enhanced rainwater infiltration, improved water quality, reduced wind velocity, carbon sequestration, recreation spaces for humans, and habitat for a variety of wildlife, including arthropod predators and parasitoids that reduce crop pests.

## Important Groups of Pollinators

The great majority of pollinators are insects, including bees, wasps, flies, beetles, butterflies, and moths (Table 1; Allen-Wardell et al. 1998; Kevan 1999; Kearns 2001), but some bird and bat species pollinate as well (Grant 1994; Valiente-Banuet et al. 2004). Bees are considered the most important group of pollinators for agricultural crops (McGregor 1976; Morse and Calderone 2000; Garibaldi et al. 2013) as well as for wild plants in temperate climates (Michener 2007). Bees are such efficient pollinators of many plants because 1) they actively collect both pollen and nectar; 2) they make many trips to flowers as they are foraging to collect nest provisions for their offspring; and 3) they have more flower constancy, i.e., once they find a good forage source they visit that type of flower over and over.

The domesticated European honey bee (*Apis mellifera*) is the most widely recognized bee worldwide and is an important managed crop pollinator. Studies indicate that honey bee pollination accounts for more than $15 billion in crop production annually in the United States (Morse and Calderone 2000; Calderone 2012).

Based on Ascher and Pickering (2020), there are over 5200 species of native bees in North America, many of which are important crop pollinators. Native bees are important in the production of crops worth an estimated $3 billion annually to the US economy (Losey and Vaughan 2006), though this may be an underestimate of their contribution. A recent analysis of 41 crop systems worldwide found that managed honey bees do not replace the pollination services provided by a diverse community of native bees (Garibaldi et al. 2013). Native bees provide pollination services in colder, windier weather (Brittain et al. 2013) and are more efficient than honey bees on an individual bee basis at pollinating particular crops, such as squash, berries, and tree fruits (e.g., Tepedino 1981; Bosch and Kemp 2001; Javorek et al. 2002; Garibaldi et al. 2013).

Most native bees live solitary lives, with each female working alone to build her nests and collect and provide food for her offspring. Some solitary bees visit a diversity of flowers to collect pollen, and others collect from flowers of a particular plant species or group of species. Bumble bees and some sweat bees are the only native bees that form social colonies. Their colonies usually have fewer than 200 bees, and are much smaller than a honey bee hive which may house up to 30,000 individuals. Bumble bees are particularly important pollinators. They are able to fly in cooler temperatures and lower light levels than many other bees, which extends their workday and improves the pollination of crops during inclement weather (Corbet et al. 1993). In addition to commercially important crops, bumble bees also play a vital role as generalist pollinators of native flowering plants (Memmott et al. 2004). They and many native bees also possess the ability to "buzz pollinate," dislodging pollen

**Table 1** Common insect pollinator groups

| | | |
|---|---|---|
|  |  |  |
| **Honey bee** | **Bumble bees** | **Ground-nesting bees** |
| Order: Hymenoptera<br>Family: Apidae<br>Genus and species: *Apis mellifera* | Order: Hymenoptera<br>Family: Apidae<br>Genus: *Bombus* | Order: Hymenoptera<br>Families: Andrenidae, Apidae, Colletidae, Halictidae |
| The European honey bee (native to Europe, Africa, and Asia) is a domesticated species that lives in large perennial social colonies (hives), with division of labor within the colony. Only the queen reproduces, while others gather nectar and pollen to feed brood (larvae) and store food (honey) for the winter. Feral colonies in the United States are somewhat rare; most hives are managed by beekeepers | Bumble bees form annual social colonies. Queen bumble bees that mated the previous fall start nests in spring and by mid-summer colonies can have dozens or hundreds of workers. They nest in insulated cavities such as under clumps of bunch grass or in old rodent nests. There are 46 recognized bumble bee species in North America | Most native bees live solitary lives, with each female working alone to build her nests and collect and provide food for her offspring. About 70% of our solitary bee species nest underground, digging slender tunnels in which they build individual cells for each egg and its provisions |
|  |  |  |
| **Tunnel-nesting bees** | **Flower-visiting flies** | **Flower-visiting beetles** |
| Order: Hymenoptera<br>Families: Apidae, Colletidae, Halictidae, Megachilidae | Order: Diptera<br>Families: Anthomyiidae, Bombyliidae, Syrphidae, Tachinidae, others | Order: Coleoptera<br>Families: Cantharidae, Coccinellidae, Scarabaeidae, others |
| Approximately 30% of solitary bee species nest in tunnels, inside already hollow stems or by chewing into the pithy center of stems, or in existing holes in wood, sometimes man-made. Most tunnel-nesting bees are solitary species | Flower-visiting flies consume nectar and sometimes pollen. Many hover flies (family Syrphidae) resemble bees or wasps in coloration. Larvae of some species are voracious predators of small insects, like aphids | Flower-visiting beetles consume nectar and pollen, and may also chew on flower parts. Larvae of some species are predatory, hunting other insects (including crop pests) as food, while others are herbivorous or are decomposers |

(continued)

**Table 1** (continued)

|  |  |  |
|---|---|---|
| **Flower-visiting wasps** | **Flower-visiting moths** | **Butterflies** |
| Order: Hymenoptera<br>Families: Sphecidae, Vespidae, Tiphiidae, Scoliidae, others | Order: Lepidoptera<br>Families: Sphingidae, Noctuidae, Arctiidae | Order: Lepidoptera<br>Families: Papilionidae, Hesperiidae, Pieridae, Lycaenidae, Nymphalidae |
| Predatory wasps, most of which are solitary, hunt for prey to bring back to their nest as food for their young. They build nests in cavities or in the ground, and may utilize pieces of grass, mud, or resin in construction of their nest. Adults maintain their energy by consuming nectar and/or pollen, and in the process may also transfer pollen between flowers | Moths, which are often subdued in color and tend to fly at dusk or night, are less visible than other groups, but many are important specialist pollinators of wild plants, while some also pollinate crops. Moths as a group form a critical food source for other wildlife | With their striking transformation from a chubby plant-chewing caterpillar to a delicate pupa to a graceful nectar-drinking adult, butterflies are some of the most beloved insects. Some species have narrow host plant needs for their caterpillars while others feed on a wide variety of plants |

Source: Flower-visiting beetle image by Jennifer Hopwood and remaining images by Nancy Lee Adamson

with a vibration that forces release from poricidal anthers found in flowers such as blueberries, cranberries, tomatoes, and peppers (Buchmann 1983).

Of the other orders of pollinating insects, flies (Diptera) also provide substantial pollination services (Kearns 2001; Larson et al. 2001; Inouye et al. 2015), especially in alpine areas and tundra. Other insects such as beetles (Coleoptera) and wasps (Hymenoptera) provide pollination services, though to a lesser extent (e.g., Frankie et al. 1990; Kevan 1999). The contribution of most butterfly and moth species (Lepidoptera) to pollination services is not well known (e.g., Frankie et al. 1990; Allen-Wardell et al. 1998; Westerkamp and Gottsberger 2000; MacGregor et al. 2015), but there are instances where butterflies have been documented pollinating wild plant species, including some flowering plants specially adapted for butterfly pollination (e.g., *Russelia, Phlox*, and *Lantana*) (Fallon et al. 2014). Ollerton (2017) estimate that more than 140,000 species of moths and butterflies visit flowers. Many butterfly species fly great distances between flowers and may carry pollen for a long time, and thus they may be effective as dispersers of pollen.

In addition to insect pollinators, there are two groups of nectar-feeding vertebrates that play an important role in pollination: hummingbirds (Trochilidae) and bats (Phyllostomidae). There are 12 species of nectar-feeding bats that are known pollinators in North America (National Research Council 2007). The known ranges for these bats correspond closely with the distribution of columnar cacti (e.g., saguaro [*Carnegiea gigantea*], *Pachycereus* spp., *Stenocereus* spp., *Lophocereus* spp.) and agaves (*Agave* spp.), the main species they are known to pollinate (Valiente-Banuet et al. 2004), primarily in the deserts of Arizona, California, Nevada, New Mexico, and Texas. Hummingbirds, which pollinate about 130 native plant species with flowers adapted for hummingbird pollination, make long migratory journeys in North America and depend on nectar corridors to sustain their long-distance movements (Nabhan et al. 2004).

## Pollinator Status and Threats

Globally, pollinators are in decline (Biesmeijer et al. 2006; National Research Council 2007; Potts et al. 2010; Sánchez-Bayo and Wyckhuys 2019), with some estimates that 40% of invertebrate pollinator species may be at risk of extinction worldwide (IPBES 2016). Threats such as the loss, degradation, and fragmentation of habitat (e.g., Kremen et al. 2002; Williams and Kremen 2007; Potts et al. 2010); introduced species (e.g., Tallamy and Shropshire 2009; Fiedler et al. 2012); use of pesticides (e.g., Dover et al. 1990; Kearns and Inouye 1997; Kevan 1999; Whitehorn et al. 2012); and diseases and parasites (e.g., Altizer and Oberhauser 1999; Colla et al. 2006; Cameron et al. 2011) all contribute to pollinator decline.

In the United States, the number of honey bee colonies has been in decline over the past half-century due to diseases, parasites, lack of floral resources, insecticides, and other factors (National Research Council 2007). Since 2012, beekeepers have experienced record high annual hive losses of 33% or more; an average of 40% of managed colonies were lost in the 2018–2019 season (Bee Informed Partnership 2019).

Much less is known about the status of most of North America's native pollinators, though what data does exist suggests that numerous species are experiencing declines similar to or more severe than the declines seen in honey bees. One-quarter of North America's bumble bees have experienced significant declines (Hatfield et al. 2014), including declines in species that were formerly some of the most common species (Cameron et al. 2011). In 2017, the once common rusty patched bumble bee (*Bombus affinis*) was added to the US Fish and Wildlife list of endangered species (US Fish and Wildlife Service 2019).

In the United States, some butterflies are also in decline. NatureServe assessed all of the country's roughly 800 butterfly species and found that 19% are at risk of extinction (NatureServe 2018). A number of generalist butterfly species have seen significant declines in recent years (Forister et al. 2011). In particular, monarch butterflies (*Danaus plexippus*) in North America are now vulnerable to extinction, according to a recently completed assessment (Semmens et al. 2016). The population of monarchs has dropped by over 80% east of the Rocky Mountains (Rendón-Salinas and Tavera-Alonso 2014) and by over 90% to the west (Schultz et al. 2017).

The loss of milkweeds (*Asclepias* spp.), the monarch's larval host plants, has been significant, particularly within agricultural fields (Pleasants and Oberhauser 2012).

The populations of both hummingbirds and nectar-feeding bats throughout the southwestern United States have also experienced declines (National Research Council 2007). Hummingbirds face disruption of migratory routes and loss of habitat (Calder 2004), while nectar-feeding bats face disturbance of their roost sites and removal of foraging habitat and nectar sources (US Fish and Wildlife Service 2006).

The loss of pollinators negatively affects plant reproduction and plant community diversity (Bawa 1990; Fontaine et al. 2005; Brosi and Briggs 2013). Threats to pollinators may have profound consequences for ecosystem health as well as our food systems (Kearns et al. 1998; Spira 2001; Steffan-Dewenter and Westphal 2008). Concerns about pollinator decline and its repercussions have led to increased efforts to reduce threats to pollinators. Managing existing habitat for insect pollinators and restoring additional habitat have been demonstrated to increase pollinator abundance and diversity (e.g., Fiedler et al. 2012; Klein et al. 2012; Morandin and Kremen 2013). By adding structural and functional diversity in landscapes, agroforestry may provide habitat and other benefits for insect and other pollinators and pollination services.

## Agroforestry's Role

Based on a review of available scientific literature, agroforestry practices can confer three key benefits for insect pollinators and pollination services: 1) providing habitat including foraging resources and nesting or egg-laying sites, 2) enhancing site and landscape connectivity, and 3) reducing pesticide exposure (Bentrup et al. 2019). Current research on supporting pollinators in agricultural landscapes has focused primarily on honey bees and native bees but general concepts may apply across other pollinator groups.

### *Providing Habitat*

#### Foraging Resources

Pollinators require a diversity of flowers to provide nectar and pollen resources to meet their nutritional needs. Nectar is an aqueous solution of sugars, amino acids, and other secondary metabolites that provides a rich source of energy for bees, butterflies, hummingbirds, bats, and some moths, wasps, beetles, and flies. Pollen is a protein-rich resource that is used by native bees, honey bees, and some wasps to feed their brood or to provision their eggs or by some adult flies and beetles as a food source. Agroforestry practices can be important sources of nectar and pollen for pollinators when appropriate plants are used (Table 2). If the agroforestry

**Table 2** North American trees and shrubs that provide abundant nectar and/or pollen[a]

| Scientific name | Common name | Bloom time[b] | Height[c] | Region[d] |
|---|---|---|---|---|
| *Acer* spp.[e] | Maple | Spring to early summer | T | WCE |
| *Amelanchier* spp.[f] | Serviceberry | Early spring to summer | SM | WCE |
| *Amorpha* spp. | Leadplant, false indigo | Spring to summer | S | WCE |
| *Arbutus* spp.[a,g] | Madrone | Early spring to summer | MT | WC |
| *Aronia* spp.[f] | Chokeberry | Spring to summer | S | C[h]E |
| *Atriplex canescens* | Four-wing saltbush | Spring to fall | SM | W |
| *Baccharis* spp.[a] | Baccharis | Summer to fall | S | WCE |
| *Callicarpa americana* | Beautyberry | Early summer | S | CE[h] |
| *Ceanothus* spp. | Native lilac, NJ tea | Early spring to summer | SM | WCE |
| *Cephalanthus occidentalis* | Buttonbush | Summer | SM | WCE |
| *Cercis* spp. | Redbud | Spring | M | WCE |
| *Chrysothamnus* spp. | Rabbitbrush | Summer-fall | SM | W |
| *Clethra alnifolia* | Sweet pepperbush | Summer | S | E |
| *Crataegus* spp. | Hawthorn | Spring | M | WCE |
| *Dasiphora* spp. | Cinquefoil | Spring | S | WCE |
| *Diospyros* spp.[e,f] | Persimmon | Spring | T | WCE |
| *Ericameria* spp. | Rabbitbrush | Summer-fall | SM | WC |
| *Eriogonum* spp. | Buckwheat | Summer | S | WC |
| *Gaylussacia* spp.[f] | Huckleberry | Early spring | S | CE |
| *Gleditsia* spp.[f] | Honey locust | Spring | T | WCE |
| *Halesia* spp. | Silverbell | Early spring | MT | E[h] |
| *Holodiscus* spp. | Cliff spirea | Summer | S | WC |
| *Hypericum* spp. | Shrubby St.-John's-wort | Late spring | S | WCE |
| *Ilex* spp.[a,g] | Holly, inkberry | Spring | SMT | WCE |
| *Itea virginica* | Virginia sweetspire | Spring | S | CE |
| *Krascheninnikovia lanata* | Winterfat | Summer | S | W |
| *Liriodendron tulipifera*[e] | Tulip tree | Spring | T | CE |
| *Mahonia* spp.[a] | Oregon grape | Spring to early summer | S | WCE |
| *Nyssa* spp.[f] | Black gum | Spring | MT | CE |
| *Oxydendrum arboreum* | Sourwood | Summer | T | E |
| *Parkinsonia* spp. | Palo Verde | Spring | M | WCE[h] |
| *Philadelphus* spp. | Mock orange | Spring | S | WCE |
| *Physocarpus* spp. | Ninebark | Spring to summer | S | WCE |
| *Prunus* spp.[e,f] | Cherry, plum, peach, apricot | Spring | M | WCE |
| *Purshia tridentata* | Antelope bitterbrush | Spring | S | W |

(continued)

**Table 2** (continued)

| Scientific name | Common name | Bloom time[b] | Height[c] | Region[d] |
|---|---|---|---|---|
| *Rhododendron* spp.[a] | Rhododendron, azalea | Early spring | SM | WCE |
| *Rhus* spp.[f] | Sumac | Spring to summer | M | WCE |
| *Robinia pseudoacacia* [e,f] | Black locust | Spring | T | E[i] |
| *Rosa* spp.[f] | Rose | Summer | S | WCE |
| *Rubus* spp.[f] | Blackberry, raspberry | Spring to fall | S | WCE |
| *Salix* spp.[f] | Willow | Early spring | MT | WCE |
| *Sambucus* spp.[f] | Elderberry | Spring to summer | S | WCE |
| *Sassafras albidum* | Sassafras | Spring | MT | CE |
| *Shepherdia* spp. | Buffaloberry | Spring | SM | WC |
| *Spiraea* spp. | Spirea | Summer | S | WCE |
| *Tilia* spp.[e] | Basswood | Spring to summer | T | CE |
| *Umbellularia californica* | California laurel | Fall to spring | T | W |
| *Vaccinium* spp.[f,g] | Blueberry, huckleberry | Early spring | S | WCE |

[a]Includes some or all evergreen species
[b]Flowering times depend on species, location, and environmental conditions, varying from year to year. Consult with local native plant experts to plan for overlapping bloom times
[c]Short (S), medium (M), tall (T)
[d]West (W), Central (C), East (E)
[e]Added value as timber
[f]Added value of fruit or other culinary crops
[g]Added value of decorative cut twigs for the floral industry
[h]Southern distribution only
[i]This species is invasive in some parts of the country and should not be planted in those regions
Source: Modified from Adamson et al. (2011)

practice lacks pollinator-suitable floral resources, pollinator use can be limited. For instance, Macdonald et al. (2018) found limited pollinator use of shelterbelts in New Zealand that were predominantly comprised of Monterey pine (*Pinus radiata* D. Don) and Monterey cypress (*Hesperocyparis macrocarpa* (Hartw.) Bartel) (wind-pollinated exotic species).

Many woody species offer abundant nectar with relatively high sugar contents such as maple (*Acer* spp.), horse chestnut (*Aesculus* spp.), basswood (*Tilia* spp.), willow (*Salix* spp.), brambles (*Rubus* spp.), cherry and plum (*Prunus* spp.), and serviceberry (*Amelanchier* spp.) (Batra 1985; Stubbs et al. 1992; Loose et al. 2005; Ostaff et al. 2015; Baude et al. 2016; Somme et al. 2016; Donkersley 2019). For example, sugar content in horse chestnut (*Aesculus hippocastanum* L.) ranges from 0.58 to 3.57 mg/flower/24 h, while black locust (*Robinia pseudoacacia* L.) ranges from 0.76 to 4.0 mg/flower/24 h (Crane and Walker 1985). For comparison, white clover (*Trifolium repens* L.) ranges from 0.01 to 0.20 mg/flower/24 h and alfalfa (*Medicago sativa* L.) ranges from 0.07 to 0.25 mg/flower/24 h (Crane and Walker 1985).

Willow, maple, cherry and plum, brambles, chestnut (*Castanea* spp.), and ash (*Fraxinus* spp.) are woody species that can provide pollen with high concentrations of amino acids, sterols, trace minerals, and other nutritionally important compounds for bees and other pollinators (Batra 1985; Tasei and Aupinel 2008; Di Pasquale et al. 2013; Ostaff et al. 2015; Russo and Danforth 2017; Filipiak 2019). Some bees are pollen specialists (oligolectic), wholly dependent on specific shrubs and trees in certain families, such as willows, dogwoods (Cornaceae), heaths like blueberry and huckleberry (Ericaceae), buckthorns such as New Jersey tea (Rhamnaceae), and roses (Rosaceae) (Dötterl and Vereecken 2010; Fowler 2016). These nutritionally rich pollen sources are often sought out by native bees (Stubbs et al. 1992; Ostaff et al. 2015) and have been shown to result in higher reproductive success and better immunity in bumble bees (e.g., Tasei and Aupinel 2008; Di Pasquale et al. 2013).

Tree and shrub plantings with overlapping bloom times provide nectar and pollen resources throughout the growing season and are key for sustaining diverse pollinator populations (Loose et al. 2005; Hannon and Sisk 2009; Miñarro and Prida 2013). Many temperate-zone trees and shrubs flower early in spring and can deliver some of the first pollen and nectar resources of the season, boosting early-season pollinator populations (Table 2) (Dirr 1990; Batra 1985; Ostaff et al. 2015; Somme et al. 2016). In Michigan, United States, Wood et al. (2018) determined that willows, maples, and *Prunus* spp. provided over 90% of the pollen collected in April by social and solitary bees. When forage is available early in the growing season, freshly emerged bumble bee queens are more successful in establishing their colonies (Carvell et al. 2017).

Plantings that include a diversity of flowers of various sizes, shapes, and colors can support a rich and abundant community of bees and other pollinators (Potts et al. 2003; Roulston and Goodell 2011; Nicholls and Altieri 2013). Flower density and subsequent nectar availability can be higher in some tree and shrub species compared to herbaceous species (Crane and Walker 1985; Loose et al. 2005). For example during peak flowering season, gray willow *(Salix cinerea* L.) can produce 334,178 flowers/m$^2$ and oneseed hawthorn (*Crataegus monogyna* Jacq.) 19,003 flowers/m$^2$ compared to sea aster (*Aster tripolium* L.) 9565 flowers/m$^2$ and buttercup (*Ranunculus acris* L.) 688 flowers/m$^2$ (Baude et al. 2016). Respectively, nectar productivity for these species is 3612, 584, 169, and 50 kg/ha cover/year. Spatially, agroforestry practices that incorporate a diversity of flowering woody and herbaceous species can deliver a high density of floral resources relative to the land area occupied due to vertical layering (Miñarro and Prida 2013; Morandin and Kremen 2013; Ponisio et al. 2016; Somme et al. 2016; Donkersley 2019). Timberlake et al. (2019) documented approximately two and four times greater nectar per unit area in hedgerows compared to woodlands and pasture, respectively.

Bees also collect resins and oils from trees and other plants (Wcislo and Cane 1996; Cane et al. 2007; Policarová et al. 2019). Some tunnel-nesting native bees use tree resins to seal off their nests while honey bees use plant resins mixed with saliva and beeswax to make propolis to seal unwanted holes in their hives. Propolis has antibacterial properties that help prevent disease transmission or pest/parasite invasion (Simone-Finstrom et al. 2017). Poplar trees (*Populus* spp.) are a common source for these resins (Greenaway et al. 1990; Bankova et al. 2000; König 1985; Drescher et al. 2019). Other tree species including pine (*Pinus* spp.), birch (*Betula* spp.), elm (*Ulmus* spp.), alder (*Alnus* spp.), beech (*Fagus* spp.), and horse chestnut can provide resin sources when poplar species are not present (Ghisalberti 1979; König 1985; Drescher et al. 2019).

Pollinator behavior, foraging, and resulting pollination services are strongly influenced by weather conditions (e.g., ambient temperature, wind speed, precipitation) (Corbet 1990; Vicens and Bosch 2000). Temperature and wind speed are two primary weather variables that agroforestry practices can influence.

Windbreaks, alley cropping, and other agroforestry practices can reduce air movement and modify temperatures in the cropped area. Daytime air temperatures are several degrees warmer within a certain distance downwind of windbreaks (8–10 times the windbreak height) (McNaughton 1988). These elevated temperatures can increase pollinator activity and pollination, particularly in vegetable- and fruit-growing regions where air temperatures at pollination time can often be below optimum (Norton 1988). The vertical structure and shaded conditions provided by agroforestry practices can offer niches that allow pollinators to find suitable sites for thermal regulation, which is becoming increasingly important under climate change (Kjøhl et al. 2011). Papanikolaou et al. (2017) found that agricultural landscapes that had a higher proportion of hedgerows and seminatural habitats (i.e., 17% compared to 2%) decreased the detrimental effects of warmer temperatures on native bee species richness and abundance.

Agroforestry plantings can address additional thermoregulation considerations for managed honey bees. Honey bees expend energy to cool themselves and their hives during hot weather. If the hives are shaded, that energy can be diverted to honey production and hive maintenance activities (Nye 1962). Trees and shrubs are useful for shading beehives, especially if the hives are placed on the north or northeast sides of the woody plantings to receive maximum shading during the summer heat (Hill and Webster 1995). Windbreaks and other woody buffers can also provide protection from winter temperatures and winds if the hives are located on the leeward side, helping reduce winter mortality (Haydak 1958). In Kansas, Merrill (1923) documented that populations in overwintered hives can be up to 52% higher when protected by windbreaks.

Foraging in moderate to high winds increases energetic costs for pollinators and reduces pollination efficiency (Vicens and Bosch 2000; Brittain et al. 2013). Agroforestry practices designed to reduce wind speeds can increase pollinator efficiency and allow pollinators to forage during wind events that would normally reduce or prohibit foraging. The protective effect on insect flight extends up to a distance equal to about nine times the height of the windbreak (Pinzauti 1986) and the sheltered zone can contain higher numbers of pollinating insects (Pasek 1988), increasing pollination and fruit set (Norton 1988). Agroforestry practices that reduce winds, such as windbreaks and hedgerows, can enhance pollination by reducing flower shedding and increasing overall flowering (Peri and Bloomberg 2002).

## Nesting and Egg-Laying Sites

The availability of nesting and egg-laying sites is also key for successful pollinator conservation (Potts et al. 2005; Steffan-Dewenter and Schiele 2008; Sardiñas et al. 2016b). The short foraging and dispersal distances of many pollinator species require that, along with food resources, nesting resources should be available within a localized area (Gathmann and Tscharntke 2002).

Solitary tunnel-nesting bees build their nests aboveground in hollow tunnels in the soft pithy centers of twigs of some plants, in abandoned wood-boring beetle tunnels, or in tunnels they excavate themselves into wood, especially rotting logs and snags (e.g., Potts et al. 2005; Cane et al. 2007). Hedgerows and other agroforestry practices that incorporate woody species with soft pithy centers can increase the availability of nesting sites (Table 2) (Morandin and Kremen 2013; Kremen and M'Gonigle 2015). A modeling study calculated a higher nesting potential for cavity-nesting species in landscapes with agroforestry compared to landscapes without agroforestry (Kay et al. 2019). Dead trees and branches left within an agroforestry practice can also provide nesting sites (Brown 2002).

Solitary ground nesters excavate underground tunnels for nesting that can be negatively impacted by tillage in agricultural fields (Shuler et al. 2005; Kim et al. 2006). The presence of trees and shrubs provides protected nesting areas that have limited soil disturbance. Hedgerows have been documented to provide suitable ground-nesting habitat and increase diversity of ground-nesting bees (Morandin and Kremen 2013; Kremen and M'Gonigle 2015; Ponisio et al. 2016); however, another study did not find enhanced nesting rates for ground-nesting bees in hedgerows (Sardiñas et al. 2016a).

Bumble bee queens often hibernate under trees in leaf litter. Upon emerging in early spring, bumble bee queens seek rodent burrows and other insulated cavities in which to start their colonies and rear their brood or offspring. Bumble bees often select nest sites at the interface between fields and linear woody habitat such as hedgerows and windbreaks (Svensson et al. 2000; Kells and Goulson 2003). One study documented bumble bee nest densities twice as great in these linear woody habitats when compared with grassland and other woodland habitats (Osborne et al. 2008b) while another study found hedgerows to be less preferred when compared to herbaceous field margins and grasslands for nest-searching bumble bee queen (Lye et al. 2009). Non-cropped habitat suitable for nesting may also facilitate movement of queens into the wider landscape (Carvell et al. 2017).

Agroforestry practices can provide egg-laying sites, larval host plants, and overwintering sites for lepidopteran (butterfly and moth) species (Dover and Sparks 2000; Maudsley 2000; Merckx et al. 2012). Woody species were found to support ten times more lepidopteran species than herbaceous plants in the US mid-Atlantic region (Tallamy and Shropshire 2009). This work also documented that lepidopteran species used native woody plant species as larval hosts 14 times more than nonnative ornamental woody species. Some of the most highly used plant genera by lepidopteran species include poplar, willows, cherry, plum, birch, and oaks (*Quercus* spp.) (Tallamy and Shropshire 2009; Dumroese and Luna 2016). Lepidopteran species and other pollinators including beetles overwinter under bark and leaf litter found in hedgerows (Dover and Sparks 2000; Maudsley 2000; Pywell et al. 2005).

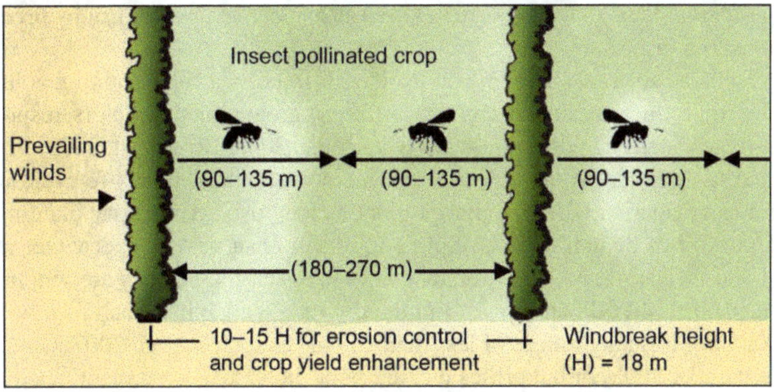


**Fig. 1** Windbreaks are typically planted at intervals of 10–15 times windbreak height (H) for reducing erosion and enhancing crop yields through microclimate modification. Using an H of 18 m as an example, the windbreaks would be spaced at 180–270 m across a field. This would place pollinator habitat within 90–135 m from the center of the cropped area, well within the foraging range of most pollinators as well as within the range of predatory and parasitoid insects to prey on crop pests. Within a 1 $km^2$ field, a 20 m wide and 18 m tall windbreak could provide 10% non-cropped habitat area to support pollinators

## *Enhancing Connectivity*

Habitat is becoming increasingly fragmented due to agricultural intensification, urban expansion, and other human activities (Saunders et al. 1991). Pollination services at the farm and landscape scale are impacted by this fragmentation (e.g., Aizen and Feinsinger 1994; Sipes and Tepedino 1995). For example, Garibaldi et al. (2011) estimated fruit set of pollinator-dependent crops decreased by 16% at 1 km distance from the nearest pollinator habitat.

Based on field level studies and modeling efforts, agroforestry practices can provide pollinator habitat close to crops and at a scale that benefits foraging and crop pollination (e.g., Morandin and Kremen 2013; Kremen and M'Gonigle 2015; Morandin et al. 2016; Sutter et al. 2018; Graham and Nassauer 2019). For example, the spatial distribution of windbreak and alley cropping plantings across fields to achieve other nonpollinator-related services places habitat within the foraging range of many pollinators, including short-distance foragers (Fig. 1) (Gathmann and Tscharntke 2002; Benjamin et al. 2014; Moisan-DeSerres et al. 2015). The benefits of agroforestry practices for pollination services are often higher when this semi-natural habitat is added to structurally simple fields and landscapes (e.g., Carvell et al. 2011; Klein et al. 2012; Ponisio et al. 2016; Ponisio et al. 2019). This distribution of habitat also supports other insect-based services in agricultural fields such as pest management by natural predatory insects. For instance, Morandin et al. (2014) documented pest control by beneficial insects extending 100 m into crop fields from hedgerows while Tscharntke et al. (2002) demonstrated that maintaining diverse habitat on more than 20% of a farm helps ensure effective pest control by predatory and parasitoid insects.

At the landscape scale, habitat connectivity is important for sustaining pollinator diversity, reproduction, and dispersal. Different groups of pollinators respond to habitat fragmentation in different ways (Cane et al. 2006; Brosi et al. 2008; Boscolo et al. 2017). Although some pollinators can complete their entire life cycle within hedgerows or riparian buffers, other pollinators may use agroforestry plantings for only a portion of their life cycle. Some pollinators can nest or overwinter in one habitat and forage in another if the distances between the patches are within their flight capabilities. Pollinators with limited dispersal capability, such as tiny sweat bees that have foraging ranges of less than 250 m (Greenleaf et al. 2007; Gathmann and Tscharntke 2002) or butterflies that are poor fliers, may need plantings directly connected to habitat to aid their dispersal. In contrast, bumble bees can forage up to 2 km or more (Osborne et al. 2008a). Habitats with greater connectivity allow pollinators to travel more safely between patches to find resources, disperse to new habitat, and encounter potential mates.

Agroforestry practices can serve as habitat corridors connecting larger patches of habitat that facilitate movement of organisms between habitat fragments, aid in establishing or maintaining populations, promote greater genetic flow among populations, and increase species diversity within isolated areas (Tewksbury et al. 2002). Experimental corridors have been found to increase the movement of pollinators (Haddad 1999) as well as facilitate pollination (Tewksbury et al. 2002; Townsend and Levey 2005). Evidence documenting pollinator use of agroforestry habitat as corridors includes hedgerow-promoted movement of butterflies (Ouin and Burel 2002), moths (Couthard et al. 2016), and bees (Cranmer et al. 2012; Klaus et al. 2015) and butterfly travel along windbreaks (Dover and Fry 2001) and riparian buffers (Meier et al. 2005). Corridors may not always need to directly connect habitat areas to help organisms to disperse (Fried et al. 2005) as patches of habitat can serve as stepping stones between isolated fragments in otherwise inhospitable landscapes (Ottewell et al. 2009).

Agroforestry plantings extending across rural and urban landscapes often contain greater plant diversity than adjacent lands, are longer term in nature, and are generally protected from further development and major disturbances. In developed landscapes, like intensively managed agricultural lands or cities, agroforestry plantings are particularly valuable (Senapathi et al. 2017). Additionally, agroforestry corridors are likely to be particularly beneficial in agricultural landscapes where natural or seminatural habitat benefits pollinator populations (e.g., Klein et al. 2012; McKechnie et al. 2017) as well as crop pollination (Morandin and Winston 2006; Blaauw and Isaacs 2014; Klatt et al. 2014). Hedgerows in intensively managed agricultural landscapes, for example, increase bee, syrphid fly, and other beneficial insect abundance and diversity in adjacent crop fields (Morandin and Kremen 2013; Morandin et al. 2014).

However, agroforestry plantings may act as barriers to some pollinators, inhibiting movement between habitats. Pollen flow can also potentially be reduced across hedgerows (Klaus et al. 2015) and possibly other tree-row plantings. Krewenka et al. (2011) found that bee foraging was not impacted by hedgerows; however, another study found that bombyliid flies had reduced pollen transfer (Campagne et al. 2009). Windbreaks and hedgerows can act as barriers for butterfly movement (Dover and Fry 2001). Hedgerows may channel pollinator movement, which could enhance connectivity but restrict movement across hedgerows, isolating some plant populations (Klaus et al. 2015). The orientation of plant rows may influence hedgerows' abilities to promote movement or act as barriers (Ouin and Burel 2002).

Climate change impacts pollinators and their relationships with plants by driving shifts in the ranges of pollinators or their host plants (Forister et al. 2010; Chen et al. 2011; Kerr et al. 2015), altering plant and pollinator phenology (Parmesan 2007; Bartomeus et al. 2011), decreasing protein concentration in floral pollen (Ziska et al. 2016), and increasing the impacts of other drivers of pollinator decline (Settele et al. 2016). Increasing landscape connectivity is one proposed strategy to reduce negative impacts of climate change on pollinators by enhancing the ability of species to move into new regions as climate shifts (Krosby et al. 2010; Gilchrist et al. 2016). Agroforestry may help enhance connectivity across rural and urban landscapes, thereby helping species extend their ranges and have some resiliency in the face of a changing climate.

## *Reducing Pesticide Exposure*

Pesticides can have acute toxicity leading to pollinator mortality and sublethal effects on growth and development, behavior, and other activities (Stanley and Preetha 2016). Sublethal effects of pesticide exposure at very low concentrations are reported on homing and foraging, larval development and adult emergence, and visual and olfactory learning (Desneux et al. 2007; Sánchez-Bayo and Goka 2014). Among social insects like honey bees and bumble bees, pesticides carried back to the nest may also impact larvae, nestmates, and the queen, and delay emergence of new adults (Wu et al. 2011). Pesticides can also suppress the immune system, making bees (and likely other organisms) more susceptible to disease and parasites (e.g., Sánchez-Bayo et al. 2016; Czerwinski and Sadd 2017; Evans et al. 2018).

On farms and in other landscapes, pollinators may come into contact with pesticides through several exposure pathways (Fig. 2) (e.g., Krupke et al. 2012; Botías et al. 2015; Chagnon et al. 2015; Johnson 2015; Hladik et al. 2016; Long and Krupke 2016; Stanley and Preetha 2016). Pollinators may also be exposed to multiple pesticides over time (with higher cumulative levels of toxicity than an individual pesticide or synergistic effects) (Sánchez-Bayo and Goka 2014). USDA (2014) provides additional information on pesticide exposure pathways and methods for preventing and mitigating potential negative impacts of pesticides on pollinators.

Agroforestry practices can help reduce pollinator exposure to pesticides that are used for managing crop pests and diseases or reducing weed competition (Vaughan et al. 2017). By understanding potential pesticide exposure pathways, farmers and land managers can better design plantings such as windbreaks, hedgerows, and

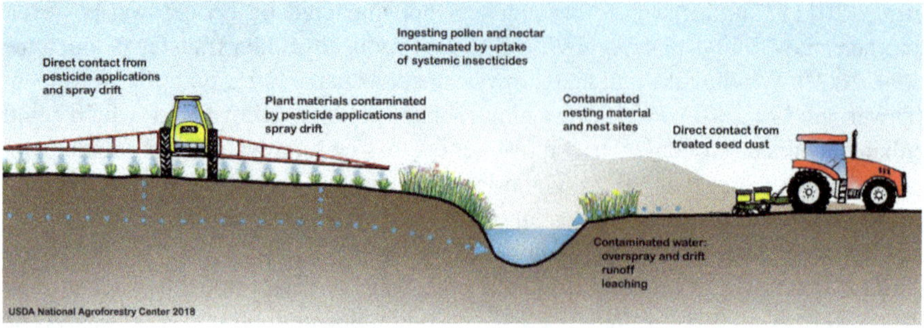


**Fig. 2** Potential pesticide exposure pathways encountered by pollinators in an agricultural landscape

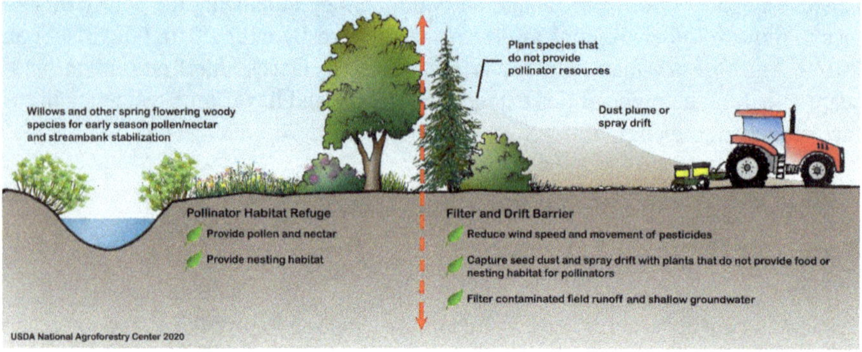


**Fig. 3** Using agroforestry practices to mitigate potential negative impacts of pesticides on pollinators

riparian buffers that help reduce or mitigate potential negative impacts of pesticides (Fig. 3). The same agroforestry practices aimed at protecting pollinators can also help reduce pesticide use and associated costs by supporting natural enemies of crop pests, such as predatory and parasitic insects and other arthropods that reduce pest populations (Morandin et al. 2016; Staton et al. 2019).

Windbreaks, hedgerows, and other linear plantings can reduce spray drift by up to 80–90% by reducing wind speeds and trapping particles (Ucar and Hall 2001; Otto et al. 2015). Buffers slow wind speeds, and the porosity of plant buffers lets wind move through the vegetation (vs. pushing up and over a solid barrier). At slower wind speeds, particles are more likely to fall out and become trapped in foliage. Agroforestry buffers that are 2.5–3 m tall, with 40–50% porosity and fine, evergreen foliage (large surface area), are generally the most effective for drift prevention (Ucar and Hall 2001; Wenneker and Van de Zande 2008; Mercer 2009; Otto et al. 2015; Chen et al. 2017). Yet, even hedgerows with porosity of nearly 75% have been found to be effective in reducing drift by more than 80% (Lazzaro et al. 2008).

In orchards or other crop systems being sprayed early in the growing season, buffers comprised of evergreen species can substantially reduce potential pollinator exposure risk from spray drift (Wenneker and Van de Zande 2008; Felsot et al. 2010). Fine, evergreen, coniferous foliage can capture 2–4 times that of broadleaf species, with the additional benefit of trapping air pollutants in winter (Chen et al. 2017). Leaf roughness, hairiness, waxiness, and other factors can affect foliage capture of particulate matter and some research indicates that the arrangement of a filter strip (with trees, shrubs, and grasses, and of an adequate length) is more important than species composition (Terzaghi et al. 2013; Chen et al. 2016).

Agroforestry buffers can also help capture pesticide runoff, prevent or slow pesticide movement through soil, and help to break down some pesticides (Chaudhry et al. 2005; Jose 2009; Pavlidis and Tsihrintzis 2017). A meta-analysis by Zhang et al. (2010) highlights how sediment captured by vegetative buffers helps improve pesticide removal, particularly those pesticides that are strongly hydrophobic such as pyrethroids and many organophosphates. Based on a review of available studies, Pavlidis and Tsihrintzis (2017) documented a 40–100% reduction of pesticides (including herbicides) in runoff using agroforestry systems. Plants and rhizosphere microorganisms vary in their ability to degrade or immobilize pesticides. Poplar, willow, birch, alder, black locust, and sycamore (*Platanus* spp.) are North American native trees with documented effectiveness in capturing pesticide runoff or immobilizing pesticides within their woody tissue (Pavlidis and Tsihrintzis 2017; Pavlidis and Tsihrintzis 2018).

However, the same factors making agroforestry practices effective buffers can also lead to pesticide accumulation and pose danger for pollinators, particularly from systemic pesticides and those with long residual activity such as neonicotinoids (Krupke et al. 2012; Hopwood et al. 2016). Nectar and pollen of early-flowering tree and shrub species may become contaminated by systemic action of neonicotinoids or through nontarget drift of treated seed-coating dust during crop planting (Long and Krupke 2016). Pesticide droplets and particles or pesticides adhering to dust can also accumulate in the foliage or at the base of agroforestry buffers (Zaady et al. 2018). Pollinators may ingest or carry back to the nest particles contaminated with pesticides (Krupke et al. 2012). If the pesticides or their metabolites have long residual activity and/or are systemically taken up into the plants, the accumulated levels could mean chronic and increased exposure over time. Pesticides accumulating in soil pose higher risks for the approximately 70% of native bees that nest in the ground.

Increasing the proportion of non-cropped habitat in agricultural landscapes has been shown to buffer the effects of pesticide on native bees (Park et al. 2015). Agroforestry practices can provide this habitat, especially when the plantings are protected from pesticide exposure. No-spray buffer zones can be used to protect agroforestry plantings that provide pollinator refuge (Davis and Williams 1990; Ucar and Hall 2001). Spray drift deposition in hedgerows was reduced by 72% when a 12 m no-spray buffer zone was used next to the hedgerows (Kjær et al. 2014). Depending on the cropping systems (and their potential spray regimes), it may be important to use plants that do not provide pollinator forage in the first rows adjacent to a field (Fig. 3).

## *Crop Pollination Services*

Available scientific evidence demonstrates the conservation benefits that agroforestry practices can provide to insect pollinators, including greater pollinator abundance and richness. Although these benefits should translate into enhanced pollination services leading to increased crop yields and quality, few studies have been conducted to document this direct agronomic benefit (Klein et al. 2012; Staton et al. 2019). Studies have shown positive effects on canola (*Brassica napus* L.) yields due to hedgerows (Morandin et al. 2016; Dainese et al. 2017) while another study showed no effects on crop pollination in sunflower (*Helianthus annuus* L.) (Sardiñas and Kremen 2015). In apple orchards, researchers found increased pollinator abundance adjacent to an artificial windbreak, which led to a 20–30% increase in fruit set with no reduction in fruit size (Smith and Lewis 1972). While the artificial windbreak was created out of coir netting, this study may suggest potential yield increases due to pollinator activity in apple orchards with planted windbreaks.

Many factors are likely to influence the ability of agroforestry practices to promote crop pollination services, including specific pollinator attributes, field size, crop type, plant composition of the agroforestry practice, and landscape context (IPBES 2016). The diversity of interacting variables makes it challenging to conduct studies and develop guidelines for producers. For instance, the ratio of agroforestry practice to crop area in order to supply sufficient pollination service is largely unexplored (Venturini et al. 2017). One study demonstrated that native bees can provide full pollination services for watermelon (*Citrullus lanatus* Thunb.) when around 30% of the land within 1.2 km of a field is in natural habitat (Kremen et al. 2004), which could be an approximate analog to an agroforestry practice. Regarding landscape context, one study found an increase in quality and quantity of strawberries grown adjacent to forest-connected hedgerows, as compared to isolated hedgerows or grass margins (Castle et al. 2019). Plants placed at forest-connected hedgerows produced more high-quality strawberries with 90% classified as "marketable." In comparison, only 75% of strawberries from plants at isolated hedgerows, 48% of strawberries from plants on grassy margins, and 41% of strawberries from self-pollinated control plants were classified as marketable. Based on market prices of 2016, the increase in economic value between strawberries produced at grassy margins and forest-connected hedgerows amounted to 61% (Castle et al. 2019). Cost-benefit studies that assess the benefits of an agroforestry practice for pollination services compared to the costs of installation and maintenance, opportunity costs, and costs of potential unintended negative effects are also very limited. Morandin et al. (2016) estimated that 7 years would be required for farmers to recover hedgerow implementation costs based on the estimated yield benefits from both pollination and pest control to the crop (Morandin et al. 2016). Future cost-benefit analyses should consider the range of agronomic effects in order to provide comprehensive economic assessment of ecosystem services.

## Summary

Agroforestry is a multifunctional land-use approach that provides a range of ecosystem services in support of production and environmental stewardship goals (Nair 2007). Capitalizing on insect-based ecosystem services, agroforestry offers opportunities to benefit pollinators and other beneficial insects and their services including crop pollination and biological pest management. Based on the available scientific literature, agroforestry practices in temperate regions can aid pollinators and pollination services by providing habitat, including foraging resources and nesting or egg-laying sites, enhancing site and landscape connectivity, and mitigating pesticide exposure.

**Table 3** General considerations for promoting pollinators and pollination services for each agroforestry practice

| Practice | Considerations for pollinators |
|---|---|
| Alley cropping (also called tree-based intercropping) | Alley cropping presents an opportunity to grow plants in close proximity that have complementary flowering periods. By paying careful attention to bloom periods and using multiple species, an alley cropping system can provide nearly continuous pollen and nectar forage within a single farmscape. Consider flowering trees like black cherry, black locust, or basswood along with the more typical alley cropping trees of walnut, pecan, or oak. Diverse forbs and shrubs may be planted in rows for cut flowers, berry production, or the nursery market, as well as for pollinators. A legume forage crop between rows will not only fix nitrogen and help manage weeds, but also provide nectar and pollen if allowed to flower |
| Windbreaks (also includes shelterbelts, hedgerows) | These practices help reduce wind speed, making it easier for pollinators to fly and visit flowers. When planted with diverse flowering shrubs and trees, windbreaks can provide shelter, pollen, and nectar for pollinators. Windbreaks and other linear plantings can serve as buffers against drifting pesticides. Do not use plants that will attract pollinators in windbreaks designed to intercept pesticide drift. Planting wildflowers during establishment can enhance pollinator resources and reduce weed pressure |
| Riparian forest buffers | Riparian forest buffers are especially important for pollinators during hot summer months when upland plants may not produce nectar or pollen. Early-flowering willows, as well as fruit and nut-bearing shrubs, can provide additional farm income as cut flowers or produce, while also providing reliable food resources for pollinators. Honey bees may also visit muddy shorelines to gather water for cooling their hives. Riparian buffers are important corridors for landscape connectivity from rural to urban areas, facilitating pollinator dispersal |
| Silvopasture | Silvopastures provide an open understory where a variety of flowering forbs (forage legumes, such as alfalfa or clover, or native wildflowers) can grow. Rotational grazing practices can give these forbs an opportunity to recover from grazing or flower before being eaten. Harvestable flowering trees, such as basswood, black locust, maple, persimmon, or tulip tree, can enhance a silvopasture system. Using thinning and prescribed fire to daylight existing seedbanks can restore natural diversity and promote flowering plants that benefit pollinators |
| Forest farming (also called multistory cropping) | Many valuable overstory crop trees, like tulip tree, maple, basswood, and black cherry, provide excellent pollinator habitat. Cultivated understory plants, such as ginseng, goldenseal, and black cohosh, may benefit from pollinator visits. For example, diverse bees pollinate black cohosh. Black cohosh does not produce nectar to attract bees, but relies on nearby prolific nectar producers, such as pale touch-me-not or whiteflower leafcup. The pollination needs of many forest-farmed crops are not well understood, but providing diverse habitat niches is the best way to support diverse pollinators. Flies are likely important pollinators since some flies are active in cooler temperatures, when many of the forest crops flower |

Source: Modified from Vaughan and Black 2006

A primary advantage for using agroforestry to support pollinators is that these practices inherently provide some pollinator benefits and with additional considerations during design and management, the effectiveness of agroforestry practices for pollinators can be enhanced. Due to common landscape settings and spatial configurations, each agroforestry practice provides different options and advantages for providing pollinator habitat, enhancing connectivity, and protecting against pesticides (Table 3).

Typically, agroforestry practices are planned for sites on individual farms and ranches. Pollinator-friendly agroforestry plantings on a single farm can have important benefits for pollinators. Even greater impact can be achieved through these plantings when planning and design are combined with other nearby farms and ranches which are using pollinator-friendly practices. While it may be uncommon for pollinators and other beneficial insects to be considered in landscape planning efforts (Steingröver et al. 2010), there are many potential benefits from broadening existing large-scale planning efforts to include pollinator issues. Working across site and landscape scales, agroforestry practices can support pollinator abundance and richness, protect biodiversity, enhance pollination, and increase food security.

**Acknowledgements** We would like to thank Justin Runyon, Judy Wu-Smart, Michele Schoeneberger, Aimee Code, and two anonymous reviewers whose comments and feedback helped to improve the chapter. This work was supported by funding provided by a contribution agreement with the United States Department of Agriculture (USDA)-Natural Resources Conservation Service (NRCS) to the Xerces Society.

## References

Adamson NL, Ward T, Vaughan M (2011) Designed with pollinators in mind. Inside Agroforestry 20(1):8–10

Aizen MA, Feinsinger P (1994) Habitat fragmentation, native insect pollinators, and feral honey bees in Argentine 'Chaco Serrano'. Ecol Appl 4(2):378–392

Allen-Wardell G, Bernhardt P, Bitner R, Burquez A, Buchmann S, Cane J, Cox P, Dalton V, Feinsinger P, Ingram M, Inouye D, Jones CE, Kennedy K, Kevan P, Koopowitz H, Medellin R, Medellin-Morales S, Nabhan G, Pavlik B, Tepedino V, Torchio P, Walker S (1998) The potential consequences of pollinator declines on the conservation of biodiversity and stability of food crop yields. Conserv Biol 12(1):8–17

Altizer SM, Oberhauser KS (1999) Effects of the protozoan parasite Ophryocystis elektroscirrha on the fitness of monarch butterflies (Danaus plexippus). J Invertebr Pathol 74:76–88

Ascher JS, Pickering J (2020) Discover life bee species guide and world checklist (Hymenoptera: Apoidea: Anthophila). http://wwwdiscoverlifeorg/mp/20q?guide=Apoidea_species Accessed (27 Feb 2020)

Bankova VB, De Castro SL, Marcucci MC (2000) Propolis: recent advances in chemistry and plant origin. Apidologie 31(1):3–15

Bartomeus I, Ascher JS, Wagner DL, Danforth BN, Colla S, Kornbluth S, Winfree R (2011) Climate-associated phenological advances in bee pollinators and bee-pollinated plants. Proc Natl Acad Sci U S A 108(51):20645–20649

Batra SWT (1985) Red maple (*Acer rubrum* L.), an important early spring food resource for honey bees and other insects. J Kans Entomol Soc 58(1):169–172

Baude M, Kunin WE, Boatman ND, Conyers S, Davies N, Gillespie MA, Morton RD, Smart SM, Memmott J (2016) Historical nectar assessment reveals the fall and rise of floral resources in Britain. Nature 530(7588):85–88

Bawa KS (1990) Plant-pollinator interactions in tropical rain forests. Annu Rev Ecol Syst 21:399–422

Bee Informed Partnership (2019) Colony Loss 2018-2019: Preliminary Results. https://beeinformed.org/wp-content/uploads/2019/11/2018_2019-Abstract.pdf. Accessed 30 Jan 2020

Benjamin FE, Reilly RJ, Winfree R (2014) Pollinator body size mediates the scale at which land use drives crop pollination services. J Appl Ecol 51(2):440–449

Bentrup G, Hopwood J, Adamson NL, Vaughan M (2019) Temperate agroforestry systems and insect pollinators: a review. Forests 10(11):981. https://doi.org/10.3390/f10110981

Biesmeijer JC, Roberts SPM, Reemer M, Ohlemüller R, Edwards M, Peeters T, Schaffers AP, Potts SG, Kleukers R, Thomas CD, Settele J, Kunin WE (2006) Parallel declines in pollinators and insect-pollinated plants in Britain and the Netherlands. Science 313(5785):351–354

Blaauw BR, Isaacs R (2014) Flower plantings increase wild bee abundance and the pollination services provided to a pollination-dependent crop. J Appl Ecol 51:890–898

Bosch J, Kemp WP (2001 How to manage the blue orchard bee as an orchard pollinator. http://agris.fao.org/agris-search/search.do?recordID=US201300072439. Accessed 30 Jan 2020

Boscolo D, Tokumoto PM, Ferreria PA, Ribeiro JW, dos Santos JS (2017) Positive responses of flower visiting bees to landscape heterogeneity depend on functional connectivity levels. Persp Ecol Conserv 15(1):18–24

Botías C, David A, Horwood J, Abdul-Sada A, Nicholls E, Hill E, Goulson D (2015) Neonicotinoid residues in wildflowers, a potential route of chronic exposure for bees. Environ Sci Technol 49(21):12731–12740

Brittain C, Kremen C, Klein AM (2013) Biodiversity buffers pollination from changes in environmental conditions. Glob Chang Biol 19(2):540–547

Brosi BJ, Briggs HM (2013) Single pollinator species losses reduce floral fidelity and plant reproductive function. Proc Natl Acad Sci U S A 110(32):13044–13048

Brosi BJ, Daily GC, Shih TM, Oviedo F, Duran G (2008) The effects of forest fragmentation on bee communities in tropical countryside. J Appl Ecol 45(3):773–783

Brown TK (2002) Creating and maintaining wildlife, insect, and fish habitat structures in dead wood. In: Proceedings of the symposium on the ecology and management of dead wood in western forests. PSW-GTR-181. USDA Forest Service, Pacific southwest Research Station, Albany, p 883–892

Buchmann SL (1983) Buzz pollination in angiosperms. In: Jones CE, Little RJ (eds) Handbook of experimental pollination biology. Van Nostrand Reinhold, New York, pp 73–113

Buehler DM, Norris DR, Stutchbury BJM, Kopysh NC (2002) Food supply and parental feeding rates of hooded warblers in forest fragments. Wilson Bull 114(1):122–127

Calder WA (2004) Rufous and broad-tailed hummingbirds—pollination, migration, and population biology. In: Nabhan P (ed) Conserving migratory pollinators and nectar corridors in Western North America. University of Arizona Press, Tucson, AZ, pp 59–79

Calderone NW (2012) Insect pollinated crops, insect pollinators and US agriculture: trend analysis of aggregate data for the period 1992–2009. PLoS One 7(5):e37235. https://doi.org/10.1371/journal.pone.0037235

Cameron SA, Lozier JD, Strange JP, Koch JB, Cordes N, Solter LF, Griswold TL (2011) Patterns of widespread decline in North American bumble bees. Proc Natl Acad Sci U S A 108(2):662–667

Campagne P, Affre L, Baumel A, Roche P, Tatoni T (2009) Fine-scale response to landscape structure in Primula vulgaris Huds.: does hedgerow network connectedness ensure connectivity through gene flow? Popul Ecol 51(1):209–219

Cane JH, Griswold T, Parker FD (2007) Substrates and materials used for nesting by North American Osmia bees (Hymenoptera: Apiformes: Megachilidae). Ann Entomol Soc Am 100(3):350–358

Cane JH, Minckley RL, Kervin LJ, Roulston TH, Williams NM (2006) Complex responses within a desert bee guild (Hymenoptera: Apiformes) to urban habitat fragmentation. Ecol Appl 16(2):632–644
Carvell C, Bourke AF, Dreier S, Freeman SN, Hulmes S, Jordan WC, Redhead JW, Sumner S, Wang J, Heard MS (2017) Bumblebee family lineage survival is enhanced in high-quality landscapes. Nature 543(7646):547–549
Carvell C, Osborne JL, Bourke AFG, Freeman SN, Pywell RF, Heard MS (2011) Bumble bee species' responses to a targeted conservation measure depend on landscape context and habitat quality. Ecol Appl 21(5):1760–1771
Castle D, Grass I, Westphal C (2019) Fruit quantity and quality of strawberries benefit from enhanced pollinator abundance at hedgerows in agricultural landscapes. Agric Ecosyst Environ 275:14–22
Chagnon M, Kreutzweiser D, Mitchell EA, Morrissey CA, Noome DA, Van der Sluijs JP (2015) Risks of large-scale use of systemic insecticides to ecosystem functioning and services. Environ Sci Pollut R 22(1):119–134
Chaudhry Q, Blom-Zandstra M, Gupta SK, Joner E (2005) Utilising the synergy between plants and rhizosphere microorganisms to enhance breakdown of organic pollutants in the environment. Environ Sci Pollut R 12(1):34–48
Chen I-C, Hill JK, Ohlemüller R, Roy DB, Thomas CD (2011) Rapid range shifts of species associated with high levels of climate warming. Science 333(6045):1024–1026
Chen L, Liu C, Zhang L, Zou R, Zhang Z (2017) Variation in tree species ability to capture and retain airborne fine particulate matter (PM2.5). Sci Rep 7:3206. https://doi.org/10.1038/s41598-017-03360-1
Chen L, Liu C, Zou R, Yang M, Zhang Z (2016) Experimental examination of effectiveness of vegetation as bio-filter of particulate matters in the urban environment. Environ Pollut 208:198–208
Colla SR, Otterstatter MC, Gegear RJ, Thomson JD (2006) Plight of the bumble bee: pathogen spillover from commercial to wild populations. Biol Conserv 129:461–467
Corbet SA (1990) Pollination and the weather. Isr J Bot 39:13–30
Corbet SA, Fussell M, Ake R, Fraser A, Gunson C, Savage A, Smith K (1993) Temperature and the pollinating activity of social bees. Ecol Entomol 18(1):17–30
Couthard E, McCollin D, Littlemore (2016) The use of hedgerows as flight paths by moths in intensive farmland landscapes. J Insect Conserv 20:345–350
Crane E, Walker P (1985) Some nectar characteristics of certain important honey sources. Pszczel Zesz Naukowe 29:29–45
Cranmer L, McCollin D, Ollerton J (2012) Landscape structure influences pollinator movements and directly affects plant reproductive success. Oikos 121(4):562–568
Czerwinski MA, Sadd BM (2017) Detrimental interactions of neonicotinoid pesticide exposure and bumblebee immunity. J Exp Zool Part A 327(5):273–283
Dainese M, Montecchiari S, Sitzia T, Sigura M, Marini L (2017) High cover of hedgerows in the landscape supports multiple ecosystem services in Mediterranean cereal fields. J Appl Ecol 54:380–388
Davis BNK, Williams CT (1990) Buffer zone widths for honeybees from ground and aerial spraying of insecticides. Environ Pollut 63(3):247–259
Desneux N, Decourtye A, Delpuech JM (2007) The sublethal effects of pesticides on beneficial arthropods. Annu Rev Entomol 52:81–106
Di Pasquale G, Salignon M, Le Conte Y, Belzunces LP, Decourtye A, Kretzschmar A, Suchail S, Brunet JL, Alaux C (2013) Influence of pollen nutrition on honey bee health: do pollen quality and diversity matter? PLoS One 8(8):e72016. https://doi.org/10.1371/journal.pone.0072016
Dirr MA (1990) Manual of woody landscape plants: their identification, ornamental characteristics, culture, propagation and uses. Stipes Publishing Co., Champaign
Donkersley P (2019) Trees for bees. Agric Ecosyst Environ 270–271:79–83
Dötterl S, Vereecken NJ (2010) The chemical ecology and evolution of bee–flower interactions: a review and perspectives. Can J Zool 88:668–697

Dover JW, Fry GLA (2001) Experimental simulation of some visual and physical components of a hedge and the effects on butterfly behaviour in an agricultural landscape. Entomol Exp Appl 100(2):221–233

Dover J, Sparks T (2000) A review of the ecology of butterflies in British hedgerows. J Environ Manag 60(1):51–63

Dover J, Sotherton N, Gobett K (1990) Reduced pesticide inputs on cereal field margins: the effects on butterfly abundance. Ecol Entomol 15(1):17–24

Drescher N, Klein AM, Schmitt T, Leonhardt SD (2019) A clue on bee glue: new insight into the sources and factors driving resin intake in honeybees (*Apis mellifera*). PLoS One 14:e0210594. https://doi.org/10.1371/journal.pone.0210594

Dumroese K, Luna T (2016) Growing and marketing woody species to support pollinators: an emerging opportunity for forest, conservation, and native plant nurseries in the Northeastern US Tree Planters' Notes 59(2):49–60 https://www.fs.usda.gov/treesearch/pubs/52882. Accessed 30 Jan 2020

Eilers EJ, Kremen C, Greenleaf SS, Garber AK, Klein AM (2011) Contribution of pollinator-mediated crops to nutrients in the human food supply. PLoS One 6(6):e21363. https://doi.org/10.1371/journal.pone.0021363

Evans AN, Llanos JE, Kunin WE, Evison SE (2018) Indirect effects of agricultural pesticide use on parasite prevalence in wild pollinators. Agric Ecosyst Environ 15(258):40–48

Fallon C, Black SH, Shepherd M (2014) Butterflies and moths as pollinators. Wings 37(2). The xerces Society for Invertebrate Conservation

Felsot AS, Unsworth JB, Linders JB, Roberts G, Rautman D, Harris C, Carazo E (2010) Agrochemical spray drift; assessment and mitigation—a review. J Environ Sci Heal B 46(1):1–23

Fiedler AK, Landis DA, Arduser M (2012) Rapid shift in pollinator communities following invasive species removal. Restor Ecol 20(5):593–602

Filipiak M (2019) Key pollen host plants provide balanced diets for wild bee larvae: a lesson for planting flower strips and hedgerows. J Appl Ecol 56:1410–1418

Fontaine C, Dajoz I, Meriguet J, Loreau M (2005) Functional diversity of plant–pollinator interaction webs enhances the persistence of plant communities. PLoS Biol 4(1):e1. https://doi.org/10.1371/journal.pbio.0040001

Forister ML, Jahner JP, Casner KL, Wilson JS, Shapiro AM (2011) The race is not to the swift: Long-term data reveal pervasive declines in California's low-elevation butterfly fauna. Ecology 92(12):2222–2235

Forister ML, McCall AC, Sanders NJ, Fordyce JA, Thorne JH, O'Brien J, Waetjen DP, Shapiro AM (2010) Compounded effects of climate change and habitat alteration shift patterns of butterfly diversity. PNAS 107(5):2088–2092

Fowler J (2016) Specialist bees of the northeast: host plants and habitat conservation. Northeast Nat 23:305–320

Frankie GW, Vinson SB, Newstrom LE, Barthell JF, Haber WA, Frankie JK (1990) Plant phenology, pollination ecology, pollinator behaviour and conservation of pollinators in Neotropical dry forest. In: Bawa KS, Hadley M (eds) Reproductive ecology of tropical forest plants. Parthenon Publishing Group, Paris, pp 37–47

Fried JH, Levey DJ, Hogsette JA (2005) Habitat corridors function as both drift fences and movement conduits for dispersing flies. Oecologia 143(4):645–651

Garibaldi LA, Steffan-Dewenter I, Kremen C, Morales JM, Bommarco R, Cunningham SA, Carvalheiro LG, Chacoff NP, Dudenhöffer JH, Greenleaf SS, Holzschuh A (2011) Stability of pollination services decreases with isolation from natural areas despite honey bee visits. Ecol Lett 14(10):1062–1072

Garibaldi LA, Steffan-Dewenter I, Winfree R, Aizen MA, Bommarco R, Cunningham SA, Kremen C, Carvalheiro LG, Harder LD, Afik O, Bartomeus I (2013) Wild pollinators enhance fruit set of crops regardless of honey bee abundance. Science 339(6127):1608–1611

Gathmann A, Tscharntke T (2002) Foraging ranges of solitary bees. J Anim Ecol 71(5):757–764

Ghisalberti EL (1979) Propolis: a review. Bee World 60:59–84
Gilchrist A, Barker A, Handley JF (2016) Pathways through the landscape in a changing climate: the role of landscape structure in facilitating species range expansion through an urbanised region. Landsc Res 41(1):26–44
Graham JB, Nassauer JI (2019) Wild bee abundance in temperate agroforestry landscapes: assessing effects of alley crop composition, landscape configuration, and agroforestry area. Agrofor Syst 93:837–850
Grant V (1994) Historical development of ornithophily in the western North American flora. Proc Natl Acad Sci U S A 91:10407–10411
Greenaway W, Scaysbrook T, Whatley FR (1990) The composition and plant origins of propolis: a report of work at Oxford. Bee World 71(3):107–118
Greenleaf SS, Williams NM, Winfree R, Kremen C (2007) Bee foraging ranges and their relationship to body size. Oecologia 153(3):589–596
Haddad NM (1999) Corridor and distance effects on interpatch movements: a landscape experiment with butterflies. Ecol Appl 9:612–622
Hannon LE, Sisk TD (2009) Hedgerows in an agri-natural landscape: potential habitat value for native bees. Biol Conserv 142(10):2140–2154
Hatfield RG, Colla SR, Jepsen S, Richardson LL, Thorp RW (2014) International Union for the Conservation of nature (IUCN) assessments for North American *Bombus* spp. for the North American IUCN bumble bee specialist group. The Xerces Society for Invertebrate Conservation, Portland. https://xerces.org/sites/default/files/publications/14-065.pdf. Accessed 6 Feb 2020
Haydak MH (1958) Wintering of bees in Minnesota. J Econ Entomol 51(3):332–334
Hill DB, Webster TC (1995) Apiculture and forestry (bees and trees). Agrofor Syst 29(3):313–320
Hladik ML, Vandever M, Smalling KL (2016) Exposure of native bees foraging in an agricultural landscape to current-use pesticides. Sci Total Environ 542:469–477
Hopwood J, Code A, Vaughan M, Biddinger D, Shepherd M, Black SH, Lee-Mäder E, Mazzacano, C (2016) How neonicotinoids can kill bees. Xerces Society for Invertebrate Conservation, Portland. http://www.xerces.org/neonicotinoids-and-bees. Accessed 30 Jan 2020
IPBES (2016) Assessment report on pollinators, pollination and food production. Intergovernmental science-policy platform on biodiversity and ecosystem services, Bonn, Germany. https://ipbes.net/assessment-reports/pollinators. Accessed 30 Jan 2020
Inouye DW, Larson BM, Ssymank A, Kevan PG (2015) Flies and flowers III: ecology of foraging and pollination. J Pollinat Ecol 16(16):115–133
Javorek SK, Mackenzie KE, Vander Kloet SP (2002) Comparative pollination effectiveness among bees (Hymenoptera: Apoidea) on lowbush blueberry (Ericaceae: Vaccinium angustifolium). Ann Entomol Soc Am 95:345–351
Johnson RM (2015) Honey bee toxicology. Annu Rev Entomol 60:415–434
Jose S (2009) Agroforestry for ecosystem services and environmental benefits: an overview. Agrofor Syst 76(1):1–10
Kay S, Kühn E, Albrecht M, Sutter L, Szerencsits E, Herzog F (2019) Agroforestry can enhance foraging and nesting resources for pollinators with focus on solitary bees at the landscape scale. Agrofor Syst 94(2):379–387. https://doi.org/10.1007/s10457-019-00400-9
Kearns CA (2001) North American dipteran pollinators: assessing their value and conservation status. Conserv Ecol 5(1):5. http://www.consecol.org/vol5/iss1/art5/. Accessed 30 Jan 2020
Kearns CA, Inouye DW (1997) Pollinators, flowering plants, and conservation biology. Bioscience 47(5):297–307
Kearns CA, Inouye DW, Waser NM (1998) Endangered mutualisms: the conservation of plant–pollinator interactions. Annu Rev Ecol Syst 29:83–113
Kells AR, Goulson D (2003) Preferred nesting sites of bumblebee queens (Hymenoptera: Apidae) in agroecosystems in the UK. Biol Conserv 109(2):165–174
Kerr JT, Pindar A, Galpern P et al (2015) Climate change impacts on bumblebees converge across continents. Science 349(6244):177–180

Kevan PG (1999) Pollinators as bioindicators of the state of the environment: species, activity and diversity. Agric Ecosyst Environ 74(1-3):373–393

Kim J, Williams N, Kremen C (2006) Effects of cultivation and proximity to natural habitat on ground-nesting native bees in California sunflower fields. J Kansas Entomol Soc 79(4):309–320

Kjær C, Bruus M, Bossi R, Løfstrøm P, Andersen HV, Nuyttens D, Larsen SE (2014) Pesticide drift deposition in hedgerows from multiple spray swaths. J Pestic Sci 39(1):14–21

Kjøhl M, Nielsen A, Stenseth NC (2011) Potential effects of climate change on crop pollination. Food and Agriculture Organization of the United Nations, Rome

Klatt BKA, Holzschu A, Westphal C, Clough Y, Smit I, Pawelzik E, Tscharntke T (2014) Bee pollination improves crop quality, shelf life and commercial value. P Roy Soc B-Biol Sci 281:20132440. https://doi.org/10.1098/rspb.2013.2440

Klaus F, Bass J, Marholt L, Mulle B, Klatt B, Kormann U (2015) Hedgerows have a barrier effect and channel pollinator movement in the agricultural landscape. J Landsc Ecol 8(1):22–31

Klein AM, Brittain C, Hendrix SD, Thorp R, Williams N, Kremen C (2012) Wild pollination services to California almond rely on semi-natural habitat. J Appl Ecol 49(3):723–732

Klein AM, Vaissière BE, Cane JH, Steffan-Dewenter I, Cunningham SA, Kremen C, Tscharntke T (2007) Importance of pollinators in changing landscapes for world crops. P Roy Soc B-Biol Sci 274:303–313

König B (1985) Plant sources of propolis. Bee World 66(4):136–139

Kremen C, M'Gonigle LK (2015) Small-scale restoration in intensive agricultural landscapes supports more specialized and less mobile pollinator species. J Appl Ecol 52(3):602–610

Kremen C, Williams NM, Bugg RL, Fay JP, Thorp RW (2004) The area requirements of an ecosystem service: crop pollination by native bee communities in California. Ecol Lett 7(11):1109–1119

Kremen C, Williams NM, Thorp RW (2002) Crop pollination from native bees at risk from agricultural intensification. Proc Natl Acad Sci U S A 99(26):16812–16816

Krewenka KM, Holzschuh A, Tscharntke T, Dormann CF (2011) Landscape elements as potential barriers and corridors for bees, wasps and parasitoids. Biol Conserv 144(6):1816–1825

Krosby M, Tewksbury J, Haddad NM, Hoekstra J (2010) Ecological connectivity for a changing climate. Conserv Biol 24(6):1686–1689

Krupke CH, Hunt GJ, Eitzer BD, Andino G, Given K (2012) Multiple routes of pesticide exposure for honey bees living near agricultural fields. PLoS One 7(1):e29268. https://doi.org/10.1371/journal.pone.0029268

Larson BMH, Kevan PG, Inouye DW (2001) Flies and flowers: taxonomic diversity of anthophiles and pollinators. Can Entomol 133(4):439–465

Lazzaro L, Otto S, Zanin G (2008) Role of hedgerows in intercepting spray drift: evaluation and modelling of the effects. Agric Ecosyst Environ 123(4):317–327

Long EY, Krupke CH (2016) Non-cultivated plants present a season-long route of pesticide exposure for honey bees. Nat Commun 7:11629. https://doi.org/10.1038/ncomms11629

Loose J, Drummond F, Stubbs C, Woods S, Hoffman S (2005) Conservation and management of native bees in cranberry. Tech bull 191. Maine agricultural and Forest Experiment Station, Orono ME

Losey JE, Vaughan M (2006) The economic value of ecological services provided by insects. Bioscience 56:311–323

Lye G, Park K, Osborne J, Holland J, Goulson D (2009) Assessing the value of rural stewardship schemes for providing foraging resources and nesting habitat for bumblebee queens (Hymenoptera: Apidae). Biol Conserv 142(10):2023–2032

Macdonald KJ, Kelly D, Tylianakis JM (2018) Do local landscape features affect wild pollinator abundance, diversity and community composition on Canterbury farms? N Z J Ecol 42:262–268

Macgregor CJ, Pocock MJ, Fox R, Evans DM (2015) Pollination by nocturnal Lepidoptera, and the effects of light pollution: a review. Ecol Entomol 40(3):187–198

Maudsley MJ (2000) A review of the ecology and conservation of hedgerow invertebrates in Britain. J Environ Manag 60(1):65–76

McGregor SE (1976) Insect pollination of cultivated crop plants. Agricultural Research Service, Washington, DC. https://naldc.nal.usda.gov/download/CAT76674944/PDF#page=11. Accessed 30 Jan 2020
McKechnie IM, Thomsen CJM, Sargent RD (2017) Forested field edges support a greater diversity of wild pollinators in lowbush blueberry (*Vaccinium angustifolium*). Agric Ecosyst Environ 237:154–161
McNaughton KG (1988) Effects of windbreaks on turbulent transport and microclimate. Agric Ecosyst Environ 22(23):17–40
Meier K, Kuusemets V, Luig J, Mander U (2005) Riparian buffer zones as elements of ecological networks: case study on *Parnassius mnemosyne* distribution in Estonia. Ecol Eng 24:531–537
Memmott J, Waser NM, Price MV (2004) Tolerance of pollination networks to species extinctions. Proc Roy Soc Lond B Bio 271:2605–2611
Mercer GN (2009) Modelling to determine the optimal porosity of shelterbelts for the capture of agricultural spray drift. Environ Model Softw 24:1349–1352
Merckx T, Marini L, Feber RE, Macdonald DW (2012) Hedgerow trees and extended-width field margins enhance macro-moth diversity: implications for management. J Appl Ecol 49(6):1396–1404
Merrill JH (1923) Value of winter protection for bees. J Econ Entomol 16(2):125–130
Michener CD (2007) The bees of the world. John Hopkins University Press, Baltimore, MD
Miñarro M, Prida E (2013) Hedgerows surrounding organic apple orchards in north-west Spain: potential to conserve beneficial insects. Agric For Entomol 15(4):382–390
Moisan-DeSerres J, Chagnon M, Fournier V (2015) Influence of windbreaks and forest borders on abundance and species richness of native pollinators in lowbush blueberry fields in Québec, Canada. Can Entomol 147(4):432–442
Morandin LA, Kremen C (2013) Hedgerow restoration promotes pollinator populations and exports native bees to adjacent fields. Ecol Appl 23(4):829–839
Morandin LA, Winston ML (2006) Pollinators provide economic incentive to preserve natural land in agroecosystems. Agric Ecosyst Environ 116(3):289–292
Morandin LA, Long RF, Kremen C (2014) Hedgerows enhance beneficial insects on adjacent tomato fields in an intensive agricultural landscape. Agric Ecosyst Environ 189:164–170
Morandin LA, Long RF, Kremen C (2016) Pest control and pollination cost–benefit analysis of hedgerow restoration in a simplified agricultural landscape. J Econ Entomol 109(3):1020–1027
Morse RA, Calderone NW (2000) The value of honey bees as pollinators of US crops in 2000. Bee Cult 128:1–15
Nabhan GP, Brusca RC, Holter L (eds) (2004) Conserving migratory pollinators and nectar corridors in western North America. University of Arizona Press, Tucson
Nair PR (2007) The coming of age of agroforestry. J Sci Food Agric 87:1613–1619
National Research Council (2007) Status of pollinators in North America. National Academies Press, Washington, DC
NatureServe (2018) Conservation status. http://explorer.natureserve.org/ranking.htm. Accessed 30 Jan 2020
Nicholls CI, Altieri MA (2013) Plant biodiversity enhances bees and other insect pollinators in agroecosystems: a review. Agron Sustain Dev 33(2):257–274
Norton RL (1988) Windbreaks: benefits to orchard and vineyard crops. Agric Ecosyst Environ 22(23):205–213
Nye WP (1962) Extra supering and shading as factors in honey production in northern Utah. Glean Bee Cult 90(5):396–399
Ollerton J, Winfree R, Tarrant S (2011) How many flowering plants are pollinated by animals? Oikos 120:321–326
Ollerton J (2017) Pollinator diversity: distribution, ecological function, and conservation. Annu Rev Ecol Evol Syst 48:353–376
Osborne JL, Martin AP, Carreck NL, Swain JL, Knight ME, Goulson D, Hale RJ, Sanderson RA (2008a) Bumble bee flight distances in relation to the forage landscape. J Anim Ecol 77(2):406–415

Osborne JL, Martin AP, Shortall CR, Todd AD, Goulson D, Knight ME, Hale RJ, Sanderson RA (2008b) Quantifying and comparing bumblebee nest densities in gardens and countryside habitats. J Appl Ecol 45(3):784–792

Ostaff DP, Mosseler A, Johns RC, Javorek S, Klymko J, Ascher JS (2015) Willows (Salix spp.) as pollen and nectar sources for sustaining fruit and berry pollinating insects. Can J Plant Sci 95(3):505–515

Ottewell KM, Donnellan SC, Lowe AJ, Paton DC (2009) Predicting reproductive success of insect-versus bird-pollinated scattered trees in agricultural landscapes. Biol Conserv 142(4):888–898

Otto S, Loddo D, Baldoin C, Zanin G (2015) Spray drift reduction techniques for vineyards in fragmented landscapes. J Environ Manag 162:290–298

Ouin A, Burel F (2002) Influence of herbaceous elements on butterfly diversity in hedgerow agricultural landscapes. Agric Ecosyst Environ 93(1-3):45–53

Papanikolaou AD, Kühn I, Frenzel M, Schweiger O (2017) Semi-natural habitats mitigate the effects of temperature rise on wild bees. J Appl Ecol 54(2):527–536

Park MG, Blitzer EJ, Gibbs J, Losey JE, Danforth BN (2015) Negative effects of pesticides on wild bee communities can be buffered by landscape context. Proc R Soc B 282(1809):20150299. https://doi.org/10.1098/rspb.2015.0299

Parmesan C (2007) Influences of species, latitudes and methodologies on estimates of phenological response to global warming. Glob Change Biol 13(9):1860–1872

Pasek JE (1988) Influence of wind and windbreaks on local dispersal of insects. Agric Ecosyst Environ 22:539–554

Pavlidis G, Tsihrintzis VA (2017) Pollution control by agroforestry systems: a short review. Eur Water 59:297–301

Pavlidis G, Tsihrintzis VA (2018) Environmental benefits and control of pollution to surface water and groundwater by agroforestry systems: a review. Water Res Manage 32(1):1–29

Peri PL, Bloomberg M (2002) Windbreaks in southern Patagonia, Argentina: a review of research on growth models, windspeed reduction, and effects on crops. Agrofor Syst 56(2):129–144

Pinzauti M (1986) The influence of the wind on nectar secretion from the melon and on the flight of bees: the use of an artificial wind-break. Apidologie 17(1):63–72

Pleasants JM, Oberhauser KS (2012) Milkweed loss in agricultural fields because of herbicide use: effect on the monarch butterfly population. Insect Conserv Divers 6(2):135–144

Policarová J, Cardinal S, Martins AC, Straka J (2019) The role of floral oils in the evolution of apid bees (Hymenoptera: Apidae). Biol J Linn Soc 128:486–497

Ponisio LC, M'gonigle LK, Kremen C (2016) On-farm habitat restoration counters biotic homogenization in intensively managed agriculture. Glob Chang Biol 22(2):704–715

Ponisio LC, de Valpine P, M'Gonigle LK, Kremen C (2019) Proximity of restored hedgerows interacts with local floral diversity and species′ traits to shape long-term pollinator metacommunity dynamics. Ecol Lett 22:1048–1060

Potts SG, Biesmeijer JC, Kremen C, Neumann P, Schweiger O, Kunin WE (2010) Global pollinator declines: trends, impacts and drivers. Trends Ecol Evol 25(6):345–353

Potts SG, Vulliamy B, Dafni A, Ne'eman G, Willmer P (2003) Linking bees and flowers: how do floral communities structure pollinator communities? Ecology 84(10):2628–2642

Potts SG, Vulliamy B, Roberts S, O'Toole C, Dafni A, Ne'eman G, Willmer P (2005) Role of nesting resources in organizing diverse bee communities in a Mediterranean landscape. Ecol Entomol 30(1):78–85

Pywell RF, James KL, Herbert I, Meek WR, Carvell C, Bell D, Sparks TH (2005) Determinants of overwintering habitat quality for beetles and spiders on arable farmland. Biol Conserv 123(1):79–90

Rendón-Salinas E, Tavera-Alonso G (2014) Monitoreo de la superficie forestal ocupada por las colonias de hibernación de la mariposa Monarca en diciembre de 2013. Alianza WWF-Telcel/CONANP

Roulston TAH, Goodell K (2011) The role of resources and risks in regulating wild bee populations. Annu Rev Entomol 56:293–312

Russo L, Danforth B (2017) Pollen preferences among the bee species visiting apple (*Malus pumila*) in New York. Apidologie 48:806–820

Sánchez-Bayo F, Goka K (2014) Pesticide residues and bees–a risk assessment. PLoS One 9(4):e94482. https://doi.org/10.1371/journal.pone.0094482

Sánchez-Bayo F, Goulson D, Pennacchio F, Nazzi F, Goka K, Desneux N (2016) Are bee diseases linked to pesticides?—a brief review. Environ Int 89:7–11

Sánchez-Bayo F, Wyckhuys KA (2019) Worldwide decline of the entomofauna: a review of its drivers. Biol Conserv 232:8–27

Sardiñas HS, Kremen C (2015) Pollination services from field-scale agricultural diversification may be context-dependent. Agric Ecosyst Environ 207:17–25

Sardiñas HS, Ponisio LC, Kremen C (2016a) Hedgerow presence does not enhance indicators of nest-site habitat quality or nesting rates of ground-nesting bees. Restor Ecol 24(4):499–505

Sardiñas HS, Tom K, Ponisio LC, Rominger A, Kremen C (2016b) Sunflower (*Helianthus annuus*) pollination in California's Central Valley is limited by native bee nest site location. Ecol Appl 26(2):438–447

Saunders DA, Hobbs RJ, Margules CR (1991) Biological consequences of ecosystem fragmentation – a review. Conserv Biol 5:18–32

Schultz CB, Brown LM, Pelton E, Crone EE (2017) Citizen science monitoring demonstrates dramatic declines of monarch butterflies in western North America. Biol Conserv 214:343–346

Semmens BX, Semmens DJ, Thogmartin WE, Wiederholt R, López-Hoffman L, Diffendorfer JE, Pleasants JM, Oberhauser KS, Taylor OR (2016) Quasi-extinction risk and population targets for the eastern, migratory population of monarch butterflies (Danaus plexippus). Sci Rep 6:23265

Senapathi D, Goddard MA, Kunin WE, Baldock KC (2017) Landscape impacts on pollinator communities in temperate systems: evidence and knowledge gaps. Funct Ecol 31(1):26–37

Settele J, Bishop J, Potts SG (2016) Climate change impacts on pollination. Nat Plants 2(7). https://doi.org/10.1038/nplants.2016.92

Shuler RE, Roulston TH, Farris GE (2005) Farming practices influence wild pollinator populations on squash and pumpkin. J Econ Entomol 98:790–795

Simone-Finstrom M, Borba RS, Wilson M, Spivak M (2017) Propolis counteracts some threats to honey bee health. Insects 8(2):46. https://doi.org/10.3390/insects8020046

Sipes SD, Tepedino VJ (1995) Reproductive biology of the rare orchid, Spiranthes diluvialis: breeding system, pollination, and implications for conservation. Conserv Biol 9(4):929–938

Smith BD, Lewis T (1972) The effects of windbreaks on the blossom-visiting fauna of apple orchards and on yield. Ann Appl Biol 72(3):229–238

Smith J, Pearce BD, Wolfe MS (2013) Reconciling productivity with protection of the environment: is temperate agroforestry the answer? Renew Agr Food Syst 28(1):80–92

Somme L, Moquet L, Quinet M, Vanderplanck M, Michez D, Lognay G, Jacquemart AL (2016) Food in a row: urban trees offer valuable floral resources to pollinating insects. Urban Ecosyst 19(3):1149–1161

Spira TP (2001) Plant-pollinator interactions: a threatened mutualism with implications for the ecology and management of rare plants. Nat Area J 21(1):78–88

Stanley J, Preetha G (2016) Pesticide toxicity to non-target organisms. Springer, Dordrecht

Staton T, Walters RJ, Smith J, Girling RD (2019) Evaluating the effects of integrating trees into temperate arable systems on pest control and pollination. Agric Syst 176:102676

Steffan-Dewenter I, Schiele S (2008) Do resources or natural enemies drive bee population dynamics in fragmented habitats. Ecology 89(5):1375–1387

Steffan-Dewenter I, Westphal C (2008) The interplay of pollinator diversity, pollination services and landscape change. J Appl Ecol 45(3):737–741

Steingröver EG, Geertsema W, van Wingerden WK (2010) Designing agricultural landscapes for natural pest control: a transdisciplinary approach in the Hoeksche Waard (The Netherlands). Landsc Ecol 25(6):825–838

Stubbs CS, Jacobson HA, Osgood EA, Drummond FA (1992) Alternative forage plants for native (wild) bees associated with lowbush blueberry, *Vaccinium* spp., in Maine. Tech Bull 148. Maine Agricultural Experiment Station, Orono

Sutter L, Albrecht M, Jeanneret P (2018) Landscape greening and local creation of wildflower strips and hedgerows promote multiple ecosystem services. J Appl Ecol 55:612–620

Svensson B, Lagerlöf J, Svensson BG (2000) Habitat preferences of nest-seeking bumble bees (Hymenoptera: Apidae) in an agricultural landscape. Agric Ecosyst Environ 77(3):247–255

Tallamy DW, Shropshire KJ (2009) Ranking lepidopteran use of native versus introduced plants. Conserv Biol 23(4):941–947

Tasei JN, Aupinel P (2008) Nutritive value of 15 single pollens and pollen mixes tested on larvae produced by bumblebee workers (*Bombus terrestris*, Hymenoptera: Apidae). Apidologie 39(4):397–409

Tepedino VJ (1981) The pollination efficiency of the squash bee (*Peponapis pruinosa*) and the honey bee (*Apis mellifera*) on summer squash (*Cucurbita pepo*). J Kansas Entomol Soc 54(2):359–377

Terzaghi E, Wild E, Zacchello G, Cerabolini BE, Jones KC, Di Guardo A (2013) Forest filter effect: role of leaves in capturing/releasing air particulate matter and its associated PAHs. Atmos Environ 74:378–384

Tewksbury JJ, Levey DJ, Haddad NM, Sargent S, Orrock JL, Weldon A, Danielson BJ, Brinkerhoff J, Damschen EI, Townsend P (2002) Corridors affect plants, animals, and their interactions in fragmented landscapes. Proc Natl Acad Sci U S A 99(20):12923–12926

Timberlake TP, Vaughan IP, Memmott J (2019) Phenology of farmland floral resources reveals seasonal gaps in nectar availability for bumblebees. J Appl Ecol 56:1585–1596

Townsend PA, Levey DJ (2005) An experimental test of whether habitat corridors affect pollen transfer. Ecology 86(2):466–475

Tscharntke T, Steffan-Dewenter I, Kruess A, Thies C (2002) Contribution of small habitat fragments to conservation of insect communities of grassland–cropland landscapes. Ecol Appl 12(2):354–363

Ucar T, Hall FR (2001) Windbreaks as a pesticide drift mitigation strategy: a review. Pest Manag Sci 57(8):663–675

Udawatta RP, Rankoth LM, Jose S (2019) Agroforestry and biodiversity. Sustainability 11:2879

USDA (2014) Preventing or mitigating potential negative impacts of pesticides on pollinators using IPM and other conservation practices. US Department of Agriculture, Washington, DC. https://www.nrcs.usda.gov/Internet/FSE_DOCUMENTS/stelprdb1261468.pdf. Accessed 30 Jan 2020

US Fish and Wildlife Service (2006) Lesser long-nosed bat (*Leptonycteris curasoae yerbabuenae*) 5-year review: summary and evaluation. US fish and wildlife service. Phoenix, Arizona

US Fish and Wildlife Service (2019) Rusty patched bumble bee listed as endangered. https://www.fws.gov/midwest/endangered/insects/rpbb/FAQsFinalListing.html. Accessed 30 Jan 2020

Valiente-Banuet A, Molina-Freaner F, Torres A, Arizmendi MC, Casas A (2004) Geographic differentiation in the pollination system of the columnar cactus *Pachycereus pecten-aboriginum*. Am J Bot 91:850–855

Vaughan M, Black SH (2006) Improving forage for native bee crop pollinators. Agroforestry Notes #33. USDA National Agroforestry Center, Lincoln

Vaughan M, Adamson N, MacFarland K (2017) Using agroforestry practices to reduce pesticide risks to pollinators & other agriculturally beneficial insects. Agroforestry Notes #35. USDA National Agroforestry Center, Lincoln

Venturini EM, Drummond FA, Hoshide AK, Dibble AC, Stack LB (2017) Pollination reservoirs for wild bee habitat enhancement in cropping systems: a review. Agroecol Sustain Food Syst 41:101–142

Vicens N, Bosch J (2000) Weather-dependent pollinator activity in an apple orchard, with special reference to Osmia cornuta and Apis mellifera (Hymenoptera: Megachilidae and Apidae). Environ Entomol 29(3):413–420

Wcislo WT, Cane JH (1996) Floral resource utilization by solitary bees (Hymenoptera: Apoidea) and exploitation of their stored foods by natural enemies. Annu Rev Entomol 41:257–286
Wenneker M, Van de Zande JC (2008) Spray drift reducing effects of natural windbreaks in orchard spraying. Asp Appl Biol 84:25–32
Westerkamp C, Gottsberger G (2000) Diversity pays in crop pollination. Crop Sci 40:1209–1222
Whitehorn PR, O'Connor S, Wackers FL, Goulson D (2012) Neonicotinoid pesticide reduces bumble bee colony growth and queen production. Science 336(6079):351–352
Williams NM, Kremen C (2007) Resource distribution among habitats determine solitary bee offspring production in a mosaic landscape. Ecol Appl 17:910–921
Wood TJ, Kaplan I, Szendrei Z (2018) Wild bee pollen diets reveal patterns of seasonal foraging resources for honey bees. Front Ecol Evol 6:210. https://doi.org/10.3389/fevo.2018.00210
Wu JY, Anelli CM, Sheppard WS (2011) Sub-lethal effects of pesticide residues in brood comb on worker honey bee (Apis mellifera) development and longevity. PLoS One 6:e14720. https://doi.org/10.1371/journal.pone.0014720
Zaady E, Katra I, Shuker S, Knoll Y, Shlomo S (2018) Tree belts for decreasing aeolian dust-carried pesticides from cultivated areas. Geosciences 8:286
Zhang X, Liu X, Zhang M, Dahlgren RA, Eitzel M (2010) A review of vegetated buffers and a meta-analysis of their mitigation efficacy in reducing nonpoint source pollution. J Environ Qual 39(1):76–84
Ziska LH, Pettis JS, Edwards J, Hancock JE, Tomecek MB, Clark A, Dukes JS, Loladze I, Polley HW (2016) Rising atmospheric CO2 is reducing the protein concentration of a floral pollen source essential for North American bees. Proc Roy Soc B-Biol Sci 283(1828):20160414

# Flood Control and Air Cleaning Regulatory Ecosystem Services of Agroforestry

**Ranjith P. Udawatta**

## Abbreviations

| | |
|---|---|
| AF | Agroforestry |
| AQ | Air quality |
| BD | Biodiversity |
| C | Carbon |
| CAFO | Concentrated animal feeding operations |
| CS | Carbon sequestration |
| ES | Ecosystem services |
| PM | Particulate matter |
| UFF | Urban food forests |
| VOC | Volatile organic chemicals |

## Introduction

In recent years an increasing number of large flood events have been reported globally. Climate models predict increases of high-intensity large rains for some regions of the world and some regions within countries (Corringham and Cayan 2019; Kirchmeier-Younga and Zhang 2020). According to Porter et al. (2014), climate change will increase the number and strength of natural hazards such as floods and droughts, and these effects are most severe at the local scale and affect livelihoods (Shaw 2006). Increasing rains over the last two decades have caused billions of weather-related damages in many countries and loss of human lives, and these losses continue to rise. Globally, floods have caused more than 500,000 deaths from 1980 to 2009 (Hirsch and Ryberg 2012). In the USA, 4586 deaths between 1959 and 2009 were caused by floods (Hirsch and Ryberg 2013).

---

R. P. Udawatta (✉)
The Center for Agroforestry, School of Natural Resources, University of Missouri,
Columbia, MO, USA
e-mail: UdawattaR@missouri.edu



R. P. Udawatta, S. Jose (eds.), *Agroforestry and Ecosystem Services*,
https://doi.org/10.1007/978-3-030-80060-4_12

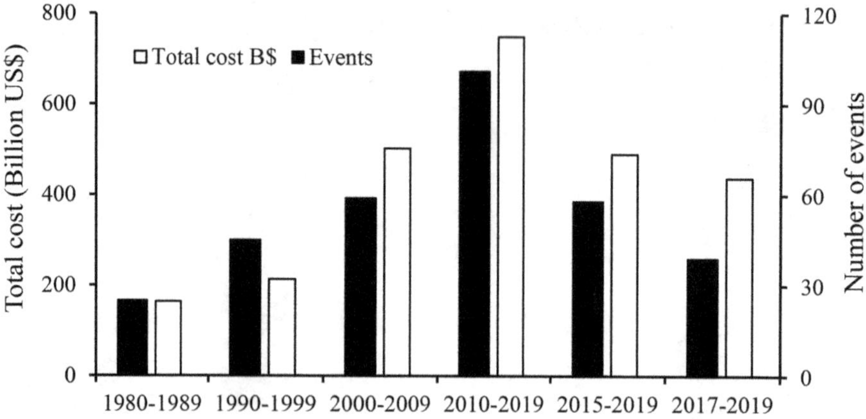


**Fig. 1** Number of weather-related damages and the monetary values for USA (source: NCDC/NOAA 2020)

NOAA has shown increasing number of weather-related damages with time in the USA (Fig. 1; NCDC/NOAA 2020). The study also showed more frequent and larger losses in recent years than in the previous two decades. According to a recent analysis of US floods from 1930 to 2000, Pielke et al. (2002) showed that the annual damage from foods has increased from ~1 billion US$ in 1930 to 4 billion US$ (1995 value) in 2000. Similar changes have been reported for other countries and the average annual damage from floods in the UK was 365 million US$ (FFF n.d.). These events also reduce the available cropland acreage and thereby affect agricultural productivity and food security.

Global temperature has increased over the last century and climate models have predicted further increases with some large, regionalized surges (climate.nasa.gov/effects/ n.d.). Increasing temperatures contribute to higher ocean temperatures and larger rain events. On the land, increasing temperatures contribute to poor air quality (AQ) by increasing the amount of pollution in the air including particulate matter (PM, dust, and various size particles), gases, vapor, and volatile organic chemicals (VOC). The effects of increasing temperature and dust are especially important for the Middle East and Northern Africa where dust is a major factor for poor AQ. In Australia, increased air temperature and reduced precipitation have affected the AQ by dust, smoke from bushfires, and other gases. Once these particles enter the higher atmosphere they will continue to travel worldwide affecting the AQ of other regions.

Air quality is also affected by anthropogenic activities and natural causes. Industry, construction, day-to-day activities, and agriculture can deteriorate AQ. For example, concentrated animal operations (CAFO) have been often criticized for poor AQ caused by PM, gases, and VOCs. Natural processes like volcanoes can emit various gases, dust, PM, and lava affecting AQ. For instance, the eruption of Eyjafjallajökull volcano in Iceland in 2010 forced many European countries to shut down air traffic. Agroforestry trees cannot stop natural disasters; however, perennial vegetation can help reduce the damage and recover faster.

Agroforestry (AF) with trees can improve regulatory ES like flood control, air cleaning, climate regulation, water cleaning, and pollination. More advanced landscape-scale AF approaches create stronger links between AF and flood control as well as AF and air cleaning (Dwyer et al. 2007; Tyndall and Colletti 2007). In AF, perennial vegetation is intentionally integrated for desired benefits arising from interactions among the components in the system (Gold and Garrett 2009). Trees, grasses, shrubs, and forbs within AF change above- and belowground structures and microclimate and add structural and spatial diversity to the landscape. Additionally, careful selection of the most suitable tree species to meet landowner objectives that are compatible with soil-site conditions and strategic placement of each component and advanced management practices can yield numerous production, environmental, and economic benefits. For example, adoption of AF has been proposed to build resilient livelihood to floods and droughts for Africa and some countries (Kandji et al. 2006; Verchot et al. 2007; Garrity et al. 2010; Simelton et al. 2015). This chapter is focused on riparian buffers, windbreaks, and urban food forest (UFF) AF practices on flood control and air cleaning regulatory ES. It also presents findings from various countries to better design AF systems for enhanced ES.

## Flood Control

Protecting farms with trees and using the flood-deposited nutrients for crop production are not new ideas. However, climate change and associated high-intensity large rains are causing severe floods across the globe destroying farmlands, cities, roads, and industry. In recent years, rain and flooding events have become more frequent, severe, unpredictable, and chaotic and produce higher peaks than a year or two ago. In some instances, these events take lives of humans, farm animals, and wildlife.

Many regions of the USA are experiencing precipitation events with increasing intensity and greater total rainfall amounts than previously recorded. From 1958 to 2012 the amount of heavy precipitation has increased from 5 to 71% in the USA (Fig. 2; NCA 2014). These climate change-induced rain events are causing greater soil erosion and more frequent flooding, especially in the Northeast and Midwest. Recent climate projections forecast an increase in heavy precipitation, even in regions where total precipitation is projected to decrease, such as the Southwest. Increasing rainfall amounts and intensities will produce higher peaks, greater velocities, and large flow volumes in streams and rivers (Vose et al. 2016).

Globally, heavy rains and flooding events have increased in recent years and will increase by 5–51% according to climate models (Nearing 2001; Nearing et al. 2004). In Pakistan, where the extreme climate can go from scorching hot to cold pounding rains throughout the year, flooding in 1950, 1992, 1993, and 2010 killed 2910, 1834, 3084, and 1781 people. Severe weather events and landslides are increasing in other countries. In Colombia, the National Unit for the Management of Disaster Risk (UNGRD) recorded 57 severe weather events during the first 13 days of the rainy season. These severe weather events have caused floods,

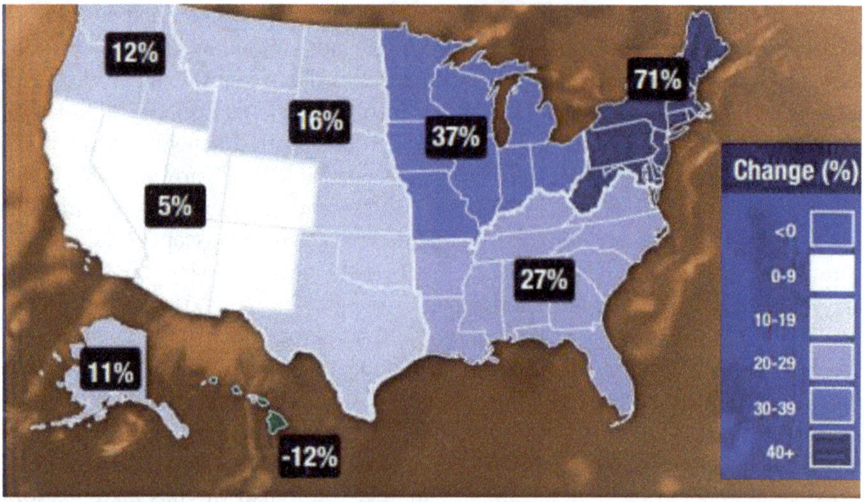


**Fig. 2** Percent increase in the amount of precipitation falling in heavy events from 1958 to 2012 for the USA (source: National Climate Assessment 2014)

landslides, and loss of lives and properties. Excessive soil water content on steep slopes with weak soils has contributed to landslides (FAO).

El Niño, La Niña, and the neutral condition alter weather globally and produce mild weather, severe storms, droughts, and flooding. For example, in the Southern USA, El Niño usually causes increased rainfall and sometimes destructive flooding during the fall through spring. Although La Niña causes drier weather in the South, it increases the number of Atlantic hurricanes, while the Northwest tends to be colder and wetter than average. Ward et al. (2014) noticed significant changes in 100-year flood risks during El Niño or La Niña years at 44% of river basins worldwide. According to the study, the Southwest USA, parts of southern South America, and the Horn of Africa will experience the biggest increases in flooding risks. These events can cause significant losses and deaths. For instance, the economic damage of the 1997–1998 El Niño was US$34 billion and caused 23,000 deaths globally (USAID n.d.).

## *The Processes of Flood Control*

The changes we do to the landscape, by removing forest canopy and replacing it with impervious surfaces like roads, parking lots, industry, homes, driveways, and annual crops, not only increase the volume of water that goes to the stream, but also shorten the amount of time it takes the water to get to the stream and increase the peak flow. The above changes reduce water infiltration and storage in soils as well

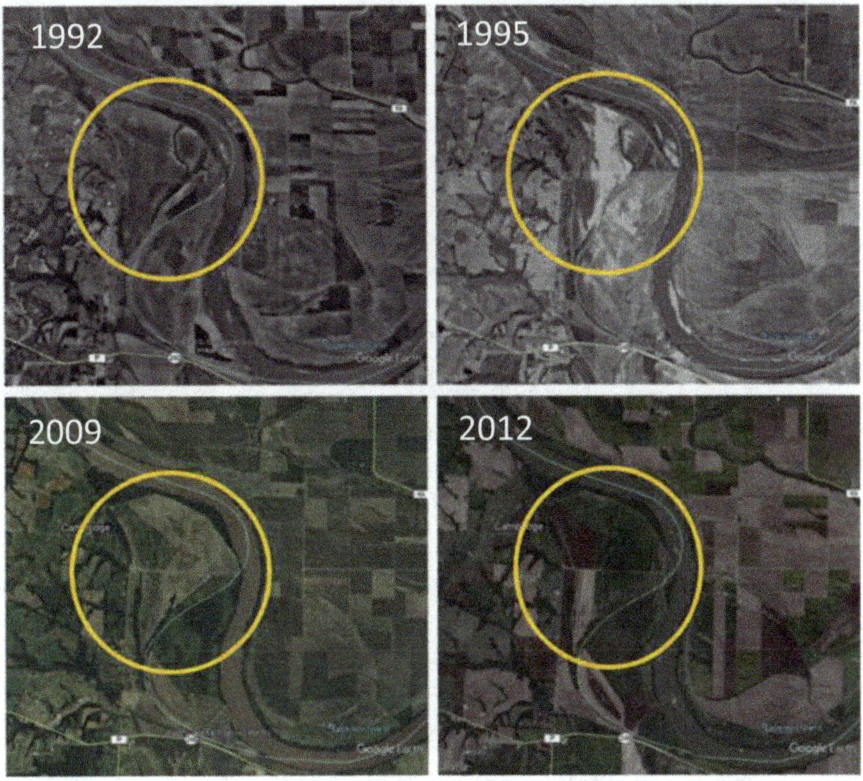


**Fig. 3** Google Earth images before the 1993 flood (22 March, 1992), levee break and sand deposition after the flood (22 February, 1995), progress of cropping and riparian buffers (7 August, 2009, and 9 September, 2012) near Cambridge, Missouri, USA

as water use by the vegetation. For example, the runoff from a 1 ha parking lot is equal to a 36 ha forest (Sammis and Herrera 1999).

Hydrology is modified when riparian forests along water bodies are cleared for croplands and other uses. Soils become less effective in storing water, with no roots to strengthen stream banks and no barriers on the levee to reduce the flow velocity. Removal of trees affects water interception, evapotranspiration, soil water dynamics, and soil stability, and these changes contribute to greater flow velocity, peak flow, and flow volumes. These changes also contribute to stream bank erosion, stream widening, levee failure, flooding, and sand deposition on farmlands (Fig. 3). Other damages with flooding include declining water quality, biodiversity, wildlife habitat, availability of productive land, lives, and financial security. However, these negative effects can be reversed by reestablishing perennial vegetation.

Studies from tropical and temperate zones prove that trees can be used to combat flooding, levee failures, landslides, and financial losses by establishing new and/or strengthening the existing riparian buffers along creeks, streams, and rivers (Allen et al. 2003; Geyer et al. 2000; Dwyer et al. 2007). Urban food forests also can help reduce large volumes of water coming to water bodies. According to Schoeneberger et al. (2012) windbreaks help address flooding intensity and frequency and help reduce losses from flood events. Strategically placed windbreaks moderate water flows similar to the way windbreaks moderate wind flows (Wallace et al. 2000). Alley cropping practices also help reduce flow volumes and flooding throughout the watershed (USDA-NAC 2016). Trees within silvopasture and forest farming can also have positive effects with regard to flood control (USDA-NAC 2016). Therefore, AF practices can minimize flooding and associated risks by changing the hydrology, increasing the evapotranspiration, strengthening banks/levees, and creating flow resistance barriers (Allen et al. 2003; Geyer et al. 2000; Dwyer et al. 2007).

Trees have been used in many countries for programs like windbreaks. For example, governments of the USA, Russia, and Canada implemented tree planting programs to combat drought, dust, declining crop yields, human malnutrition, and mass exodus and to reduce deaths. The role of trees for flood protection and water cycle modification will need extra attention in addition to carbon sequestration (CS) and other ES as flood-related damages occur more frequently now and will occur more often in the future in our increasingly wetter and drier world.

Trees intercept, consume, and store water in their structure including the stem, branches, and leaves, thereby reducing the total amount of runoff and peak flow. Over 30% of the precipitation in tropical forests can be captured by interception (Brooks et al. 1991). In North Vancouver, British Columbia, the average canopy interception for Douglas fir (*Pseudotsuga menziesii*) and Western red cedar (*Thuja plicate*) was 49% and 61%, respectively (Asadian and Weiler 2009). The interception by the urban trees in Santa Monica, California, ranged from 15% for a small *Jacaranda mimosifolia* to 66% for a mature *Tristania conferta* (Xiao and McPherson 2002). Interception values ranging from 20% to 52% were recorded among 19 evaluated trees in Australia (Venkatraman and Ashwath 2016). Removing ~400 mature trees from a forested 1 ha riparian buffer significantly reduces interception and increases the streamflow.

Trees in AF play an incredible role in reducing runoff in several ways and removing or filtering pollutants that would otherwise end up in waterways. A mature pecan (*Carya* spp.) tree transpires 565–950 Lt of water daily during hot summer (Sammis and Herrera 1999). Cottonwood (*Populus deltoides*) evapotranspiration of 127 cm $yr.^{-1}$ was closely followed by willow (*Salix* spp. 120 cm $yr.^{-1}$) while values for mesquite ranged from 49 to 76 cm $yr.^{-1}$ in the USA (Fig. 4; Scott et al. 2000). Another study in Arizona also showed that a single cottonwood tree transpires 120 cm $yr.^{-1}$ (Jetton 2008). Even narrow strips of trees on an alley cropping AF practice can change soil water dynamics and help reduce runoff volumes and flow velocities (Anderson et al. 2009; Udawatta et al. 2011; Sahin et al. 2016; Alagele et al. 2020). Trees begin to transpire before the crop is established and reduce the soil water content. During recharge periods AF areas store more water due to improvements in soil parameters including C, hydraulics, and other physical properties (Udawatta et al. 2011). The mean interception and transpiration vary by tree species, age, season, and characteristics of the rain event (intensity and volume). When these trees are removed, all the water will be available for surface runoff, and

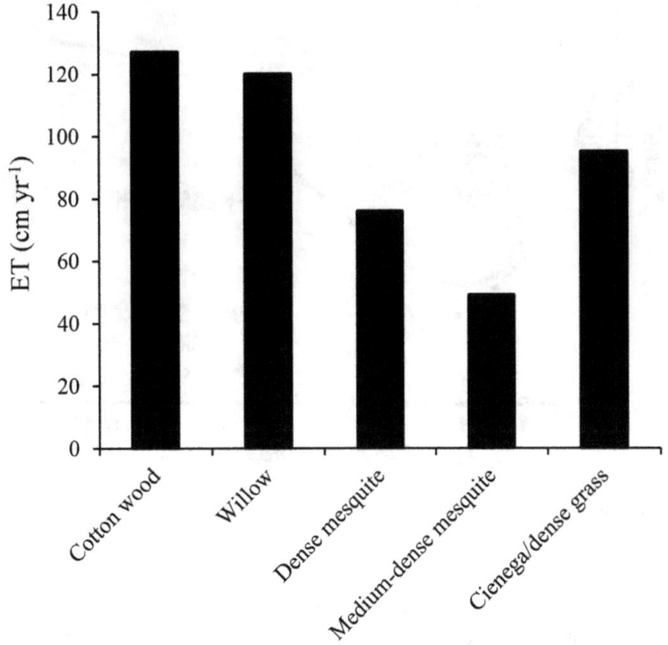


**Fig. 4** Evapotranspiration (ET) of riparian vegetations in the Sierra Vista sub-catchment of the San Pedro River Basin, Arizona, USA (adapted from Scott et al. 2000)

will eventually end up in water bodies. The worst situation is 100% impervious surfaces like parking lots, roofs, buildings, and roads.

Riparian buffers along the streams impose resistance to flow and considerably slow the release of water. Trees of AF vegetation provide better protection on stream banks than annual crops and herbaceous vegetation. This service reduces peak flow level, flow velocity, flash flooding, bank erosion, levee breaks, and inundation of larger areas along rivers. Because of these services, forested banks record lower rates of erosion than unforested (Geyer et al. 2000; Allen et al. 2003; Zaimes et al. 2004, 2006). Diverse multispecies vegetative buffers are more effective than sparsely vegetated non-intact buffers. A riparian buffer with several canopy layers and a dense ground vegetation can impose greater resistance to water flow and reduce the flow velocity.

Trees help increase water infiltration into the soil and soil water storage. Infiltration rate was 3 and 14 times greater in grass and AF buffers compared to row crop areas for an AF alley cropping practice in northern Missouri (Seobi et al. 2005). Infiltration rates in buffer treatments have increased with time as the buffer vegetation matures and occupies more soil volume (Akdemir et al. 2016). Similar findings have been reported in grazing practices with AF than without AF (Kumar et al. 2008a). Soil porosity values were also greater in row crop practices and silvopasture with AF than cropping and grazing managements alone, thus increasing water infiltration and storage. Bharati et al. (2002) in Iowa also found similar benefits of AF

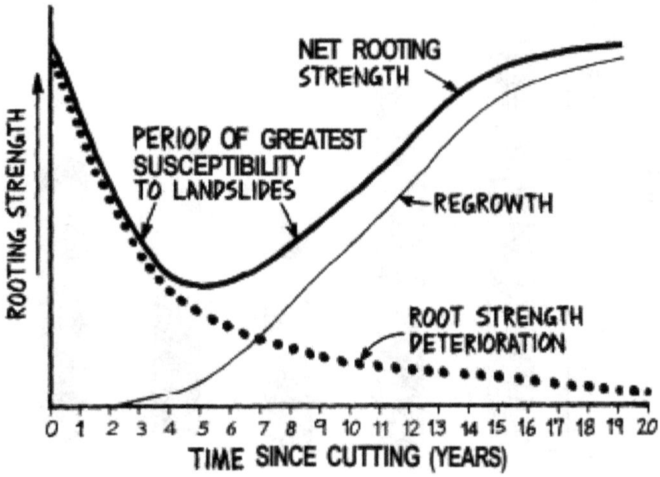


**Fig. 5** Root strength and time since tree harvesting (source: Sidle 1985)

compared to cropping systems. Many other soil- and plant-related benefits on soil water dynamics that help flood control can be found in chapter "Water Quality and Quantity Benefits of Agroforestry and Processes: Long-Term Case Studies from Missouri, USA" of this book. These changes imply improved soil water storage potential when AF is integrated. Findings of studies show that AF practices improve soil water dynamics, infiltration, water storage, and porosity and thereby reduce flow volumes and velocity of surface runoff and subsurface flow to streams and rivers.

Planting trees stabilizes the soil (Hairiah et al. 2020). This service of roots and trees acts as a woven fence that holds soils in place. When trees are harvested, the stability is lost and it takes years for roots to reestablish and strengthen the bank and soils (Fig. 5). The soil is significantly vulnerable during the first ~10 years after the harvest (Sidle 1985). As the new vegetation regenerates and roots develop, soil stability increases gradually. Faster growing trees can stabilize the soil sooner as they develop above- and belowground biomass in a shorter period than slow-growing trees. According to Twedt et al. (2002) 5- to 9-year-old cottonwood had twice the conservation value of oak (*Quercus* spp.) trees. Other management practices like weed control, sapling protection from browsing, and supply of deficient nutrients during early stages of growth would promote even faster tree growth of fast-growing species.

All six AF practices like riparian buffers, windbreaks, alley cropping, silvopasture, forest farming, and UFF help reduce runoff water volumes, flow velocity, and flow peaks and thereby help flood control. Contributions of AF trees include rainfall interception, water storage, water use, flow resistance, and soil modifications, and these processes occur when trees are integrated into regular farming practices. Specifically, UFF and homegardens which are common in many tropical countries

**Table 1** Levee failures and the width of the woody corridor in Missouri, USA (adapted from Dwyer et al. 1997)

| Corridor width | | Failures |
|---|---|---|
| Feet | Meters | Number |
| 0 | 0 | 45 |
| 100 | 30.5 | 15 |
| 200 | 61.0 | 7 |
| 300 | 91.4 | 0 |
| 400 | 121.9 | 2 |
| 500 | 152.4 | 2 |
| 600 | 182.9 | 1 |
| 700 | 213.4 | 1 |
| 800 | 243.8 | 0 |
| 900 | 274.3 | 1 |
| 1000 | 304.8 | 1 |
| 1100 | 335.3 | 0 |
| 1200 | 365.8 | 1 |

can reduce the flow velocity and volume in urban areas and thereby reduce the peak flow and flow velocities in streams and rivers.

A study conducted by Dwyer et al. (1997) along a 40-mile stretch of Missouri River, USA, showed that wide and intact riparian buffers protected levees, farmland, and adjacent areas. The study showed a significantly strong relationship between the buffer width and levee failure; as the width of the woody corridor decreased, the length of the levee failure increased (Table 1; Dwyer et al. 1997). Narrower buffers were the main reason for levee failures while buffers wider than 100 m significantly controlled levee failures. The study also showed that failure length was significantly larger for no buffers or narrower buffers than wider buffers. The greatest number of failures and longer breaks were reported for buffers narrower than 100 m. However, data are limited and more data for small and large catchments will help improve predictions and design criteria for effective flood control riparian buffers (Table 1).

Another similar study conducted by Geyer et al. (2000) in Central Kansas examined streambank changes after the flood of 1993. Their study used aerial photography to determine the relationship of streambank stability to natural riparian vegetation, stream channel morphology, and soil type. Image analysis of this 100-500-year flood has showed more erosion on cropland than forested streambanks and the erosion was greater on sandy soils than silty soils. This flood has caused greater water flow over great distances much of the time. Other studies in Tennessee (Scheifele 1928) and California (Shields Jr. and Gray 1992) also reported similar findings between streambank riparian vegetation and levee protection.

## Air Cleaning

Climate models predict increasing droughts and fires for some parts of the world and their potential negative effects on the air quality (AQ). Several record-breaking large fires in 2020 in California, Colorado, Nevada, Arizona, Oregon, and Washington states have significantly damaged the AQ, and some state agencies have requested their citizens stay at home as much as possible and wear masks, especially for elderly and people with predisposed health conditions (NYTimes 2020). Similarly, Australia experienced more bushfires in 2020 than the normal (Wheeling 2020). The number of fires erupted in Europe in 2019 and 2020 was also above normal.

Industry, transportation, day-to-day activities, prescribed burning, waste incineration, and backyard burning can also deteriorate the AQ. Agricultural and forestry operations also affect the AQ by machinery use, chemicals used in agriculture, and dust, and due to poor land management practices. According to USEPA (2020)'s National Emissions Inventory for 2017, agriculture contributes to 0.8 and 4 million tons $yr.^{-1}$ $PM_{2.5}$ and $PM_{10}$ emissions, approximately 14% and 24% of the total US emissions. Desert dust is also contributing to declining AQ, especially in the Middle East and North Africa. With the predicted climate change, airborne dust concentrations will increase due to drying soil and blowing wind (Zobeck and Van Pelt 2006). Declining AQ can affect rural communities and their economy.

Changes in demographics like urbanization and expansion of cities emphasize the importance of AQ. For instance, more than 2 billion additional people will migrate to cities in the twenty-first century and by 2050 the majority of humans will live in cities, towns, and other urban areas (McDonald et al. 2016). According to WHO-UN currently over 3.5 billion people live in cities and the number will increase to 6.5 billion by 2050: about 60% of the world population (UN 2019). As the population concentrates in cities, greater amount of pollution is released to the atmosphere, water, and soil. Global AQ maps show deteriorating AQ in big cities and the need for immediate measures for AQ improvements (Fig. 6).

Air pollution is a growing global threat, affecting human health and ES; with the rising emission rates it could be worse. Approximately one in nine deaths worldwide is attributed to poor AQ (WHO 2016). Annually, PM contributed to an estimated 3.2 million premature deaths and other nonfatal health issues like coughing, asthma, bronchitis, irregular heartbeat, and nonfatal heart attacks (Mcdonald et al. 2016). Two recent studies have predicted 6.2 million and 260,000 annual global deaths due to PM and heat stress by 2050 (Daniels et al. 2000; WHO 2005). Poor AQ can aggravate asthma, emphysema, rashes, nausea, or headaches and increase breathing difficulties, hospitalization, and premature birth to name a few. Fine particles can reduce visibility by scattering or absorbing light, and create a haze even over large areas that may include several states and may contribute to accidents and traffic fatalities. The declining AQ positively correlates with increasing health issues, health cost, and deaths, and other contributing factors like heat and pathogens.

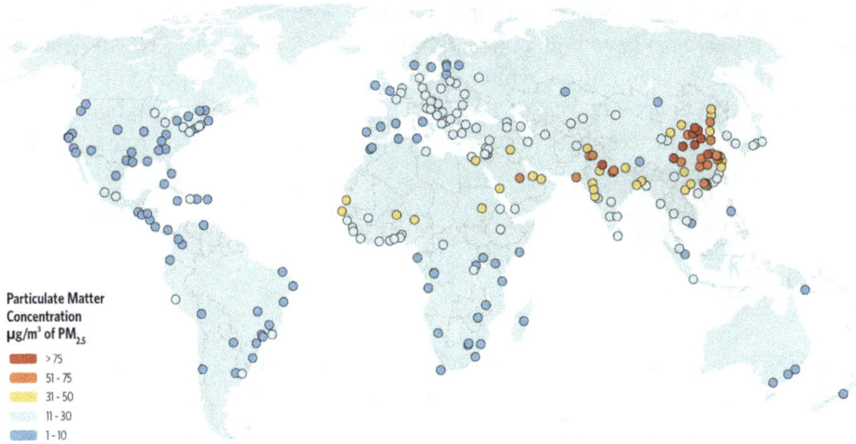


**Fig. 6** Mean particulate matter (PM) concentrations for 2010–2014 for cities around the world (source: McDonald et al. 2016)

Degradation of AQ has been reported in China, India, and many other countries and some countries have taken special measures to reduce further degradation. For example, colors and digits on vehicle registration/license plates in China indicate on which dates they can be on roads. Despite numerous efforts by the European Union (EU) to reduce air pollution, urban populations are still exposed to higher levels of pollution than EU standards (Selmi et al. 2016). Recent studies have shown that tens of millions of Americans live in areas that exceed the national health standards for AQ. In the Midwest USA, more than 20 million people experience AQ that does not meet national AQ standards (Pryor and Barthelmie 2013). USEPA has established primary standards to protect health and secondary standards to prevent environmental and property damage under the Clean Air Act to reduce damages caused by declining AQ (https://www.epa.gov/agriculture/agriculture-and-air-quality). Air quality degradation can also affect water quality.

Numerous sources and pollution types affect AQ. Pollutants from natural and man-made activities include gases like carbon monoxide (CO), nitrogen oxides ($NO_y$), sulfur oxides ($SO_x$), hydrogen sulfide ($H_2S$), volatile organic chemicals (VOC), dioxins, and ozone ($O_3$). Particulate matter refers to solid and liquid aerosolized particles suspended in the atmosphere and these can be divided into three main categories: large particles measured up to 10 micrometers (µm) across ($PM_{10}$), up to 2.5 µm across ($PM_{2.5}$), and nanoparticles. These include tiny organic particles, acids, metals, dust, soot emitted by various industries, vehicles, and construction sites. The list can be expanded and depends on sources like industry, transportation, fossil fuel burning, fracking, field activities (lawn, agriculture, and forestry), livestock operations, and natural processes. For example, volcanic eruptions can release a significant amount of dust and other particles as well as numerous gases and heat to the atmosphere in addition to lava on the earth surface. The concentration also varies by location, region, cities, countries, and geographical area and the highest concentrations of $PM_{2.5}$ were found in China and Northern India. North Africa and the Middle East also have high concentrations of dust. Europe and North America have moderate concentrations while Australia, South Africa, and South America have the lowest concentrations of PM (Fig. 6). Types of pollution and concentrations also vary by meteorological factors. Additionally, minerals, pathogens, spores, and toxins also affect AQ.

## *Windbreaks*

Livestock industry including poultry, swine, and cattle generate odor-emitting compounds like PM, gases, and VOC and these pollutants affect AQ, human health, standard of living, and economics. Although studies are limited on AF's air cleaning service, the limited number of studies have confirmed positive effects of AF for air cleaning services and odor mitigation from CAFOs. Pollutants generated from CAFOs include PM (feed dust, mammalian cell debris, aeroallergens, and bioaerosols), gases and vapors (VOCs, vapors and gases, odoriferous microbial compounds, phenolic compounds, and nitrogen-containing compounds), and odors (hydrogen sulfide, dimethyl sulfide, butyric acid, isobutyric acid, valeric acid, isovaleric acid, skatole, and indole) (Thorne 2002).

A comprehensive review by Tyndall and Colletti (2007) emphasizes the importance of strategic placement of multispecies tree windbreaks for effective odor mitigation. According to NRCS (2004) six main processes of windbreaks reduce livestock odor and improve the visual perception and AQ. Physical processes like reduction of wind speed, dilution, filtration, deposition, and immobilization reduce further spread and origination of pollutions from fields (Tibke 1988; Dochinger 1980). Approximately 40% of the wind should go through the canopy of the trees and the remaining 60% should be deflected up and over the windbreak (NRCS 2004). Other services include reduction of the area exposed to wind (Asbjornsen et al. 2014). The vegetation also collects and stores chemicals and PM within wood and on leaves (NRCS 2004). Windbreaks create a physical barrier to wind and intercept and deposit material and chemicals within the windbreak and the downward of the windbreak (Brandle et al. 2004; NRCS 2004). Windbreaks can also lift the odor plume into the lower atmosphere and facilitate dilution and dispersion over a larger area (Lin et al. 2006). Windbreaks also improve aesthetics while blocking the visibility of the farm buildings and thereby changing the perception of livestock operations (Tyndall and Colletti 2007). The article also explains sources, mitigation mechanisms, and design criteria for optimum air cleaning.

In northeast Missouri, USA, Lin et al. (2021) established a three-row windbreak in 2007 to quantify windbreak effects on AQ of a CAFO operation. The study design consisted of pitch loblolly pine (*Pinus rigida x P. Taeda*) as the inner row, alternately planted red maple (*Acer rubrum*) and pin oak (*Quercus palustris*) as the middle row, and Viburnum "Allegheny" (*Viburnum rhytidophyllum x V. Lantana*) as the outer row. They evaluated concentrations of $NH_3$, $H_2S$, and >20 VOCs. Results of the study showed that the multispecies windbreak significantly reduced PM and odor from the CAFO operation due to physical, chemical, and biological processes. Windbreaks reduced the wind speed, filtered air, absorbed gases, and promoted deposition of PM within the windbreak and these processes reduced degradation of AQ. They have also simulated the American Meteorological Society/Environmental Protection Agency Regulatory Model (AERMOD) and evaluated plume characteristics. The model simulation showed that fully developed windbreaks reduced 27% of the $NH_3$ concentration. Windbreaks along roads established to reduce snow accumulation can help reduce pollution from vehicle traffic and road materials. They also filter minerals and gases emitted by vehicles (Singh et al. 2020).

## *Urban Food Forest (UFF)*

Humans have been clearing forests for cities, roads, and industrial sites; however, cities are now investing to create green spaces as it makes cities healthier, air safer, habitat favorable, aesthetically pleasing, and microclimate favorable. The Mayor of London announced a 7000-tree planting program in January 2019 and slated to be completed by the end of 2020 (bbc.com/future/article/20200504-which-trees-reduce-air-pollution-best). Paris, France, began an urban tree planting program to improve AQ, reduce pollution, protect iconic landmarks, and adapt to climatic change. In order to improve the air quality of Beijing, Hebei Province is establishing a green belt to reduce the pollution from surrounding factories.

Urban food forests are somewhat similar to homegardens in tropical countries. These are gaining attention in Western countries for several benefits. Trees in UFF serve as an air cleaning agent and help improve AQ similar to any other urban forestry program. Currently, literature is limited on the role of UFF on AQ. However, findings on AQ improvements by urban forestry, roadside tree planting (Fig. 7), and tress on parking lots, and lawns, can be used to explain AQ benefits of UFF and homegardens. The number of trees and the area covered by UFF may be smaller than urban forests and roadside plantings. The limited number of trees and locations will perform the same air cleaning service. Benefits of urban forestry have been reported by studies conducted in New York, Atlanta, and Baltimore (Nowak 2002). For instance, trees in New York removed 1821 metric tons of air pollution in 1994 with an estimated value of $9.5 million (Nowak 2002). During daytime of in-leaf season AQ improvement by trees in New York was 0.47% for PM, 0.45% for $O_3$, 0.43% $SO_2$, 0.30% for $N_2O$, and 0.002% for $CO_2$. Air quality improvements vary by pollution (type, amount, and concentration), vegetation (species, density, distribution, and length of in-leaf season), and meteorological characteristics (precipitation, wind, temperature, and humidity) of the area (Nowak 2002). Trees in parking lots or edges of UFF also improve air quality by reducing the surface temperature and absorbing gases emitted by vehicles.

McDonald et al. (2016) evaluated effects of trees on PM removal in 245 large cities that house about 910 million people across the globe. The findings of the

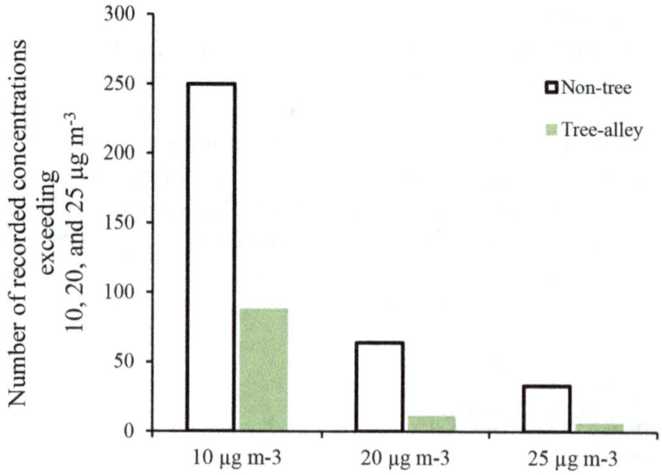


**Fig. 7** Number of recorded concentrations exceeding 10, 20, and 25 µg m-3 concentration for non-tree and tree alley in Dublin, Ireland (adapted from Riondato et al. 2020)

study showed that trees are currently providing on average 1.3 million people with at least a 10 µg $m^{-3}$ reduction in $PM_{2.5}$, 10.2 million people with at least a 5 µg $m^{-3}$ reduction, and 52.1 million people with at least a 1 µg $m^{-3}$ reduction. Similarly, trees are already providing 68.3 million people with a roughly 0.5–2.0 °C (0.9–3.6 °F) reduction in summer maximum air temperatures and thus reducing propagation of air pollutions (dust, spores, etc.). They also noticed that the majority of the effects occur within 300 m from the trees and the effectiveness varies from city to city. However, the rate of return was greater for best neighborhoods than for the less favorable neighborhoods.

## *The Processes of Air Cleaning*

Green vegetation including trees, shrubs, and grasses improves AQ. The simplest example is that green plants serve like "ecosystem lungs" taking atmospheric $CO_2$ through the stomata and releasing $O_2$ back to the atmosphere. Another clear example is dust-coated trees along unpaved and paved roads. Trees can also serve as "ecosystem livers" by taking $SO_2$ and other gases from the atmosphere. Various tree species of AF directly and indirectly improve AQ.

Air purification services of AF include several processes: (a) deposition of particles on the vegetation; (b) filtration, interception, and dilution of airborne particle as wind blows through the windbreak; (c) absorption of particles by rough surfaces of the vegetation; (d) adherence of pollutants to wax and wet surface of leaves, stem, and branches; and (e) intake of certain gases by stomate, and wet solubility (Dochinger 1977; Nowak 2002; NRCS 2004). The physical process of air cleaning involves dispersion, dilution, and removal of airborne particles as wind blows through a tree buffer. The extent to which each species performs the desired function is determined by canopy shape and size, leaf size, leaf density, leaf position, angle, and composition. Bigger canopies and leaves can trap more pollution than smaller ones. Larger trunks and branches with more rough surfaces trap or intercept more particulate pollutants than a small smooth bark. According to Nowak et al. (2002), some particles are absorbed into the trees while the majority is intercepted and retained on the plant surface. These pollutants can be washed off by rain and released to the soil; therefore, this trapping mechanism is a temporary air cleaning process.

In general, leaves are the primary receptor of pollutants. Some species have spines or hair on leaf surfaces to trap particles. In a wind tunnel study at the University of Lancaster, Wang et al. (2019) evaluated PM capture of nine tree species and reported that silver birch (*Betula pendula*), yew (*Taxus baccata*), and elder (*Sambucus* spp.) trees captured 79%, 71%, and 70% of the particles and it was the hairs of their leaves that contributed to the reduction of pollution. Some leaves have wax or moisture films where particles can be adhered to. Leaves of some species have a small electrical charge and these surfaces can adsorb pollutants with an opposite charge (Dochinger 1977). Rough, rugged, waxy, and hairy leaf surfaces can serve as better filters than smooth, non-waxy, and non-hairy leaf surfaces.

Through the stomata, gases like $CO_2$ and $SO_2$ enter leaves and thereby remove gaseous pollution. Inside the leaf, these gases diffuse into intercellular space, absorbed by water films, react with inner leaf surfaces, and may form acids (Smith 1990). Therefore, the most important character is the number of stomata and their distribution on a leaf. The arrangement of leaves, their angle, density, structure, and size are important factors that determine the filtration efficiency. Larger leaves and canopies can trap more pollutants than small leaves and small canopies.

Air cleaning process of trees varies by season and months. Pollutant removal rates were greater from May to September and were lower during October to April in Strasbourg, France (Selmi et al. 2016). Approximately 81% of the removal occurred during the leaf-on season due to greater leaf area and higher concentrations of pollution. The amount of pollutants removed by tree cover varies by location (Nowak et al. 2006; Selmi et al. 2016). These differences could be due to location, measurement procedure, trees, and meteorological parameters. A study conducted in India showed that roadside trees accumulate a significant amount of metals (Cu) and pollution (dust) than the trees in a forest (Singh et al. 2020).

According to Tyndall and Colletti (2007), younger trees with more complex branch architect and more leaves have been more effective than older trees with fewer branches and leaves. Larger trees with many branches are more effective in reducing the wind velocity and thereby enhancing greater deposition of particles within an AF buffer. Deciduous trees may appear to be better cleaners as they have larger leaves; however, conifers keep their leaves year-around and may have more leaf area than deciduous trees. Although conifers appear to be better than deciduous, many conifer species are sensitive to salt levels and therefore salts used for deicing can affect these trees. Year-round leaves could block sunlight and melting of ice which could cause traffic problems. Their review presents information on the effects of tree species, design criteria, and management of windbreaks for maximum air cleaning efficiency.

The efficiency of air cleaning is different among species and, in some instances, within a species. For example, one of the early studies conducted by Dochinger (1977) reported that some bigtooth aspen (*Populus grandidentata*) grew vigorously for $SO_2$ fumigation while others became stunted in a laboratory study. The tolerance level for pollutants and their reactions also vary by trees. Some trees can survive longer, while others may lose their air cleaning effectiveness earlier than others. Trees exposed to higher doses of pollution produced more proline than trees exposed to lower doses (Singh et al. 2020). Proline is a biochemical produced by trees to adjust to the stress level. Therefore, AF needs to be maintained by removing ineffective trees, replanting new saplings, and conducting other required maintenance like branch removal for effective air cleaning service. This also emphasizes the importance of selection of more tolerant, effective, and long-living species that require minimum maintenance.

Windbreaks/shelterbelts and UFF are two main AF practices that can help clean air. Riparian buffers and alley cropping also clean air although their effects are less quantified. Alley cropping and riparian buffers can reduce dust and agrichemical dispersion across large areas. A wide and dense riparian buffer system can reduce airflow velocity and increase filtration, and thereby improve the AQ similar to processes with windbreaks.

Model simulations also have shown air cleaning and other benefits of trees in urban and rural areas. Selmi et al. (2016) simulated the i-Tree Eco model to quantify tree effects on pollutant removal of Strasbourg, France. The field study used 228 plots to collect the required data to calibrate the model. Their study showed that urban forests and parks removed a total of 88 t of pollutants between July 2012 and June 2013. Removal rates varied by months, tree cover, and concentrations. Riondato et al. (2020) compared tree effects on air cleaning in Ireland. They used a 150 m length with old-growth Norway maple (*Acer platanoides*) and another 150 m length with no trees on Drumcondra Road, Dublin. The study demonstrated that trees were significantly effective in removing $PM_{2.5}$ particles from the atmosphere. In their study 3, 5, and 5 times more 10, 20, and 25 μg $m^{-3}$ $PM_{2.5}$ particle concentrations were observed from the treeless roads than the road section with trees.

## Practical Implications

The total river length in the USA is about 5.7 million km (USEPA 2016). There is a great potential for expanding the area under riparian forest buffers on many rivers and streams in the USA to protect farmlands and other properties. For example, about 6.1 million ha (15 million ac) of farmland was flooded at least once during 1993 in nine Midwest states (Larson 1995). Once a cropland is flooded and sand deposited, the recovery will take years and cost a significant capital to remove sand

and revegetate the affected area (Fig. 2). The 1993 flood carried away >600 billion tons of topsoil and deposited great amounts of sand and silt on valuable farmlands (dnr.ne.gov/floodplain/PDF_Files/FloodUpdateStory_Rev3.pdf). During this time, the landowners/farmers will have no cropland and income from sand-deposited farmlands. Even though the land can be converted for crops, no one can predict the next damage. Therefore, it is worth allocating riverfront land for intact riparian buffers. This should be a stakeholder or community effort as one landowner cannot control flooding. Landowners with insufficient buffer widths need to work together to establish an intact buffer along the river length.

States and federal governments can purchase land out of annual crop production for permanent vegetation. After 1993, 1995, and 2011 floods, Missouri Department of Conservation (32,000 ha) and Missouri Department of Natural Resources purchased much of the flood-damaged land. This purchase has helped to protect farmlands, levees, cities, and other structures from flood risk and damages. There is a lot of interest in restoration of riparian buffers but there is a lag time of 5–15 years to establish a strong intact buffer. This would mean planning ahead for future rains and floods and changing the incentive programs prioritizing to reduce flood risks. The program like purchasing flood-damaged and vulnerable riverfront land from landowners by Missouri Department of Conservation/Natural Resources after floods is a good example of preventing and reducing future flooding risks by developing proper policies and investing for flood control.

Dosskey et al. (2012) have proposed establishment of AF for Lower Mississippi River forest restoration. They have stated multiple benefits of restoration including restoration of hydrology, improvements in flood control, and reduction of risks of crop failures by selecting AF tree species that are more compatible with wetness and periodic flooding. The selected tree species should not compete for resources with companion crops within the system. Deep-rooted trees with greater proportion of roots in the subsurface soil in the tree-crop interphase can have less competition for resources (Udawatta et al. 2005). Tree species that can tolerate several days or extended periods of flooding reduce maintenance and replanting costs. Tree species with faster growth rates, better survival rates, resistance to pest and diseases, and herbicide tolerance would be ideal for AF practices intended for enhanced ES. These trees also should be drought hardy, winter hardy, long-living, and aesthetically pleasing. Other benefits including short- and long-term marketable products, enhanced biodiversity, improved wildlife habitat, and leasing for hunting can enhance adoption of flood control measures by landowners and farmers.

Riparian buffers, windbreaks, and UFF must be designed to perform its main function with optimum efficiency. Ideally, more species will increase BD; some studies suggest not to include >10% of the same species. Strategic planting of selected species can also optimize other benefits. For example, selection of fast-growing species like poplar can enhance C storage in above- and belowground biomass than slow-growing species. However, integration of slow-growing species can extend the life span of the practice. These species should be easy to maintain and live under these settings for 3–4 decades. The practice should contain tall and short trees. Farmers and landowners could select trees and shrubs and understory plants that have higher market value, survive under various conditions, or require little attention. Proper combinations can be environmentally friendly, economically profitable, and sustainable, and may require less inputs. This can assure continuous supply of food for wildlife, better habitat, and additional income while providing various ecosystem services. Trees should be fully foliated during times of rainy seasons and high wind periods for optimum flood control and air cleaning. Established intact vegetation can also help reduce downstream flooding while protecting valuable farmlands. This integration, in turn, benefits the trees and makes it a win-win situation.

Small holdings and tenure type can affect the establishment of perennial trees. Although the effectiveness of flood control and air cleaning of trees are well known, the establishment of a riparian buffer, windbreak, or UFF on croplands or in urban areas may not be that easy especially on rented lands. Other incentives or some form of compensation can convince short- and long-term tenured farmers to allocate land for these practices. Incorporation of perennial trees into agricultural ecosystems, through AF for the protection of riverbanks, provides security for their fields and their neighbors and may help diversify their revenue stream. These practices could also help reduce expansion of the hypoxia zone in the Gulf of Mexico. Similarly, AF programs can help reduce flooding and hypoxia zones globally.

Since landscape and tree parameters determine the cleaning efficiency those should be evaluated very carefully, especially for selecting trees for UFF. Some tree species emit volatile organic compounds and may contribute to formation of ozone which is harmful for human health (Nowak 2002). For example, a heat wave in Berlin in 2006 formed ozone and reacted with other pollutants to reduce the AQ. Proper planning can reduce the release of these compounds, as the release of these compounds is temperature dependent and trees reduce the temperature. According to Nowak et al. (2002) mulberry (*Morus* spp.), cherry (*Prunus* spp.), linden (*Tilia* spp.), and honey locust (*Gleditsia* sp.) had the greatest effect on lowering ozone in Brooklyn, New York, USA. The most effective native or exotic species should be incorporated with a greater diversity that most suits for the location, whether a city with tall buildings and narrow roads or medium-size buildings with wide roads and open areas. Because large canopy trees on narrow roads can trap pollutants similar to what happened in Beijing, China. Planners should also consider the survival of these trees in those areas and maintenance requirements as well as wind direction and opinions of general public about those selected species. Some species can be allergic and generate more pollen and host harmful or unpleasant organisms.

Creation of buffers increases land value, aesthetics, wildlife habitat, and numerous other ecosystem services. Andy Mason, the former director of the Forest Service's National Agroforestry Center, said "*agroforestry is not converting farms to forest. It's the right tree at the right place for the right reason*" (http://www.nytimes.com/2011/11/22/science/quiet-push-for-agroforestry-in-us.html Accessed 11 19 20). Although AF can provide multiple benefits and links between many services and AF has been established in the recent literature, there remains much work for promoting and adopting AF by large (Schoeneberger et al. 2008).

Aerial images, decision support tools, and models can help identify vulnerable riparian areas where bank erosion and failures are likely to occur and where riparian forest buffers would be effective at flood controlling. The same tools can be used to evaluate best locations and species for a windbreak, UFF, and other AF practices for optimum air cleaning services. These tools can help urban planners and AF experts develop location-appropriate AF for optimum regulatory services. For example, according to Kumar et al. (2008b, 2009, 2011), short vegetation is more suitable for narrow roads with tall buildings while typical tall trees are suitable for wide roads with low-rise buildings. Urban planning, tree growth, and 3-D model software may be used to visualize time series progression of the selected system. For example, a free software provided by the US Forest Service, iTree species, ranks species based on a set of variables including air pollution removal capability, carbon storage, and VOC emissions. Information on locally available, exotic, and effective tree species as well as input from local citizens should be used to finalize the tree species list because the software would select species that will not perform well under local conditions. If local tree manuals are available, those can be used to identify suitable species for specific conditions and purpose. There are several undesirable characteristics on some species and one should pay attention to factors like production of pollen, VOCs, and requiring high maintenance.

Abhijith et al. (2017) reviewed existing literature on urban forestry and air quality to recommend a design criterion. According to the review, high-level vegetation canopies deteriorate air quality, while low-level vegetation (hedges) improves AQ of narrow roads with skyscrapers. Although generic recommendations can be adopted to open roads, designers must pay attention to the landscape of urban and other areas for optimum air cleaning. The review provides a design criterion to consider for urban forests for optimum air cleaning. Matocha et al. (2012) suggested to pay attention on selecting species. This will be more critical for UFF where selected trees should be ideal for the situation and not impact health and aesthetics of the urban areas. Climate-smart designs should not conflict with interests of the community.

Extension, education, and workshops can help promote adoption of AF for flood control and air cleaning services. Although streambank erosion is a natural river adjustment process, the rate of damage can be accelerated by poor management of the floodplain, adjacent lands, and streambanks. Anthropogenic activities can influence the ratio of their height to bank full stage, density and depth of roots, and type of bank vegetation which determine the bank erosion rates (Rosgen 1996). Therefore, increased public awareness through workshops, demonstrations, published material, and social media may help improve the role of AF practices for enhanced ES.

Earlier studies have shown that riparian vegetation is effective for stabilizing streambanks and reducing erosion under normal stream flow conditions (Nunnally 1978; Turner 1 Once established, timely management can improve various ES and reduce the potential negative effects. Trees and other vegetation may need protection during the early stages of growth to promote the establishment of the stand and enhance the early growth. Trees may also need some fertilizers for early growth on nutrient-poor degraded sites. Schultz et al. (2009, 2021) have provided details on buffer management for production, environmental, and aesthetic benefits. These include removal of older trees, replanting, and removing soil berm and removal of invasive species and several other important maintenance procedures.

Integration of AF practices brings significant co-benefits for local ecosystems (Matocha et al. 2012). Planting trees in riparian buffers, windbreaks, alley cropping, and UFF and maintaining those for optimum ES provide a series of secondary benefits (Jose 2009). Because of the diversity, AF systems are considered more resilient than monocropping. These benefits are not limited to aesthetics but also provisional, regulating, supportive, and cultural ES. Perennial vegetation of riparian buffers, windbreaks, alley cropping, and UFF enhances land productivity, CS, biodiversity, wildlife, water quality, soil health, and economics while reducing flooding risks (Kort and Turnock 1999; Jose 2009; Nair et al. 2009a, b; Udawatta and Jose 2012; Udawatta et al. 2017, 2019, 2021; Shi et al. 2018). AF buffers, windbreaks, and UFF can reduce the air-conditioning cost for homes. For instance, windbreaks reduced fuel cost by 18–25% for heating when homes on protected and unprotected sites were compared (AAF Canada 2020). Agroforestry practices help increase crop yields and livestock performance (Brandle et al. 2004; Kallenbach et al. 2006). These various services bring additional income to landowners.

## Summary and Future Directions

The increasing number of floods and degrading air quality emphasize the importance of cost-effective mitigation measures to reduce flooding and improve AQ. Agroforestry can be implemented to reduce flood risks and improve AQ. The reduction of flood risks can be attributed to water use by trees, changing soil water dynamics, stabilizing soils, reducing bank erosion, and strengthening levees. The air quality improvement can be attributed to vegetation characteristics including tree architecture, stem features, leaf surface features, and leaf parameters. The species composition, configuration, tree density, age, and size of trees determine the effectiveness of flood control and air cleaning efficiency.

The selection of soil-site-climate-suitable species combinations for the desired purpose is important. These species combination should meet landowner objectives. Effective design criteria and maintenance of the AF are vital for the efficiency of the system. The review also emphasizes the importance of multispecies integration for greater heterogeneity of the landscape. However, proper planning must be conducted before the practices are adopted. Design criteria should consider meteorological parameters, landscape features, flood-related parameters, and air quality-related parameters. Regular maintenance like removal of inefficient trees, replanting saplings, and soil modification can enhance many ES in addition to flood control and air cleaning.

Educational programs to increase the public awareness of AF practices on various ES may help promote the adoption of these practices. State and federal level support and changes in policies as well as incentives help promote the establishment of these practices for flood control and air cleaning regulatory ES.

Simulation of flood control and air cleaning models may enhance our understanding and predictability of short- and long-term benefits of adopted AF practices. Additionally, landscape design criteria can be used to design aesthetically optimized designs for urban areas while many other benefits can occur naturally. Evaluation of long-term benefits may help longevity of the riparian buffer, windbreak, and UFF. Future studies could evaluate the effectiveness of flood control and air cleaning services of exotic and high-value trees with various marketable products. Properly established and well-maintained systems can offer many other provisional, regulating, supporting, and cultural ecosystem services in addition to flood control and air cleaning.

## References

AAF Canada (Agriculture and Agri-Food Canada) (2020). https://www.agr.gc.ca/eng/agriculture-and-the-environment/agricultural-practices/agroforestry/benefits-of-agroforestry/?id=1344633257343. Accessed 11 30 Nov 2020

Abhijith KV, Kumar P, Gallagher J, McNabola A, Baldauf R, Pilla F, Broderick B, Sabatino SD, Pulvirenti B (2017) Air pollution abatement performances of green infrastructure in open road and built-up street canyon environments e a review. Atmos Environ 162:71–86

Akdemir E, Anderson SH, Udawatta RP (2016) Influence of agroforestry buffers on soil hydraulic properties relative to row crop management. Soil Sci 181:368–376. https://doi.org/10.1097/SS.0000000000000170

Alagele SM, Anderson SH, Udawatta RP (2020) Agroforestry, grass, biofuel, and row crop management effects on soil water dynamics for claypan landscapes. Soil Sci Soc Am J 84:203–219

Allen SB, Dwyer JP, Wallace DC, Cook EA (2003) Missouri river flood of 1993: role of woody corridor width in levee protection. J Am Water Res Assoc 39:923–933

Anderson SH, Udawatta RP, Seobi T, Garrett HE (2009) Soil water content and infiltration in agroforestry buffer strips. Agrofor Syst 75:5–16. https://doi.org/10.1007/s10457-008-9128-3

Asadian Y, Weiler M (2009) A new approach in measuring rainfall interception by urban trees in coastal British Columbia. Water Qual Res J 44(1):16–25. https://doi.org/10.2166/wqrj.2009.003

Asbjornsen H, Hernandez-Santana V, Liebman L, Bayala J, Chen J, Helmers M, Ong CK, Schulte LA (2014) Targeting perennial vegetation on agricultural landscapes for enhancing ecosystem services. Renew Agric Food Syst 29:101–125

Bharati L, Lee KH, Isenhart TM, Schultz RC (2002) Soil water infiltration under crops, pasture, and established riparian buffer in Midwestern USA. Agrofor Syst 56:249–257

Brandle JR, Hodges L, Zhou X (2004) Windbreaks in sustainable agriculture. Agrofor Syst 61:65–78

Brooks KN, Ffolliott PF, Gregersen HM, Thames JL (1991) Hydrology and the management of watersheds. Iowa State Univ Press, Ames IA, pp 13–36

Climate.nasa.gov/effects/ (n.d.) The effects of climate change. Accessed 20 Dec 2020

Corringham TW, Cayan DR (2019) The effect of El Nino on flood damages in the Western United States. Am Meteorol Soc 11:489–504

Daniels MJ, Dominici F, Samet JM, Zeger SL (2000) Estimating particulate matter-mortality dose-response curves and threshold levels: an analysis of daily time-series for the 20 largest US cities. Am J Epidemiol 152(5):397–340. https://doi.org/10.1093/aje/152.5.397

Dochinger LS (1977) How tree help clean the air? USDA, Forest Service Agricultural Information Bulletin No 412
Dochinger LS (1980) Interception of airborne particles by tree planting. J Environ Qual 9:265–268
Dosskey MG, Bentrup G, Schoeneberger M (2012) A role for agroforestry in forest restoration in the lower Mississippi alluvial valley. J For 48:55. https://doi.org/10.5849/jof.10-061
Dwyer J, Wallace D, Larsen DR (1997) Value of woody river corridors in levee protection along the Missouri River in 1993. J Am Water Res Asso 33(2):481–489
Dwyer JP, Wallace D, Larsen DR (2007) Value of woody river corridors in levee protection along the Missouri River in 1993 First published: 08 June 2007 https://doi.org/10.1111/j.1752-1688.1997.tb03526.x Paper No. 96184 of the Journal of the American Water Resources Association (formerly Water Resources Bulletin)
FFF (Foresight Future Flooding) (n.d.) Flood control defense project. www.foresight.gov.uk
Garrity DP, Akinnifesi FK, Ajayi OC, Weldesemayat SG, Mowo JG, Kalinganire A, Larwanou M, Bayala J (2010) Evergreen agriculture: a robust approach to sustainable food security in Africa. Food Secur 2(3):197–214. https://doi.org/10.1007/s12571-010-0070-7
Geyer WA, Neppl T, Brooks K, Carlisle J (2000) Woody vegetation protects streambank stability during the 1993 flood in Central Kansas. J Soil Water Conserv 55(4):483–486
Gold MA, Garrett HE (2009) Agroforestry nomenclature, concepts, and practices. In: Garrett HE (ed) North American agroforestry, an integrated science and practice, 2nd edn. American Society of Agronomy, Madison, WI, pp 45–55
Hairiah K, Widianto W, Suprayogo D, Van Noordwijk M (2020) Tree roots anchoring and binding soil: reducing landslide risk in Indonesian agroforestry. Land 9:256. https://doi.org/10.3390/land9080256
Hirsch RM, Ryberg KR (2012) Has the magnitude of floods across the USA changed with global $CO_2$ levels? Hydrol Sci J 57:1–9. https://doi.org/10.1080/02626667.2011.621895
Hirsch RM, Ryberg KR (2013) United States flood loss report - water year 2011. National Oceanic and Atmospheric Administration, National Weather Service. p 10
Jetton AJ (2008) Estimation of evapotranspiration of cottonwood trees in the Cibola National Wildlife Refuge, Cibola, Arizona. MS Thesis, Wright University. https://corescholar.libraries.wright.edu/etd_all/825
Jose S (2009) Agroforestry for ecosystem services and environmental benefits: an overview. Agrofor Syst 76(1):1–10
Kallenbach RL, Kerley MS, Bishop-Hurley GJ (2006) Cumulative forage production, forage quality and livestock performance from an annual ryegrass and cereal rye mixture in a pine walnut silvopasture. Agrofor Syst 66:43–53. https://doi.org/10.1007/s10457-005-6640-6
Kandji ST, Verchot LV, Mackensen J, Boye A, van Noordwijk M, Tomich TP, Ong C, Albrecht A, Palm C (2006) Opportunities for linking climate change adaptation and mitigation through agroforestry systems. In: Garrity D, Okono A, Grayson M, Parrott S (eds) World agroforestry into the future. World Agroforestry Centre, Nairobi, pp 113–122
Kirchmeier-Younga MC, Xuebin Zhang (2020) Human influence has intensified extreme precipitation in North America. PNAS p 6. www.pnas.org/cgi/doi/10.1073/pnas.1921628117
Kort J, Turnock R (1999) Carbon reservoir and biomass in Canadian Prairie Shelterbelts. Agrofor Syst 44:175–186
Kumar S, Anderson SH, Bricknell LG, Udawatta RP, Gantzer CJ (2008a) Soil hydraulic properties influenced by agroforestry and grass buffers for grazed pasture systems. J Soil Water Conserv 63:224–232
Kumar P, Fennell P, Britter R (2008b) Effect of wind direction and speed on the dispersion of nucleation and accumulation mode particles in an urban street canyon. Sci Total Environ 402:82–94
Kumar P, Garmory A, Ketzel M, Berkowicz R, Britter R (2009) Comparative study of measured and modelled number concentrations of nanoparticles in an urban street canyon. Atmos Environ 43:949–958

Kumar P, Ketzel M, Vardoulakis S, Pirjola L, Britter R (2011) Dynamics and dispersion modelling of nanoparticles from road traffic in the urban atmospheric environment - a review. J Aerosol Sci 42:580e603

Larson LW (1995) The great USA flood of 1993. U.S.- Italy Research Workshop on the Hydrometeorology, Impacts, and Management of Extreme Floods. Perugia (Italy), November 1995

Lin XJ, Barrington S, Nicell J et al (2006) Influence of windbreaks on livestock odour dispersion plume in the field. Agric Ecosys Environ 116:263–272

Lin CH, Weber EE, Walter WD, Lim T, Garrett HE (2021) Vegetative environmental buffers for air quality benefits. In: Garrett HE, Jose S, Gold MA (eds) North American agroforestry, 3rd edn. Agronomy Society of America, Madison, WI

Matocha J, Schroth G, Hole THD (2012) Integrating climate change adaptation and mitigation through agroforestry and ecosystem conservation agroforestry - the future of global land use. Adv Agrofor:105–126

McDonald R, Kroeger T, Boucher T, Longzhu W, Salem R (2016) Planting healthy air: A global analysis of the role of urban trees in addressing particulate matter pollution and extreme heat. Nat Conserv 136

Nair PKR, Kumar BM, Nair VD (2009a) Agroforestry as a strategy for carbon sequestration. J Plant Nutr Soil Sci 172:10–23

Nair PKR, Nair VD, Gama-Rodriguez EF, Garciad R, Haile SG, Howlett DS, Kumar BM, Mosquera-Losada MR, Saha SK, Takimoto ANG, Tonucci RG (2009b). Soil carbon in agroforestry systems: an unexplored treasure?) Nat Proc. https://doi.org/10.1038/npre.2009.4061.1

NCA (2014). http://nca2014.globalchange.gov/highlights/overview/overview/graphics/observed-change-very-heavy-precipitation. Accessed 10 Dec 2020

NCDC/NOAA (2020). https://www.ncdc.noaa.gov/billions/summary-stats/US/2000-2020>. Accessed 10 Dec 2020

Nearing MA (2001) Potential changes in rainfall erosivity in the U.S. with climate change during the 21st century. J Soil Water Conserv 56:229–232

Nearing MA, Pruski FF, O'Neal MR (2004) Expected climate change impact on soil erosion rates: a review. J Soil Water Conserv 59:43–50

Nowak DJ (2002) The effect of urban trees on air quality. USDA Forest Service Northern Research Station, Syracuse, NY, p 4

Nowak DJ, Crane DE, Stevens JC, Ibarra M (2002) Brooklyn's urban forest. USDA For Ser Gen Tech Rep NE-290, p 107

Nowak DJ, Crane DE, Stevens JC (2006) Air pollution removal by urban trees and short shrubs in the United States. Urban For Urban Greening 4:115–123

NRCS (2004) Windbreak/shelterbelt-odor control, conservation practice information sheet (IS-MO380)

NYTimes (2020). https://www.nytimes.com/interactive/2020/09/24/climate/fires-worst-year-california-oregon-washington.html#:~:text=With%20more%20than%20a%20month,the%20largest%20fires%20ever%20recorded. Accessed 10 Dec 10 2020

Pielke RS, Downton MW, Miller JZB (2002) Flood damage in the United States, 1926-2000: a reanalysis of national weather service estimates. Environmental and societal impacts group. National Center for Atmospheric Research PO Box 3000 Boulder Colorado 80307-3000

Porter JR, Xie L, Challinor AJ et al (2014) Food security and food production systems. In: Field CB, Barros VR, Dokken DJ, Mach KJ, Mastrandrea MD, Bilir TE, Chatterjee M, Ebi KL, Estrada YO, Genova RC, Girma B, Kissel ES, Levy KM, MacCracken S, Mastrandrea PR, White LL (eds) Climate change 2014: impacts, adaptation, and vulnerability. Part A: global and sectoral aspects. Contribution of Working Group II to the Fifth Assessment Report of the Intergovernmental Panel on Climate Change. Cambridge University Press, Cambridge, pp 485–533. http://www.ipcc.ch/report/ar5/wg2/

Pryor SG, Barthelmie RJ (2013) The Midwestern United States: socio-economic context and physical estimate. In: Pryor SC (ed) Climate change in the Midwest: impacts, risks, vulnerability, and adaptation. Indiana University Press, Bloomington, IN, pp 12–47
Riondato E, Pilla F, Basu AS, Basu B (2020) Investigating the effect of trees on urban quality in Dublin by combining air monitoring with i-tree eco model. Sustain Cities Soc 61:102356
Rosgen DL (1996) Applied river morphology. Elsievier Science, Amsterdam, p 30
Sahin H, Anderson SH, Udawatta RP (2016) Water infiltration and soil water content in claypan soils influenced by agroforestry and grass buffers compared to row crop management. Agrofor Syst 90:839–860. https://doi.org/10.1007/s10457-016-9899-x
Sammis T, Herrera E (1999) Estimating water needs for pecan trees guide H-636. New Mexico State University, Las Cruces, NM, p 2
Scheifele OS (1928) Protecting riverbanks and levees. The Canadian Engineer, January 10, 1928: 1-7
Schoeneberger M, Bentrup G, Current D, Wight B, Simpson T (2008) Building bigger better buffers for bioenergy. Water Res Impact 10:22–26
Schoeneberger M, Bentrup G, de Gooijer H, Soolanayakanahally R, Sauer T, Brandle J, Zhou X, Current D (2012) Branching out: agroforestry as a climate change mitigation and adaptation tool for agriculture. J Soil Water Conserv 67:128–136
Schultz RC, Isenhart TM, Colletti JP, Simpkins WW, Udawatta RP, Schultz PL (2009) Riparian and upland buffer practices. In: Garrett HE (ed) North American agroforestry: an integrated science and practice, 2nd edn. American Society of Agronomy Inc., Madison, WI, pp 163–218
Schultz RC, Udawatta RP, Isenhart TM, Colletti JP, Simpkins WW, Schultz PL (2021) Riparian and upland buffer practices. In: Garrett HE, Gold MA, Jose S (eds) North American agroforestry: an integrated science and practice, 3rd edn. American Society of Agronomy Inc., Madison, WI
Scott RL, Shuttleworth WI, Goodrich DC, Maddock T (2000) The water use of two dominant vegetation communities in a semiarid riparian system. Agric For Meteor 105:241–256
Selmi W, Weber C, Rivière E, Blonda N, Mehdi L, Nowak D (2016) Air pollution removal by trees in public green spaces in Strasbourg city, France. Urban For Urban Green 17:192–201
Seobi T, Anderson SH, Udawatta RP, Gantzer CJ (2005) Influences of grass and agroforestry buffer strips on soil hydraulic properties. Soil Sci Soc Am J 69:893–901
Shaw R (2006) Community-based climate change adaptation in Vietnam: inter-linkages of environment, disaster, and human security. In: Sonak S (ed) Multiple dimensions of global environmental changes. TERI Publications, New Delhi, pp 521–547. http://citeseerx.ist.psu.edu/viewdoc/download?doi=10.1.1.475.8279&rep=rep1&type=pdf
Shi L, Feng W, Xu J, Kuzyakov Y (2018) Agroforestry systems: meta-analysis of soil carbon stocks, sequestration processes, and future potentials. Land Degra Devel 29:3886–3897. https://doi.org/10.1002/ldr.3136Citations
Shields FD Jr, Gray DH (1992) Effects of woody vegetation on sandy levee integrity. Water Res Bull 28(5):917–931
Sidle RC (1985) Factors influencing the stability of slopes. In Swanston D (ed) Proceedings of a Workshop on Slope Stability: Problems and Solutions in Forest Management USDA For Serv Pacific Northwest Forest and Range Exp Station General Technical Report PNW-180 April 1985. pp 17–25
Simelton E, Dam BV, Catacutan D (2015) Trees and agroforestry for coping with extreme weather events: experiences form northern and central Vietnam. Agrofor Syst 89(6):1065–1082. https://doi.org/10.1007/s10457-015-9835-5
Singh H, Yadav M, Kumar N, Kumar A, Kumar M (2020) Assessing adaptation and mitigation potential of roadside trees under the influence of vehicular emissions: a case study of Grevillea robusta and Mangifera indica planted in an urban city of India. PLoS One 15(1):e0227380. https://doi.org/10.1371/journal.pone.0227380
Smith WH (1990) Air pollution and forests. Springer-Verlag, New York, p 618
Thorne PS (2002) Chapter 3.0. Air quality issues. In Iowa concentrated animal feeding operations air quality study. Final Report. Iowa State University and University of Iowa. pp 35–43

Tibke G (1988) Basic principles of wind erosion control. Agric Ecosyst Environ 22(23):103–122
Twedt DJ, Wilson RR, Henne-Kerr JL, Grosshuesch DA (2002) Avian response to bottomland hardwood reforestation: the first 10 years. Restor Ecol 10:645–655
Tyndall J, Colletti J (2007) Mitigating swine odor with strategically designed shelterbelt systems: a review. Agrofor Syst 69:45–65
Udawatta RP, Jose S (2012) Agroforestry strategies to sequester carbon in temperate North America. Agrofor Syst 86:225–242. https://doi.org/10.1007/s10457-012-9561-1
Udawatta RP, Nygren PO, Garrett HE (2005) Growth of three oak species during establishment in an agroforestry practice for watershed protection. Can J For Res 35:602–609. https://doi.org/10.1139/x04-206
Udawatta RP, Anderson SH, Motavalli PP, Garrett HE (2011) Calibration of a water content reflectometer and soil water dynamics for an agroforestry practice. Agrofor Syst 82:61–75. https://doi.org/10.1007/s10457-010-9362-3
Udawatta RP, Gantzer CJ, Jose S (2017) Agroforestry practices and soil ecosystem services. In: Al-Kaisi MM, Lowery B (eds) Soil health and intensification of Agroecosytems. USA, Elsevier AP, pp 305–334
Udawatta RP, Rankoth LM, Jose S (2019) Agroforestry and biodiversity. Sustainability 11:2879. https://doi.org/10.3390/su11102879
Udawatta RP, Anderson SH, Kremer RJ, Garrett HE (2021) Agroforestry and soil health. In: Garrett HE, Gold MA, Jose S (eds) North American agroforestry: an integrated science and practice, 3rd edn. American Society of Agronomy Inc., Madison, WI
UN (2019) World population prospects: the 2018 revision. United Nations Development of Economics and Social Affairs Population Division, New York
USAID (n.d.) El Nino-related disasters. How USAID's office of U.S. foreign disaster assistance is preparing and responding. https://2012-2017.usaid.gov/sites/default/files/documents/1866/1.13.16_El%20Ni%C3%B1o.pdf
USDA-NAC (USDA-National Agroforestry Center) (2016) How can agroforestry help landowners adapt to increased rain intensity? 402-437-5178 ext. 4011, fax 402-437-5712, 1945 N. 38th St., Lincoln, Nebraska 68583-0822. http://nac.unl.edu
USEPA (2016) U.S. Environmental Protection Agency. Office of Water and Office of Research and Development. National Rivers and Streams Assessment 2008-2009: A collaborative survey (EPA/841/R-16/007). Washington, DC March 2016. http://www.epa.gov/national-aquatic-resource-surveys/nrsa
USEPA (2020) 2017 National Emissions Inventory, Complete Release, Technical Support Document. U.S. Environmental Protection Agency Office of Air Quality Planning and Standards Air Quality Assessment Division Emissions Inventory and Analysis Group Research Triangle Park, North Carolina
Venkatraman K, Ashwath N (2016) Canopy rainfall intercepted by nineteen tree species grown on a phytocapped landfill. Int J Waste Resour 6:1. https://doi.org/10.4172/2252-5211.1000202
Verchot LV, Van Noordwijk M, Kandji S, Tomich T, Ong C, Albrecht A, Mackensen J, Bantilan C, Anupama KV, Palm C (2007) Climate change: linking adaptation and mitigation through agroforestry. Mitig Adapt Strateg Glob Chang 12(5):901–918. https://doi.org/10.1007/s11027-007-9105-6
Vose JM, Clark JS, Luce C, Patel-Weynard T (2016) Effects of droughts and on forests and rangelands in the United States: a comprehensive science synthesis. Gen Tech Rep WO-93b. USDA, Forest Service Washington Office, Washington DC, p 289
Wallace DC, Geyer WA, Dwyer JP (2000) Waterbreaks: managed trees for floodplain. Agroforestry notes: special applications #4. http://www.unl.edu/nac/agroforestrynotes/an19sa04.pdf
Wang H, Maher BA, Ahmed IA, Davison B (2019) Efficient removal of ultrafine particles from diesel exhaust by selected tree species: implications for roadside planting for improving the quality of urban air. Environ Sci Technol 53(12):6906–6916. https://doi.org/10.1021/acs.est.8b06629

Ward PJ, Jongman B, Kummu M, Dettinger MD, Weiland FCS, Winsemius HC (2014) Strong influence of El Niño southern oscillation on flood risk around the world. PNAS 111(44):15659–15664. https://doi.org/10.1073/pnas.1409822111
Wheeling K (2020) Australia's most extreme bushfire season, statistically speaking. Eos. 101., https://doi.org/10.1029/2020EO151949. Accessed 10 Dec 2020
WHO (2005) Air quality guidelines for particulate matter, ozone, nitrogen dioxide and sulfur dioxide. World Health Organization, Geneva
WHO (2016) Ambient air pollution: a global assessment of exposure and burden of disease. World Health Organization, Development of Economics and Social Affairs Population Division
Xiao Q, McPherson EG (2002) Rainfall interception by Santa Monica's municipal urban forest. Urban Ecosyst 6:291–302. https://doi.org/10.1023/B:UECO.0000004828.05143.67
Zaimes GN, Schultz RC, Isenhart TM (2004) Streambank erosion adjacent to riparian forest buffers, rowcrop fields and continuously-grazed pastures along Bear Creek in Central Iowa. J Soil Water Conserv 59:19–27
Zaimes GN, Schultz RC, Isenhart TM (2006) Riparian land uses and precipitation influences on streambank erosion in Central Iowa. J Am Water Res Assoc 42:83–97
Zobeck TM, Van Pelt RS (2006) Wind-induced dust generation and transport mechanics on a bare agricultural field. J Hazard Mater 132:26–38

# Establishing Agroforestry Conservation Buffer Zones to Protect Tropical Peatland Forests of Indonesia

**Haryo A. Dewanto, Hong S. He, Shibu Jose, and Ranjith P. Udawatta**

## Introduction

Peatlands in Indonesia are estimated at 20.7 Mha of the total geographic area (47% of total tropical peatland) and 1138 $Gm^3$ of volume (65% of the total tropical peatlands), with an average 5.5 m thick peat layer (Page et al. 2011). Peatland in Indonesia occurs on three big islands. They are Sumatra 43% or 6.44 Mha, Kalimantan 32% or 4.78 Mha, and Papua 25% or 3.69 Mha (Fig. 1) (Osaki and Tsuji 2016).

These peatland ecosystems are threatened by deforestation and forest degradation. For example, Kalimantan has been affected heavily by both human and climate change impact. In Kalimantan, trees from *Dipterocarpaceae* family are dominated in this tropical peatland ecosystem. *Dipterocarpaceae* tree species are highly commercial, especially in Indonesia. Trees such as *Shorea balangeran*, *Shorea parvifolia*, *Calophyllum* spp., and *Alstonia* spp. are favorites to build houses, boats, and furniture.

Tropical peat swamp forest is crucial not only for its wealth of diverse bioresources, but also for its huge carbon pool (Tawarya et al. 2003). Peat soil is considered as organic soil (Histosols). The unique properties of Histosols are a very high content of organic matter in the upper 80 cm (32 in.) of the soils and no permafrost. The amount of organic matter is at least 20–30% in more than half of this thickness, or the horizon that is rich in organic matter rests on rock or rock rubble

H. A. Dewanto (✉)
HCV Network, Oxford, UK
e-mail: haryo@hcvnetwork.org

H. S. He · S. Jose · R. P. Udawatta
University of Missouri, Columbia, MO, USA
e-mail: heh@missouri.edu; joses@missouri.edu; udawattar@missouri.edu



R. P. Udawatta, S. Jose (eds.), *Agroforestry and Ecosystem Services*,
https://doi.org/10.1007/978-3-030-80060-4_13

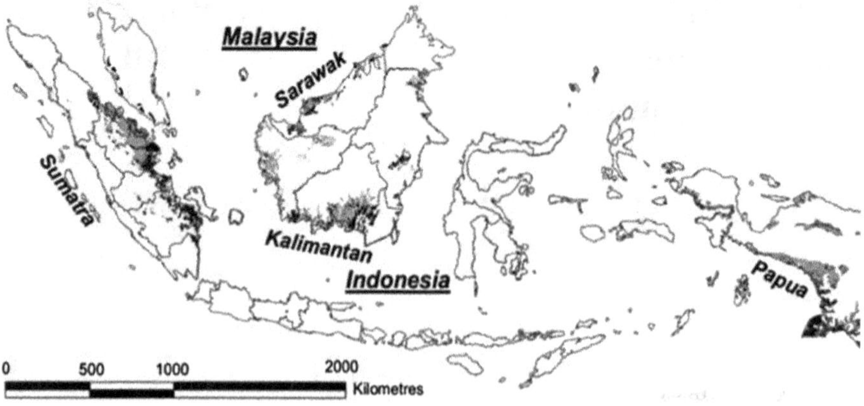


**Fig. 1** Peatland in Indonesia (Hooijer et al. 2010; reprinted under the Creative Commons Attribution 4.0 License)

(Soil Survey Staff, USDA 1999). Peat soil can be found in transitional ecosystems between aquatic and terrestrial ecosystems.

## *Conservation Challenges for Indonesian Peatland*

Peat soil is vulnerable if it is overexposed and cultivated without considering the conservation rules. Destruction of peat soil damages its physical, biological, and chemical properties (Noor 2010). Peat soil is hydrophilic which means that it can bind water. There are three categories of peat soil: first, fibric peat (raw) that can bind water as much as 850% of its dry weight, and sapric peat (processed) that only binds water as much as 450% of its dry weight (Noor 2010). If the physical, chemical, and biological properties of hydrophilic soils are damaged, it will become hydrophobic, where the peat soil can no longer bind water.

Peatland utilization for agriculture is not new in Indonesia, especially in Sumatra and Kalimantan. It was the indigenous people who looked and utilized peatland as a resource to produce traditional food crops, fruits, and spices. However, traditional peatland utilization changed when the mega rice project was initiated. The project was established in 1995 and discontinued in 2000 in the Central Kalimantan Province. It was a project that proposed to open one million hectare of peatland as paddy fields to produce rice in support of food security in Indonesia. The opened peatland areas in Kalimantan for this project are also known as the three largest greenhouse gas (GHG) emission zones in the world (Noor 2010).

In the 1990s, the local communities relied on the forest to make a living. They opened the land for ladang (farm), paddy fields, palm oil, and rubber plantations by using the slash and burn method, but most of the land cultivations did not manage the fire well. The communities also logged timber inside the forest for cash revenue.

After the 1990s, Indonesian Government introduced a *Manajemen Hutan Lestari* (Sustainable Forest Management) operation to regulate logging activity. This policy made logging activity the least favored livelihood option for the local communities. However, these communities still harvest timber on a smaller scale. Besides the government policy, the decrease in the number of commercial timber species in these forests also led to the reduction in logging activities. Lately, most of the villagers earn their living from river fishing (Lemons et al. 2011).

## *Agroforestry as a Conservation Option*

Agroforestry systems can be defined as an agroecosystem approach to land use that incorporates trees and shrubs into farming practice, in which both trees and agricultural crops or livestock are combined on the same field (Nair 1993). Agroforestry systems combine the potential to provide a variety of non-marketed ecosystems while maintaining a high agricultural production (Clough et al. 2011). Agroforestry can contribute to a number of ecosystem services such as water quality improvement, biodiversity enhancement, and soil conservation (Jose 2009). Agroforestry systems can be classified as windbreaks, silvopastoral systems, forest farming systems, integrated riparian forest systems, and tree-based intercropping system—also known as alley cropping (Jose 2009).

Agroforestry systems can provide the security of a diversified source of products, usually by combining food crops, cash crops, timber, and various non-timber products, and they are very resilient to the economic and ecological crisis (Feintrenie et al. 2010). As a strategy for reducing tropical forest destruction and degradation, agroforestry systems are becoming important in tropical areas (Perfecto and Vandermeer 2008).

Agroforestry is considered as a viable alternative to provide income while protecting peatlands in Indonesia by not only the government but the local communities as well. Lemons et al. (2011) indicated that the goal of agroforestry implementation was to achieve restoration and reforestation through integrated natural forest regrowth with a community-based cash crop, multistory mixed agroforestry, and low-impact aquaculture programs that alleviate hunger, poverty, and pressures on the surrounding primary and secondary forests. Lemons et al. (2011) also suggested executing agroforestry in cooperation and participation with the palm oil concessionaires (as joint venture partners) to address leakage. Leakage in this context consists of canal opening, illegal logging, peat burning, and other ecological disturbance that threaten the ecological balance of the peatlands. The idea is that agroforestry practices would become the buffer zones of the conservation areas and protect the forest from future encroachment and deforestation, thus offering great value to conservation.

The main goal of this research was to assist the establishment of agroforestry areas as buffer zones (adaptive management zone) in Rimba Raya Biodiversity Reserve. To accomplish this goal, the following objectives were constructed:

1. Determine suitable locations for agroforestry practices as buffer zones in the conservation area.
2. Determine whether agroforestry areas could reduce the pressure on natural forests from deforestation or forest conversion.

## Materials and Methods

### *Study Area*

The study area is located in Seruyan district, Central Kalimantan Province, Indonesia. It is bounded by Tanjung Puting National Park to the west, Java Sea to the south, Seruyan River to the east, and KUCC oil palm plantation to the north (Fig. 2) (Lemons et al. 2011). Rimba Raya is divided into a three-unit management, Northern unit (10,978.53 ha), Central unit (25,022.13 ha), and Southern unit (28,161.33 ha).

Mean annual rainfall in the Project Zone is approximately 2500–2700 mm. Based on Oldeman classification, the Project Zone falls into B1 and C1 zones. Zone B1 has >200 mm/month of precipitation for long-term averages of 7–9 months per year and <100 mm/month for <2 months per year (Lemons et al. 2011). Furthermore,

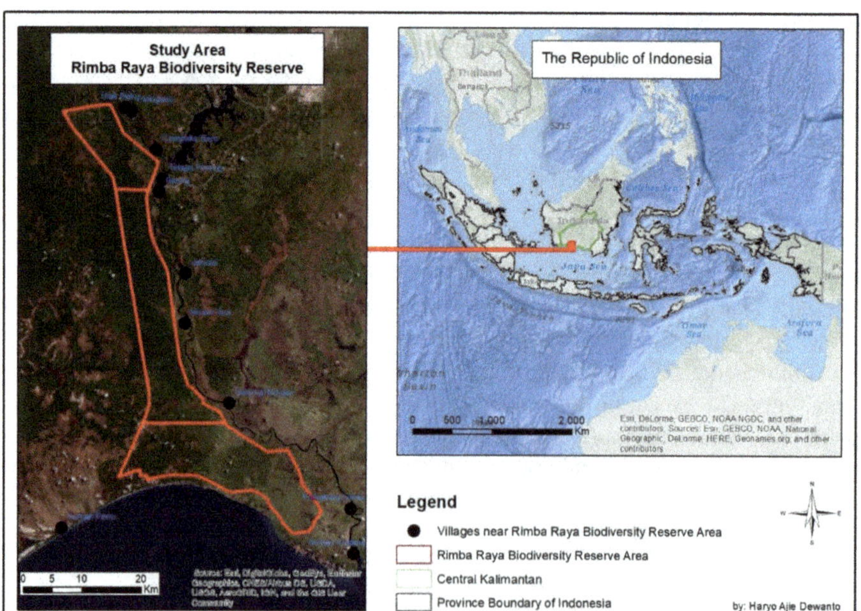


**Fig. 2** Study area location of Rimba Raya Biodiversity Reserve in Central Kalimantan Province, Indonesia

C1 has >200 mm of precipitation/month for 5–6 months and <100 mm per month for <2 months per year (Oldeman et al. 1980). The mean annual temperature is 20–32 °C with the average relative humidity of 75% (seruyankab.go.id).

## *Rimba Raya Biodiversity Reserve as a Case Study*

Indonesia is one of the best places to implement the REDD+ project. One of the ecosystem restoration concessions (ERC) held in Central Kalimantan to implement the REDD+ project in peat swamp forest is PT Rimba Raya Conservation, one of the private sector companies in Indonesia that get the license from the Ministry of Forestry to manage a restoration concession. This concession has been carrying out a project named Rimba Raya Biodiversity Reserve Project that started in 2008 (Indriatmoko et al. 2014), which has a crediting period of 30 years. The total areal of the project is 64.162 ha (Project Management Area) with 47.237 ha set aside as a Carbon Accounting Area (Lemons et al. 2011).

The location was chosen because it is a restoration concession in peat swamp forests which has experienced forest degradation, deforestation, fires, illegal logging, and other threats. Also, the location is critical to protecting the habitat of endangered species, especially Bornean orangutans. Moreover, this concession has received recognition from VCS (Verified Carbon Standard) and CCBA (Climate, Community, and Biodiversity Alliance) standards (www.rimba-raya.com). For this reason, the location is considered as a good representative site to describe the condition of tropical peat swamp forest restoration. The Rimba Raya project eliminates many of the incentives driving illegal logging and unnecessary conversion of forest to agricultural land. The project also trains project-zone community members and offers them priority employment in all the key project activities (Lemons et al. 2011).

## *Data*

Data used for this research included ecological factors including peat soil depth, landcover data and NDVI data (Normalized Difference Vegetation Index), and disturbance data such as fire hotspot data, access data, and traditional land-use data. All data sets were obtained from either direct measurements or secondary data provided by Rimba Raya Biodiversity reserve.

### Soil

A soil map for the Project Zone was produced using the Soil Resource Exploration Map (Pontianak MA49, Centre for Soil and Agroclimatic Research, Bogor, Indonesia) at a scale of 1:1,000,000.

**Table 1** Soil type of Rimba Raya biodiversity reserve (adapted from Lemons et al. 2011)

| Soil type | General description | Parent material | Sub-landform | Relief |
|---|---|---|---|---|
| Haplohemist, sulfihemists | Moderately decomposed peat soils, some of which are sulfic | Organic | Peat dome | Flat |
| Endoaquepts, sulfaquents | Saturated inceptisols and saturated sulficentisols | Alluvium | Delta or estuary | Flat |
| Endoaquepts, dystrudepts | Saturated inceptisols and acidic inceptisols | Alluvium | Alluvial floodplain | Flat |
| Quartzipsamments, durorthods | Quartzicentisols and Spodosols with a cemented hardpan | Sediment | Terraces | Flat rolling |

Soil types and its description were derived from Soil Taxonomy (Soil Survey Staff, USDA 1999)

Associated soil types in each mapping unit are summarized in Table 1. The great groups and general descriptions are derived from Soil Taxonomy (Soil Survey Staff, USDA 1999).

Peat soil in Sumatra is more fertile than peat soil in Kalimantan depending on the basic material, mineral materials, and soil depth. Based on the depth, peat soil can be classified as (Noor 2010):

- Shallow peat (50–100 cm)
- Medium shallow peat (101–200 cm)
- Deep peat (201–300 cm)
- Very deep peat (>300 cm)

Rimba Raya Biodiversity Reserve conducted a massive work to map the peat depth inside their concession. According to referrals of Department of Agriculture (BB Litbang SDLP 2008), the peatland area that can be used for agriculture is in the shallow peatland depth (<2 m). Peat soil with >2 m depth is used as conservation areas. These peat soils are fragile if converted to farmland. Rimba Raya Biodiversity Reserve has various peat depths ranging from 0 to 7.2 m.

## Land Cover

Land cover map of 2017 was produced from Landsat 8 OLI (Fig. 3). To classify the land cover type in Rimba Raya area, the unsupervised and supervised classifications were assigned. Unsupervised method was done using object-based tool (Feature Extraction) in ENVI 5.3 with the value of edge feature set at 10 and merge level at 90. This method was used because there was a haze that covered the study area. The haze would disrupt the pixel recognition if we utilized the pixel-based tool. There are 11 classes from the land cover interpretation that consist of coastal forest, farmland (plantation), fishpond, grassland, low sparse vegetation, peat swamp forest, palm oil plantation, riparian forest, shrubland, wetland, and water body (Table 2).

Land cover map validation was conducted only for the latest image, 2017. There were 136 random sample plots that were visited during ground check in May–July

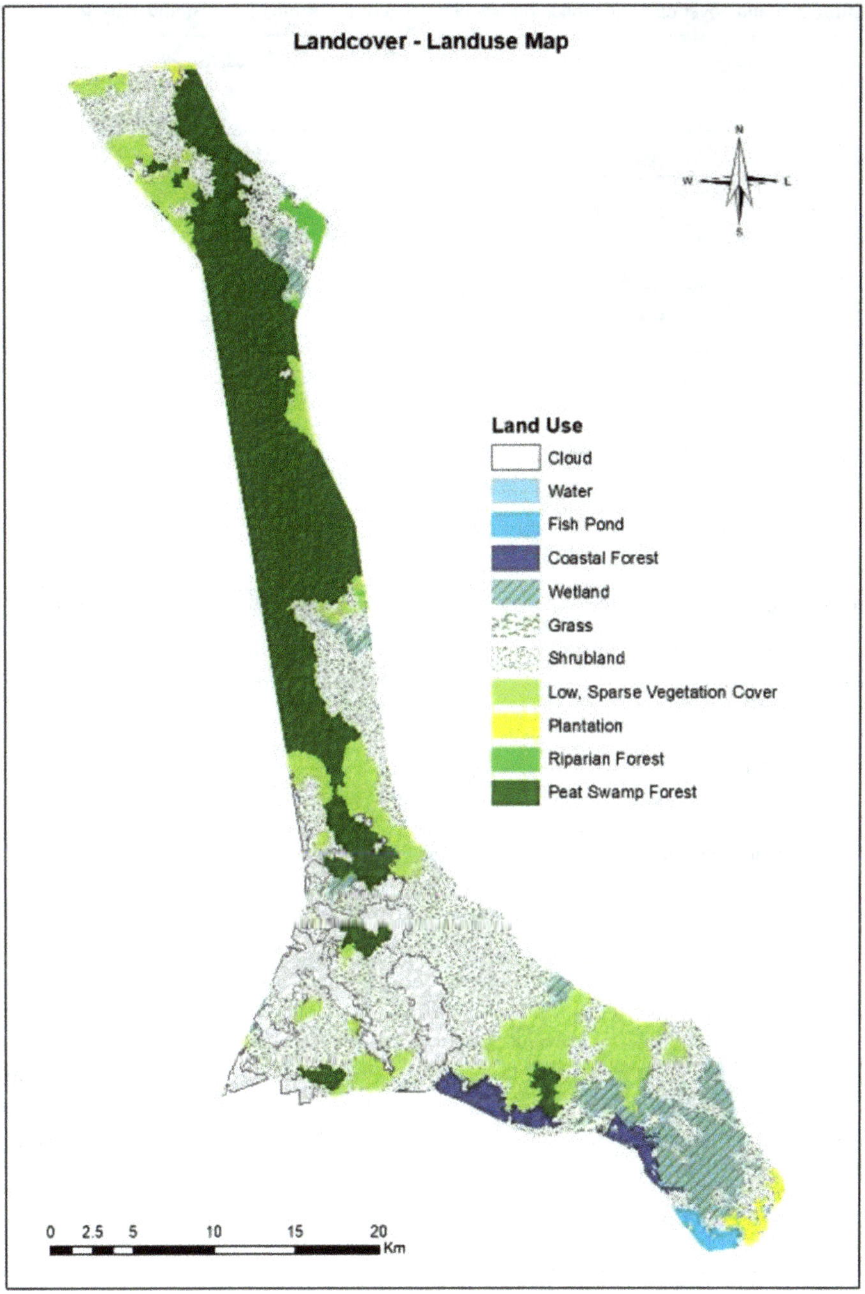


**Fig. 3** 2017 Landcover Map of Rimba Raya Biodiversity Reserve, Kalimantan Province, Indonesia

**Table 2** Land cover classification and associated area of the Rimba Raya Biodiversity Reserve

| Land cover | Area (ha) |
|---|---|
| Coastal forest | 1418.14 |
| Fishpond | 395.11 |
| Grass | 329.41 |
| Low, sparse vegetation cover | 9068.94 |
| Cloud | 5518.18 |
| Peat swamp forest | 18,601.25 |
| Plantation | 550.3 |
| Riparian forest | 383.81 |
| Shrubland | 23,246.34 |
| Water | 44.12 |
| Wetland | 4623.68 |

**Table 3** Formula of accuracy assessment used in the analysis

| Name | Formula[a] | References |
|---|---|---|
| Overall accuracy | $\frac{1}{N}\sum_{i=1}^{c} n_{ii}$ | Story and Congalton (1986) |
| User's accuracy | $\frac{n_{ii}}{N_i}$ | Story and Congalton (1986) |
| Producer's accuracy | $\frac{n_{ii}}{M_i}$ | Story and Congalton (1986) |
| Kappa coefficient | $\frac{P_o - P_e}{1 - P_e}$ | Congalton et al. (1983) |

[a]$N$ is the total number of samples, $n_{ii}$ is the number of samples that correctly classified, $N_i$ is the row totals for class $i$, $M_i$ is the column totals for class $i$, $P_o$ is $\frac{1}{N}\sum_{i=1}^{c} n_{ii}$ , and $P_e$ is $\frac{1}{N^2}\sum_{i=1}^{c} N_i M_i$

2017. After that, error matrix was constructed to calculate overall accuracy, user's accuracy, and producer's accuracy and kappa coefficient of agreement (Table 3).

Based on the error matrix table (Table 4), the Kappa coefficient was 0.91 that meant the accuracy of the classification was high. This high value of accuracy assessment revealed that the classification result had good quality and was close to the reality.

## NDVI (Normalized Difference Vegetation Index)

Vegetation cover of the study area was mapped from Landsat 8 image using NDVI, which is the normalized ratio between the reflected radiance in the red channel and the reflected radiance in the infrared channel (Svoray et al. 2005). Values close to zero represent rock and bare soil and negative values represent water and clouds.

**Table 4** Error matrix for land cover map 2017 for Rimba Raya Biodiversity Reserve

| | | Class types determined from reference source | | | | | | | | | | Total | User's |
|---|---|---|---|---|---|---|---|---|---|---|---|---|---|
| | | PSF | LSV | SL | RF | PL | WB | WL | FP | G | CF | (Ni) | accuracy |
| Class types determined from classified map | PSF | 49 | | | | | | | | | | 49 | 100.00 |
| | LSV | | 3 | | | | | | | | | 3 | 100.00 |
| | SL | 5 | | 27 | | | | 1 | | | | 33 | 81.82 |
| | RF | | | | 14 | | | | | | | 14 | 100.00 |
| | PL | | | | | 8 | | | | | | 8 | 100.00 |
| | WB | | | | | | 13 | | | | | 13 | 100.00 |
| | WL | | | 1 | | | 1 | 5 | | | | 7 | 71.40 |
| | FP | | | | | | | | 2 | | | 2 | 100.00 |
| | G | | | 1 | 1 | | | | | 1 | | 3 | 33.30 |
| | CF | | | | | | | | | | 2 | 2 | 100.00 |
| | Total | 54 | 3 | 29 | 15 | 8 | 14 | 6 | 2 | 1 | 2 | 134 | |
| | Producer's accuracy | 90.74 | 100.00 | 93.10 | 93.33 | 100.00 | 92.86 | 83.33 | 100.00 | 100.00 | 100.00 | | 92.60 |
| | Kappa | 0.91 | | | | | | | | | | | |

The formula for NDVI is (NIR band − RED band)/(RED band + NIR band), where NIR is near infrared (band 5) and red band is band 4 in Landsat 8 images. The NDVI values were grouped into five classes using natural break, with high score for high NDVI range value. The value of NDVI ranged from −0.066 to 0.583 (Fig. 4). Areas with high NDVI values were located at the center part of project area.

### Fire

Disturbance events were common in the project area. Fire, illegal logging, illegal hunting, and property trespassing had often occurred. The land cover which was affected by the most significant number of fire events (hot spot) was shrubland, followed by wetland, and farmland (plantation) (Fig. 5). A different pattern took place between 2014 and 2015 due to a significant number of hot spots also affecting the peat swamp forest (Table 5). In the peat swamp forest, there were only two hot spots in 2014 and became significantly higher in 2015 (99 hot spots); the peak fire was in early November which was the end of a rainless period. There was no fire in 2016 and only two hot spots occurred in 2017 on shrubland (Table 5). In 2015, the biggest number of hot spots occurred in shrubland. This condition was due to the low level of groundwater at the end of the rainless period.

In a dry season, shrubland is vulnerable because the shrubland is the source of fuel when it becomes scorched (easy to ignite by small fire). If the communities cannot control the fire, it could spread to the other land cover like peat swamp forest. When the peat dries and ignites the fire, the ground fire will occur and become difficult to extinguish.

Wildfire brought negative impact to the location including destruction of peat soil. Physical damages on soil include the subsidence on peat soil, irreversible drying, enhanced hydrophobic properties, inability of peat soil to bind nutrients, and decreased water supply horizontally because of increased porosity (Noor 2010). Fires were more common in the south (in the shrublands and grasslands) (Fig. 5) and posed threats to the long-term sustainability of peat soil.

Wildfire occurred because fires that were set by the communities were not well managed. Some traditional farmers thought that burning peat soil would give benefits to them because of the peat ash that brought back nutrients in the peat soil (Noor 2010). The use of peat ash is popular among traditional farmers in Kalimantan because the ash could bring back the nutrients and make the soil more fertile (Table 6).

Fire events also happened because of access. New road, hunter path, small rivers, and canals mostly facilitated access to everyone, particularly in the north and south where fires were more prevalent.

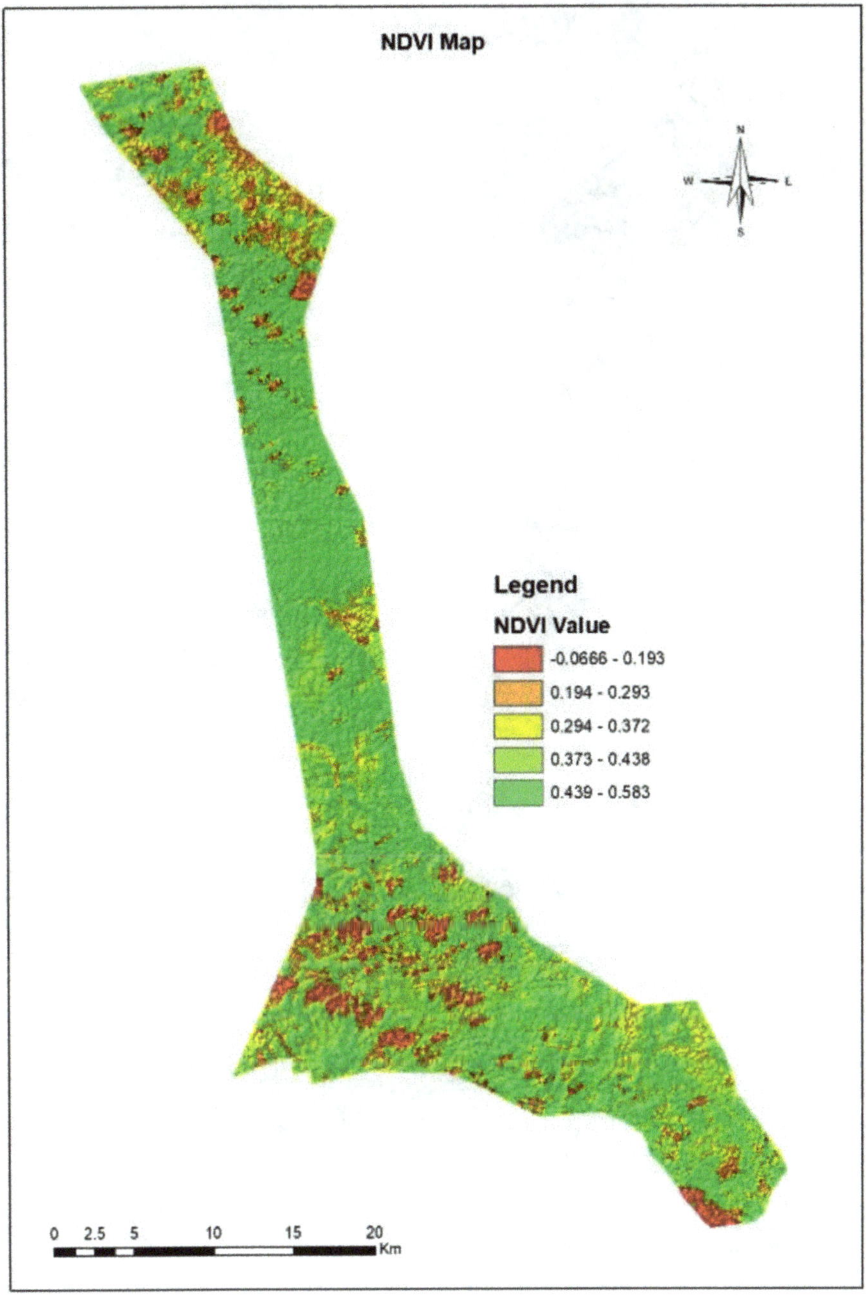


**Fig. 4** Normalized Difference Vegetation Index (NDVI) map of Rimba Raya Biodiversity Reserve

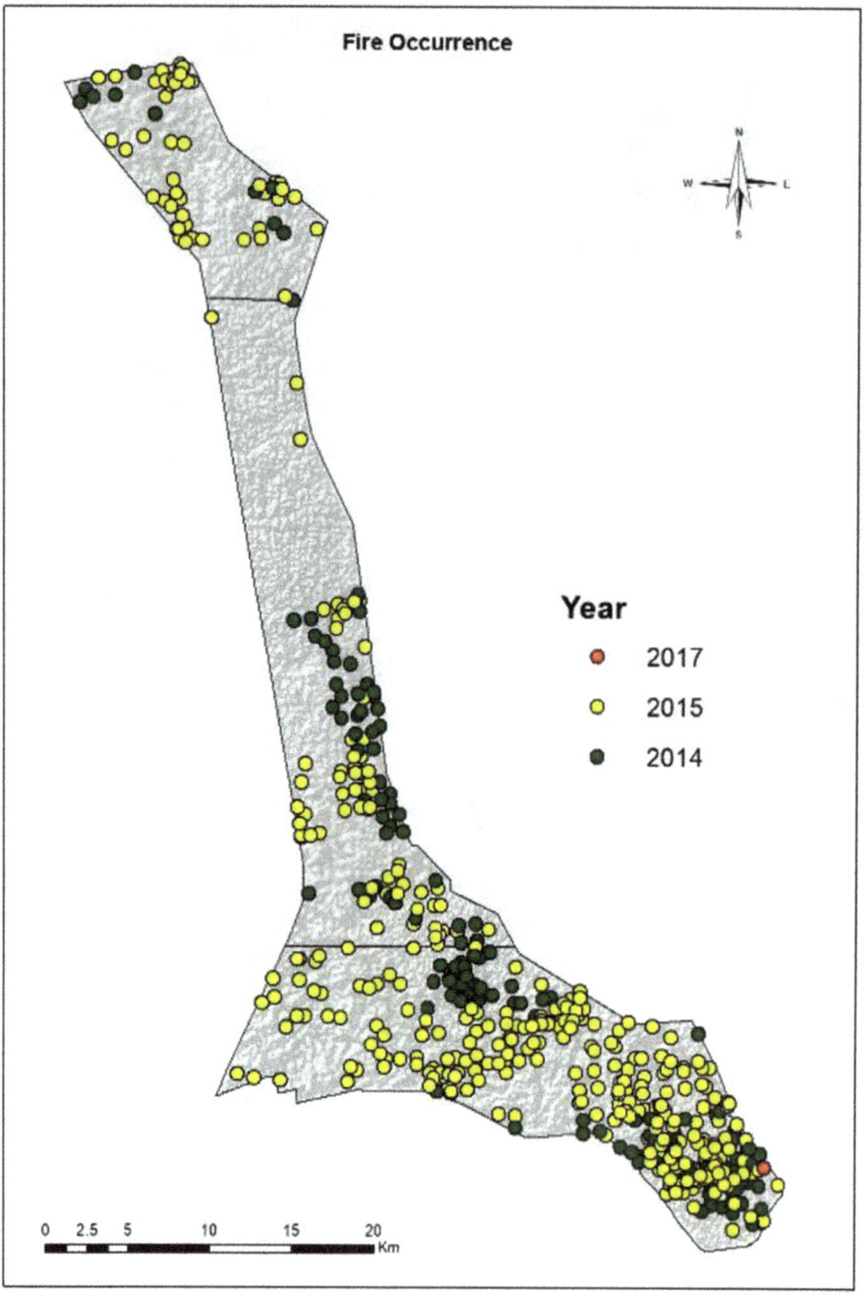


**Fig. 5** Fire occurrence map for 2014, 2015, and 2017 of Rimba Raya Biodiversity Reserve

**Table 5** Hot spots per land cover type at Rimba Raya Biodiversity Reserve

| Land use | Hot spot 2014 | Hot spot 2015 | Hot spot 2016 | Hot spot 2017 |
|---|---|---|---|---|
| Cloud | 4 | 32 | 0 | 0 |
| Coastal forest | 5 | 16 | 0 | 0 |
| Grass | 5 | 1 | 0 | 0 |
| Low, sparse vegetation | 10 | 2 | 0 | 0 |
| Peat swamp Forest | 2 | 99 | 0 | 0 |
| Plantation | 2 | 14 | 0 | 0 |
| Riparian forest | 1 | 6 | 0 | 0 |
| Shrubland | 90 | 1 | 0 | 2 |
| Wetland | 34 | 121 | 0 | 0 |

Data derived from Nasa Fire Information for Resource Management System (FIRMS)

**Table 6** Nutrients from burned peat (adapted from Noor 2010)

| Nutrients and pH | Peat soil ash |
|---|---|
| pH | 6.33 |
| Phosphate (%) | 1.2 |
| Nitrogen (%) | 1.22 |
| Kalium (%) | 0.02 |
| Calcium (%) | 0.16 |
| Magnesium (%) | 0.01 |

## Village Demographic and Spatial Planning

There are nine villages located in the east part of Rimba Raya within the potential sphere of influence of the project (Table 7). Most of the villagers work as traditional farmers, fishermen, palm oil plantation laborers, and wood crafters.

Undefined boundary of villages near Rimba Raya Project led to encroachment to the conservation area. That is why participatory mapping program was conducted by Rimba Raya Conservation to reduce the pressure to the forest area and help the village to develop spatial planning.

Integration of all data (biophysical and socioeconomic) from stakeholders was necessary for a participatory mapping. Involvement of government agencies, private sector, and community is very central in this activity. A participatory rural appraisal is an approach used by nongovernmental organizations (NGOs) and other agencies involved in international development. The process aims to incorporate the knowledge and opinions of rural people in the planning and management of development projects and programs. In the Rimba Raya Biodiversity Reserve area, the government does not have spatial planning in every village. Rimba Raya Biodiversity Reserve intensively communicated with local government and community to prepare the participatory map. In this case, communication was essential to gather stakeholders' opinion. Participants responded to the following questions:

**Table 7** Population data of Rimba Raya Biodiversity Reserve, 2016

| No. | Village | No. of people | No. of women | No. of men | Predominant tribe |
|---|---|---|---|---|---|
| 1 | Ulak Batu | 181 | 89 | 92 | Dayak Nadju and Banjar |
| 2 | Palingkau | 168 | 77 | 91 | Dayak Nadju and Banjar |
| 3 | Cempaka Baru | 566 | 216 | 250 | Dayak and Banjar |
| 4 | Telaga Pulang | 2313 | 1008 | 1305 | Dayak and Banjar |
| 5 | Baung | 2015 | 992 | 1223 | Dayak and Banjar |
| 6 | Jahitan | 477 | 208 | 269 | Dayak and Banjar |
| 7 | Muara Dua | 523 | 236 | 287 | Dayak and Banjar |
| 8 | Tanjung Rangas | 1406 | 641 | 765 | Dayak and Banjar |
| 9 | Pematang Limau | 3575 | 1658 | 1917 | Dayak and Banjar |
| **Total** | | **11,224** | **3467** | **4282** | |

1. Which policies or methods should be established in response to forest management?
2. How can different people who are involved in management coordinate their actions where some conflicts arise that overlap between needs and policies?
3. How can policies be enforced when voluntary efforts are the most important thing to consider?

A participatory approach implies that the perceptions, views, and concerns of all relevant stakeholders or actors must be accommodated.

### Access and Traditional Land-Use Data

Fire, illegal logging, and any other illegal activity occurred because of easy access (Erten et al. 2004). A combined map of access (showing river, canals, hunter path, and distance from village) and traditional land use (generated from village spatial planning data) is given in Fig. 6. These maps were created in ArcGIS 10.5.1 using administrative data from the Rimba Raya Biodiversity Reserve data.

## *Index-Based GIS Modeling*

The analysis used in this research was GIS-based analytic hierarchy process (AHP), a multicriteria analysis/evaluation technique that is often used in land suitability analysis. This approach required a process to determine the weights of the variables. The qualitative variables must be translated into numerical data and all variables must have the same standard of making weight and score.

In this technique, a criterion of high suitability value receives a high score, whereas a criterion of low fitness receives a low score (Malczewski 1999). All criteria indexes should be assigned with a score. To determine the score and weight of each criterion in this study, the analysis was built using land-use (agriculture

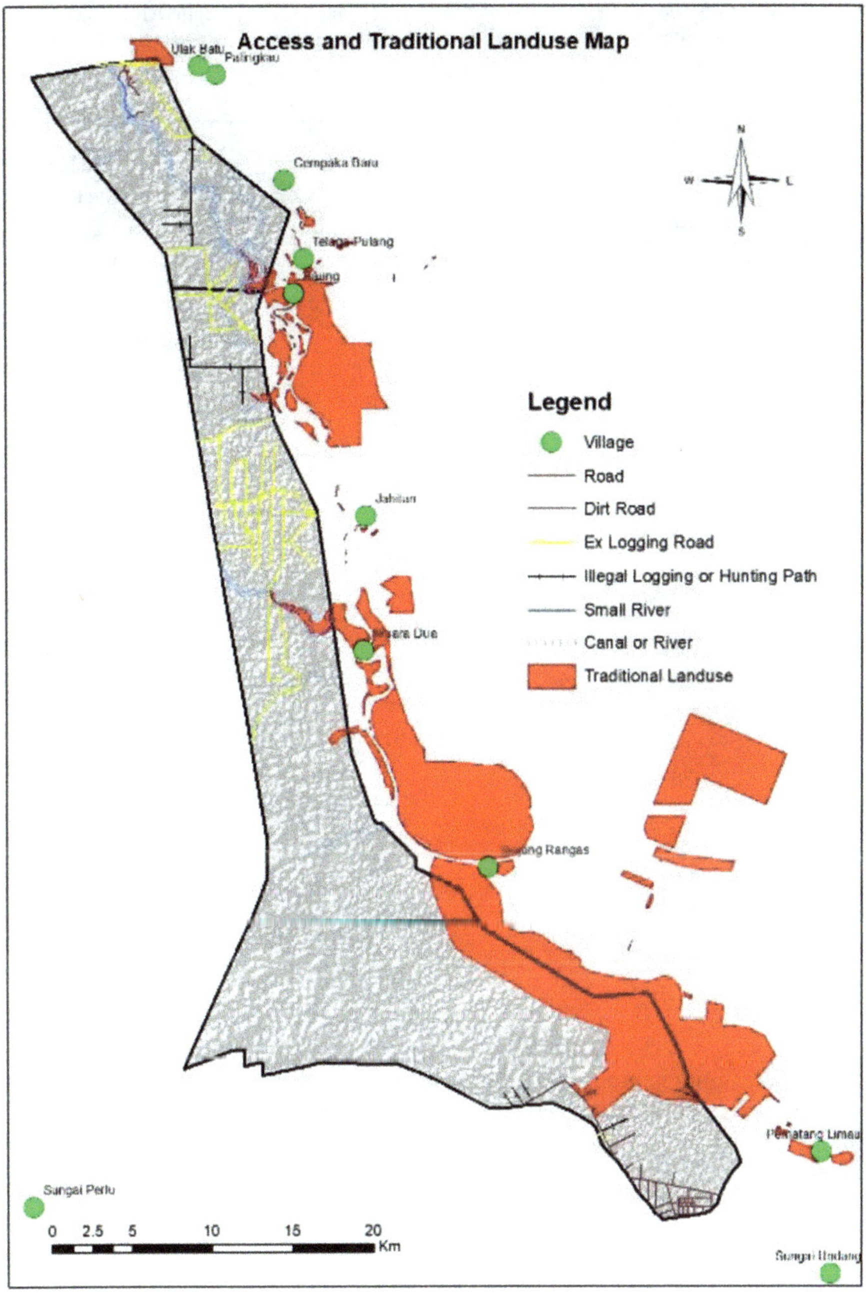


**Fig. 6** Access and traditional land-use map for the Rimba Raya Biodiversity Reserve

**Table 8** Suitability class (a range of values 1–5, except for peat soil depth, where each range represents the suitability level of agriculture and agroforestry land use) for each parameter chosen for the analysis at Rimba Raya Biodiversity Reserve

| Parameter | Range value | Score | Suitability |
|---|---|---|---|
| Peat soil depth | 0–100 cm | 5 | Very suitable |
| | 101–200 cm | 3 | Medium |
| | >201 cm | 1 | Unsuitable |
| Land cover | Peat swamp forest, fishponds, wetland, water, and coastal forest | 1 | Unsuitable |
| | Riparian forest | 2 | Low |
| | Cloud (undetermined land cover) | 3 | Medium |
| | Palm oil plantation, low sparse vegetation | 4 | High |
| | Shrubland and grassland | 5 | Very suitable |
| NDVI | −0.067 to 0.19 | 5 | Very suitable |
| | 0.2–0.29 | 4 | High |
| | 0.3–0.37 | 3 | Medium |
| | 0.38–0.44 | 2 | Low |
| | 0.45–0.58 | 1 | Unsuitable |
| Fire density | 0–4876.74 | 1 | Unsuitable |
| | 4876.75–11,661.78 | 2 | Low |
| | 11,661.79–22,051.37 | 3 | Medium |
| | 22,051.38–36,681.61 | 4 | High |
| | 36,681.62–54,068.27 | 5 | Very suitable |
| Access density | 0–0.0292 | 1 | Unsuitable |
| | 0.0293–0.0743 | 2 | Low |
| | 0.0744–0.125 | 3 | Medium |
| | 0.126–0.206 | 4 | High |
| | 0.207–0.338 | 5 | Very suitable |

criteria) capability in production. Each variable was classified into five conformity types that had a range of values from 1 to 5 where each range represented the suitability level for agriculture or agroforestry (Table 8).

## Data Analysis

The scoring processes for each criterion were done using ArcGIS 10.5.1 software, which involved reclassifying raster data based on suitability score (Fig. 7).

The procedure for this method consisted of the following phases:

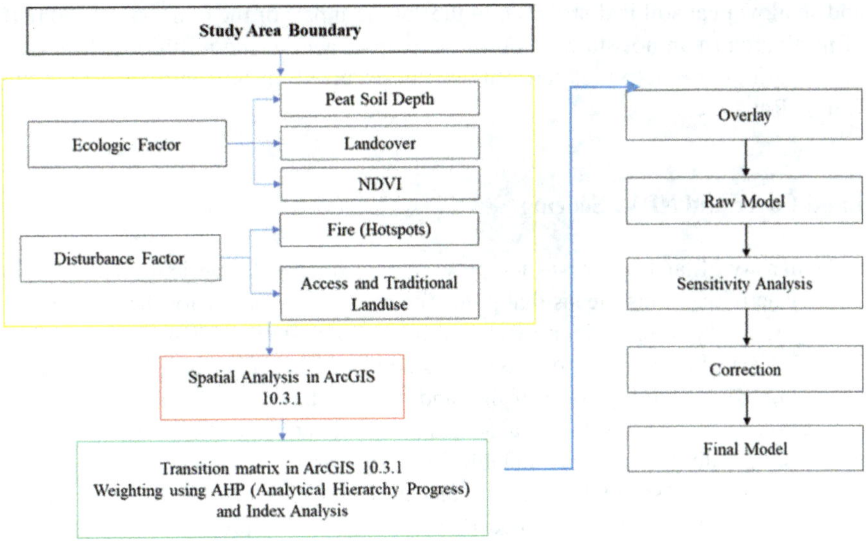


**Fig. 7** Data analysis flowchart to establish agroforestry areas as buffer zone

1. Assessment of the suitability structure: Choosing the land-use factors and defining their importance and how these factors impact the suitability priority
2. Producing layers of map: Data acquisition and making appropriate GIS data sets
3. Modeling/spatial analysis: Defining viable area and combining the suitability factors
4. Sensitivity analysis: Signifying the effect of different criteria weights on the spatial pattern of suitability index (Store and Kangas 2001)

## *Ecological Factors*

### Peat Soil Depth Scoring

The peat soil depth parameter used three scores (Table 8). These were based on the Agriculture Department statement for agricultural cultivation in peat soil, where peat soil depth within 100 cm is possible to do the cultivation. In 101–200 cm, it is still possible, but the maintenance is much more expensive. Peat soil with depth >200 cm was considered deep peat soil and used for conservation purposes. Deep peat soil is accumulated in the central part of the Rimba Raya area and shallow peat soil is distributed in the southern part of the concession and part of north area (map not shown). From the map of access and traditional land use, many agricultural activities have massively occurred in the southern part of the Rimba Raya area.

### Land Cover and NDVI Scoring

If the density of land cover was less, it was more suitable for agriculture and agroforestry activities. This means that in the future it will be easier for the community to prepare the land for agriculture and agroforestry activities. There are ten classes of land cover. The land cover data is reclassified into five classes based on suitability level (Fig. 8). Suitability level of the land cover is based on the accessibility for cultivation or openness of the area, access to water, and vegetation density (similar with NDVI value). The most suitable location was on shrubland and grassland (score 5) and the least suitable area was on peat swamp forest (Table 8). The description of land cover types is provided to give more comprehensive information (Table 9).

NDVI values were positively linked with canopy cover, with higher values corresponding to denser canopy cover. Using natural break, values of NDVI were determined and also given five classes (Table 8).

## *Disturbance Factors*

### Fire Density Scoring

Fire density was analyzed using a kernel estimation method to figure out fire-affected areas in the last 10 years. Moreover, a kernel density estimation method was used to map distribution and concentration of fires within the study area and to indicate areas that had high frequency (and possibly greater risk) of fires (Asgary et al. 2010). The kernel density estimation was used as an analytical method to derive fire disturbance density (Rui et al. 2013).

All active fires or hot spots from 2014 to 2017 were combined into one GIS layer and then the spatial distribution of fire points was modeled as density kernel function (Takahata et al. 2010). The kernel density was drawn on default setting in ArcGIS software.

The kernel density estimation is simply modeled as:

$$\hat{f}_h(x) = \frac{1}{n}\sum_{i=1}^{n} K_h(x - x_i) = \frac{1}{nh}\sum_{i=1}^{n} K\left(\frac{x - x_i}{h}\right),$$

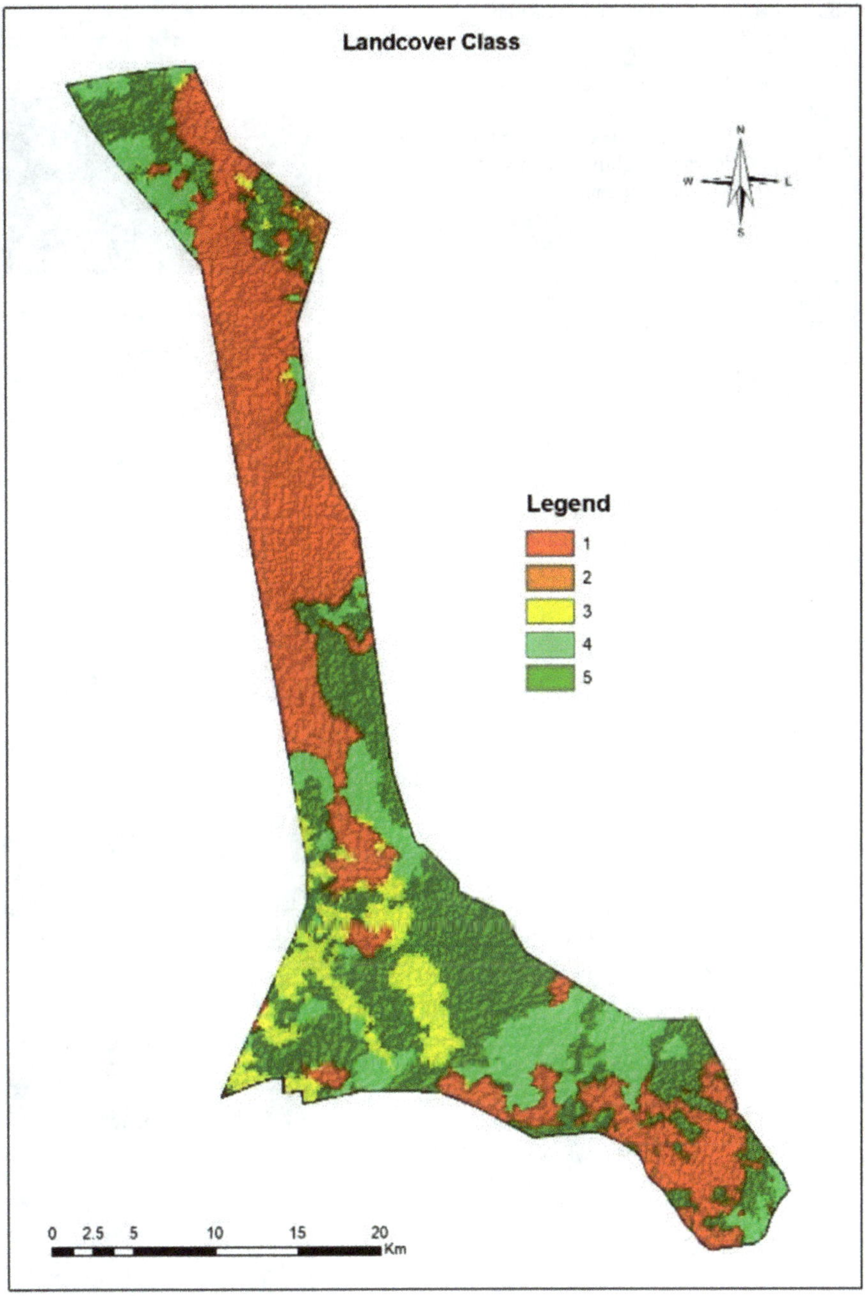


**Fig. 8** Peat swamp forest, fishponds, wetland, water, and coastal forest (*1*); riparian forest (*2*); cloud undetermined land cover (*3*); palm oil plantation; low sparse vegetation (*4*); and shrubland and grassland (*5*) land cover class map of Rimba Raya Biodiversity Reserve

**Table 9** Land cover description for Rimba Raya Biodiversity Reserve

| Name class | Description | Location |
|---|---|---|
| Peat swamp forest | Peat swamp forest, locally "hutan rawa" denoting seasonally wet forests on peat substrate. All peat swamp forests in Rimba Raya are lightly to highly degraded by selective logging | Project zone and boundary area  |
| Riparian forest | Forest areas adjacent or near to a river. The areas have mineral soil | Boundary area  |
| Coastal forest | It is generally found above the high-tide mark on sandy soil and may merge into agricultural land or upland forest | Project zone  |
| Shrubland | Formerly peat swamp forests, these areas were deforested by fire in the last 10 years. Seasonally wet areas characterized by shrubby regrowth and scattered remnant trees | Project zone and boundary area  |
| Grassland | Ground covered by vegetation dominated by grasses, with little or no tree cover | Project zone  |
| Oil palm plantation | In the Rimba Raya vicinity, all plantation agriculture is oil palm plantation and is currently confined to the WSSL concession in the north, with some recent expansion into the Project Management Zone | Boundary area and project zone  |

(continued)

**Table 9** (continued)

| Name class | Description | Location |
|---|---|---|
| Fishpond | Repeatedly burned cultivation land, locally "ladang," often abandoned after several years of cultivation. Active cultivation land may appear bright green on imagery from postfire herbaceous growth. Old ladang often has woody shrubs and scattered trees | Boundary area  |
| Wetland | Locally "danau" or seasonal lake, most of these areas were formerly peat swamp forests that have been logged and burned. Where these are adjacent to rivers, flooding may be semipermanent. Most are sedge dominated | Boundary area and project zone  |
| Low sparse vegetation | Areas with sparse grass or herb cover or bare ground, usually associated with recent, severe, or frequent burning in areas of human activity. Most of these areas have been cleared by fire but are interpreted to be outside cultivation lands | Boundary area and project zone  |
| Water body | Deep water with no vegetation | Boundary area and project zone  |

(Based on Lemons et al. 2011)

where $f_h(x_t)$ is the density distribution which is used to estimate the weights, $x$ is the observed value, $K$ is the Kernel function, and h is the width of the Kernel function. The values of the fire density were between 0 and 54,068 and were classified into five classes using natural break in ArcGIS (Table 8). The higher the value, the more suitable the location was for agroforestry because those areas that burned were mostly changed into farmland already. The area heavily damaged by fire occurred in the southern part of Rimba Raya area and experienced farming activities including rice fields.

**Table 10** Weighted variables of agroforestry suitability areas

| Goal | | Weight | | Weight |
|---|---|---|---|---|
| Suitable agroforestry areas | Variable | W1 | Sub-variable | W2 |
| 1 | Ecologic | 0.5 | Peat soil depth | 0.33 |
| | | | Land cover | 0.33 |
| | | | NDVI | 0.33 |
| | Disturbance | 0.5 | Fire | 0.5 |
| | | | Access | 0.5 |

### Access Density Scoring

Access and traditional land-use data were used to generate the access (line density) density map. The density values ranged from 0 to 0.338 square kilometers (Table 8). These values were divided into five classes by using natural break in ArcGIS software.

The higher the value, the denser was the access in the particular area, meaning that the area was easily accessible by the community and more feasible to cultivate. The access was mostly facilitated by ex-logging road, dirt road, illegal logging path, and canal inside the project area and more importantly the traditional use by the communities.

We reclassified each criterion into scores and weighted the criteria also using the ArcGIS software in spatial analysis tools. The total weight for the suitability area was equal to 1 or 100%. Each variable first must be weighted. For the ecological and disturbance factors used, equal weight was given and sub-variables varied depending on their importance to the suitability analysis (Table 10). In this project, the ecological and disturbance elements received the same weight because the assumption was that all of the variables had the same influence to determine the suitability of the locations for the establishment of agroforestry areas.

## Result and Discussion

### *Suitability Map*

The suitability map of agroforestry areas shows that the most suitable area for agroforestry was located in the southern part of the project area and some parts at central and northern parts of the project area. To analyze the suitability area, the result was classified into suitability classes (Fig. 9).

In this model, the total area of very suitable (shown as Suitable in Fig. 9) areas for agroforestry was about 1.62% (1044 ha) of the total project area. The high suitability class was 24.1% (15,404 ha) of the total area. Most of the unsuitable area was located in the central part of the project area. The total unsuitable area was 20.9% (13,431 ha) of the total project area. The low suitability also had a relatively

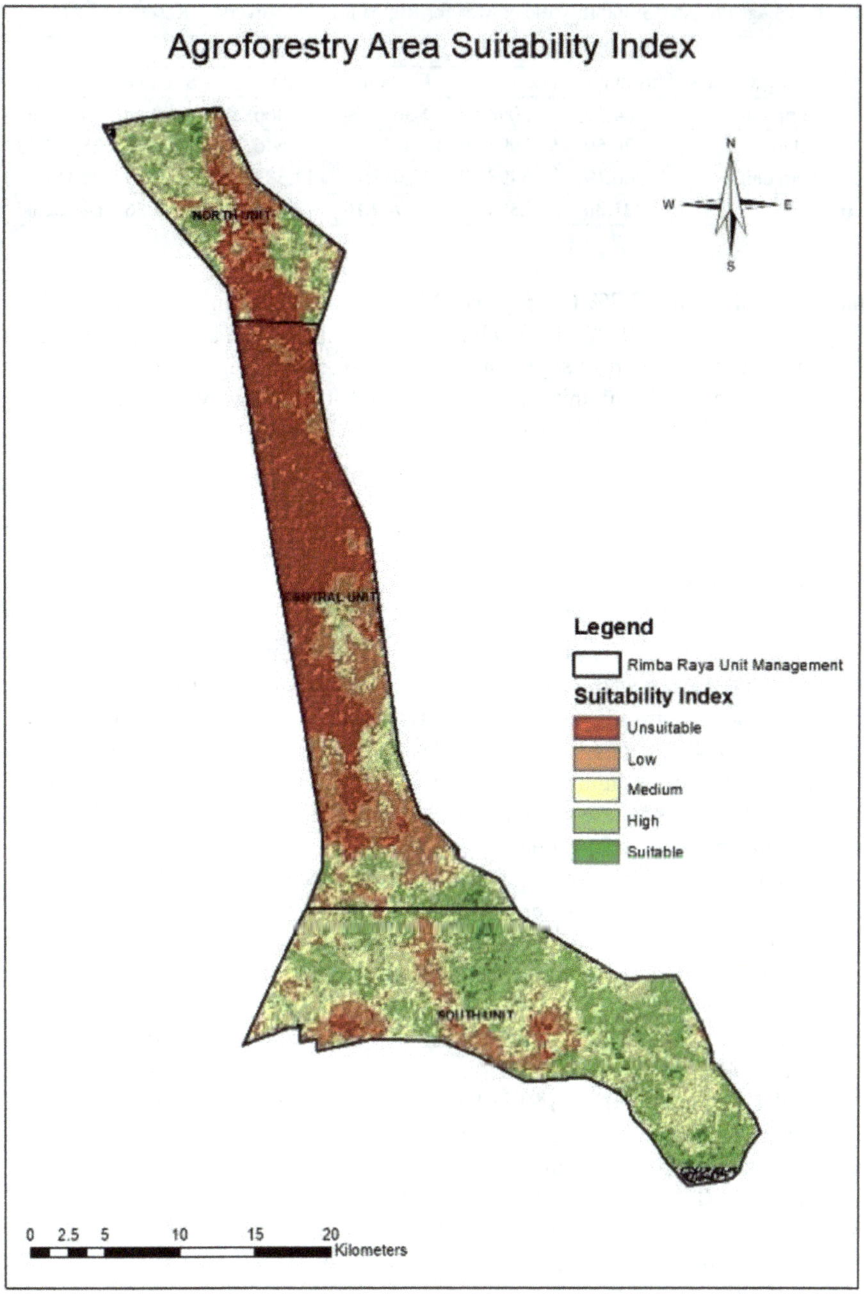


**Fig. 9** Agroforestry Suitability Index Map (model 1) of Rimba Raya Biodiversity Reserve

**Table 11** Agroforestry Suitability Index Area for Rimba Raya Biodiversity Reserve

| | Suitability index area (ha) | | | | | |
|---|---|---|---|---|---|---|
| Unit management | Unsuitable | Low | Medium | High | Suitable | Total (ha) |
| Northern unit | 1784.59 | 3098.59 | 3627.79 | 2306.59 | 160.99 | **10,978.53** |
| Central unit | 11,270.59 | 7998.19 | 4038.19 | 1546.99 | 168.19 | **25,022.13** |
| Southern unit | 366.19 | 3498.19 | 12,048.19 | 11,533.39 | 715.39 | **28,161.33** |
| **Total** | **13,421.36** | **14,594.96** | **19,714.16** | **15,386.96** | **1044.56** | **64,162.00** |

large area of about 22.7% (14.583 ha). Medium suitability in this model had the largest area of about 30.7% (19,699 ha). These numbers were also derived for each management unit. The most significant allocation for potential agroforestry use was located in the southern unit of Rimba Raya Biodiversity reserve where 74% (11,533.39 ha) belonged to high suitable and 60% (715.39 ha) to suitable categories (Table 11).

## *Sensitivity Analysis*

Sensitivity analysis can be used to analyze the influence of different criteria weights on the spatial pattern of the suitability index, and can also help to find alternatives for conservation purposes. Sensitivity analysis was accomplished by applying different weighting schemes for the variables. The results illustrate how changes in weighting affect the optimal choice of areas allocated as agroforestry area (Table 12). The suitability maps for the weighted system were created and the sensitivity analysis was done by making 25 models.

There is a changing pattern between these 25 models. Various kinds of output area for each suitability level were obtained with these 25 models. The least area for a very suitable location is found in model A35N35 (Fig. 10a) with only 378.8 ha. Model A35N35 represents that the weight for access and NDVI is 35% of all parameters, which means that peat depth, land cover, and NDVI parameter do not have a significant effect on the model. The most area for a suitable location is found in model LC20P50N20 (Fig. 10b) (land cover weight: 20%, peat depth weight: 50%, and NDVI weight: 20%) with 9154 ha. This is because of the proportion of shrubland in the land cover parameter, which has a vast area of about 20,000 ha, and also the shallow peat proportion in peat depth parameter. If compared with the model LC20P20N50 (Fig. 10c), which uses the same parameter, the weight for NDVI is higher than land cover and peat depth (the most suitable area was only 2887.6 ha, meaning that the higher the density or canopy cover, the less area that can be used for agroforestry or agriculture). The smallest unsuitable area is found in model A35F35 (Fig. 10d) (access weight: 35% and fire weight: 35%) with 9069.7 ha. It can be assumed that high access and fire occurrences were located in shallow peat and under less dense canopy cover. The largest area for unsuitable location is found in model P50N30 (Fig. 10e) (peat depth weight: 50% and NDVI weight: 30%) with

**Table 12** Matrix of sensitivity analysis for Rimba Raya Biodiversity Reserve

| | Suitability level | | | | |
|---|---|---|---|---|---|
| Model | Unsuitable | Low suitability | Medium | High suitability | Most suitable |
| All balance | 13,421.3 | 14,594.9 | 19,714.1 | 15,386.9 | 1044.5 |
| P50 | 19,414.6 | 6614.5 | 8270.8 | 25,017.6 | 4844.5 |
| N50 | 17,641.0 | 20,682.5 | 18,274.9 | 6467.9 | 1095.7 |
| A50 | 10,810.3 | 29,429.7 | 18,062.4 | 5441.8 | 417.8 |
| F50 | 16,149.5 | 20,788.8 | 18,157.7 | 8476.0 | 590.0 |
| P0 | 12,195.5 | 19,447.6 | 25,006.7 | 7101.8 | 410.4 |
| A30F30 | 11,264.7 | 19,070.1 | 23,295.3 | 9674.3 | 857.5 |
| A35F35 | 9069.7 | 29,019.3 | 17,190.2 | 8146.2 | 736.6 |
| A35N35 | 13,203.2 | 27,542.5 | 18,949.2 | 4093.3 | 373.8 |
| A35P35 | 12,807.5 | 15,057.5 | 22,412.2 | 12,100.2 | 1784.6 |
| A45P25 | 9454.4 | 24,753.8 | 21,213.9 | 7471.9 | 1267.9 |
| A30N25 | 11,206.1 | 21,932.1 | 24,251.8 | 6368.9 | 403.1 |
| N30LC10 | 14,031.4 | 17,179.2 | 22,870.2 | 9289.5 | 791.5 |
| P50N30 | 20,125.5 | 6980.9 | 8373.4 | 23,196.4 | 5485.8 |
| P25N50 | 17,721.6 | 15,977.3 | 18,472.8 | 9656.0 | 2334.3 |
| P25N45 | 17,124.3 | 14,790.0 | 19,825.0 | 10,366.9 | 2055.8 |
| P45N25 | 18,531.5 | 8047.3 | 9824.6 | 23,933.0 | 3825.8 |
| A30P30F30 | 13,859.2 | 16,585.6 | 20,374.7 | 10,553.8 | 2788.7 |
| A30P30N30 | 13,104.3 | 19,249.7 | 22,353.5 | 8904.8 | 549.7 |
| A30LC30P30 | 13,265.5 | 12,976.0 | 16,406.0 | 18,447.2 | 3067.2 |
| LC50P20N20 | 17,106.0 | 8245.2 | 11,202.4 | 19,454.9 | 8153.5 |
| LC20P20N50 | 15,651.1 | 15,790.4 | 20,568.9 | 9263.9 | 2887.6 |
| LC20P50N20 | 17,025.3 | 8498.0 | 9304.2 | 20,180.5 | 9154.0 |
| LC25P25N25 | 15,610.8 | 11,774.1 | 19,660.1 | 15,893.0 | 1223.9 |
| LC30P30N30 | 15,951.6 | 10,407.2 | 14,412.5 | 19,227.7 | 4162.9 |
| **Mean** | **14,627.9** | **16,627.1** | **17,947.8** | **12,556.7** | **2402.6** |
| **St. dev** | **3058.042** | **6691.622** | **5016.943** | **6294.130** | **2398.131** |

A: Access, F: Fire, LC: Land cover, N: NDVI, and P: Peat
Number after the abbreviation is the weight of each variable

20,125.5 ha. Deep peat and dense canopy cover contributed to this result as such areas are unsuitable location for agroforestry or agricultural activity.

The standard deviation values of the model outcomes are high, meaning that the data is dispersed or spread, and it can be explained that each model has a different implication for the total area in each suitability level.

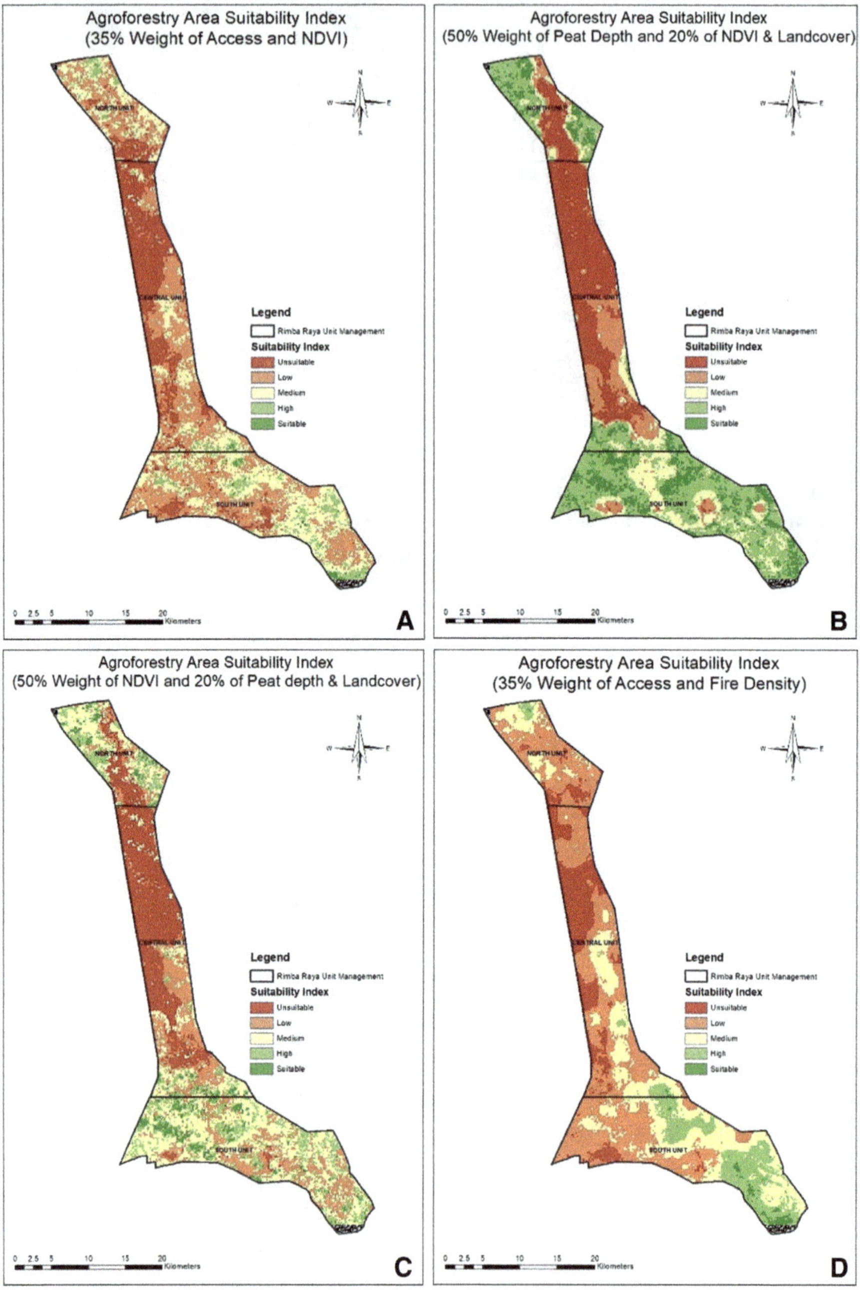


**Fig. 10** Sensitivity analysis using different models

**Fig. 10** (continued)

## Overall Conclusion and Future Directions

Agroforestry implementation is important to achieve restoration and reforestation goals through an integrated natural forest regrowth plan combined with a community-based cash crop, multistory mixed agroforestry, and low-impact aquaculture. Such an approach will alleviate hunger, poverty, and pressure on the surrounding primary and secondary forests. The establishment of agroforestry practices would become conservation areas' buffer zones and protect the remaining natural forests from future encroachment and deforestation.

From the modeling exercise, the most area with high suitability area is in the southern unit and some parts of the northern unit. In the central unit, there are a few locations that could be used for agroforestry, but conservation should be the priority here due to the dense forest canopy and deep peat soil.

These models were created for consideration by Rimba Raya Biodiversity Reserve to protect the core area (conservation area) by allocating appropriate areas as buffer zones for agroforestry so that pressure on the core can be alleviated. A collaborative buffer zone (the collaboration between company and communities) should be a barrier to the conservation area.

Recommendations and next steps of this research are the following:

1. Develop other models using additional parameters including the forest inventory data to know the potential forest location for high carbon stock.
2. Add more comprehensive socioeconomic data for communities to produce more detailed results in the forest pressure valuation.
3. Develop more sensitivity analysis using different weights in each parameter to predict various conditions that occur in the Rimba Raya Biodiversity Reserve area.
4. Use high-resolution imagery to get a more precise model including drone mapping for affordability reason.

## Conclusion and Management Implications

Mapping areas suitable for agroforestry buffers are crucial for conservation management especially in the Rimba Raya Biodiversity Reserve, which needs to protect the remaining natural forests. The agroforestry suitability index in this study was analyzed spatially using scoring, weighting, and overlaying methods. GIS software ArcGIS (Version 10.5.1) was used to produce and examine the suitability for agroforestry at the Rimba Raya Biodiversity area. The information about agroforestry area selection was transformed into score values. The score values ranged from 1 to 5, with 1 for the unsuitable area and 5 for very highly suitable for agroforestry. The critical step for mapping suitability area was determining the weight value between variables. This study shows that analytical hierarchy process is an excellent tool for land suitability analysis and subsequent decision-making. To examine the robustness of the weight value, sensitivity analysis was performed using two weighting schemes. Most of the suitable locations were placed in locations that have easy access, less dense canopy cover, and shallow peat depth. Results show that patterns differ from model to model (model A35N35, LC20P50N20, LC20P50N20; model A35F35; and model P50N30). Each management unit in the project area will have a different approach to managing their territory. The most suitable location appears to be in the southern unit compared to the northern and central units.

Mapping suitability area using GIS spatial analysis is more effective and efficient compared to traditional surveys. Compared to the index-based modeling, the traditional approach is harder and less reliable due to lack of access to specific locations and lack of resources.

The approach used in this study also has some limitations. For example, accuracy and resolution of satellite image data can limit the usefulness of the GIS model. In this study, both Landsat-8 and DEM data had a 30 m spatial resolution. Using satellite imagery with high spatial resolution can improve the results, but image preprocessing is also needed to improve the accuracy of satellite image data. This study is beneficial for decision-making and is expected to improve the participatory conservation management with local communities in the study area.

## References


Asgary A, Ghaffari A, Levy J (2010) Spatial and temporal analyses of structural fire incidents and their causes: a case of Toronto, Canada. Fire Safety J 45:44–57. https://doi.org/10.1016/j.firesaf.2009.10.002

Clough Y, Barkmann J, Juhrbandt J, Kessler M, Wanger TC, Anshary A, Buchori D, Cicuzza D, Darras K, Putra DD, Erasmi S, Pitopang R, Schmidt C, Schulze CH, Seidel D, Steffan-Dewenter I, Stenchly K, Vidal S, Weist M, Wielgoss AC, Tscharntke T (2011) Combining high biodiversity with high yields in tropical agroforests. Proc Natl Acad Sci 108:8311–8316. https://doi.org/10.1073/pnas.1016799108

Congalton RG, Oderwald RG, Mead RA (1983) Assessing landsat classification accuracy using discrete multivariate statistical techniques. Remote Sens 49:1671–1678

Erten E, Kurgum V, Musaoglu N (2004) Forest fire risk zone mapping from satellite imagery and GIS: a case study. In: Proceedings of 20th Congress of ISPRS, Istanbul, Turkey

Feintrenie L, Schwarze S, Levang P (2010) Are local people conservationist? Analysis of transition dynamics from agroforests to monoculture plantation in Indonesia. Ecol Soc 15(4):37

Hooijer A, Page S, Canadell JG, Silvius M, Kwadijk J, Wosten H, Jauhiainen J (2010) Current and future $CO_2$ emissions from drained peatlands in Southeast Asia. Biogeosciences 7(5):1505–1514. https://doi.org/10.5194/bg-7-1505-2010

Indriatmoko Y, Atmadja SS, Ekaputri AD, Komalasari M (2014) Rimba Raya biodiversity reserve project. Central Kalimantan, Indonesia. Retrieved from http://www.cifor.org/redd-case-book/case-reports/indonesia/rimba-raya-biodiversity-reserve-project-central-kalimantan indonesia/

Jose S. (2009) Agroforestry for ecosystem services and environmental benefits: An overview Agrofor Syst 76:1–10. https://doi.org/10.1007/s10457-009-9229-7

Lemons T, Bolick L, Reece J, Procanik J, Fuad F, Stanley S, Paoli G, McDermott S (2011) Project design document: The Rimba Raya biodiversity reserve project. PT. Rimba Raya Conservation

Malczewski J (1999) GIS and multicriteria decision analysis. John Willey & Sons, Inc., Canada

Nair PKR (1993) An Introduction to agroforestry. Kluwer Academic Publisher, Dordrecht, NL

Noor M (2010) Peatland: development, conservation and climate change. Gadjah Mada University Press, Yogyakarta. Print

Oldeman LR, Irsal, Muladi (1980) Contribution: The Agroclimatic Maps of Kalimantan, Maluku, Irian Jaya and Bali, West and East Nusa Tenggara. Bogor: Central Research Institute for Agriculture

Osaki M, Tsuji N (2016) Tropical peatland ecosystems. Springer, Tokyo. Print

Page SE, Rieley J, Banks C (2011) Global and regional importance of the tropical peatland carbon pool. Glob Change Biol 17(2):798–818

Perfecto I, Vandermeer J (2008) Biodiversity conservation in tropical agroecosystems, a new conservation paradigm. Ann N Y Acad Sci 1134:173–200

Rui T, Yang Z, Zhou Y, Fang H, Zhu J (2013) Target detection based on kernel density estimation combined with correlation coefficient. In: Huet B et al (eds) PCM 2013, vol 8294. LNCS, pp 769–778

Soil Survey Staff (1999) Soil taxonomy: A basic system of soil classification for making and interpreting soil surveys. 2nd edition. Natural Resources Conservation Service. U.S. Department of Agriculture Handbook 436

Store R, Kangas J (2001) Integrating spatial multi-criteria evaluation and expert knowledge for GIS-based habitat suitability modelling. Landsc Urban Plan 55:79–93. https://doi.org/10.1016/S0169-2046(01)00120-7

Story M, Congalton R (1986) Accuracy assessment: a user's perspective. Remote Sens 52:397–399

Svoray T, Bar KP, Bannet T (2005) Urban land-use allocation in a Mediterranean ecotone: habitat heterogeneity model incorporated in a GIS using a multi-criteria mechanism. Landsc Urban Plan 72:337–351. https://doi.org/10.1016/j.landurbplan.2004.05.001

Takahata C, Amin R, Sarma P, Banerjee G, Oliver W, Fa JE (2010) Remotely-sensed active fire data for protected area management: eight-year patterns in the Manas National Park, India. Environ Manag 45:414–423. https://doi.org/10.1007/s00267-009-9411-8

Tawarya K, Takarya Y, Turjaman M, Tuah SJ, Limin SH, Tamai Y, Cha JY, Wagatsuma T, Osaki M (2003) Arbuscular mycorrhizal colonization of tree species grown in peat swamp forest of Central Kalimantan, Indonesia. For Ecol Manag 182:381–386

# Cultural Ecosystem Services in Agroforests

**Tomasz B. Falkowski and Stewart A. W. Diemont**

## Abbreviations

| | |
|---|---|
| CES | Cultural ecosystem services |
| ES | Ecosystem services |
| NOAA | United States National Oceanic and Atmospheric Administration |
| spp. | Species |
| TEK | Traditional ecological knowledge |
| VAC | *Vườn Ao Chuồng* (garden-pond-livestock in Vietnamese) |

## Introduction

### *Cultural Ecosystem Service Definition and History*

The Millennium Ecosystem Assessment defined cultural ecosystem services (CES) as "the nonmaterial benefits people obtain from ecosystems through spiritual enrichment, cognitive development, reflection, recreation, and aesthetic experiences" (Millennium Ecosystem Assessment 2005). Many other ecosystem service typologies include CES or some variant on the concept (Costanza et al. 1997; De Groot et al. 2002; Boyd and Banzhaf 2007; Costanza 2008; Kumar 2012). For example, Fish et al. (2016) wrote that CES are "the contributions ecosystems make to human well-being in terms of the identities they help frame, the experiences they help

T. B. Falkowski (✉)
Department of Forestry, New Mexico Highlands University, Las Vegas, NM, USA
e-mail: tbfalkowski@nmhu.edu

S. A. W. Diemont
Department of Environmental and Forestry Biology, State University of New York, College of Environmental Science and Forestry, Syracuse, NY, USA
e-mail: sdiemont@esf.edu

 
R. P. Udawatta, S. Jose (eds.), *Agroforestry and Ecosystem Services*,
https://doi.org/10.1007/978-3-030-80060-4_14

**Table 1** Examples of CES adapted from the Millennium Ecosystem Assessment (2005)

| Cultural ecosystem service | Description |
|---|---|
| Cultural diversity | Cultural diversity, or the diversity of cultural expressions, is affected by and affects biological diversity (Pilgrim et al. 2009). It can contribute to resiliency of human societies (Turner et al. 2003) |
| Spiritual and religious value | Natural elements, ecosystems, and/or landscapes may hold religious or spiritual value(s) for many cultures (Albanese 1991) |
| Knowledge systems and education | Many cultures have developed complex knowledge, practice, and belief systems from observing local ecosystems (Berkes 2008). Ecosystems can also serve as a "living classroom" for study and scientific research (Falkowski et al. 2015) |
| Inspiration | Ecosystems can provide inspiration for art, architecture, technology, etc. (Carlson 2000; Shu et al. 2011) |
| Aesthetic value | Individuals and cultures perceive aesthetic beauty in natural elements, ecosystems, and/or landscapes (Cooper et al. 2016) |
| Social relationships, identity, and cohesion | Ecosystems define relationships between individuals both within and between communities. The identities individuals use to define themselves and their communities can be associated with resource management, religious belief, and cultural heritage, which are also provided by ecosystems (Clayton and Opotow 2003) |
| Recreation and tourism | Ecosystems provide opportunities for recreation and for nature tourism. Examples include bird-watching, photography, stargazing, camping, climbing, hiking, hunting, and fishing (Ceballos-Lascuráin 1996) |
| Therapeutic benefits; mental and emotional health | Ecosystems provide emotional and mental health benefits, including stress reduction (Buzzell and Chalquist 2009) |
| Cultural heritage and sense of place | Many cultures are closely tied to the places where they developed. These natural landscapes become imbued with cultural and historical meanings that are passed between generations and maintained through customary practices and social institutions, thereby contributing to their cultural identity (Berkes 2008) |

enable, and the capabilities they help equip." Alternately, Chan et al. (2012) phrase their definition as "ecosystems' contributions to the nonmaterial benefits … that humans derive from human-ecological interactions."

Despite differences in classification schemes and definitions, CES are generally considered to provide humans with intangible, constructed benefits resulting from the interactions between sociocultural and environmental systems, such as social cohesion, cultural identity, mental and physical health, and intellectual and spiritual stimulation (Daniel et al. 2012; Milcu et al. 2013) (Table 1). The interconnectedness of these two spaces makes a socioecological framework useful for understanding CES. The environment enables development, expression, and maintenance of cultural practices, which in turn shapes the structure and function of the environment (Fish et al. 2016). For example, ecosystems support wildlife which local people can hunt. Hunting can be a valuable cultural practice that creates community, shapes identity, and provides recreational opportunities, or hunting can be the focus of other cultural practices, such as dances and legends. These practices can shape the environment by affecting wildlife populations. Environmental changes, such as population shifts, can affect practice, as is in the case of taboos that limit hunting during particular seasons to maintain resources.

The Millennium Ecosystem Assessment reported that 70% of CES are being degraded or used unsustainably worldwide, largely as a result of land cover change. Many of these CES are not substitutable or replaceable. This decline reduces the benefits humans obtain from CES and negatively influences other ES, such as regulating and supporting ES (Millennium Ecosystem Assessment 2005). For example, sacred groves in India are being deforested as perceived economic value of land exceeds its religious value (Chandrakanth et al. 2004; Osuri et al. 2014). This conversion not only impacts spiritual benefits from the groves, which may be preserved as spaces for deities, but also reduces water retention (regulating) and wildlife habitat (supporting) services (Chandran and Hughes 1997; Bhagwat et al. 2005a, b). This conversion further reduces CES such as religious identity and social cohesion in communities (Kandari et al. 2014; Tilliger et al. 2015; Wehi and Lord 2017).

Long-term traditions associated with CES, such as cultural identity and aesthetic appreciation, can slow environmental degradation and land conversion (Sneed et al. 2013). CES tend to hold deep value for stakeholders and thereby serve as an important way of relating to nature, facilitating support for environmental protection and stewardship (Chan et al. 2012; Daniel et al. 2012; Fish et al. 2016). For example, indigenous land management practices and traditional governance structures have consistently and effectively limited deforestation in the Brazilian Amazon. Even though they are often located in frontier zones with high deforestation rates, indigenous reserves have inhibited deforestation within their traditional lands as effectively as strict, non-extractive reserves and parks, underscoring the importance of maintaining traditional cultural practices and sovereignty (Schwartzman et al. 2000; Nepstad et al. 2006).

## *Assessing CES*

CES are rarely measured directly because they are typically intangible. Instead, proxy indicators provide indirect CES assessments. Hirons et al. (2016) provide a comprehensive overview of CES assessments, including shadow and hedonic pricing, anthropological methods and participatory GIS, and narrative and artistic methods. These methods can be quantitative or qualitative, be monetary or nonmonetary, ignore or involve stakeholders, and be spatially explicit or implicit. Care must be taken to choose the appropriate CES method for a particular socioecological context and research objective. For instance, using photos to consider the value of peoples' visual perceptions of landscapes as ecosystem services is useful for gauging aesthetic cultural ecosystem services. This qualitative method can be spatially explicit if photos are georeferenced. However, this approach may be biased toward sites that are easily accessible, and while it can be made quantitative by counting the number of photographs taken of a particular site, it cannot assess the quality or importance of the aesthetic ecosystem service to different stakeholders. It is also difficult to quantify this metric monetarily, which may be desirable in some decision-making frameworks (Kelemen et al. 2015; Hirons et al. 2016).

Several barriers have limited the integration of CES into decision-making. First, the concept of "culture" itself is fluid and open to interpretation (Satz et al. 2013). While this does not preclude the incorporation of CES into comprehensive frameworks, decision makers must be clear as to how they are defining CES and their benefits. The abstract and intangible nature of CES makes them difficult to classify and measure for decision-making. Furthermore, while CES have value, most are not easily monetized. It is also debated whether they should be quantified in economic terms even if it were possible to do so with precision. Economically valuing CES may result in the commodification and undervaluation of services that are often described as indescribable and priceless (Milcu et al. 2013; Satz et al. 2013; Fish et al. 2016). This risk leads to concerns of incommensurability between CES and other ES. Although this problem can be addressed using deliberative approaches, it precludes tidy decision-making procedures (Chan et al. 2012).

The complex feedback between environmental spaces and cultural practices makes implementing CES assessments difficult (Fish et al. 2016). The distinction between benefits, services, and values can be tenuous (Milcu et al. 2013). The values associated with CES may change over time and vary among stakeholder groups. Furthermore, CES may differ across spatiotemporal scales (Satz et al. 2013). As a result, they may hold different values within and between scales of social organization (e.g., individual, community, and society) (Chan et al. 2012). For example, while a backcountry hiker and farmer may have different perceptions of the aesthetics of a particular landscape, both may share a similar appreciation for the aesthetic CES provide to the society of which they are a part. Additionally, many CES overlap, which may lead to double counting. For example, traditional ecological knowledge (TEK) can be considered an education service or a cultural heritage service (Chan et al. 2012; Daniel et al. 2012).

The diversity of CES frameworks makes comparing CES results difficult (Costanza 2008). While some have argued that this lack of consistent and concrete frameworks has precluded their integration into holistic assessments of ES, many CES frameworks exist that could be used for this purpose (Chan et al. 2012; Gould et al. 2015; Felipe-Lucia et al. 2015; Fish et al. 2016). Limited CES assessment implementation may be due to perceived imprecision and intangibility or limited understanding of CES assessment methods. Several authors raised the second point, noting that research in CES tends to be based on social science methods such as ethnographic interviews and participatory mapping, underscoring the importance of collaborations between biophysical and social scientists (Milcu et al. 2013; Fish et al. 2016).

As a result, CES are considered less frequently than other ecosystem service categories in research (Hernández-Morcillo et al. 2013). Furthermore, CES tend to not be the primary focus of projects; more commonly, they are a secondary component of broader analyses. The difficulties associated with quantifying CES make valuing them in an economic context particularly challenging (Milcu et al. 2013).

More than half of the assessments that have considered CES focused on recreation and tourism, which is unsurprising given that it may be the most easily quantifiable and economically valued metric. Other CES, such as inspiration and religious and spiritual services, were only considered in a combined 10% of cases (Milcu et al. 2013; Hernández-Morcillo et al. 2013).

## *Cultural Ecosystem Services in Agroforests*

Proponents of agroforestry often argue that agroforests are sustainable in part because they are managed to provide multiple ecosystem services (Zhang et al. 2007; Jose 2009; Power 2010; Letcher et al. 2015). For example, Altieri and Toledo (2011) point out that peasant agroecosystems in Latin America place a high degree of importance on traditional knowledge, empower smallholder communities, serve as an opportunity for expression of often marginalized cultures, and integrate biophysical and social processes into management. Moreno et al. (2017) show that agroforests throughout Europe provide recreation, tourism, education, aesthetic beauty, and cultural heritage.

Indigenous peoples often note the importance of CES in their land management systems. TEK is a knowledge-practice-belief complex. Therefore the natural history of the region, their environmental management systems (e.g., agroforests), social institutions, and cultural practices are all nested and inextricably linked. In fact, the concept of natural gifts in many indigenous cultures as discussed by Kimmerer (2014) closely reflects the ecosystem service concept. Although indigenous worldviews may reject an anthropogenic perspective of nature solely as service provider, they do recognize themselves as part of a web of reciprocity between themselves and nature (Kimmerer 2011; Chan et al. 2012). Perhaps coincidentally, Díaz et al. (2015) use the term "natural gifts" to describe ecosystem services in establishing the framework for the Intergovernmental Platform on Biodiversity and Ecosystem Services.

That said, it is important to note that CES are not exclusive to traditional or indigenous agroforestry systems. Because traditional and nontraditional agroforestry systems apply the same principles of socioecological organization and management, they both may offer the benefits of social cohesion, heritage, recreation, aesthetic beauty, education, and inspiration to communities around the world. A distinction between CES provided by traditional and nontraditional agroforests is the spiritual component, which may not be prioritized in the latter.

Despite their importance, CES are rarely considered in ecosystem service assessments in agroforestry systems, mirroring the trends described in section "Cultural Ecosystem Services in Agroforests" (Tengberg et al. 2012; Tilliger et al. 2015). For example, in a special issue dedicated to ecosystem services in the journal *Agroforestry Systems*, only 1 of 19 articles considered CES (Jose 2009). The ecosystem service assessment tool (Tsonkova et al. 2014) for agroforests lacks any mention of CES. We were able to identify only three manuscripts that explicitly assessed CES in agroforests (Langenberger et al. 2009; Calvet-Mir et al. 2012; Moreno et al. 2017).

This limited consideration of CES is particularly distressing given the positive feedback between environmental degradation in agroecosystems and loss of the CES they provide (Tilliger et al. 2015). CES are often one of the primary management objectives in agroforests due to their high value for individual land managers and communities (Barrena et al. 2014). Calvet-Mir et al. (2012) found that CES, such as relaxation, aesthetic beauty, and cultural heritage, are the most valued ecosystem services for both scientists and practitioners in their study of homegardens in Spain. Cultural, regulating, and supporting ecosystem services provided by agroforests are also highly correlated, so the loss of CES will likely deleteriously affect other ES (Calvet-Mir et al. 2012; Tilliger et al. 2015).

## Case Studies

We have selected agroforests from around the world that have been developed in a wide range of sociocultural and environmental contexts in order to illustrate CES of agroforestry systems. After describing each agroforestry system and how its function influences the CES it provides, we describe how socioecological changes have affected how it is managed, its CES, and the feedback between the two. We do not advocate for or evaluate any particular framework for assessing CES as it would be inappropriate to do so without primary data and a deeper understanding of each of these systems. After presenting these case studies, we will highlight some common themes elucidated from this overview, which can inform recommendations about future work regarding CES in agroforestry research.

### *Lacandon Maya Milpa: Chiapas, Mexico*

#### Description and History

The Lacandon Maya likely settled in the humid lowlands of southwestern Mexico more than 2–3 centuries ago after fleeing the Yucatan Peninsula of Mexico after the Spanish conquest (Palka 2005). They adapted to local environmental conditions by developing a swidden, successional agroforestry system, which has been called the *milpa* cycle (Ford and Nigh 2016).

At the end of the dry season around March and April, the Lacandon agroforestry cycle (Fig. 1) is initiated by the farmer slashing patches of vegetation typically measuring less than two hectares. Farmers will preferentially clear secondary forest vegetation as opposed to mature forest, which is maintained as a source of many ecosystem services, including seed rain and wildlife habitat and provisioning game, timber, firewood, and wild edible plants (Nations and Nigh 1980). Farmers then

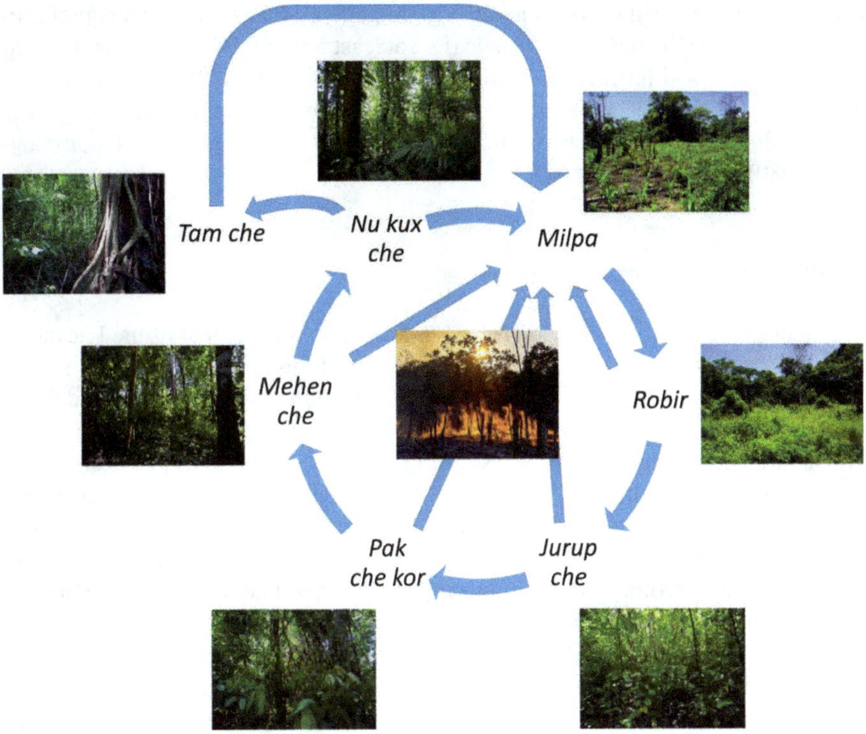


**Fig. 1** Diagram of the Lacandon Maya milpa cycle

burn the slash just before the onset of the rainy season in May. This burn creates biochar, which serves as a soil amendment (Nigh and Diemont 2013).

Lacandon farmers cultivate a diverse polyculture of crops in the first stage of the *milpa* cycle. This stage, called *milpa* in Spanish or *kor* in Lacandon Mayan, is dominated by *Zea mays* but can contain between 50 and 100 different crop species and cultivars, including both annual herbaceous and perennial tree crops (Falkowski et al. 2019b). After 3–5 years, *milpa* production begins to decline due to declining soil fertility and increased weed populations. At this point, farmers allow the plot to go fallow (Diemont and Martin 2009). While fallowing connotes a lack of management, Lacandon farmers actively manage these stages, although not to the same degree as *milpas*. For example, Lacandon farmers plant or clear ruderal herbaceous vegetation around naturally occurring tree seedlings of slow-growing species just before fallowing a *milpa* plot. They encourage the growth of these species because they are valuable for either provisioning (e.g., timber) or regulating (e.g., soil fertility enhancement) services they provide in latter successional stages. Lacandon farmers recognize several distinct stages in the fallow period based upon a suite of physical characteristics, such as canopy cover, biomass, dominant plant species, and light transmission. These include, in order, *robir, jurup che, pak che kor, mehen che,* and *nu kux che.* Fallow periods can last from 2 to 60 years before the plot is slashed and burned again (Falkowski et al. 2019a). In general, Lacandon farmers prefer to wait until at least the *pak che kor* stage (i.e., at least 5 years) to make *milpa* again in order to restore soil fertility after cultivation (Falkowski et al. 2016). In addition to actively managing fallow succession, Lacandon farmers hunt, fish, and gather medicinal and edible plants from these secondary forest stages (Nations and Nigh 1980).

## Cultural Ecosystem Services

The management of traditional agroforestry systems, such as that of the Lacandon Maya, is often imbued with cultural meaning. Agroforests have historically served as infrastructure for TEK education in Lacandon communities. TEK is passed down from generation to generation as parents and grandparents guide their children and grandchildren in managing agroforests. In the process, Lacandon youth learn about the natural history of the region, as well as traditional agroecological management (Falkowski et al. 2015). In addition to the education CES provide to Lacandon communities, Lacandon farmers have actively collaborated with researchers to study ecology in their agroforests (Diemont and Martin 2009; Falkowski et al. 2016).

Lacandon agroforests also provide therapeutic and aesthetic CES. Lacandon farmers sometimes plant particular flowering species in their *milpas* in part because they are beautiful. Farmers remark that they enjoy spending time in the forest because it is enjoyable and "tranquil" (Adolfo Chankin, personal communication, July 2017). Tourists are attracted to the region due to its natural beauty and the unique cultural history of the Lacandon. They learn about forest ecosystems and Lacandon Maya history while visiting. Many Lacandon families are increasingly reliant upon the income associated with ecocultural tourism (van den Berghe 1995).

According to Alcorn and Toldeo (1998), *milpa* is not exclusively—or even primarily—a spatial concept defined as an agricultural production system. It is a social institution and a process that is encoded in a cultural script, or an internalized plan used to make decisions given cultural and social constraints. These cultural scripts are transmitted between generations through legends, beliefs, and social events. Thus, traditional Maya culture influences land management systems and vice versa. Rodas et al. (1940) said that the Maya "do not raise maize to live, they live to raise maize." Nigh (1976) noted that "… the making of *milpa* is the central, most sacred act, one which binds together the family, the community, the universe … *milpa* forms the core institution of Indian society in Mesoamerica and its religious and social importance often appear to exceed its nutritional and economic importance." Maintaining *milpa* agroforests integrates Lacandon smallholders into a network of reciprocity that ensures assistance in times of social, economic, or ecological stress. It is also associated with social status and a fundamental component of Lacandon cultural identity and heritage (Alcorn and Toldeo 1998).

Particular agroforestry management events are marked by religious ceremonies (Alcorn and Toldeo 1998). Many of the materials used for these events are obtained from the agroforestry system itself. For example, *balche* is a ceremonial beverage made by fermenting honey and the sap from *Lonchocarpus* spp. trees. Copal—an aromatic resin from the *Protium copal* tree—was traditionally burned as an offering to the gods. Many gods in the traditional Lacandon pantheon were associated with nature and with agroforests (e.g., the god of corn). Given that legends and stories are often metaphors encoding these scripts, elements of *milpa* agroforestry management permeate Maya mythology and cosmology. For example, according to the Popol Vuh—the Maya creation story—humans are made from maize. *Ceiba pentandra* is the axis mundi that connects the underworld (*Xibalba*), terrestrial world, and celestial world, as well as being tree species that grew at the site where humans were created. To this end, *C. pentandra* trees are often maintained by Lacandon in advanced forest (*tam che*) stands (McGee 2002). In a legend indicative of the connection of gods to Maya agroforestry, the wind god, Chäk Ik Al, rendered a strong wind that destroyed the forest. The creator god, Hachäkyum, who was displeased with his creation, then burned the felled trees. Chäk Ik Al brought a strong storm that inundated the world. The only survivors were the people and plants Akinchob, the god of the *milpa*, placed in a canoe. This people, the ancestors of the Lacandon, then repopulated the earth and planted their *milpas* (McGee 1990). This legend mirrors the process for making a *milpa*, wherein farmers fell vegetation, burned the slash, and planted their crops at the onset of the rainy season. Thus, cultivating *milpa* is a sacred act commemorating creation itself (McGee 2002).

## Socioecological Changes

Socioecological changes in the last decades have fundamentally altered the way Lacandon Maya value CES, and led to similar changes in their agroforestry management and livelihood strategies. Immigration to the Lacandon region spurred by land reforms increased population density and development of the Lacandon rainforest throughout much of the twentieth century. Lacandon territory decreased because of deforestation, expanding from logging roads and newly established agricultural settlements, as well as population declines caused by outbreaks of diseases to which the Lacandon had not been exposed. Due to these reasons and government resettlement, they were clustered together in more centralized communities (Perera and Bruce 1986; Boremanse 1998; McGee 2002).

The Lacandon largely abandoned their traditional religion by the early 1990s as missionaries converted young people and the older generation died, taking their traditions and rituals with them (McGee 2002; Palka 2005). McGee (2002) points to three main causes for the decline of the traditional Lacandon Maya religion: a decline in the necessity of healing rituals with increasing access to modern medicine, a shift from traditional subsistence agriculture to a tourism-based economy, and the introduction of Western institutions and technologies, namely television and primary schools, which facilitated a growing divide between younger and older generations.

The weakness of the peso made Mexico an attractive destination for international tourists. Some Lacandon capitalized on this tourist boom by selling souvenirs at the nearby ruins of Palenque and Bonampak. Tourists also began to travel to the Lacandon communities along newly constructed roads. Some residents built lodges and restaurants to meet the growing demand for tourist infrastructure (McGee 2002). This shift away from subsistence agriculture to a tourism-based market economy has had profound changes on Lacandon Maya culture and TEK.

Lacandon communities transitioned from a subsistence-oriented economy to one based on income from tourists. The income from providing souvenirs, room, and board for tourists is far greater than can be earned by maintaining traditional agroforests, so many people have abandoned agroforestry. The shift away from traditional healing and agricultural practices obviated the need for healing rituals and asking for bountiful harvests, so younger Lacandon saw little need to practice them (McGee 2002).

The income from tourism allowed Lacandon to purchase nonlocal goods and changed the socioecological structure. After attending school (where they are taught in Spanish, but not Lacandon Mayan), children often play on the computer or watch television instead of working alongside their parents in agroforests. The proliferation of purchased goods in Lacandon communities has resulted in shifting perspectives on social standing. Increasingly, material wealth is the indicator of social status as opposed to effective *milpa* cultivation and wisdom acquired with age. As opposed to older Lacandon who tried to maintain their traditional lifeways, younger generations tend to seek material wealth, providing them with more social capital. This change has led to tensions between the younger and older generations (Valle-García 2014). Men typically earn more money through tourism. The ability to purchase products such as store-bought clothes and commercial food products depreciated products made by women, such as clothing and food, increasing the gap in power between the genders (McGee 2002). Finally, certain families have profited more from tourism than others, leading to tension between families and socioeconomic disparity (Valle-García 2014).

The increasing role of tourism in the local economy has changed Lacandon traditional agroforestry management. Few farmers still manage traditional *milpas*. Even if they did not abandon agriculture altogether, they have less time to manage their *milpas*. To compensate for lost labor, farmers may reduce the diversity of their *milpas* to ease management; hire additional workers; or add chemical fertilizers, herbicides, and pesticides (McGee 2002).

This case study exemplifies how changing socioeconomic conditions drive changes in agroforestry management and CES valuation, accelerating and/or magnifying socioecological change. That said, the integrity of traditional Lacandon culture, and *milpa* agroforestry in particular, has helped maintain practices which may have eroded faster. Tourism serves as a double-edged sword in this situation. On the one hand, tourism in Lacandon communities has encouraged traditional practices to be maintained, such as wearing traditional dress. On the other, it has also contributed to a commodification and abandonment of some cultural practices at the expense of others (van den Berghe 1995).

## *VAC Homegardens: Vietnam*

### Description and History

Many Vietnamese smallholders cultivate diverse homegarden agroforests. Households manage homegardens to mimic the structure of the surrounding natural ecosystems. While their primary function is to provide provisioning ecosystem services such as foods and medicines, they also often hold cultural and social significance. Homegardens are an example of a socioecological system that includes a household, the community to which it belongs, surrounding ecosystems, and the plants and animals incorporated into the homegarden itself (Kumar and Nair 2010).

One of the most common homegardens in Vietnam is the *Vườn Ao Chuồng* (VAC), which translates to garden-pond-livestock. This system likely originated in the rich soils of the Red River Delta in northern Vietnam and subsequently spread throughout Vietnam in the later half of the twentieth century. Homegardens served a critical role in ensuring subsistence for rural smallholders during the wars with France and Vietnam during this period. The spread of VAC systems can be largely attributed to the communist government's support for small-scale integrated agroecosystems in an attempt to improve food security for rural smallholders (Luu 1992). The campaign resulted in a dramatic increase in homegarden cultivation. Today, up to 90% of rural families maintain some form of homegarden, and approximately 44% of all households maintain the complete VAC system, which consists of gardens, livestock, and aquaculture. On average, these systems provide 30–60% of rural families' income and most of their subsistence (Mohri et al. 2013) (Fig. 2).

In general, homes are situated near the pond for easy disposal of domestic and kitchen waste, which is drained into the water to support stocked fish populations and aquatic vegetation which in turn supports ducks. Households plant a diverse polyculture of crops in the garden, including annual crops (e.g., sweet potato and sugarcane) and fruit trees (e.g., orange, banana, and apricot trees). Many of these plants are cultivated either for food or for traditional medicine. Families fertilize their crops with livestock manure and pond silt. They use kitchen scraps and weeds to feed poultry and pigs (Luu 1992; Mohri et al. 2013).

VAC system design and management are adapted to local conditions, including topographical, economic, ecological, and cultural factors. Trinh et al. (2003)

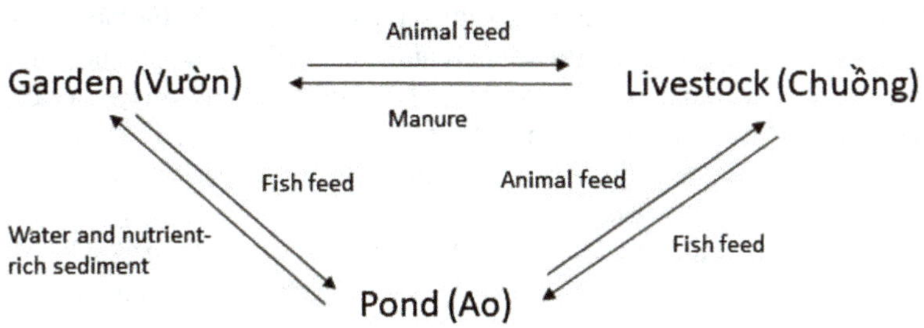


**Fig. 2** Diagram of VAC system components and exchanges. Adapted from Thanh (2010)

identified four different kinds of VAC systems depending on their location and how resources are managed. These include fruit trees in southern Vietnam, aquaculture ponds and livestock in the Red River Delta and central Vietnam, vegetables in the Red River Delta and central Vietnam, and forest trees throughout the country.

## Cultural Ecosystem Services

In addition to their importance for rural smallholders as sources of income and subsistence, VAC systems provide many CES. While still nascent, there is a growing body of research regarding the ecological function and impacts of VAC agroecosystems (Trinh et al. 2003; Sekhar 2007; Vlkova et al. 2011; Nguyen et al. 2013b; Mohri et al. 2013). Furthermore, these VAC systems are the result of and embody centuries of accumulated traditional ecological and cultural knowledge through maintenance of the agroecosystem itself.

Prior to the *Đổi Mới* Renovation in 1986, which led to a drastic and rapid loosening of government regulations, farmers were not given a choice in crop selection and farmland was collectivized. However, farmers were permitted to maintain small, private homegardens for subsistence (Mohri et al. 2013). As such, crop selection in VAC homegardens is mediated strongly by cultural practices. Trinh et al. (2003) describe how *Areca catechu* fruit, *Piper betle* leaf, and lime (*Citrus* spp.) are often grown and consumed together, particularly at weddings when the combination represents the union between husband and wife. This combination of crops has been metaphorically encoded into folklore and legends, both underscoring their importance and describing how they are to be managed and used. They are also important for religious and spiritual experiences due to their role as gifts and offerings during festivals and weddings. During the Vietnamese New Year (*Tết*) celebrations, household shrines must include five fruits that symbolize elements of Vietnamese Buddhism: *Musa* spp. *Citrofortunella microcarpa* or *Citrus* spp., *Capsicum* sp. or *Vitis vinifera*, *Citrus grandis*, and *Pyrus pyrifolia* (He 1991). Other crops utilized for cultural purposes during *Tết* festivities include *Momordica cochinchinensis* (used to dye rice red), *Phrynium parviflorum* (used to wrap rice cakes), and *Saccharum officinarum* (placed by doorways to prevent evil spirits from entering the home). These crops are widespread and commonly cultivated in traditional VAC homegardens (Hodel et al. 1999).

VAC agroecosystems provide therapeutic and aesthetic CES. Farmers plant certain trees because they are beautiful (e.g., orange tree and acacia) and note that they like to spend time socializing and relaxing in their homegardens (Vlkova et al. 2011; Nguyen et al. 2013b). The growing importance of tourism to the Vietnamese economy may spur further development of ethnotourism in regions where homegarden cultivation is central to the cultural expression ethnic minority groups (Sekhar 2007; Vlkova et al. 2011; Shih and Do 2016). This change may be underway given the prevalence of homestays throughout rural tourist areas in Vietnam. Ecotourism development itself can lead to changes in agroecological management and cultural expression (Cochrane 2008; Kontogeorgopoulos et al. 2015).

VAC homegarden cultivation and management are a source of cultural identity for rural smallholders in Vietnam, including many ethnic minority groups such as the Nung and H'mong (Sekhar 2007). Farmers express pride in their management acumen. Ethnic minority homegardens are also typically managed differently from those of the Kinh majority, illustrating the cultural differences between the two groups. Homegardens of ethnic minorities typically contain more medicinal plants, but the vegetable crops are less diverse than Kinh homegardens. This difference is in part due to the residence patterns of ethnic minority groups, most of which live near forests in the midland and mountainous areas of northern and central Vietnam, allowing them to collect food from these unmanaged ecosystems. The homegardens of ethnic minorities generally contain fewer commercial crop varieties and more crops that are used in regional traditional cuisine, in part because their communities are commonly located further from market hubs and dense population centers (Trinh et al. 2003).

## Socioecological Changes

VAC management continues to evolve as new crops and forms of resource management are integrated into the system. Rice cultivation, forestry, and biogas production are increasingly incorporated into the VAC. Changes in management, composition, structure, and function come from local government support to foreign investment in agriculture in Vietnam. VAC systems have consequently shifted from subsistence and culturally important crops to market-oriented resource management (Mohri et al. 2013). This shift has been attributed to the *Đổi Mới* renovation and decollectivization policies of the late 1980s and early 1990s, which opened Vietnamese markets to increased foreign investment. Before these policies, homegardens were more necessary for subsistence; farmers diversified their VAC to increase their resilience to environmental stochasticity. With open markets homegarden cultivation is increasingly becoming an opportunity to produce economically valuable crops, such as *Arachis hypogaea, Acacia* spp., and *Hevea brasiliensis* (Sekhar 2007).

Despite these market changes, the biodiversity within VACs today still remains high. It is unclear, however, whether traditions will be maintained along with culturally important crops, or reductions in these crops will erode cultural practice. Sekhar (2007) found that commercialized homegarden agroforests contained less than half the plant species of traditional systems. He also observed that they were managed more intensively with shorter fallow periods and increased fertilizer inputs. Alternatively, Fey (1989) found that commercialized homegardens were more diverse than subsistence homegardens, suggesting that homegardens may still be a refugia for culturally important species that are not economically valuable. However, this finding only takes into account species richness, not plant community composition. Increased commercialization of VAC systems is associated with exacerbated economic inequality, reduced use of traditional medicinal plants, and increased land privatization and fragmentation (Trinh et al. 2003; Mohri et al. 2013).

Increasing population density and growing economic inequality between rural and urban populations have also resulted in urban migration as rural families abandon their farms and homegardens to pursue more economically profitable activities in cities. While many former rural residents maintain customs and cultural practices, their cultural expressions are shifting with changes in livelihoods, physical environment, and social community. Traditional medicine is still practiced in cities (Albala 2014). Religious beliefs are generally transferable, so culturally important crops are still used in traditional rites and ceremonies in urban environments (Mazumdar and Mazumdar 2012). However, urbanization has reconfigured family structures in Vietnam, commonly leading to the breakdown of support networks (Barbiéri and Bélanger 2009). Urbanization, cultural shifts, and increasing income have increased meat consumption in Vietnam, thereby increasing the value and management intensity of the livestock component of VAC systems outside cities (Albala 2014; Hansen and Jakobsen 2020). Fast food is also becoming an increasingly staple component of urban Vietnamese diets, replacing traditional home-cooked food (Baumann 2006). In turn, this has resulted in marked increases in cardiovascular disease, obesity, and other diet-related health problems which compound health issues associated with poor air and water quality in urban environments (Cuong et al. 2007; Lâm et al. 2011; Nguyen et al. 2013a; Kien et al. 2017).

## *Rubber Homegardens: Brazil*

### Description and History

Rubber tappers who reside in the Brazilian Amazon share a cultural identity originally centered on their common history as peasant laborers for rubber estate owners during the nineteenth century (Weinstein 1983). Many rubber tappers were and are *caboclos*: Amazonian mestizos of mixed indigenous and European ancestry. While many rubber barons abandoned their land following the collapse of the Brazilian rubber industry after World War II, rubber tappers, or *seringueiros*, remained and continued small-scale rubber tapping operations.

Rubber cultivation in the Brazilian Amazon ranges in management intensity from forests with a high percentage of naturally occurring *Hevea brasiliensis* trees to intentionally planted rubber agroforests (Murrieta and Rueda 1995). Leaf blight (*Microcyclus ulei*), which is endemic to South America and decimates *H. brasiliensis* plantations, precludes the development of extensive rubber plantations in Amazonia (Gouyon et al. 1993). Despite this, smallholders have long planted relatively small rubber tree agroforest groves or supplemented natural *H. brasiliensis* stands with additional trees for their latex and edible seeds (Schurz et al. 1925) (Fig. 3).

*H. brasiliensis* is also often a dominant component of smallholder homegardens. Rubber agroforests are generally swidden agroecosystems in which *H. brasiliensis* seeds are planted between annual crops. The annual crops are typically cultivated

**Fig. 3** Brazilian homegarden with cultivated rubber trees (*H. brasiliensis*), cupuaçu (*Theobroma grandiflorum*), and açai (*Euterpe oleracea*). *H. brasiliensis* is the large stem at the center of the photograph. Image courtesy of Goetz Schroth and originally published in Schroth et al. (2003)

for approximately 2 years, after which point the plot is left fallow. Many farmers also extract timber and non-timber forest products besides rubber from these agroforests (Schroth et al. 2003). Other forms of agroforestry management are also common in the region, such as intercropped black pepper/orange agroforests (Smith et al. 1996).

## Cultural Ecosystem Services

Rubber tapping has historically been at the core of the local economy and has indelibly influenced the culture (e.g., music and legends) and social structures (Vadjunec et al. 2011; Gomes et al. 2012). In the 1960s and 1970s, the Brazilian Government enacted policies to encourage colonization and development of the Amazon frontier, including selling lands informally owned by *seringueiros* to wealthy ranchers from southern Brazil. In response, the rubber tappers' identity shifted to emphasizing the sustainability of their resource management systems, especially relative to cattle ranching. They formed trade unions which allowed them to collectively fight for their rights to the land and continued resource management. These social institutions were both a product and source of common rubber tapper identity. In this way, identity as *seringueiros*—once a source of socioeconomic stigma—became a globally recognized symbol of environmental stewardship and badge of honor (Gomes et al., 2012).

Agroforest homegarden plant communities are seen as being parts of kin networks, and different plants have unique histories which are in turn tied to particular uses and characteristics. While primarily cultivated for provisioning ecosystem services, *caboclo* homegardens are also maintained for their aesthetic beauty. Ornamental plants such as *Rosa* spp. are often included in homegardens. Homegarden management is also an expression of the syncretism that typifies *caboclo* religious worldviews (WinklerPrins and De Souza 2005). Other plants, such as *Jatropha gossypiifolia*, are cultivated for use in traditional plant-based medicines. Others still are selected for their uses in syncretic religious practices that incorporate elements of traditional native religions and Catholicism (Miller et al. 2006).

*Seringueiro* culture and social institutions, which are both predicated upon historic rubber tapping, represent a deep and long-term understanding of *H. brasiliensis* physiology, ecology, and management which has been passed down from generation to generation for centuries (Schroth et al. 2004). More recently, there has been some renewed research interest in traditional *seringueiro* management because their cultivation and extraction techniques seem to sustain rubber production to a greater degree than industrial methods in Southeast Asia. While most research regarding rubber agroforestry has centered on agroecosystems in southeast Asia, which dominates global rubber production, there has been a great deal of research on rubber cultivation and management in Brazil historically given that *H. brasiliensis* is native to the region.

Finally, the region is well known for its biodiversity and protected areas. This conservation ethic, combined with the global support for rubber tappers during clashes with cattle ranchers in the 1980s, can help draw tourists to the region's extractive reserves and rubber agroforests (Schroth et al. 2004).

## Socioecological Changes

Rubber tapper heritage and identity continue to adapt to social, political, and economic changes. The Brazilian federal government has recently cut rubber subsidies, global rubber prices have been declining steadily for at least the past two decades, and the Amazon frontier is becoming increasingly integrated with national and international markets (Hoelle 2011; Gomes et al. 2012). As the economic viability of rubber tapping declines, *seringueiros* have increasingly adopted agriculture and cattle ranching. This shift contravenes their own identity as forest stewards of their extractive reserves, as it is associated with environmental degradation and deforestation, as well as being at the root of their conflict with cattle ranchers in the 1970s and 1980s (Hoelle 2011; Gomes et al. 2012). Because the activities of rubber tapping and forest management are fundamental to *seringueiro* identity, many *seringueiros* were emotionally impacted by this change. One rubber tapper commented, "We all became sad and didn't know what to do." However, this reaction was mixed. "Life is better now … I would do exactly what I am doing now if the price of rubber improved," said another former rubber tapper (Salisbury and Schmink 2007).

Traditional definitions of *seringueiro* identity no longer apply as a result of these lifestyle shifts. While many local residents still identify as rubber tappers, this self-identification is not inherently associated with resource management, but rather historical occupancy and participation in social organizations (Salisbury and Schmink 2007; Vadjunec et al. 2011; Hoelle 2011). Only 33% of households Vadjunec et al. (2011) interviewed stated that rubber tapper identity is contingent upon practicing rubber tapping management. These views are not necessarily homogeneous, and a great diversity exists in individuals' reasoning for self-identifying as rubber tappers (Gomes et al. 2012). The increase in cattle ranching among *seringueiros* has caused tensions in communities historically unified by a common resource management identity. While some of those who still tap rubber understand the motives of community members who have transitioned to cattle ranching, others see it as a betrayal. Cattle ownership is generally seen as a status symbol, exacerbating tensions between community members as rubber tapping is associated with poverty and lack of education (Salisbury and Schmink 2007; Vadjunec et al. 2011; Hoelle 2011).

In addition to changing parameters of identity, the changes in land management have impacted the expression of *seringueiro* culture. For example, country music, rodeos, and Western cowboy attire are increasingly common and popular in historically rubber tapping regions. Meat is increasingly central to the diet (Hoelle 2011; Gomes et al. 2012). Thus, CES of identity and heritage provided by rubber agroforests are being replaced by those provided by cattle ranches and cowboy culture.

In addition to the changes in rural livelihoods, urbanization is shifting demographics in the Brazilian Amazon. Many *caboclo* rural migrants continue to manage homegardens in cities as a tie to their cultural heritage and to supplement their diets and incomes. They exchange garden products in a "network of giving" that is more than just an informal market that ensures food security. It also strengthens social ties and promotes a feeling of well-being and affection. For example, individuals who remain in rural areas but visit family in urban environments bring goods that cannot be produced in urban homegardens. This exchange ensures dietary diversity for urban residents and a sense of familial connection for isolated rural residents (WinklerPrins and De Souza 2005). Thus, urban residents have adopted agroforest homegarden management to provide them with CES in a new environment.

Cultural traditions and social institutions allow for the enforcement of rules governing resource management. Therefore, it is questionable whether *seringueiro* communities will continue sustainable forest extractivism. Although permitting economic development and resource use through rubber tapping management has been shown to limit deforestation, continued socioeconomic pressures may combine with cultural trends to facilitate further expansion of cattle ranching among *seringueiro* communities. While CES associated with rubber tapping and rubber agroforestry management, such as cultural identity, may be replaceable with cattle culture, this shift could be associated with reductions in regulating and supporting ecosystem services. The flexible cultural boundary of this group makes them more open to changes that can either increase or decrease their ecological and cultural resilience (Berkes and Folke 1998).

## *Tree-Vine Vineyards: Portugal*

### Description and History

Ancient Greek viticulturalists grew grapevines along trees and trellises as high as 15 m using a technique known as "high vine," which was thought to be the source of the best wine (Thompson 1937). The Greeks introduced high vine viticulture to the Etruscans (Surico 2000), and the Romans subsequently transplanted their traditional vineyard management as they conquered other cultures throughout what is now Europe, including Portugal (Anderson 2000).

Modern-day Portuguese traditional vineyards are a mix of high vine management types that have trees and those that do not have trees to support the vines. High vines that are supported by concrete poles and metal wire rather than trees can be found throughout much of northwestern Portugal (Altieri and Nicholls 2002), while vineyard agroforestry management is patchily distributed and generally restricted to an area within 20 miles of the city of Braga in the Minho region (Altieri and Koohafkan 2004; Koohafkan and Altieri 2017). Even within this region, vineyards are dominated by high vines that do not include trees and conventional forms of viticulture that are not high vine at all (Fig. 4).

**Fig. 4** A vineyard agroforest in the Minho region of northwest Portugal

Vineyard agroforests are typically found in multifaceted family farms, most of whose land is dedicated to cultivating a polyculture of crops surrounded by a perimeter of trees. Farmers space these trees approximately 10 m apart, string several wires between them, and cultivate three or four grapevines at the base of each tree. As the vines grow, farmers interweave them within the tree branches and festoon them along the wire between the two trees, creating mixed foliage of grape and tree leaves. Historically, farmers have used numerous tree species, many of which also provide fruits or nuts, such as cherry, chestnut, and oak (Stanislawski 1970; Altieri and Nicholls 2002). Farmers pollard trees once a year between February and March, encouraging young branches to support new grapevine growth as the growing season progresses. Tree foliage fills in and becomes a dense cover for the grapevines before the hot and sunny summer months of July and August. Despite high temperatures of 23 °C and monthly precipitation in these months averaging only 40 mm over 2 days (NOAA n.d.), farmers do not irrigate these grapes, even as modern row single-species viticulture in the region requires regular irrigation.

Farmers employ tall ladders to reach the grapes for harvest in September. Traditionally, farmers and their families would stomp the grapes by foot to make wine. These traditional vineyard agroforests can produce more than 1000 L of low-proof effervescent wine, from what is essentially a living fence surrounding a 2 ha farm. Families consume the resulting table wine throughout the entire year.

## Cultural Ecosystem Services

Viticulture agroforestry has been traditionally a central part of family activity in the Minho region, and is integral to the family economy. The home, while not located within the agroforest, is typically within an easy walking distance. Daily management activities, such as weeding field crops within the vineyard agroforest, often involve the entire family, while annual activities, such as planting, pollarding, harvesting, and winemaking, involve extended families. Even as land is divided, extended families will share equipment and human resources for larger annual activities, such as pollarding, tilling the field, grape harvest, and winemaking. These agroforestry management practices serve to unify family around a shared activity, providing an opportunity for bolstering relationships within the nuclear family and with distant relations.

Vineyard agroforest landscapes are fairly open. Parcels have no divisions beyond the living fence trees and intertwined grapes that surround field crops. This openness contrasts with other private conventional vineyards in the area, many of which are surrounded by perimeter fencing and guarded by dogs. As a result, traditional vineyard agroforests provide space for relaxation and recreation for the general public. Visitors can stroll through between vineyard agroforest parcels, despite not being community members or members of the farming family.

The landscape of vineyard agroforests provides an important and unique agroecological aesthetic (Stanislawski 1970). During winter, the pollarded trees offset by the bare fields accentuate the quiet and cold of the season. After this bareness comes the activity of spring, as families begin to prepare the fields, and then the summer months, during which families and wildlife are active daily.

#### Socioecological Changes

Agroforestry viticulture around Braga has changed markedly over the past few decades. Although traditional family farms remain, they are increasingly sold to landowners from outside the community because most of the agroforests are maintained by older farmers, and members of younger generations are less interested in farming than their parents. Although new owners do not typically remove the trees after purchasing these farms, they rarely focus on traditional vineyard agroforest maintenance. Many fields are abandoned, and trees are rarely pollarded. Older grapevines are neither pruned nor replaced and no longer provide abundant grapes for winemaking as a result. Traditional winemaking techniques have been largely replaced by machinery, if not discontinued entirely. These changes are not entirely the result of local social changes, but also result from economic incentives.

Vineyard agroforests have been converted to row cropping vineyards with drip irrigation. According to interviewed farmers, various European Union agricultural incentive programs provide benefits to landowners who wish to modernize their traditional vineyards, which are considered to be less productive than those employing commercial grape-growing techniques. Trees are absent from these conventional monoculture viticulture systems. While traditional agroforestry vineyards require considerable labor, grapevines under row cropping require irrigation, infrastructure, and fuel resources (far above traditional viticulture). As a result, these row systems may sacrifice other ecosystem services (e.g., recreation, aesthetic), and may reduce the system's resilience to environmental change (including climate change) (Costa et al. 2016; Hannah et al. 2013; Viers et al. 2013).

## Conclusions and Recommendations

### *Common Themes*

Agroforests consistently provide a wide range of CES. These services are perceived as among the most valuable ES provided by agroforest ecosystems (Martín-López et al. 2012). While production-oriented rationales for agroforestry are no doubt important to agriculture, it is imperative to consider culture and other social factors as well.

The desire to maintain CES can promote sustainable agroecosystem management and limit environmental degradation. Just as biodiversity loss and environmental degradation are pressing global concerns, so too is the loss of cultural diversity. Indeed, many have argued that the two are inextricably linked (Díaz et al.

2006; Clark et al. 2014). Any attempts to address the former must engage the latter to be successful, which poses challenges associated with interdisciplinary and intercultural work involving multiple stakeholders. It also implies that addressing one issue offers an opportunity to address the other.

While CES are central to many cultures, the socioecological systems that create value for CES are open and must adapt to changes, which are not necessarily normative (Berkes and Folke 1998). Cultural heritage is a product of not only the past, but also how it is maintained, expressed, valued, and transmitted in modern society (Tengberg et al. 2012). While it can be argued that CES are not replaceable, the case studies presented here indicate that cultures find cultural value in changing ecosystems as well (Hirons et al. 2016). That said, while different ecosystems may both provide similar CES, more research is necessary to assess the nature of these ES and whether changes have also impacted the quantity or quality of services.

## *Framework for Assessing Cultural Ecosystem Services in Agroforests*

Brown et al. (2014) provide a general framework for developing ecosystem service indicator frameworks. The first step is identifying and consulting with stakeholders to determine management objectives. It is then useful to develop a conceptual model and determine key questions regarding potential indicators. Data acquisition can also be done collaboratively to provide local stakeholders with a vested interest in the work. This step is critical in cases of CES which are the product of stakeholder interactions with the environment. After indicators are calculated, findings should be broadly communicated so that the indicators can be evaluated and refined with stakeholders to ensure accuracy and comprehensiveness. Because all of these steps involve local stakeholders, building positive relationships is critical.

Calvet-Mir et al. (2012), Barrena et al. (2014), Nahuelhual et al. (2014), Tilliger et al. (2015), and Tengberg et al. (2012) describe additional methods that aim to explicitly assess CES in agroecosystems. The interdisciplinary concept of cultural landscapes, which is well established in land-use science, social sciences, humanities, and paleoecology, may offer a useful framework for integrating cultural services into broader assessments of ecosystem services. Cultural landscape research includes methods for assessing and valuing CES at multiple spatiotemporal scales using participatory research, historical land-use analysis, ethnographic surveys, and spatial analysis. However, the cultural landscape research community seems primarily focused on historical assessments of cultural services, which risks overlooking how persisting systems are adapting to modern changes (Schaich et al. 2010).

Researchers must undertake CES assessments with cultural sensitivity and attention to nuance. The CES framework has the potential to integrate multiple disciplines and epistemologies in identifying important factors that sustain socioecological systems However, if implemented carelessly and callously, it can also be used to further marginalize the stakeholders, ecosystems, and services they aim to assess and protect (Hirons et al. 2016).

Although CES are currently relegated to the periphery of most ecosystem service assessments, the number of studies incorporating CES is growing. This new body of literature offers many new frameworks for assessing CES and incorporating them into decision-making processes. While this change is admirable, an overemphasis on quantification and placing multiple ecosystem services into a single scale for ease of comparison may obfuscate that ecosystem services are a conceptual tool that facilitate holistic exploration of socioecological systems and the way humans relate to nature. While CES present certain challenges to incorporation into comprehensive ES assessments, their fundamental role in socioecological systems makes them critical to consider in some way, even if it is imprecise or indefinite.

## *The Future of Cultural Ecosystem Services in Agroforests*

A future scenario in which CES are increased according to the Millennium Ecosystem Assessment (2005) is the "adapting mosaic," in which watershed-scale landscapes are the basic socioeconomic unit. Local institutions are strengthened so as to improve the collective understanding of local ecosystem function and sustainable management. Emphasis of economic growth is replaced with steady-state economics focused on decreasing economic inequality, stabilizing population, and restricting economic expansion (Daly 1991). Sociocultural and biological diversity is emphasized, maintained, and celebrated in order to ensure resilience of socioecological systems in the face of change. Local institutions are connected through socioeconomic networks to share knowledge and resources in addressing socioecological problems (Millennium Ecosystem Assessment 2005).

This narrative underscores the importance of maintaining ecocultural diversity. Agroforestry management offers a way in which all three pillars of sustainability—social, economic, and environmental—can be achieved. While CES are under threat due to cultural, economic, and environmental homogenization associated with globalization, they also offer a potential way to minimize the negative impacts associated with these trends.

## References

Albala K (ed) (2014) From famine to fast food: nutrition, diet, and concepts of health around the world. Greenwood, Santa Barbara, CA

Albanese CL (1991) Nature religion in America: from the Algonkian Indians to the new age. Nachdr. Univ. of Chicago Press, Chicago, IL

Alcorn JB, Toldeo VM (1998) Resilient resource management in Mexico's forest ecosystems: the contribution of property rights. In: Berkes F, Folke C (eds) Linking social and ecological systems: management practices and social mechanisms for building resilience. Cambridge University Press, Cambridge, pp 216–249

Altieri MA, Koohafkan P (2004) Globally important ingenious agricultural heritage systems (GIAHS): extent, significance, and implications for development. In: Proceedings of the second international workshop and steering committee meeting for the globally important agricultural heritage systems (GIAHS) project. FAO, Rome, Italy, pp 7–9

Altieri MA, Nicholls CI (2002) The simplification of traditional vineyard based agroforests in northwestern Portugal: some ecological implications. Agrofor Syst 56(3):185–191

Altieri MA, Toledo VM (2011) The agroecological revolution in Latin America: rescuing nature, ensuring food sovereignty and empowering peasants. J Peasant Stud 38:587–612

Anderson JM (2000) The history of Portugal. Greenwood Press, Westport, CT

Barbiéri M, Bélanger D (eds) (2009) Reconfiguring families in contemporary Vietnam. Stanford University Press, Stanford, CA

Barrena J, Nahuelhual L, Báez A et al (2014) Valuing cultural ecosystem services: agricultural heritage in Chiloé Island, southern Chile. Ecosyst Serv 7:66–75

Baumann LC (2006) A training program for diabetes care in Vietnam. Diabetes Educ 32:189–194

Berkes F (2008) Sacred ecology, 2nd edn. Taylor & Francis, New York, NY

Berkes F, Folke C (eds) (1998) Linking social and ecological systems: management practices and social mechanisms for building resilience. Cambridge University Press, Cambridge

Bhagwat S, Kushalappa CG, Williams PH, Brown ND (2005a) The role of informal protected areas in maintaining bio-diversity in the Western Ghats of India. Ecol Soc 10(1):8

Bhagwat SA, Kushalappa CG, Williams PH, Brown ND (2005b) A landscape approach to biodiversity conservation of sacred groves in the Western Ghats of India. Conserv Biol 19:1853–1862

Boremanse D (1998) Hach Winik: the Lacandon Maya of Chiapas, Southern Mexico. Albany, NY, Institute for Mesoamerican Studies, University of Albany

Boyd J, Banzhaf S (2007) What are ecosystem services? The need for standardized environmental accounting units. Ecol Econ 63:616–626

Brown C, Reyers B, Ingwall-King L et al (2014) Measuring ecosystem services: guidance on developing ecosystem service indicators. UNEP-WCMC Cambridge, United Kingdom

Buzzell L, Chalquist C (eds) (2009) Ecotherapy: healing with nature in mind. Sierra Club Books, San Francisco, CA

Calvet-Mir L, Gómez-Baggethun E, Reyes-García V (2012) Beyond food production: ecosystem services provided by home gardens. A case study in Vall Fosca, Catalan Pyrenees, Northeastern Spain. Ecol Econ 74:153–160

Carlson A (2000) Aesthetics and the environment: the appreciation of nature, art, and architecture. Routledge, London, United Kingdom

Ceballos-Lascuráin H (1996) Tourism, ecotourism, and protected areas: the state of nature-based tourism around the world and guidelines for its development. International Union for the Conservation of Nature, Gland, Switzerland

Chan KMA, Guerry AD, Balvanera P et al (2012) Where are cultural and social in ecosystem services? A framework for constructive engagement. Bioscience 62:744–756

Chandrakanth MG, Bhat MG, Accavva MS (2004) Socio-economic changes and sacred groves in South India: protecting a community-based resource management institution. Nat Resour Forum 28:102–111

Chandran MDS, Hughes JD (1997) The sacred groves of South India: ecology, traditional communities and religious change. Soc Compass 43:413–427

Clark NE, Lovell R, Wheeler BW et al (2014) Biodiversity, cultural pathways, and human health: a framework. Trends Ecol Evol 29:198–204

Clayton SD, Opotow S (eds) (2003) Identity and the natural environment: the psychological significance of nature. MIT Press, Cambridge, MA

Cochrane J (ed) (2008) Asian tourism: growth and change. Elsevier, Amsterdam, The Netherlands

Cooper N, Brady E, Steen H, Bryce R (2016) Aesthetic and spiritual values of ecosystems: recognising the ontological and axiological plurality of cultural ecosystem 'services'. Ecosyst Serv 21:218–229

Costa JM, Vaz M, Escalona J, Egipto R, Lopes C, Medrano H, Chaves MM (2016) Modern viticulture in southern Europe: vulnerabilities and strategies for adaptation to water scarcity. Agric Water Manag 164:5–18

Costanza R (2008) Ecosystem services: multiple classification systems are needed. Biol Conserv 141:350–352

Costanza R, d'Arge R, de Groot R et al (1997) The value of the world's ecosystem services and natural capital. Nature 387:253–260

Cuong TQ, Dibley MJ, Bowe S et al (2007) Obesity in adults: an emerging problem in urban areas of Ho Chi Minh City, Vietnam. Eur J Clin Nutr 61:673–681

Daly HE (1991) Steady-state economics, with new essays, 2nd edn. Island Press, Washington, DC

Daniel TC, Muhar A, Arnberger A et al (2012) Contributions of cultural services to the ecosystem services agenda. Proc Natl Acad Sci 109:8812–8819

De Groot RS, Wilson MA, Boumans RM (2002) A typology for the classification, description and valuation of ecosystem functions, goods and services. Ecol Econ 41:393–408

Díaz S, Fargione J, Chapin FS, Tilman D (2006) Biodiversity loss threatens human well-being. PLoS Biol 4:e277

Díaz S, Demissew S, Carabias J et al (2015) The IPBES conceptual framework—connecting nature and people. Curr Opin Environ Sustain 14:1–16

Diemont SAW, Martin JF (2009) Lacandon Maya ecosystem management: sustainable design for subsistence and environmental restoration. Ecol Appl 19:254–266

Falkowski TB, Martinez-Bautista I, Diemont SAW (2015) How valuable could traditional ecological knowledge education be for a resource-limited future? An energy evaluation in two Mexican villages. Ecol Model 300:40–49

Falkowski TB, Diemont SAW, Chankin A, Douterlungne D (2016) Lacandon Maya traditional ecological knowledge and rainforest restoration: soil fertility beneath six agroforestry system trees. Ecol Eng 92:210–217

Falkowski TB, Chankin A, Diemont SAW (2019a) Successional changes in vegetation and litter structure in traditional Lacandon Maya agroforests. Agroecol Sust Food Syst 44(6):747–767

Falkowski TB, Chankin A, Diemont SAW, Pedian RW (2019b) More than just corn and calories: a comprehensive assessment of the yield and nutritional content of a traditional Lacandon Maya milpa. Food Security 11(2):389–404

Felipe-Lucia MR, Comín FA, Escalera-Reyes J (2015) A framework for the social valuation of ecosystem services. Ambio 44:308–318

Fey RL (1989) Diversity index usage in the geographic study of home gardens in Quintana Roo, Mexico. PhD Dissertation, University of Colorado

Fish R, Church A, Winter M (2016) Conceptualising cultural ecosystem services: a novel framework for research and critical engagement. Ecosyst Serv 21:208–217

Ford A, Nigh R (2016) The Maya forest garden: eight millennia of sustainable cultivation of the tropical woodlands. Left Coast Press, Inc., Walnut Creek, CA

Gomes CVA, Vadjunec JM, Perz SG (2012) Rubber tapper identities: political-economic dynamics, livelihood shifts, and environmental implications in a changing Amazon. Geoforum 43:260–271

Gould RK, Klain SC, Ardoin NM et al (2015) A protocol for eliciting nonmaterial values through a cultural ecosystem services frame: analyzing cultural ecosystem services. Conserv Biol 29:575–586

Gouyon A, de Foresta H, Levang P (1993) Does 'jungle rubber' deserve its name? An analysis of rubber agroforestry systems in Southeast Sumatra. Agrofor Syst 22:181–206

Hannah L, Roehrdanz PR, Ikegami M, Shepard AV, Shaw MR, Tabor G, Zhi L, Marquet PA, Hijmans RJ (2013) Climate change, wine, and conservation. Proc Natl Acad Sci 110(17):6907–6912

Hansen A, Jakobsen J (2020) Meatification and everyday geographies of consumption in Vietnam and China. Geogr Ann: SerB 102(1):21–39

He PH (1991) The plants of Vietnam. Mekong Printing, Santa Ana, CA

Hernández-Morcillo M, Plieninger T, Bieling C (2013) An empirical review of cultural ecosystem service indicators. Ecol Indic 29:434–444

Hirons M, Comberti C, Dunford R (2016) Valuing cultural ecosystem services. Annu Rev Environ Resour 41:545–574

Hodel U, Gessler M, Cai HH et al (1999) In situ conservation of plant genetic resources in home gardens of southern Vietnam. IPGRI, Rome, Italy

Hoelle J (2011) Convergence on cattle: political ecology, social group perceptions, and socioeconomic relationships in acre, Brazil: convergence on cattle. Cult Agric Food Environ 33:95–106

Jose S (2009) Agroforestry for ecosystem services and environmental benefits: an overview. Agrofor Syst 76:1–10

Kandari LS, Bisht VK, Bhardwaj M, Thakur AK (2014) Conservation and management of sacred groves, myths and beliefs of tribal communities: a case study from North India. Environ Syst Res 3(1):1–10

Kelemen E, Barton D, Jacobs S et al (2015) Preliminary guidelines for integrated assessment and valuation of ecosystem services in specific policy contexts. In: Operationalisation of natural capital and ecosystem services. European Commission FP7

Kien VD, Van Minh H, Giang KB et al (2017) Socioeconomic inequalities in self-reported chronic non-communicable diseases in urban Hanoi, Vietnam. Glob Public Health 12:1522–1537

Kimmerer R (2011) Restoration and reciprocity: the contributions of traditional ecological knowledge. In: Egan D, Hjerpe EE, Abrams J (eds) Human dimensions of ecological restoration. Island Press, Washington, DC

Kimmerer RW (2014) Braiding sweetgrass: indigenous wisdom, scientific knowledge, and the teachings of plants. Milkweed Editions, Minneapolis, MN

Kontogeorgopoulos N, Churyen A, Duangsaeng V (2015) Homestay tourism and the commercialization of the rural home in Thailand. Asia Pac J Tour Res 20:29–50

Koohafkan P, Altieri MA (2017) Forgotten agricultural heritage: reconnecting food systems and sustainable development. Routledge, Taylor & Francis Group, London, New York

Kumar P (ed) (2012) The economics of ecosystems and biodiversity: ecological and economic foundations. Routledge, London, United Kingdom

Kumar BM, Nair PKR (eds) (2010) Tropical homegardens: a time-tested example of sustainable agroforestry. Springer, Dordrecht, The Netherlands

Lâm HT, Rönmark E, Văn Tường N et al (2011) Increase in asthma and a high prevalence of bronchitis: results from a population study among adults in urban and rural Vietnam. Respir Med 105:177–185

Langenberger G, Prigge V, Martin K et al (2009) Ethnobotanical knowledge of Philippine lowland farmers and its application in agroforestry. Agrofor Syst 76:173–194

Letcher SG, Lasky JR, Chazdon RL et al (2015) Environmental gradients and the evolution of successional habitat specialization: a test case with 14 Neotropical forest sites. J Ecol 103:1276–1290

Luu LT (1992) Farmer proven integrated agriculture-aquaculture: a technology information kit. International Institute for Rural Reconstruction and International Center for Living Aquatic Resources Management

Martín-López B, Iniesta-Arandia I, García-Llorente M et al (2012) Uncovering ecosystem service bundles through social preferences. PLoS One 7:e38970

Mazumdar S, Mazumdar S (2012) Immigrant home gardens: places of religion, culture, ecology, and family. Landsc Urban Plan 105:258–265

McGee RJ (1990) Life, ritual, and religion among the Lacandon Maya. Wadsworth Publishing Co., Belmont, CA

McGee RJ (2002) Watching Lacandon Maya lives. Allyn and Bacon, Boston, MA

Milcu A, Hanspach J, Abson D, Fischer J (2013) Cultural ecosystem services: a literature review and prospects for future research. Ecol Soc 18(3)
Millennium Ecosystem Assessment (2005) Ecosystems and human well-being: synthesis. Washington, DC
Miller RP, Penn JW, van Leeuwen J (2006) Amazonian homegardens: their ethnohistory and potential contribution to agroforestry development. In: Kumar BM, Nair PKR (eds) Tropical homegardens. Springer Netherlands, Dordrecht, The Netherlands, pp 43–60
Mohri H, Lahoti S, Saito O et al (2013) Assessment of ecosystem services in homegarden systems in Indonesia, Sri Lanka, and Vietnam. Ecosyst Serv 5:124–136
Moreno G, Aviron S, Berg S et al (2017) Agroforestry systems of high nature and cultural value in Europe: provision of commercial goods and other ecosystem services. Agrofor Syst 92(4):877–891
Murrieta JR, Rueda RP (eds) (1995) Extractive reserves. International Union for the Conservation of Nature, Gland, Switzerland
Nahuelhual L, Carmona A, Laterra P et al (2014) A mapping approach to assess intangible cultural ecosystem services: the case of agriculture heritage in Southern Chile. Ecol Indic 40:90–101
Nations JD, Nigh RB (1980) The evolutionary potential of Lacandon Maya sustained-yield tropical forest agriculture. J Anthropol Res 36:1–30
Nepstad D, Schwartzman S, Bamberger B (2006) Inhibition of Amazon deforestation and fire by parks and indigenous lands. Conserv Biol 20:65–73
Nguyen PVN, Hong TK, Hoang T et al (2013a) High prevalence of overweight among adolescents in Ho Chi Minh City, Vietnam. BMC Public Health 13(1):1–7
Nguyen Q, Hoang MH, Öborn I, Noordwijk M (2013b) Multipurpose agroforestry as a climate change resiliency option for farmers: an example of local adaptation in Vietnam. Clim Chang 117:241–257
Nigh R (1976) Evolutionary ecology of Maya agriculture in Highland Chiapas, Mexico. PhD dissertation, Stanford University
Nigh R, Diemont SAW (2013) The Maya milpa: fire and the legacy of living soil. Front Ecol Environ 11:45–54
NOAA National Oceanic and Atmospheric Administration, USA https://www.ncdc.noaa.gov/
Osuri AM, Madhusudan MD, Kumar VS et al (2014) Spatio-temporal variation in forest cover and biomass across sacred groves in a human-modified landscape of India's Western Ghats. Biol Conserv 178:193–199
Palka JW (2005) Unconquered Lacandon Maya: ethnohistory and archaeology of indigenous culture change, 1st edn. University Press of Florida, Gainesville, FL
Perera V, Bruce RD (1986) The last lords of palenque: the Lacandon Mayas of the Mexican
Pilgrim S, Pretty J, Adams B et al (2009) The intersections of biological diversity and cultural diversity: towards integration. Conserv Soc 7:100
Power AG (2010) Ecosystem services and agriculture: tradeoffs and synergies. Philos Trans R Soc B Biol Sci 365:2959–2971
Rodas N, Flavio O, Rodas C, Hawkins LF (1940) Chichicastenango: The Kiche Indians, Their History and Culture. Union Typografica, Guatemala City, Guatemala
Salisbury DS, Schmink M (2007) Cows versus rubber: changing livelihoods among Amazonian extractivists. Geoforum 38:1233–1249
Satz D, Gould RK, Chan KMA et al (2013) The challenges of incorporating cultural ecosystem services into environmental assessment. Ambio 42:675–684
Schaich H, Bieling C, Plieninger T (2010) Linking ecosystem services with cultural landscape research. Gaia-Ecol Perspect Sci Soc 19:269–277
Schroth G, Coutinho P, Moraes VHF, Albernaz AL (2003) Rubber agroforests at the Tapajós river, Brazilian Amazon—environmentally benign land use systems in an old forest frontier region. Agric Ecosyst Environ 97:151–165
Schroth G, Moraes VHF, da Mota MSS (2004) Increasing the profitability of traditional, planted rubber agroforests at the Tapajós river, Brazilian Amazon. Agric Ecosyst Environ 102:319–339

Schurz WL, Hargis OD, Marbut CF, Manifold CB (1925) Rubber production in the Amazon Valley. Department of Commerce, Washington, DC
Schwartzman S, Moreira A, Nepstad D (2000) Rethinking tropical forest conservation: perils in parks. Conserv Biol 14:1351–1357
Sekhar NU (2007) Traditional versus improved agroforestry systems in Vietnam: a comparison. Land Degrad Dev 18:89–97
Shih W, Do NTH (2016) Impact of tourism on long-run economic growth of Vietnam. Mod Econ 07:371–376
Shu LH, Ueda K, Chiu I, Cheong H (2011) Biologically inspired design. CIRP Ann 60:673–693
Smith NJ, Falesi IC, de Alvim PT, EAS S (1996) Agroforestry trajectories among smallholders in the Brazilian Amazon: innovation and resiliency in pioneer and older settled areas. Ecol Econ 18:15–27
Sneed P, Nietschmann B, Dean TD et al (2013) Conservation through cultural survival indigenous peoples and protected areas. Island Press, Washington, DC
Stanislawski D (1970) Landscapes of Bacchus: the vine in Portugal. University of Texas Press
Surico G (2000) The grapevine and wine production through the ages. Phytopathol Mediterr 39(1):3–10
Tengberg A, Fredholm S, Eliasson I et al (2012) Cultural ecosystem services provided by landscapes: assessment of heritage values and identity. Ecosyst Serv 2:14–26
Thanh PV (2010) VAC integrated system with entire energy chain in Vietnam. In: Technical consultation: how to make 'integrated food energy systems' work for small-scale farmers and rural people. FAO, Rome, Italy
Thompson DB (1937) The garden of Hephaistos. Hesperia: J Am School Classical Stud Athens 6(3):396–425
Tilliger B, Rodríguez-Labajos B, Bustamante J, Settele J (2015) Disentangling values in the interrelations between cultural ecosystem services and landscape conservation—a case study of the Ifugao Rice terraces in the Philippines. Land 4:888–913
Trinh LN, Watson JW, Hue NN et al (2003) Agrobiodiversity conservation and development in Vietnamese home gardens. Agric Ecosyst Environ 97:317–344
Tsonkova P, Quinkenstein A, Böhm C et al (2014) Ecosystem services assessment tool for agroforestry (ESAT-A): an approach to assess selected ecosystem services provided by alley cropping systems. Ecol Indic 45:285–299
Turner NJ, Davidson-Hunt IJ, O'flaherty M (2003) Living on the edge: ecological and cultural edges as sources of diversity for social—ecological resilience. Hum Ecol 31:439–461
Vadjunec JM, Schmink M, Gomes CVA (2011) Rubber tapper citizens: emerging places, policies, and shifting rural-urban identities in acre, Brazil. J Cult Geogr 28:73–98
Valle-García SE (2014) Ecotourism: sustainable indigenous policies and its effects in Mayan communities, southern Mexico. In: Brebbia CA, Favro S, Pineda FD (eds) Sustainable tourism VI. WIT Press, Southampton, United Kingdom, pp 239–250
van den Berghe PL (1995) Marketing Mayas: ethnic tourism promotion in Mexico. Ann Tour Res 22:568–588
Viers JH, Williams JN, Nicholas KA, Barbosa O, Kotzé I, Spence L, Webb LB, Merenlender A, Reynolds M (2013) Vinecology: pairing wine with nature. Conserv Lett 6:287–299
Vlkova M, Polesny Z, Verner V et al (2011) Ethnobotanical knowledge and agrobiodiversity in subsistence farming: case study of home gardens in Phong My commune, Central Vietnam. Genet Resour Crop Evol 58:629–644
Wehi PM, Lord JM (2017) Importance of including cultural practices in ecological restoration: biocultural restoration. Conserv Biol 31:1109–1118
Weinstein B (1983) The Amazon rubber boom, 1850–1920. Stanford University Press, Stanford, CA
WinklerPrins AM, De Souza PS (2005) Surviving the city: urban home gardens and the economy of affection in the Brazilian Amazon. J Lat Am Geogr 4:107–126
Zhang W, Ricketts TH, Kremen C et al (2007) Ecosystem services and dis-services to agriculture. Ecol Econ 64:253–260

# Carbon Sequestration Potential of Agroforestry Systems in India: A Synthesis

**B. Mohan Kumar and T. K. Kunhamu**

## Abbreviations

| | |
|---|---|
| AFOLU | Agriculture, forestry, and other land uses |
| AFS | Agroforestry systems |
| C | Carbon |
| CSP | Carbon sequestration potential |
| GHG | Greenhouse gases |
| MPT | Multipurpose tree |
| SCS | Soil carbon sequestration |
| SOC | Soil organic carbon |
| SOM | Soil organic matter |
| UNFCCC | United Nations Framework Convention on Climate Change |

## Introduction

India is a physiographically diverse and geographically large country with varied ecologies. Rich natural resource endowments in terms of soil, plant, animal, and fish wealth make India and the contiguous areas of South Asia a mega-biodiverse region. The National Bureau of Soil Survey and Land Use Planning (India), based on soil, bioclimatic, and physiographic features (Sehgal et al. 1992), has divided the country into 20 agroecological regions (Fig. 1), which broadly fall under arid, semiarid, subhumid, humid-perhumid, and coastal ecosystems. Land-use systems differ profoundly across these regions but agroforestry dominates in most parts. The Indian

B. M. Kumar (✉)
Arunachal University of Studies, Knowledge City, Namsai, Arunachal Pradesh, India
e-mail: bmkumar@arunachaluniversity.ac.in

T. K. Kunhamu
Department of Silviculture and Agroforestry, College of Forestry, Kerala Agricultural University, KAU (P.O.), Thrissur, Kerala, India
e-mail: kunhamu.tk@kau.in



R. P. Udawatta, S. Jose (eds.), *Agroforestry and Ecosystem Services*,
https://doi.org/10.1007/978-3-030-80060-4_15

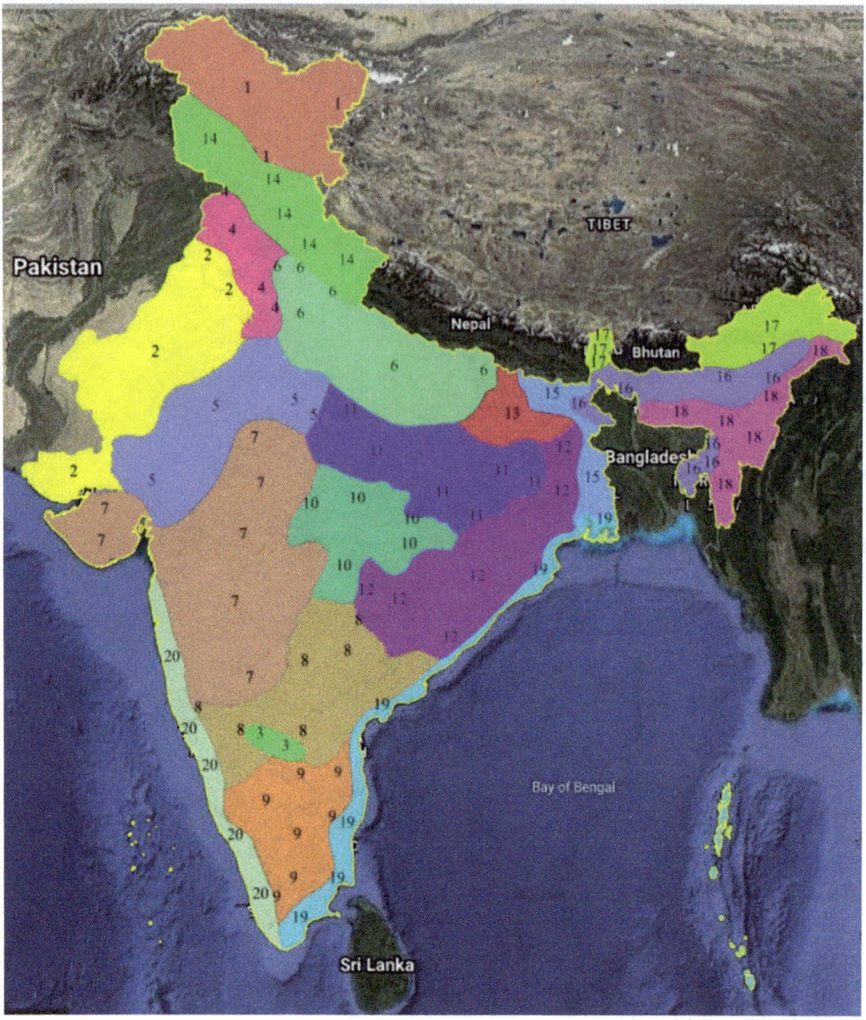


**Fig. 1** Agroecological regions of India. 1. Western Himalayas (cold arid), 2. Western Plains and Kutch Peninsula (hot arid), 3. Deccan Plateau (hot arid), 4. Northern Plains (Upper Gangetic; semiarid to subhumid), 5. Northern Plains (Rajasthan Upland and Gujarat Plains; hot semiarid), 6. Northern Plains (Middle Gangetic Plains; hot semiarid to subhumid), 7. Deccan Plateau (Malwa Plateau, Gujarat Plains, and Kathiawar peninsula; hot, semiarid with moderately deep black soils and length of growing period (LGP) 120–150 days), 8. Deccan Plateau (hot semiarid with mixed red and black soils and LGP 120–180 days), 9. Deccan Plateau (hot semiarid with red loamy soils and LGP 150–210 days), 10. Eastern Plateau (Satpura Range and Mahanadi Basin; hot subhumid), 11. Eastern Plateau (Bundelkhand Upland; hot subhumid with red and yellow soils and LGP 120–180 days), 12. Eastern Plateau (hot subhumid with red and lateritic soils and LGP 150–210+ days), 13. Northern Plains (Lower Gangetic; hot, subhumid), 14. Western Himalayas (warm to hot subhumid to humid), 15. Bengal basin (hot, subhumid), 16. Assam and North Bengal Plains (warm humid to perhumid), 17. Eastern Himalayas (warm perhumid), 18. North Eastern hills (Purvanchal; warm perhumid), 19. Eastern Coastal Plains and Islands of Andaman and Nicobar (hot subhumid), and 20. Western Ghats (Coastal Plains and Western Hills; hot humid to perhumid). Reprinted/adapted by permission from the National Bureau of Soil Survey and Land Use Planning, Nagpur (source: http://www.bhoomigeoportal-nbsslup.in/)

**Fig. 2** A Kerala homegarden with a multistrata arrangement of coconut palms (*Cocos nucifera*), banana (*Musa* spp.), and other species (photo: BM Kumar)

farmers, as their counterparts elsewhere, have domesticated fruit trees and other agricultural crops over millennia, primarily to meet their subsistence requirements. The tropical homegardens, which represent a complex integration of diverse trees (Fig. 2) with understory crops performing several production and service functions, are a case in point (Kumar et al. 2012). Indeed, the biophysical heterogeneity and climatic variability of the country affect the choice of tree and crop species and their productivity, implying profound variability in the nature and composition of agroforestry practices in India (Tejwani 1994; Puri and Panwar 2007). India is also one of the early countries to launch a national initiative on agroforestry research; indeed, as early as in 1983, it started the All India Coordinated Research Project on Agroforestry (Chinnamani 1993).

Since the late twentieth century, the phenomenon of "climate change" or "global warming" has been attracting global attention at a scale unparalleled in the history of humankind. Scientists, policy makers, and the general public continue to grapple with the adverse impacts of climate change and in figuring out strategies for mitigating the same. It is very likely that climate change may cause unprecedented shifts in global weather patterns producing a range of effects from threats to food security to rising sea levels that increase the risk of catastrophic flooding. India's average temperature has risen by around 0.7 °C during the 1901–2018 period and it is likely to increase further by approximately 4.4 °C by 2100 (relative to the 1976–2005 average; Krishnan et al. 2020). It is widely recognized that climate change is caused by rise in the atmospheric concentrations of the so-called greenhouse gases (GHGs) such as carbon dioxide ($CO_2$), methane ($CH_4$), and nitrous oxide ($N_2O$). The atmospheric concentration of $CO_2$, a prominent GHG, which accounts for 76% of the total global GHG emissions, has increased at unprecedented rates from the pre-industrial concentration of about 280 ppm to the current level of approximately 410 ppm (https://www.esrl.noaa.gov/gmd/ccgg/trends/gl_trend.html). The principal anthropogenic factors contributing to the increase in atmospheric $CO_2$ levels include the burning of fossil fuels such as coal, gas, and oil for industrial and other purposes, and agriculture, forestry, and other land uses (AFOLU), including deforestation. The average decadal growth rate of $CO_2$, which was 2.0 ppm per year in the 2000s, had surged to 2.4 ppm per year during the 2010–2019 period (https://www.co2.earth/co2-acceleration). Significantly, India is the third largest emitter of GHGs and accounts for 7% of total GHG emissions in the world as per the 2018 emission data (https://www.ucsusa.org/resources/each-countrys-share-co2-emissions).

Carbon sequestration is a key strategy for reducing atmospheric concentrations of $CO_2$, and thereby mitigating global warming. It is a process of storing atmospheric $CO_2$ or other forms of carbon (C) in long-standing pools. The United Nations Framework Convention on Climate Change (UNFCCC) describes it as "the process of removing C from the atmosphere and depositing it in a reservoir, or the transfer of atmospheric $CO_2$ to secure storage in long-lived pools" (UNFCCC 2007). Green plants—especially woody perennials—and soil play a central role in this. Dubbed as biological carbon sequestration, plants assimilate atmospheric $CO_2$ through photosynthesis and store the products of photosynthesis in their parts. The soil also is a major C sink as organic matter can remain in the soil for extended periods. Forestry and agroforestry systems (AFS) play a major role in biological carbon sequestration and stabilization of atmospheric GHG levels. Ever since climate change became a matter of stark global concern, agroforestry has received immense importance as a land management strategy with considerable potential for reducing atmospheric $CO_2$ levels. The average carbon sequestration potential (CSP) of agroforestry in India has been estimated to be 25 Mg C $ha^{-1}$ over 96 million ha (Sathaye and Ravindranath 1998) and agroforestry figures prominently in the country's climate change mitigation strategies (https://www4.unfccc.int/sites/ndcstaging/PublishedDocuments/IndiaFirst/INDIAINDCTOUNFCCC.pdf). There are, however, considerable variations in the CSP of agroforestry across different regions and land-use systems and based on the method of estimation. This chapter examines the range of AFS by agroecological regions of India and their potential to sequester atmospheric $CO_2$ and thus mitigate global warming. Such information can help focus attention on promising AFS and in adopting appropriate stand management practices including choice of species for enhancing the potential of biological carbon sequestration and for evolving national climate change mitigation strategies, which are cost effective.

## Agroforestry: A Cardinal Feature of the Indian Landscape

India is regarded as the cradle of agroforestry with diverse kinds of AFS (Kumar et al. 2012). These include the tropical, subtropical, and temperate AFS. India, with a geographical area of 329 million hectares, features 20 diverse agroecological regions each with an array of AFS (Table 1). Many of these are indeed traditional systems, practiced since time immemorial. For instance, homegardening and rearing of silkworm (*Bombyx* spp.) and lac insect (*Kerria lacca*) were practiced in the Indian subcontinent during the epic era of *Ramayana* and *Mahabharat* (7000 and 4000 BCE, respectively; Puri and Nair 2004). The travelogue of *Ibn Battuta* (Persian traveler; 1325–1354 CE) provides the earliest literary evidence of agroforestry from peninsular India and it mentions that in the densely populated and intensively cultivated landscapes of Malabar Coast, coconut (*Cocos nucifera*) and black pepper (*Piper nigrum*) were prominent around the houses (Randhawa 1980). The ecoclimatic situations under which agroforestry is practiced in India are also correspondingly diverse and range from the humid tropical valleys through to the high-elevation temperate regions and from humid tropical forests to the semiarid and arid drylands, including both irrigated and rain-fed ecosystems.

The predominant Indian AFS include agrisilviculture involving poplar (*Populus deltoides*; Fig. 3); *Eucalyptus* spp.; plantation agriculture involving coffee (*Coffea* spp.; Fig. 4), tea (*Camellia sinensis*; Fig. 5), cacao (*Theobroma cacao*), and spices (e.g., black pepper, cardamom, or *Elettaria cardamomum*) in association with a wide spectrum of trees (planted as well as trees in the natural forests); betel vine (*Piper betel* L.) + areca palm (*Areca catechu*); intercropping systems with coconut, Para rubber (*Hevea brasiliensis*), and other trees; commercial crop production under the shade of trees in natural forests (e.g., cardamom; Fig. 6); homegarden systems; and parkland systems. Table 1 provides a detailed account on this, agroecological region-wise. Deliberate growing of trees on field bunds (risers) and in agricultural fields as scattered trees and the practice to utilize open interspaces in the newly planted orchards and forests for cultivating field crops are also widespread in the Indian subcontinent (Singh 1987). In the relatively bigger landholdings of Himachal Pradesh, agri-horticulture is widespread, and in the northern and southern aspects, apple trees (*Malus domestica*) dominate. Growing arable crops in association with alder (*Alnus nepalensis*) is a remunerative AFS in the northeastern hill region of the country. Indeed, alder-based production system is an outstanding example of sustainable land use that stood the test of time in many parts of eastern Himalayas. Kumar et al. (2018) recently reviewed the literature on agroforestry in the Indian Himalayan region.

The traditional land-use systems, however, have been transformed over time—owing to the interplay of socioeconomic and technological factors. In particular, agricultural transformations brought about by market economies in the past, especially the incorporation of exotic commercial crops (e.g., *Hevea brasiliensis*), have led to the decimation of many traditional land-use systems (Kumar 2005). For example, the homegardens that constituted a predominant land-use activity in the

**Table 1** Major agroforestry systems and practices in different agroclimatic regions of India

| Sl. no. | Agroecological region | Agroforestry systems/ practices | Major tree and crop species |
|---|---|---|---|
| 1. | Western Himalayas (cold region) | Agrisilviculture, agri-silvi-horticulture, boundary plantations, fruit tree orchards, silvopasture | **Forest trees**: Banj oak (*Quercus leucotrichophora*), birch (*Betula* spp.), black locust (*Robinia pseudoacacia*), black poplar (*Populus nigra*), brown oak (*Quercus semecarpifolia*), cherry elm (*Ulmus villosa*), Chilgoza pine (*Pinus gerardiana*), sea buckthorn (*Hippophae* spp.), chinar (*Platanus orientalis*), Chinese albizia (*Albizia chinensis*), chir pine (*Pinus roxburghii*), green oak (*Q. dilatata*), Himalayan alder (*Alnus nepalensis*), Himalayan elm (*Ulmus wallichiana*), Himalayan poplar (*Populus ciliata*), Himalayan mulberry (*Morus laevigata*), Indian horse chestnut (*Aesculus indica*), Indian willow (*Salix tetrasperma*), juniper (*Juniperus* spp.), tama bamboo (*Dendrocalamus hamiltonii*), pines (*Pinus* spp.), red cedar (*Toona ciliata*), tree of heaven (*Ailanthus altissima*), tree rhododendron (*Rhododendron arboreum*), West Himalayan alder (*Alnus nitida*), wild olive (*Olea ferruginea*), white willow (*Salix alba*)<br>**Fruit and nut trees**: Almond (*Prunus dulcis*), apple (*Malus pumila*), apricot (*Prunus armeniaca*), citrus (*Citrus* spp.), common pear (*Pyrus communis*), common or European plum (*Prunus domestica*), Indian gooseberry (*Emblica officinalis* syn. *Phyllanthus emblica*), peach (*Prunus persica*), pear (*Pyrus pyrifolia*), pomegranate (*Punica granatum*), walnut (*Juglans regia*)<br>**Crops:** Medicinal and aromatic plants, millets, mustard (*Brassica juncea*), oats (*Avena sativa*), pulses, rice (*Oryza sativa*), vegetables, wheat (*Triticum aestivum*)<br>**Grasses:** Setaria grass (*Setaria anceps*), *Panicum* spp., etc. |

(continued)

**Table 1** (continued)

| Sl. no. | Agroecological region | Agroforestry systems/ practices | Major tree and crop species |
|---|---|---|---|
| 2. | Western Plains and Kutch Peninsula (hot arid) | Agrisilvicultural system, agri-silvi-horticulture, boundary plantations, parkland systems, silvopasture | **Forest trees:** Babul (*Acacia nilotica*), cactus (*Opuntia* spp.), cassia tree (*Cassia siamea* syn. *Senna siamea*), desert teak (*Tecomella undulata*), horsebean (*Parkinsonia aculeata*), khejri tree (*Prosopis cineraria*), Persian neem (*Melia azedarach*), pongam tree (*Millettia pinnata* syn. *Pongamia pinnata*), sicklebush (*Dichrostachys cinerea*)<br>**Fruit and nut trees**: Ber or Indian jujube (*Ziziphus mauritiana*), date palm (*Phoenix dactylifera*), common fig (*Ficus carica*), jamun (*Syzygium cumini*), phalsa (*Grewia asiatica*)<br>**Crops:** Maize (*Zea mays*), pearl millet (*Pennisetum glaucum*), sorghum (*Sorghum bicolor*), sesame (*Sesamum indicum*), foxtail millet (*Setaria italica* syn. *Panicum italicum*)<br>**Vegetables**: Cluster bean (*Cyamopsis tetragonoloba*), cowpea (*Vigna unguiculata*), watermelon (*Citrullus lanatus*), round melon (*Cucumis melo*), long melon (*Cucumis melo* var. *utilissimus*) |
| 3. | Deccan Plateau 7 (hot arid) | Agri-horticulture, agrisilviculture, block planting, boundary planting, silvopasture | **Trees:** Anjan (*Hardwickia binata*), babul, casuarina (*Casuarina equisetifolia*), eucalyptus (*Eucalyptus tereticornis*), jujube (*Ziziphus nummularia*), khejri, mesquite (*Prosopis juliflora*), mahua (*Madhuca longifolia*), neem (*Azadirachta indica*), Persian neem, safed khair (*Acacia ferruginea* syn. *Senegalia ferruginea*), siris tree (*Albizia lebbeck*), white-bark acacia (*Acacia leucophloea* syn. *Vachellia leucophloea*)<br>**Fruits:** Custard apple (*Annona squamosa*), guava (*Psidium guajava*), Indian gooseberry, lotebush (*Ziziphus nummularia*), mango (*Mangifera indica*), tamarind (*Tamarindus indica*)<br>**Crops**: Cowpea, finger millet (*Eleusine coracana*), groundnut (*Arachis hypogaea*), legumes, millets, pigeon pea (*Cajanus cajan*), pearl millet, rice, seasonal grasses, sorghum |

(continued)

**Table 1** (continued)

| Sl. no. | Agroecological region | Agroforestry systems/ practices | Major tree and crop species |
|---|---|---|---|
| 4. | Northern Plains (Upper Gangetic; semiarid to subhumid) | Agri-horticulture, agrisilviculture, agri-silvi-horticultural system, silvopasture, parkland systems | **Trees**: Arjun (*Terminalia arjuna*), babul, citrus, eastern poplar (*Populus deltoides*), eucalyptus, Indian tree of heaven (*Ailanthus excelsa*), Indian gooseberry, khejri tree, mesquite, miswak (*Salvadora persica*), pongam oil tree, sesban (*Sesbania sesban*), shisham (*Dalbergia sissoo*), *tamarisk* (*Tamarix articulata*)<br>**Crops:** Barley (*Hordeum vulgare*), black gram (*Vigna mungo*), berseem (*Trifolium alexandrinum*), cowpea, cluster bean, green gram (*Vigna radiata*), lentil (*Lens culinaris*), marigold (*Tagetes erecta*), mint (*Mentha piperita*), mustard, oats, pearl millet, pigeon pea, potato (*Solanum tuberosum*), taro (*Colocasia esculenta*), sorghum, sugarcane (*Saccharum officinarum*), rice, sesame, turmeric (*Curcuma longa*), wheat<br>**Fodder crops**: Buffel grass, birdwood grass, blue panic grass (*Panicum antidotale*), butterfly pea (*Clitoria ternatea*), Caribbean stylo, cowpea, Napier grass (*Pennisetum purpureum*), Sewan grass (*Lasiurus scindicus*) |
| 5. | Northern Plains (Rajasthan Upland and Gujarat Plains; hot semiarid) | Agrisilvicultural system, parkland systems, silvipasture | **Trees:** Anjan, babul, ber, banwali (*Acacia jacquemontii* syn. *Vachellia jacquemontii*), casuarina, citrus, common bamboo (*Bambusa vulgaris*), eucalyptus, gum arabic tree (*Acacia senegal* syn. *Senegalia senegal*), Indian gooseberry, jujube, large toothbrush tree (*Salvadora oleoides*), khejri, lotebush, mango, sapota, subabul, umbrella thorn (*Acacia tortilis* syn. *Vachellia tortilis*)<br>**Crops:** Barley, black gram, cluster bean, cowpea, chickpea (*Cicer arietinum*), green gram, mustard, pearl millet, pigeon pea, sesame, sorghum<br>**Fodder species**: Buffel grass, desert grass (*Panicum turgidum*), marvel grass (*Dichanthium annulatum*), Sewan grass |
| 6. | Northern Plains (Middle Gangetic Plain; hot semiarid to subhumid) | Agri-horticulture, agrisilviculture, agri-silvi-horticultural system, silvopasture, parkland systems | **Trees:** Arjun, babul, citrus, eastern poplar, eucalyptus, Indian gooseberry, Indian tree of heaven, khejri tree, mesquite, miswak, pongam tree, sesban (*Sesbania sesban*), shisham (*Dalbergia sissoo*), tamarisk<br>**Crops:** Barley (*Hordeum vulgare*), berseem, black gram (*Vigna mungo*), cowpea, cluster bean, green gram (*Vigna radiata*), lentil (*Lens culinaris*), marigold, mint, mustard, oats, pearl millet, potato, taro (*Colocasia esculenta*), sesame, sorghum, sugarcane, turmeric, wheat<br>**Fodder crops**: Buffel grass, birdwood grass, blue panic grass, butterfly pea, Caribbean stylo, cowpea, Napier grass, Sewan grass |

(continued)

**Table 1** (continued)

| Sl. no. | Agroecological region | Agroforestry systems/ practices | Major tree and crop species |
|---|---|---|---|
| 7. | Deccan Plateau (Malwa Plateau, Gujarat Plains, and Kathiawar Peninsula; hot, semiarid with moderately deep black soils and length of growing period (LGP) 120–150 days) | Agrisilviculture, agri-silvi-horticulture, boundary plantations, fruit tree orchards, live fence, horti-silvi-pasture, parkland systems, silvi-horticulture Silvopasture | **Trees:** Anjan, babul, ber, banwali, casuarina, common bamboo, eucalyptus, gum arabic tree (*Acacia senegal* syn. *Senegalia senegal*), henna (*Lawsonia alba*), horsebean, Indian laurel (*Terminalia elliptica*), large toothbrush tree, khejri, lotebush, Manila tamarind (*Pithecellobium dulce*), *Opuntia* spp., palmyra palm (*Borassus flabellifer*), Persian neem, pongam tree, sicklebush (*Dichrostachys cineraria*), siris tree, subabul (*Leucaena leucocephala*), spotted gliricidia (*Gliricidia sepium*), teak (*Tectona grandis*), umbrella thorn<br>**Fruits trees:** Ber, common fig, custard apple, drumstick (*Moringa oleifera*), guava, Indian gooseberry, jamun, mango, orange (*Citrus reticulata*), phalsa, pomegranate, sapota (*Manilkara zapota*), tamarind<br>**Crops:** Black gram, brinjal (*Solanum melongena*), chickpea, cluster bean, cowpea, curry leaf (*Murraya koenigii*), green gram, groundnut, lathyrus (*Lathyrus sativus*), linseed (*Linum usitatissimum*), long melon, maize, okra (*Abelmoschus esculentus*), pearl millet, pigeon pea, rice, safflower (*Carthamus tinctorius*), sesame, sorghum, soybean (*Glycine max*), sunflower, sunn hemp (*Crotalaria juncea*)<br>**Fodder species**: Buffel grass, desert grass (*Panicum turgidum*), marvel grass (*Dichanthium annulatum*), Sewan grass |
| 8. | Deccan Plateau (hot semiarid with mixed red and black soils and LGP 120–180 days) | Agri-horticultural system, agrisilvicultural system, agri-silvi-horticulture, fruit tree orchards, horti-silvi-pasture, silvi-horticulture, silvopasture | **Trees:** Belliric myrobalan (*Terminalia bellirica*), eucalyptus, Indian laurel (*Terminalia elliptica*), mahua, Persian neem, pongam tree, shisham, subabul, tamarind, teak (*Tectona grandis*)<br>**Fruit trees:** Custard apple, guava, Indian gooseberry, mango, orange (*Citrus reticulata*), sapota, tamarind<br>**Crops**: Black gram, cowpea, curry leaf, finger millet, foxtail millet, groundnut, horse gram (*Macrotyloma uniflorum*), Indian aloe (*Aloe vera* syn. *Aloe barbadensis*), lathyrus, linseed, maize, pearl millet, safflower, rice, sorghum, sunn hemp<br>**Fodder crops**: Hybrid Napier (*Pennisetum glaucum* × *P. purpureum*), stylo (*Stylosanthes guianensis*), desmanthus (*Desmanthus virgatus*) |

(continued)

**Table 1** (continued)

| Sl. no. | Agroecological region | Agroforestry systems/ practices | Major tree and crop species |
|---|---|---|---|
| 9. | Deccan Plateau (hot semiarid with red loamy soils and LGP 150–210 days) | Agrisilviculture, agri-silvi-horticulture, block plantations, fruit tree orchards, horti-pastural system, horti-silvi-pasture, silvi-horticulture, silvopasture | **Trees:** Agati (*Sesbania grandiflora*), casuarina, coconut (*Cocos nucifera*), East Indian *sandalwood* (*Santalum album*), eucalyptus, gmelina (*Gmelina arborea*), Indian laurel (*Terminalia elliptica*), jackfruit (*Artocarpus heterophyllus*), kapok (*Ceiba pentandra*), Malabar neem (*Melia dubia*), mahua, mulberry (*Morus alba*), palmyra palm, teak (*Tectona grandis*), shisham, silk cotton tree (*Bombax ceiba*), white-bark acacia<br>**Fruit trees/crops:** Custard apple, guava, Indian gooseberry, mango, banana (*Musa* spp.), orange (*Citrus reticulata*), papaya (*Carica papaya*), pomegranate, lemon (*Citrus* spp.), sapota, tamarind<br>**Crops**: Black gram, curry leaf, green gram, horse gram, lathyrus, linseed, maize, pigeon pea, rice, sorghum, sunn hemp<br>**Oilseeds**: Groundnut, sesame, sunflower (*Helianthus annuus*), safflower<br>**Vegetables:** Bitter gourd (*Momordica charantia*), bottle gourd (*Lagenaria siceraria*), ridge gourd (*Luffa acutangula*), snake gourd (*Trichosanthes cucumerina*)<br>**Fodder crops:** African tall maize (*Zea mays*), buffel grass (*Cenchrus ciliaris*), birdwood grass (*Cenchrus setigerus*), Caribbean stylo (*Stylosanthes hamata*), desmanthus, hybrid Napier |
| 10. | Eastern Plateau (Satpura Range and Mahanadi Basin; hot subhumid) | Agri-silvi-horticultural system, agrisilviculture | **Trees**: Anjan, arjun, babul, ber, eucalyptus, flame of the forest (*Butea monosperma*), gmelina, neem, pongam tree, sweet orange (*Citrus aurantium*), white siris (*Albizia procera*)<br>**Fruit trees**: *Citrus* spp., guava, litchi (*Litchi chinensis*), mango, papaya<br>**Crops**: Bottle gourd, fodder species, linseed, lentil, rice, mustard, okra, pointed gourd (*Trichosanthes dioica*) |

(continued)

**Table 1** (continued)

| Sl. no. | Agroecological region | Agroforestry systems/ practices | Major tree and crop species |
|---|---|---|---|
| 11. | Eastern Plateau (Bundelkhand Upland; hot subhumid with red and yellow soils and LGP 120–180 days) | Agrisilvicultural system, agri-horti-silviculture, boundary planting, homegardens, silvopastoral system | **Trees**: Anjan, arjun, babul, ber, banwali, casuarina, common bamboo, eucalyptus, flame of the forest, gmelina, gum arabic tree (*Acacia senegal* syn. *Senegalia senegal*), Indian gooseberry, large toothbrush tree, khejri, lotebush, neem, pongam tree, shisham, sweet orange, solid bamboo (*Dendrocalamus strictus, D. hamiltonii*), white siris, subabul, umbrella thorn<br>**Fruit trees**: *Citrus* spp., guava, Indian gooseberry, Indian date (*Phoenix sylvestris*), litchi, mango, papaya, sapota<br>**Crops**: Barley, black gram, bottle gourd, chickpea, cluster bean, cowpea, green gram, green pea (*Pisum sativum*), lentil, linseed, mustard, okra, pearl millet, pigeon pea, pointed gourd, rice, sesame, sorghum, wheat<br>**Fodder species**: Buffel grass, desert grass (*Panicum turgidum*), marvel grass (*Dichanthium annulatum*), Sewan grass |
| 12. | Eastern Plateau (hot subhumid with red and lateritic soils and LGP 150–210+ days) | Agri-horticultural system, agrisilviculture, alley cropping, homegardens, silvopasture, lac cultivation, commercial forestry, windbreaks | **Trees:** Agati, Australian wattle (*Acacia auriculiformis*), belliric myrobalan (*Terminalia bellirica*), casuarina, chebulic myrobalan (*Terminalia chebula)*, coconut, eucalyptus, jackfruit tree, gmelina, guava, mangium (*Acacia mangium*), litchi, mango, mahogany (*Swietenia macrophylla*), orange, palmyra palm, papaya, shisham, som (*Machilus bombycina* syn. *Persea bombycina*), teak<br>**Crops:** Arrowroot (*Maranta arundinacea*), black gram, forages, ginger, green gram, groundnut, mango ginger (*Curcuma amada*), mustard, pigeon pea, pineapple (*Ananas comosus*), pulses, rice, turmeric, vegetables, wheat |
| 13. | Northern Plains (Lower Gangetic; hot, subhumid) | Agrisilviculture, agri-silvi-horticultural system, silvopasture, parkland systems, *Hevea* | **Trees:** Arjun, babul, citrus, eastern poplar, eucalyptus, khejri tree, mesquite, miswak, pongam tree, sesban (*Sesbania sesban*), tamarisk<br>**Crops:** Berseem, cluster bean, cowpea, green gram, marigold, mint, mustard, oats, pearl millet, potato, taro, sorghum, sugarcane, turmeric, wheat<br>**Fodder crops**: Buffel grass, birdwood grass, blue panic grass, butterfly pea, Caribbean stylo, cowpea, Sewan grass |

(continued)

**Table 1** (continued)

| Sl. no. | Agroecological region | Agroforestry systems/ practices | Major tree and crop species |
|---|---|---|---|
| 14. | Western Himalayas (warm to hot subhumid to humid) | Agri-silvi-horticulture system, agrisilviculture, agri-horticulture, agri-horti-silviculture, silvopasture | **Trees:** Arjun, axle wood tree (*Anogeissus latifolia*), babul, ber, bihul (*Grewia optiva*), cherry elm, Chinese albizia, chir pine, cutch tree (*Acacia catechu*), eastern poplar, East Indian sandalwood, eucalyptus, haldu (*Adina cordifolia*), Himalayan mulberry, Indian elm (*Holoptelea integrifolia*), Indian gooseberry, Indian willow, Indian tree of heaven, kachnar (*Bauhinia variegata*), kadam (*Neolamarckia cadamba*), lote tree or honeyberry (*Celtis australis*), mulberry, oaks (*Quercus* spp.), Persian neem, red cedar, sesbania (*Sesbania aegyptiaca*), siris tree, shisham, solid bamboo, soapberry (*Sapindus mukorossi*), subabul, teak, wild olive<br>**Horticulture trees**: Apple, citrus, guava, Indian gooseberry, jackfruit, litchi, mango, papaya.<br>**Crops:** Brinjal, cabbage (*Brassica oleracea var. capitata*), cauliflower (*Brassica oleracea var. botrytis)*, chilies (*Capsicum* spp.), French bean (*Phaseolus vulgaris*), green pea, maize, medicinal and aromatic plants, millets, mustard, oats, okra, onion (*Allium cepa*), pulses, potato, radish (*Raphanus sativus*), rice, tomato (*Solanum lycopersicum*), turnip (*Brassica rapa subsp. rapa*), wheat<br>**Grasses:** Green foxtail (*Setaria* spp.), Guinea grass (*Panicum* sp.), Napier (*Pennisetum* spp.), etc. |
| 15. | Bengal basin (hot, subhumid) | Agrisilvicultural system, agri-silvi-horticultural system, homegardens | **Trees:** Akil (*Dysoxylum binectariferum*), areca nut (*Areca catechu*), bamboo (*Bambusa balcooa, B. tulda*), coconut, kadam, Indian laurel (*Litsea glutinosa*), sal (*Shorea robusta*), solid bamboo, white siris<br>**Fruit trees**: Ber, litchi, guava, mango<br>**Crops:** Banana, bottle gourd, cabbage, cauliflower, ginger (*Zingiber officinale*), groundnut, lentil, mustard, pineapple, pointed gourd, soybean, rice, turmeric |

(continued)

**Table 1** (continued)

| Sl. no. | Agroecological region | Agroforestry systems/ practices | Major tree and crop species |
|---|---|---|---|
| 16. | Assam and North Bengal Plains (warm humid to per humid) | Agrisilvicultural system, agri-silvi-horticultural system, homegardens | **Trees:** *Acacia* spp., *Albizia* spp., akil, areca nut, bamboos (*Bambusa balcooa, B. tulda, Dendrocalamus hamiltonii*), belliric myrobalan, chebulic myrobalan, common macaranga (*Macaranga peltata*), *Ficus* spp., gmelina, Indian laurel (*Litsea glutinosa*), kadam, kapok, mulberry, palmyra palm, Persian neem, rubber (*Hevea brasiliensis*), sal (*Shorea robusta*), semul (*Bombax ceiba*), solid bamboo, som, teak, white siris<br>**Fruit trees**: Ber, *Ficus* spp., jackfruit, jamun, guava, litchi, mango, pomegranate, orange (*Citrus* spp.), papaya<br>**Crops:** Banana, black pepper (*Piper nigrum*), betel leaf (*Piper betle*), bottle gourd, brinjal, cabbage, cauliflower, cucumber (*Cucumis sativus*), French bean, ginger, green pea, groundnut, knolkhol (*Brassica oleracea*), lentil, mustard, pineapple, pointed gourd, potato, pumpkin (*Cucurbita pepo*), rice, soybean, radish, sesame, tea (*Camellia sinensis*), tomato, turmeric |
| 17. | Eastern Himalayas (warm per humid) | Agrisilviculture, hedgerow intercropping, and many traditional systems | **Trees**: Agarwood (*Aquilaria malaccensis*), belliric myrobalan, Himalayan alder, Indian tree of heaven, champa (*Michelia champaca*), rubber, southern magnolia (*Magnolia* sp.), bamboos (28 bamboo species)<br>**Medicinal plants**: Galangal (*Kaempferia galanga*), green chirayta (*Andrographis paniculata*), long pepper (*Piper longum*), patchouli (*Pogostemon cablin*), sarpagandha (*Rauwolfia serpentina*), sugandhmantri (*Homalomena aromatica*)<br>**Crops:** Large cardamom (*Amomum* spp.), ginger, maize, pineapple, potato, rice, sweet potato (*Ipomoea batatas*), tea, turmeric, vegetables<br>**Hedgerow species**: Eastern rattlepod (*Crotalaria tetragona*), gliricidia, large-leaf flemingia (*Flemingia macrophylla*), pigeon pea, true indigo (*Indigofera tinctoria*), white tephrosia (*Tephrosia candida*) |
| 18. | North Eastern hills (Purvanchal; warm perhumid) | Agri-silvi-horticulture system, jhum cultivation, upland terrace farming | Trees: Apple, Himalayan alder, coffee (*Coffea arabica, C. canephora*), *Dipterocarps* spp., oak (*Quercus* spp.), orange, peach, pear (*Pyrus communis*), pines (*Pinus* spp.)<br>**Crops**: Banana, chilies, cotton, ginger, large cardamom, maize, medicinal plants, millets, mesta (*Hibiscus sabdariffa*), pineapple, potato, rice, sweet potato, sesame, sugarcane, tea |

(continued)

**Table 1** (continued)

| Sl. no. | Agroecological region | Agroforestry systems/ practices | Major tree and crop species |
|---|---|---|---|
| 19. | Eastern Coastal Plains and Islands of Andaman and Nicobar (hot subhumid) | Agrisilviculture system, block planting, silvopasture, horti-pasture system, silvi-horticulture system | **Trees**: Australian wattle, casuarina, coconut, gliricidia, Indian tree of heaven, jackfruit, mangium, mango, white-bark acacia, subabul<br>**Bamboos:** Brandisii bamboo (*Dendrocalamus brandisii*), chivari (*Dendrocalamus stocksii*)<br>**Crops:** Black pepper, cowpea, finger millet, rice<br>**Fodder:** Indian lovegrass (*Eragrostis pilosa*), mulberry (*Morus indica*), calliandra (*Calliandra calothyrsus*), shrubby stylo (*Stylosanthes scabra*), lovegrass (*Chrysopogon* sap), Napier |
| 20. | Western Ghats (Coastal Plains and Western Hills; hot humid to perhumid) | Alley cropping, animal-based integrated farming systems, aquaculture, homegardens, improved fallows, live fences, multipurpose trees, plantation-crop combinations, rotational tree fallows | **Trees:** Areca nut, cacao (*Theobroma cacao*), cashew (*Anacardium occidentale*), coconut, gmelina, guava, Indian coral tree (*Erythrina indica*), jackfruit, Malabar tamarind (*Garcinia gummi-gutta*), mango, mahogany, maharukh (*Ailanthus triphysa*), oil palm (*Elaeis guineensis*), palmyra palm, rubber, subabul, teak, sapota, spotted gliricidia<br>**Crops:** Black pepper, cardamom (*Elettaria cardamomum*), cassava (*Manihot esculenta*), clove (*Syzygium aromaticum*), elephant foot yam (*Amorphophallus paeoniifolius*), galangal, ginger, nutmeg (*Myristica fragrans*), rice, taro, turmeric, yams (*Dioscorea* spp.), vegetables<br>**Fodder:** Mulberry, calliandra, subabul, hybrid Napier, guinea grass, stylo |

Note: This list is compiled from various sources including Handa et al. (2019) and Kumar et al. (2018) and only the major species of agroforestry relevance are mentioned here

subcontinent, of late, have been showing symptoms of decline in some localities (Guillerme et al. 2011)—owing to rising population pressure and policies oriented towards land-use intensification to meet the rising demands for food grains (e.g., promoting monospecific production systems).

Environmental concerns such as global warming, land degradation, erosion of biodiversity, loss of wildlife habitats, and increased nonpoint source pollution of ground- and surface water, however, have provided impetus for the development and adoption of agroforestry around the world. Of late, economic incentives to the land managers have also acted as a major driver for promoting agroforestry. The poplar-based agroforestry in northern India, especially in the lowland "Tarai" areas at the base of the Himalayas, is a case in point (Fig. 7). An estimated 317,800 ha has been planted with *P. deltoides* in the country, of which 60% are block plantations and 40% are boundary plantations (National Poplar Commission of India 2012–15). Woodlots of other fast-growing trees such as eucalypts (*Eucalyptus* spp.), leucaena (*Leucaena leucocephala*), casuarina (*Casuarina equisetifolia*), mangium (*Acacia mangium*), Australian wattle (*Acacia auriculiformis*), *maharukh* (*Ailanthus triphysa*), and Malabar neem (*Melia dubia*) are also becoming increasingly popular among farmers in several parts of India.

**Fig. 3** Agroforestry systems involving poplar (*Populus deltoides*), turmeric (*Curcuma longa*), mango (*Mangifera indica*; pruned trees), and litchi (*Litchi chinensis*) in Yamunanagar district, Haryana; note the systematic arrangement of different components (photo: BM Kumar)

**Fig. 4** Coffee (*Coffea* spp.) agroforestry in Wayanad, Kerala; shade-loving coffee plants are raised in the understory of areca palms (*Areca catechu*) (photo: BM Kumar)

**Fig. 5** Tea (*Camellia sinensis*) + silver oak (*Grevillea robusta*) trees (for partial shade) in Idukki district, Kerala (photo BM Kumar). Reprinted/adapted by permission from Springer (South Asian Agroforestry: Traditions, Transformations, and Prospects; Kumar et al. 2012)

**Fig. 6** Cardamom (*Elettaria cardamomum*) with diverse kinds of shade trees in Idukki district, Kerala; principal trees include *Vernonia arborea, Artocarpus heterophyllus, Actinodaphne malabarica,* and *Persea macrantha* (photo: BM Kumar)

**Fig. 7** Poplar (*Populus deltoides*) trees (leafless during winter) and understory wheat (*Triticum aestivum*) in Pantnagar, Uttarakhand (photo: BM Kumar)

## Area Under Agroforestry in India

Although AFS abound in India, precise quantitative estimates on the extent of area under agroforestry are lacking—presumably because of the nonavailability of proper procedures for delineating the area influenced by trees in a mixed stand of trees and crops (Nair et al. 2009a). While in the multistrata systems (e.g., homegardens, shaded perennial systems, and intensive tree intercropping) the entire area occupied by such tree-crop combinations can be reckoned as agroforestry, most other agroforestry systems are rather extensive, where the components, especially trees, are not planted at regular spacing or density; for example, the parkland system and extensive silvopastures in central and northern India. The problem is acute in the case of practices such as windbreaks and boundary planting where the trees are planted at wide intervals or on farm boundaries. In the sequential agroforestry systems such as improved fallows and shifting cultivation, the beneficial effect of woody vegetation (in the fallow phase) on the crops in the sequence (in the cropping phase) may last for a variable length of time (years).

Given the diversity of AFS in India and the complexity of its components, it is a formidable task to determine the area under agroforestry. Nonetheless, some attempts have been made in this direction. Dhyani et al. (2013), using the databases of agricultural, horticultural, and forestlands of the country, deduced the area under agroforestry as 25.32 m ha, or 8.2% of the total geographical area of India with Maharashtra, Gujarat, and Rajasthan ranking high among the states. In another attempt, Rizvi et al. (2014), using geospatial techniques, estimated the area under agroforestry in India as 14.46 m ha and the potential area as 17.45 m ha. Forest Survey of India (FSI 2013), using digital interpretation of remote sensing data, however, estimated it as 11.54 m ha. Given the lack of consistency among the available estimates and the need to evolve climate change mitigation strategies through land-use management, it is imperative to estimate the area under agroforestry in India more precisely; however, such efforts are still rudimentary.

## Agroforestry for Climate Change Mitigation and Adaptation

Agroforestry provides an excellent opportunity for combining the twin aims of climate change mitigation (technological changes and substitution that reduce GHG emissions by averting emissions and sequestering GHGs) and adaptation (evolving approaches to reduce the harmful effects of climate change). In addition to its potential for reducing atmospheric $CO_2$ levels, AFS play an important role in reducing vulnerability of agricultural production systems to climate change (i.e., imparting increased resilience); they also increase livelihood security of the dependent populations. Given such advantages, the importance of promoting agroforestry in the country cannot be overemphasized. In particular, there is scope for conversion of wastelands and grasslands to agroforestry, which according to IPCC (2007) has huge potential to absorb $CO_2$ from the atmosphere. There are about 120 million hectares of degraded lands in India (ICAR-NAAS 2010) and a significant chunk of that could probably be converted into agroforestry. While the potential for agroforestry in India is enormous, there are also challenges such as dearth of quality planting materials, lack of credit and marketing facilities, meager insurance cover, and weak extension, which hamper the adoption of AFS. To capitalize on the ecological and production functions of agroforestry, the Government of India launched the landmark National Agroforestry Policy in 2014 (http://www.indiaenvironmentportal.org.in/content/389156/national-agroforestry-policy-2014/), which aims to mainstream tree growing on farms and meet a wide range of developmental and environmental goals.

### *Vegetation Carbon Sequestration Potential of AFS in India*

Agroforestry systems, which occur under diverse ecological conditions in India, offer immense scope for enhancing carbon stocks in the terrestrial ecosystems. During photosynthesis, atmospheric $CO_2$ is fixed as C in vegetation, detritus, and soil pools for "secure" storage. Vegetation carbon pools include those long-lasting products derived from biomass such as timber and belowground biomass such as roots. Nair et al. (2009a, 2010) reviewed the global literature on CSP of AFS and highlighted that aboveground CSP of AFS is tremendously variable, ranging from 0.29 to 15.21 Mg $ha^{-1}$ $year^{-1}$. Dhyani et al. (2016) reviewed the Indian literature on this topic and found that the CSP values (aboveground) range from 0.25 to 19.14 Mg C $ha^{-1}$ $year^{-1}$ for the tree components; and for bamboo-based systems, it may be as high as 21.36 Mg C $ha^{-1}$ $year^{-1}$ (Nath and Das 2012). A perusal of the data in Table 2, which summarizes the relatively recent studies on this, echoes the gross variability in CSP values of Indian AFS: aboveground C sequestration ranges from 0.23 to 23.55 Mg C $ha^{-1}$ $year^{-1}$ and belowground (root) C sequestration varies from 0.03 to 5.08 Mg C $ha^{-1}$ $year^{-1}$. Given the diverse nature of tree components involved, besides variations in ecoclimatic conditions, site quality, and stand management practices adopted, this is not unusual. The following section provides a brief account of the major factors influencing aboveground CSP of AFS.

## Agroforestry Systems and the Nature of Components

As mentioned, the diverse range of ecoclimatic conditions and the disparate array of agroforestry systems and practices in India representing profound variability in species and management regimes result in enormous variability of CSP values. In general, woodlots of bamboos, *Acacia auriculiformis*, *A. mangium*, and *Populus deltoides* are characterized by relatively high CSP (Table 2). Likewise, boundary plantation of 8-year-old *P. deltoides* had lower carbon stocks (4.51 Mg $ha^{-1}$) than block plantations (28.67 Mg $ha^{-1}$) in the Central Himalayan region (Kanime et al. 2013) with carbon sequestration rates of 0.43 and 2.75 Mg C $ha^{-1}$ $year^{-1}$, respectively. Mangalassery et al. (2014) found that silvopastoral systems involving *Acacia tortilis* and *Azadirachta indica* and grasses such as *Cenchrus ciliaris* and *C. setigerus* showed higher sequestration potential compared with systems containing only trees or pastures in the arid northwestern India.

While most AFS (e.g., multipurpose trees, silvopasture, energy plantations) have great potential for C sequestration, homegardens are unique in this respect. They not only sequester C in biomass and soil, but also conserve agrobiodiversity (Kumar 2006). Tilman et al. (1997) and Kirby and Potvin (2007) have suggested that plant assemblages with high species diversity may promote more efficient use of site resources compared with those of lesser diversity. It signifies that "biodiverse" systems such as tropical homegardens can maintain greater net primary production and consequently higher CSPs than AFS with fewer species. In a case study from peninsular Indian homegardens, Kumar (2011) found that average aboveground standing stock of C ranged from 16 to 36 Mg $ha^{-1}$. Structural attributes such as size of the homegardens, however, may alter the carbon sequestration rates; for example, small homegardens in the reported study showed higher C stocks on unit area basis than large- and medium-sized ones.

**Table 2** Biomass (aboveground + roots) carbon sequestration potentials of some agroforestry systems in India

| Agroforestry/ land-use system | Major system components | Stand characteristics | Method of estimation | Carbon sequestration potential[a] ($Mg\ ha^{-1}\ year^{-1}$) | | | Source |
|---|---|---|---|---|---|---|---|
| | | | | Aboveground | Roots | Total | |
| Agri-horticulture, Tarai region, Central Himalaya | *Mangifera indica* + wheat (*Triticum aestivum*) | 278 trees $ha^{-1}$, 15 years old[b] | Allometric equation | 5.48 | 1.37 | 6.85 | Adhikari et al. (2020) |
| | *Populus deltoides* + wheat | 400 trees $ha^{-1}$; 8 years old | | 6.07 | 1.62 | 7.69 | |
| Agrisilviculture, Chhattisgarh (humid and subhumid) | *Gmelina arborea* + soybean (*Glycine max*) | 1000–2500 trees $ha^{-1}$; 4 years old | Destructive sampling | 0.28–1.13 | 0.17–0.24 | 0.46–1.36 | Swamy et al. (2003) |
| | *Gmelina arborea* + eight field crops | 592 trees $ha^{-1}$; 5 years old | Destructive sampling | 1.00 | 0.26 | 1.26 | Swamy and Puri (2005) |
| Agrisilviculture: multipurpose tree and black pepper (*Piper nigrum*), Kerala (humid tropics) | *Casuarina equisetifolia* | 1111 trees $ha^{-1}$; 22 years old | Destructive sampling | 6.12 | 0.77 | 6.89 | Kunhamu et al. (2018) |
| | *Macaranga peltata* | | | 2.83 | 0.91 | 3.75 | |
| | *Ailanthus triphysa* | | | 2.68 | 0.52 | 3.20 | |
| | *Artocarpus heterophyllus* | | | 4.91 | 1.19 | 6.09 | |
| | *Acacia auriculiformis* | | | 5.66 | 1.37 | 7.03 | |
| | *Grevillea robusta* | | | 6.35 | 1.35 | 7.69 | |
| Agrisilviculture/ silvoarable systems, arid western Rajasthan | *Prosopis cineraria* + crops | 45 trees $ha^{-1}$;19 years old | Allometric relationships | | | 0.46 | Tanwar et al. (2019) |
| | *Hardwickia binata* + crops | 145 trees $ha^{-1}$; 19 years old | Allometric relationships | | | 0.56 | |
| Homegarden, Mizoram | Mixed-species stand | Young (<20 years) and old (>20 years) homegardens | Allometric relationships | | | 2.46–4.11 | Singh and Sahoo (2018) |
| Horticultural plantations, Central Himalayan Tarai region, Uttarakhand (Indo-Gangetic region) | *Litchi chinensis* | 100 trees $ha^{-1}$; 7 years old | Nondestructive method | | | 1.20 | Kanime et al. (2013) |
| | *Mangifera indica* | 100 trees $ha^{-1}$; 15 years old | Nondestructive method | | | 1.80 | |
| | *Prunus salicina* | 400 trees $ha^{-1}$; 5 years old | Nondestructive method | | | 1.61 | |

| | | | | | | | |
|---|---|---|---|---|---|---|---|
| Hortipasture, arid western Rajasthan | *Ziziphus mauritiana* + grass | 120 trees $ha^{-1}$; 19 years old | Allometric relationships | | | 0.28 | Tanwar et al. (2019) |
| Mine spoil reclamation, dry tropical region (Singrauli coalfield), Madhya Pradesh | *Dendrocalamus strictus* plantation | 3999 green culms $ha^{-1}$; 3 years old | Harvest method | 2.73 | 5.08 | 7.81 | Singh and Singh (1999) |
| | | 10,854 green culms $ha^{-1}$; 5 years old | | 2.61 | 4.86 | 7.47 | |
| Plantation forests, Himachal Pradesh (Northwestern Himalaya) | *Quercus leucotrichophora* | 320 trees $ha^{-1}$; 2 years old | Allometric relationships | | | 3.47 | Devi et al. (2013) |
| | *Pinus roxburghii* | 875 trees $ha^{-1}$; 2 years old | | | | 2.78 | |
| | *Acacia catechu* | 342 trees $ha^{-1}$; 32 years old | | | | 0.97 | |
| | *Acacia mollissima* | 217 trees $ha^{-1}$; 32 years old | | | | 2.9 | |
| | *Albizia procera* | 603 trees ha−1; 32 years old | | | | 4.42 | |
| | *Alnus nitida* | 159 trees $ha^{-1}$; 32 years old | | | | 4.68 | |
| | *Eucalyptus tereticornis* | 2233 trees $ha^{-1}$; 32 years old | | | | 3.76 | |
| | *Ulmus villosa* | 1617 trees $ha^{-1}$; 22 years old | | | | 4.6 | |
| Plantation forests, Chhattisgarh (humid and subhumid) | *Gmelina arborea* | 525 trees $ha^{-1}$; 5 years old | Destructive sampling | 1.57 | 0.38 | 1.95 | Swamy and Puri (2005) |
| Plantations of bamboo, Barak Valley, Assam | Mixed patch of *Bambusa cacharensis*, *B. vulgaris*, and *B. balcooa* | Total density 11,030 culms $ha^{-1}$ | Allometric relationships | 18.93–23.55 | – | – | Nath and Das (2012) |

(continued)

**Table 2** (continued)

| Agroforestry/ land-use system | Major system components | Stand characteristics | Method of estimation | Carbon sequestration potential[a] (Mg $ha^{-1}$ $year^{-1}$) Aboveground | Roots | Total | Source |
|---|---|---|---|---|---|---|---|
| Shifting agriculture, Mizoram | Mixed-species stand | Young (<5 years) and old (>5 years) fallows | Allometric relationships | | | 2.50–2.77 | Singh and Sahoo (2018) |
| Silvopasture in arid western Rajasthan | Silvopasture: *Colophospermum mopane* + grass | 19-year-old trees; 98 trees $ha^{-1}$ | Allometric relationships | | | 0.28 | Tanwar et al. (2019) |
| Silvopasture, Kerala (humid tropics) | Coconut + *Calliandra calothyrsus* | Calliandra stand 17,777–27,777 plants $ha^{-1}$; 3 years old | Destructive sampling | 1.49–2.06 | 0.19–0.24 | 1.68–2.29 | Joy et al. (2019) |
| | | Coconut: 173 palms $ha^{-1}$; 25 years old | From bole volume and density; destructive sampling of bunches and fronds | 1.32–1.33 | – | – | |
| | Coconut + mulberry (*Morus* spp.) | Mulberry 27,777–49,382 plants $ha^{-1}$; 3 years old | Destructive sampling | 1.11–2.12 | 0.69–1.18 | 1.80–3.30 | John et al. (2019) |
| | | Coconut: 173 palms $ha^{-1}$; 25 years old | From bole volume and density; destructive sampling of bunches and fronds | 1.25–1.30 | – | – | |

| | | | | | | | |
|---|---|---|---|---|---|---|---|
| Silvopasture, Kurukshetra, sodic soils, semiarid region | *Acacia nilotica, Dalbergia sissoo, Prosopis juliflora* + grasses (*Desmostachya bipinnata* and *Sporobolus marginatus*) | [illegible]250 trees ha$^{-1}$; [illegible] years old | Allometric relationships | 0.23–2.55 | 0.03–0.80 | 0.26–3.35 | Kaur et al. (2002) |
| Silvopasture, hot semiarid Rajasthan | *Hardwickia binata* + *Cenchrus setigerus* | [illegible]33–666 trees ha$^{-1}$; [illegible]0 years old | Allometric relationships | | | 0.75–1.06 | Gupta et al. (2019) |
| Silvopasture, arid northwestern India, Gujarat | *Acacia tortilis* | [illegible]78 trees ha$^{-1}$;10 years old | Harvest method | 0.5 | 0.10 | 0.6 | Mangalassery et al. (2014) |
| | *Azadirachta indica* | | | 0.29 | 0.07 | 0.36 | |
| | *Acacia tortilis* + *Cenchrus ciliaris* | | | 0.51 | 0.18 | 0.69 | |
| | *A. tortilis* + *C. setigerus* | | | 0.49 | 0.12 | 0.61 | |
| | *A. indica* + *C. ciliaris* | | | 0.35 | 0.14 | 0.49 | |
| | *A. indica* + *C. setigerus* | | | 0.37 | 0.12 | 0.49 | |

(continued)

**Table 2** (continued)

| Agroforestry/ land-use system | Major system components | Stand characteristics | Method of estimation | Carbon sequestration potential[a] ($Mg\ ha^{-1}\ year^{-1}$) | | | Source |
|---|---|---|---|---|---|---|---|
| | | | | Aboveground | Roots | Total | |
| Woodlots, humid tropics, Kerala | *Acacia auriculiformis* | Block plantations; 2500 trees $ha^{-1}$; 9 years old | Destructive sampling | 19.11 | 0.99 | 20.10 | Kumar et al. (1998) |
| | *Ailanthus triphysa* | | | 2.67 | 0.41 | 3.08 | |
| | *Artocarpus heterophyllus* | | | 5.12 | 0.56 | 5.69 | |
| | *Artocarpus hirsutus* | | | 3.89 | 0.62 | 4.51 | |
| | *Casuarina equisetifolia* | | | 5.62 | 0.31 | 5.93 | |
| | *Leucaena leucocephala* | | | 1.44 | 0.18 | 1.62 | |
| | *Paraserianthes falcataria* | | | 10.96 | 0.77 | 11.72 | |
| | *Phyllanthus emblica* | | | 4.52 | 0.70 | 5.22 | |
| | *Pterocarpus marsupium* | | | 4.08 | 0.41 | 4.48 | |
| | *Acacia mangium* | 625–5000 trees $ha^{-1}$; 12 years old | Destructive sampling | 4.17–8.97 | 1.37–3.71 | 5.55–12.68 | Rocha et al. (2017) |
| | *Grevillea robusta* | 460 trees $ha^{-1}$; 21 years old | Destructive sampling | 1.49 | 0.38 | 1.87 | Thakur et al. (2015) |
| | *Acacia mangium* | 625–5000 trees $ha^{-1}$; 6.5 years old | Allometric equation | 5.53–10.22 | 0.83–2.37 | 6.37–12.59 | Kunhamu et al. (2011) |
| Woodlot, arid western Rajasthan | *Acacia tortilis* | Farm forestry; 116 trees $ha^{-1}$; 19 years old | Allometric relationships | | | 1.65 | Tanwar et al. (2019) |
| Woodlots, bamboo, Assam | *Schizostachyum dullooa* | Block plantation; 4 years old | Allometric relationships | 0.9 | | | Singnar et al. (2017) |
| | *Pseudostachyum polymorphum* | Block plantation; 4 years old | | 0.9 | | | |
| | *Melocanna baccifera* | Block plantation; 4 years old | | 2.05 | | | |

| | | | | | | | |
|---|---|---|---|---|---|---|---|
| Woodlot, Central Punjab (Trans-Gangetic plains) | *Populus deltoides* | Block plantation, 500 trees ha$^{-1}$; 6 years old | Allometric relationships | 11.0 | – | | Gera et al. (2011) |
| | | Boundary planting, 250–290 trees ha$^{-1}$; 6 years old | | 6.2 | – | | |
| Woodlot, Tarai Uttarakhand (Western Himalayan foothills) | *P. deltoides* | [illegible]00 trees ha$^{-1}$; [illegible]1 years old | | 8.19 | | | Arora et al. (2014) |
| Woodlots, Central Himalayan Tarai region, Uttarakhand (Indo-Gangetic region) | *P. deltoides* | Block plantation, 500 trees ha$^{-1}$; 8 years old | Destructive method | | | 3.58 | Kanime et al. (2013) |
| | | Boundary planting; 70 trees, 8 years old | | | | 0.56 | |
| | *Eucalyptus tereticornis* | Boundary planting; 120 trees ha$^{-1}$; 10 years old | | | | 1.05 | |
| | *Dalbergia sissoo* | Block plantation, 1666 trees ha$^{-1}$; 10 years old | | | | 4.34 | |
| Woodlot, Pantnagar, Uttarakhand | *P. deltoides* | 200–1000 trees ha$^{-1}$; 3 years old | Allometric equation | 3.63–8.99 | – | | Pingale et al. (2014) |

(continued)

**Table 2** (continued)

| Agroforestry/ land-use system | Major system components | Stand characteristics | Method of estimation | Carbon sequestration potential[a] (Mg ha$^{-1}$ year$^{-1}$) | | | Source |
|---|---|---|---|---|---|---|---|
| | | | | Aboveground | Roots | Total | |
| Woody perennial plantation agriculture, Northeast India | Para rubber (*Hevea brasiliensis*) | Monoculture; 714 trees ha$^{-1}$; 30 years old | Allometric equation | 4.22 | 0.27 | 4.49 | Brahma et al. (2018) |
| | Areca (*Areca catechu*) | Monoculture; 1560 palms ha$^{-1}$; 30 years old | | 0.73 | 0.17 | 0.90 | |
| | Betel vine (*Piper betle*)-*jhum* (slash and mulching) agroforestry | Support trees: *Oroxylum indicum*, *Artocarpus chama*, *Mangifera indica*, *Areca catechu*, *Syzygium cumini*, *Musa* sp., and *Terminalia chebula*; 1350 trees ha$^{-1}$; 30 years old | | 4.02 | 1.02 | 5.04 | |

[a]Wherever biomass values had been reported, the C stocks were deduced as 50% of the biomass stocks
[b]"Age" of the system, though not clearly defined, is assumed to be the number of years since the establishment of the tree component in the system

## Ecoregions and Site Quality

Agroforestry systems on humid and tropical sites have higher potential to sequester carbon than those on arid, semiarid, and temperate sites. For example, AFS in the Western Himalayan and humid tropical regions showed higher CSP than those in the arid and semiarid regions (Table 2). Ajit et al. (2017a) using the dynamic carbon accounting model, CO2FIXv3.1, simulated the CSP of extant AFS in 26 districts of 10 selected states in India over a 30-year period. Comparisons across districts indicate that CSP ranged from 0.05 to 1.03 Mg C $ha^{-1}$ $year^{-1}$ with a mean value of 0.21 Mg C $ha^{-1}$ $year^{-1}$. In another study involving the $CO_2$FIX model, these authors (Ajit et al. 2017b) showed that the CSP (tree, crop, and soil) of the extant AFS in Kupwara district of Kashmir valley involving species such as *Malus* (33.75%), *Populus* (29.91%), *Salix* (14.32%), *Juglans* (6.68%), and *Robinia* (4.7%) was 0.88 Mg C $ha^{-1}$ $year^{-1}$. The CSP of an AFS, apart from the nature of the species involved (section "Species and Stand Age"), is driven by stand management (section "Silvicultural Management") and the prevailing ecological quality of the site (site quality). In spite of the potential benefits of site-specific ecological conditions in enhancing stand growth, there are no studies addressing the impacts of site quality on CSP of AFS.

Altitudinal ranges as reported by some authors significantly influence carbon density (amount of carbon per unit area for a given ecosystem or vegetation type). For example, Rajput et al. (2015) showed that biomass carbon density in Kullu valley (Northwestern Himalayas) increased from 1000 to 1600 m altitude and declined thereafter, presumably because of the lower cropping intensity and shorter growing period prevailing in the upper altitudinal zones, which depress carbon density. As a result, carbon stocks/density may decline in the aboveground biomass and woody debris at high elevations (>1600 m). However, the soil organic carbon (SOC) may increase with elevation, albeit modestly, owing to the lower organic matter decay rates prevailing at higher altitudes, offsetting any net change in total carbon density (vegetation + soil) with increasing elevation.

## Species and Stand Age

Choice of species is an important criterion that determines the carbon stocks of AFS. Fast-growing species such as bamboos, acacia (*A. mangium*; *A. auriculiformis*), poplar, eucalypts, and leucaena are generally characterized by high CSPs (Table 2). Dhyani et al. (2016) also reported similar results. Russell and Kumar (2019) using the CENTURY model showed that inclusion of trees with traits that promoted C sequestration such as lignin content, along with the use of best management practices, resulted in higher biomass (and therefore higher CSP), suggesting that the nature of tree components, besides the tree and stand management practices, holds the key in this respect. While evaluating the carbon sequestration in an age series of *P. deltoides*, a short-rotation plantation crop in Tarai region of central Himalaya, Arora et al. (2014) found that the C sequestration rate (in wood products and by substitution of biomass for coal) in mature plantations (7–11 years) varied from 5.8 to 6.5 Mg C $ha^{-1}$ $year^{-1}$. They also showed that aboveground carbon stocks increased from 0.5 Mg $ha^{-1}$ in 1-year-old stands to 90.1 Mg $ha^{-1}$ at 11 years of age, implying the dominant role of stand age in determining carbon stocks. Due to fast growth rate and adaptability to a range of environments, short-rotation plantations, in addition to high carbon storage, produce biomass for energy and contribute to reduced greenhouse gas emissions (Kaul et al. 2010). They also reported that high net annual carbon sequestration rates were achieved for fast-growing short-rotation poplar (8 Mg C $ha^{-1}$ $year^{-1}$) and eucalyptus (6 Mg C $ha^{-1}$ $year^{-1}$) plantations compared to the moderately fast-growing teak (*Tectona grandis*; 2 Mg C $ha^{-1}$ $year^{-1}$) and the relatively slow-growing (long-rotation) sal (*Shorea robusta*) forests (1 Mg C $ha^{-1}$ $year^{-1}$).

## Silvicultural Management

Carbon sequestration being a function of tree growth and productivity, stand management practices (stand density regulation through thinning or through controlling initial planting density, pruning, fertilization, and weeding), apart from increasing the quality and quantity of production, may also promote C sequestration. In general, fast-growing tropical conifers and broad-leaved species respond favorably to silvicultural treatments. Information on the effect of planting density, crown pruning, and other management practices on the C accumulation potential, however, is scarce in the Indian context. In one such study, Kunhamu et al. (2011) found that biomass C stock of *A. mangium* trees was significantly altered by planting density and pruning treatments. The total tree (aboveground + roots) C sequestration was higher for the 5000 trees $ha^{-1}$ treatment (81.82 Mg $ha^{-1}$) than that for the 625 trees $ha^{-1}$ (41.39 Mg $ha^{-1}$) at 6.5 years of age. Rocha et al. (2017) using the same experimental stand reported that CSP ranged from 5.55 to 12.68 Mg $ha^{-1}$ $year^{-1}$ at 12 years of age with denser stocks having substantially higher values (Table 2). In another study involving a 30-year-old *Hardwickia binata*-based AFS in the hot semiarid environment of Rajasthan, Gupta et al. (2019) also reported a significant impact of tree population density on carbon sequestration. Average biomass carbon sequestered per tree (118.44 ± 50.26 kg C $tree^{-1}$) was significantly more (44.5%) in the low-density (333 tree $ha^{-1}$) stand compared to the high-density (666 tree $ha^{-1}$) system. However, the total biomass carbon sequestered per hectare was significantly more (40.8%) in the high-density stand (31.6 ± 12.6 Mg C $ha^{-1}$), implying the silvicultural trade-off between maximization of individual tree growth and maximization of stand growth.

## *Soil Carbon Sequestration*

Soil carbon pool refers to the relatively stable forms of organic and inorganic C in the soil, which account for about two-thirds of the total C sequestration. Biomass such as plant residues that is not removed from the site is eventually incorporated into the soil as soil organic matter (SOM). Apart from plant residues, tree roots (both coarse roots and fine roots), which represent about one-fifth to one-fourth of the total living biomass, signify another important input of organic matter into the soil. SOM plays a vital role in determining C storage in terrestrial ecosystems and in regulating atmospheric $CO_2$ fluxes. Soil C sequestration (SCS), therefore, is a significant greenhouse gas removal strategy (Lal 2008). However, literature on SCS potential of AFS in India, as it is generally the case elsewhere, is very scanty. Yet another problem is that many of the reported studies lack the required rigor (e.g., low sampling intensity, inadequate sampling depth, and/or inappropriate analytical procedures employed: section "Measurement and Estimation of C Sequestration in Agroforestry Systems"), making generalizations somewhat difficult.

Reviewing the global literature on SCS in AFS, Nair et al. (2009a) reported that the estimates vary greatly across systems, ecological regions, and soil types. The "best-bet estimates" ranged from 5–10 kg C $ha^{-1}$ in about 25 years in extensive tree-intercropping systems on arid and semiarid lands to 100–250 kg C $ha^{-1}$ in about 10 years in species-intensive multistrata shaded perennial systems and homegardens of the humid tropics (Nair et al. 2009b). In the Indian context, soil carbon stocks in AFS (0–100 cm depth) varied from 10.02 Mg C $ha^{-1}$ for *Ziziphus mauritiana* + grass system in the arid western Rajasthan to as high as 229.5 Mg C $ha^{-1}$ in the homegarden systems of Mizoram (Table 3). Like vegetation carbon stocks (Table 2), SCS potential was relatively low for the AFS in the arid and semiarid ecosystems compared to that of the humid tropical ecosystems (e.g., homegardens and woodlots; Table 3), which is consistent with the global trends mentioned above. Indeed, Saha et al. (2010) reported that soil carbon stocks of multistrata homegardens in central Kerala were next only to the adjacent tropical moist deciduous forest ecosystems. Despite the generally low SCS potential of the arid northwest Indian ecosystems, silvopastoral systems were found to be promising. For example, Mangalassery et al. (2014) reported that the SOC and net carbon sequestered were greater in the silvopastoral system in the arid parts of Gujarat, which had 36.3–60.0% more total SOC stock compared to the tree system and 27.1–70.8% more SOC than the pasture system.

The influence of AFS on SCS generally depends on the quantity and quality of biomass inputs provided by the tree and non-tree components of the system, besides soil attributes such as soil structure and aggregation. Taxa of the multipurpose tree (MPT), stand age, and stand density are key factors in this regard. Dhyani et al. (2020) reported that MPTs like *Alnus nepalensis*, *Parkia roxburghii*, *Michelia oblonga*, *Pinus kesiya*, and *Gmelina arborea* with high ground surface cover, constant leaf litterfall, and extensive root systems have huge potential for augmenting SOC levels and for enhancing soil aggregate stability. Silvicultural management of

**Table 3** Recent reports on soil carbon stocks of agroforestry systems in India

| Agroforestry system | Species | Stand age (year) | Location | Soil depth (cm) | Soil C (Mg ha$^{-1}$) | Reference |
|---|---|---|---|---|---|---|
| Agri-horticulture | Apple (*Malus pumila*) + field crops, 1900–2170 m altitude | 25 | Kinnaur district, Himachal Pradesh (high-altitude dry temperate region) | 0–100 | 146.52 | Chisanga et al. (2018) |
| | Apple + field crops, 2170–2440 m altitude | 20 | | | 122.79 | |
| | Apple + field crops, 2440–2710 m altitude | 18 | | | 186.0 | |
| | *Ziziphus mauritiana* + crops | 19 | Arid western Rajasthan | 0–100 | 11.49 | Tanwar et al. (2019) |
| Agri-horti-silviculture | *Robinia* sp. (40 years) + Apple (25 years), *Ailanthus altissima* (40 years), *Salix tetrasperma* (50 years) + field crops, 1900–2170 m altitude | – | Kinnaur district, Himachal Pradesh (high-altitude dry temperate region) | 0–100 | 122.22 | Chisanga et al. (2018) |
| | *Robinia* sp. (40 years), apple, *Populus ciliata*, *Cedrus deodara* + field crops, 2170–2440 m altitude | – | | | 128.31 | |
| | *Cedrus deodara* + apple, *Pinus gerardiana* + field crops, 2440–2710 m altitude | – | | | 125.58 | |
| Agri-horti-silviculture | Mango (*Mangifera indica*) + teak (*Tectona grandis*) + okra (*Abelmoschus esculentus*) | – | Navsari, Gujarat | 0–30 | 27.22 | Singh et al. (2019) |

| | | | | | | |
|---|---|---|---|---|---|---|
| Agrisilviculture | *Gmelina arborea* + soybean (*Glycine max*) | 5 | Raipur, Chhattisgarh | 0–60 | 27.4 | Swamy and Puri (2005) |
| | Mixed-species stands | – | Thane, Maharashtra | 0–90 | 85.24 | Newaj et al. (2017) |
| | | – | Nasik, Maharashtra | | 80.82 | |
| | | – | Chittoor, Andhra Pradesh | | 55.84 | |
| | | – | Tumkur, Karnataka | | 62.57 | |
| | | – | Bellary, Karnataka | | 51.54 | |
| | | – | Mandi, Himachal Pradesh | – | 22.28 | Ajit et al. (2017b) |
| | | – | Ludhiana, Punjab | – | 9.12 | |
| | | – | Faizabad, Uttar Pradesh | | 4.6 | |
| | | – | Nawada, Bihar | | 16.67 | |
| | | – | Upper Gangetic plain, Hisar, Haryana | | 10.31 | |
| | | – | Gujarat plains and hills, Dahod, Gujarat | | 24.13 | |
| | | – | Desert arid and hot, Sikar, Rajasthan | | 4.28 | |
| Agrisilviculture | Teak + sugarcane (*Saccharum officinarum*) | – | Navsari, Gujarat | 0–30 | 23.81 | Singh et al. (2019 |
| Agrisilviculture/ silvoarable systems | *Prosopis cineraria* + crops | 19 | Arid western Rajasthan | 0–100 | 10.33 | Tanwar et al. (2019) |
| | *Hardwickia binata* + crops | 19 | | 0–100 | 10.82 | |

(continued)

**Table 3** (continued)

| Agroforestry system | Species | Stand age (year) | Location | Soil depth (cm) | Soil C (Mg ha$^{-1}$) | Reference |
|---|---|---|---|---|---|---|
| Homegarden | Mixed species | – | Kerala (humid tropics) | 0–100 | 103.32–119.30 | Saha et al. (2010) |
| | Young (<20 years) and old (>20 years) homegardens | – | Mizoram | 0–100 | 144.6–229.5 | Singh and Sahoo (2018) |
| Homegarden | Multispecies | – | Navsari, Gujarat | 0–30 | 31.03 | Singh et al. (2019) |
| Horticulture | Apple (*Malus domestica*), 1900–2170 m altitude | 25 | Kinnaur district, Himachal Pradesh (high-altitude dry temperate region) | 0–100 | 151.15 | Chisanga et al. (2018) |
| | Apple, 2170–2440 m altitude | 20 | | | 145.91 | |
| | Apple, 2440–2710 m altitude | 18 | | | 124.47 | |
| | *Litchi chinensis* | 7 | Central Himalayan Tarai region, Uttarakhand (Indo-Gangetic region) | 0–30 | 36.30 | Kanime et al. (2013) |
| | *Mangifera indica* | 15 | | | 40.70 | |
| | *Prunus salicina* | 5 | | | 36.97 | |
| Horti-pasture | *Ziziphus mauritiana* + grass | 19 | Arid western Rajasthan | 0–100 | 10.02 | Tanwar et al. (2019) |
| Multipurpose tree-based black pepper system | *Casuarina equisetifolia* | 22 | Kerala (humid tropics) | 0–100 | 63.62 | Kunhamu et al. (2018) |
| | *Macaranga peltata* | | | | 68.64 | |
| | *Ailanthus triphysa* | | | | 65.56 | |
| | *Artocarpus heterophyllus* | | | | 64.42 | |
| | *Acacia auriculiformis* | | | | 71.39 | |
| | *Grevillea robusta* | | | | 61.26 | |

| | | | | | | |
|---|---|---|---|---|---|---|
| Silvopasture | Coconut + mulberry (*Morus* spp.) | 3 | Kerala (humid tropics) | 0–40 | 32.88–54.65 | John et al. (2019) |
| | Coconut + *Calliandra* | 5 | | 0–100 | 90.83–103.43 | Joy et al. (2019) |
| | *Pinus gerardiana*, *Artemisia indica* (50 years) + *A. brevifolia* + grasses, 1900–2170 m altitude | | Kinnaur district, Himachal Pradesh (high-altitude dry temperate region) | 0–100 | 127.33 | Chisanga et al. (2018) |
| | *Pinus gerardiana* + *A. brevifolia* (60 years) + grasses, 2170-2440 m altitude | | | | 95.35 | |
| | *Cedrus deodara* + *A. brevifolia* + *Lespedeza gerardiana* + grasses, 2440-27[illegible]0 m altitude | | | | 106.61 | |
| | *Colophospermum mopane* + grass | 19 | Arid western Rajasthan | 0–100 | 9.78 | Tanwar et al. (2019) |
| | *Hardwickia binata* + *Cenchrus setigerus* | 20 | Hot semiarid environment, Rajasthan | 0–30 | 22.94–23.25 | Gupta et al. (2019) |
| Shifting agriculture fallows | Young (<5 years) and old (>5 years) fallows | – | Mizoram | 0–100 | 102.6–144.3 | Singh and Sahoo (2018) |
| Tropical moist deciduous forest | Mixed species | – | Southern Western Ghats | 0–100 | 176.6 | Saha et al. (2010) |
| Tropical wet evergreen forest | Mixed stand | – | Barak Valley region, Assam | 0–100 | 133.08 | Brahma et al. (2018) |

(continued)

**Table 3** (continued)

| Agroforestry system | Species | Stand age (year) | Location | Soil depth (cm) | Soil C (Mg ha$^{-1}$) | Reference |
|---|---|---|---|---|---|---|
| Woodlots (block plantations) | *Acacia tortilis* | 19 | Arid western Rajasthan | 0–100 | 13.50 | Tanwar et al. (2019) |
| | *Grevillea robusta* | 21 | Kerala (humid tropics) | 0–100 | 77.56 | Thakur et al. (2015) |
| | *Acacia mangium* | 6.5 | | 0–15 cm | 27.02–34.64 | Kunhamu et al. (2011) |
| | *Quercus leucotrichophora* | 32 | Himachal Pradesh (Northwestern Himalaya) | 0–100 | 165.0 | Devi et al. (2013) |
| | *Pinus roxburghii* | 32 | | | 165.0 | |
| | *Acacia catechu* | 32 | | | 18.0 | |
| | *Acacia mollissima* | 32 | | | 195.0 | |
| | *Albizia procera* | 32 | | | 163.0 | |
| | *Alnus nitida* | 32 | | | 213.0 | |
| | *Eucalyptus tereticornis* | 32 | | | 164.0 | |
| | *Ulmus villosa* | 32 | | | 207.0 | |
| | *Gmelina arborea* | 5 | Raipur, Chhattisgarh | 0–60 | 36.1 | Swamy and Puri (2005) |
| | *Populus deltoides* | 11 | Central Himalayan Tarai region, Uttarakhand (Indo-Gangetic region) | 0–90 | 200.35 | Arora et al. (2014) |
| | *Populus deltoides* | 8 | | 0–30 | 42.17 | Kanime et al. (2013) |
| | *Dalbergia sissoo* | 10 | | | 48.99 | |
| Woodlots (boundary plantations) | *Populus deltoides* | 8 | | 0–30 | 41.83 | |
| | *Eucalyptus tereticornis* | 10 | | | 37.23 | |
| Woody perennial plantation agriculture | Coconut (*Cocos nucifera*) | 30 | Kerala (humid tropics) | 0–100 | 91.7 | Saha et al. (2010) |
| | Para rubber (*Hevea brasiliensis*) | 50 | | | 119.2 | |
| | Coconut | 25 | | 0–40 | 41.81 | John et al. (2019) |
| | Rubber | 30 | Barak Valley region, Assam | 0–100 | 101.95 | Brahma et al. (2018) |
| | Areca (*Areca catechu*) | | | | 96.18 | |
| | Betel vine (*Piper betle*)- *jhum* (slash and mulching) agroforestry | | | | 115.85 | |

stands may also increase SOM prompting improved productivity, besides providing climate change mitigation effects—signifying a win-win situation. Very little, however, is known about the changes in soil C storage of MPT stands under differing stand density management regimes. In a solitary study, Kunhamu et al. (2011) reported that high stand densities (5000 and 2500 trees $ha^{-1}$) promoted SCS in 6.5-year-old *A. mangium* stands (31.79 and 34.64 Mg C $ha^{-1}$, respectively) in the top (0–15 cm) layer of the soil profile. Intense pruning (up to 50% of tree height), however, depressed overall tree growth and soil C stocks at high (5000 tree $ha^{-1}$) and low (625 tree $ha^{-1}$) stand densities, while at intermediate densities (2500 and 1250 tree $ha^{-1}$), pruning exerted a beneficial effect, signifying the need to maintain optimal stand densities, besides adopting appropriate tree management practices, for reaping carbon sequestration benefits.

The association between biodiversity (especially plant diversity) and SCS has become a topic of considerable scientific interest. Saha et al. (2009) reported that the soil C stock was directly related to plant diversity of homegardens. They found that homegardens with higher species richness and tree density than monocultural systems had greater soil carbon stocks, especially in the top 50 cm of soil. Overall, within the 1 m profile, soil C content ranged from 101.5 to 127.4 Mg $ha^{-1}$. Furthermore, small-sized gardens (<0.4 ha) that had higher tree density and plant species diversity had relatively more soil C per unit area (119.3 Mg $ha^{-1}$) than large-sized (>0.4 ha) gardens (108.2 Mg $ha^{-1}$).

Higher species richness of tropical homegardens may also ensure greater stability of the SOM fractions, especially at lower soil depths. Undeniably, SOM represents a significant carbon store and can remain in the soil for extended periods as a part of soil aggregates. The recalcitrant fraction of SOM is "protected" from further rapid decomposition by biochemical recalcitrance, chemical stabilization, and physical protection (Christensen 1996; von Luetzow et al. 2008). Biochemical recalcitrance occurs when the chemical composition of SOM involves aromatic polymers and other structures that are difficult for microbes to break down (Christensen 1996). A familiar example is lignin, one of the main constituents of woody plants. Russell and Kumar (2019) in the modeling study mentioned earlier indicated that inclusion of trees with traits that promoted C sequestration such as lignin, along with the use of best management practices, resulted in higher soil C storage. Studies on aspects of SCS and factors leading to aggregate formation and stability are scarce in the Indian context.

## *Measurement and Estimation of C Sequestration in Agroforestry Systems*

Yet another factor that determines the magnitude of soil and vegetation carbon sequestration is the methods employed for estimating vegetation CSP and SCS. Biomass is often taken as a surrogate of total C and the aboveground CSP values are typically the direct spin-offs of biomass measurements made either through destructive procedures or by employing allometric equations (Table 2). To derive carbon stocks, the amount of harvested and standing biomass is summed up assuming that 50% of the biomass comprises C, which however is variable depending on tissue types. Whole-tree harvest procedures for biomass estimation are also cumbersome. General allometric equations (Brown 1997; Piccard et al. 2012; Chave et al. 2014) are, therefore, widely employed in forestry, and are recommended by UNFCCC (2006) for tree biomass estimation in AFS also. Biomass estimation equations, however, vary with species, age, bole shape, and/or bole wood density. This has created the dilemma of whether to use the generalized equation for tree biomass estimation in AFS or not. Clearly, there is a need to develop a robust generic allometry that accounts for the heterogeneity of tree diversity throughout the landscape (Kuyah et al. 2012a).

As mentioned, often equations built for predicting biomass of forest trees are used in AFS. Variations in tree management, however, can be a concern, which limit the use of standard allometric equations developed for forests in agroforestry; for instance, trees in AFS may be pruned depending on management objectives or may have different growth forms due to differences in spacing compared to natural (forest) systems (Nair et al. 2009a). The determination of biomass production from AFS, therefore, is a challenging task and makes extrapolation from one system to others difficult and sometimes unrealistic (Nair 2012). Biomass regression equations, generalized for a geographic region, have been developed in a few cases to minimize errors in estimated biomass that result from such variability in sampled trees (e.g., Kumar et al. 1998). However, such location-specific allometric equations are not available for many agroforestry tree species.

In addition to aboveground biomass fractions, belowground net primary productivity (biomass) is a major pool of C. However, belowground biomass is difficult to measure and only very few Indian studies have characterized that. Root-to-shoot ratio is commonly used to estimate belowground living biomass. The ratios, however, differ substantially among species and across ecological regions, posing a serious problem in estimating belowground C sequestration in living biomass. Allometric equations for predicting root biomass have been constructed internationally (e.g., Kuyah et al. 2012b), but they are yet to gain popularity.

Apart from the root biomass, organic C occurs in soils as microbial biomass, and as SOM in labile and recalcitrant forms. The intricate interactions among these different forms make the measurement of SCS also a formidable task. The Walkley-Black (WB) procedure (Walkley and Black 1934) has been parsimoniously employed for SOC determination in India and elsewhere; it involves digestion of organic matter in the sample through oxidation with potassium dichromate. Although fast, convenient, and inexpensive, it is semiquantitative in nature and does not completely recover the organic carbon in soil (Abraham 2013). In fact, complete oxidation of SOC does not take place and variable levels of carbon recoveries have been reported (e.g., 60–86%: Nelson and Sommers 1996), implying that underestimation of SOC is in the WB procedure. The problem of incomplete digestion of the organic matter in the WB method, however, has been partially resolved by supplying external heat during sample digestion in the modified WB protocol (Nelson and Sommers 1996). Dry combustion methods, widely used for routine laboratory analysis, are considered to be the "gold standard" and superior to wet digestion (Nayak et al. 2019). Spectroscopic techniques for sensing of SOC are also evolving rapidly; nevertheless, the conventional methods will continue to be used in the near future despite their limitations (Nayak et al. 2019). Another major issue is the lack of uniformity in soil sampling, especially the depth of sampling (see Table 3). Although this problem is universal in nature (Nair 2012), it is more acute in the Indian context. Most soil studies are restricted to the surface soil layers, i.e., to 20 or 30 cm depth. In view of the fact that tree roots extend to deeper soil horizons, and the role of subsoil in long-term stabilization of C, the need for sampling the deeper layers of the soil profile cannot be overemphasized. Overall, a uniform set of methods and procedures are not available for estimating C sequestration in AFS. Wide variations also exist in the procedures used for soil sampling and analysis, which can greatly affect the conclusions made when comparing the differences under various management practices, soils, environments, and social conditions (Nair 2012).

## Concluding Remarks

Agroforestry systems abound in India with profound variability in the nature of components and their dynamics. Biological carbon sequestration (in vegetation and soil) is an intrinsic feature of agroforestry. Being a low-cost strategy, it has immense scope in the national climate change mitigation debate. In general, AFS with multistrata canopy architecture are characterized by higher CSP (aboveground) than those with simpler canopy structures. Likewise, AFS in the humid regions have higher aboveground CSPs than those in the arid and semiarid regions. Aboveground CSP values of Indian AFS reported in the literature range from 0.23 to 23.55 Mg C ha$^{-1}$ year$^{-1}$. More than half of the C assimilated is also transported belowground via root growth and organic matter turnover processes (e.g., fine root dynamics, rhizodeposition, and litter dynamics), which enrich the soil organic carbon pool. Species diversity (especially plant diversity), stand age, and stocking levels, besides depth of sampling, are key determinants of SCS. Soil carbon stocks (0–100 cm depth) varied from 10.0 Mg C ha$^{-1}$ to as high as 229.5 Mg C ha$^{-1}$, signifying great variability in SCS among the various ecoregions and AFS of India. Older, densely stocked (e.g., block plantations) and biodiverse AFS (e.g., multistrata homegardens) are more efficient in SCS. Much like the aboveground CSP, AFS in the arid and semiarid regions showed much less potential for SCS than those in the humid regions. Proper choice of AFS involving rapidly growing multipurpose tree species and adopting appropriate stand management practices are, therefore, key to enhancing the prospects of biological carbon sequestration and evolving national climate change mitigation strategies, which are cost effective.

## References


Abraham J (2013) Organic carbon estimations in soils: analytical protocols and their implications. Rubber Sci 26(1):45–54

Adhikari B, Lodhiyal N, Lodhiyal LS (2020) Assessment of crop yield, productivity and carbon sequestration in agroforestry systems in Central Himalaya, India. Agrofor Syst 94:281–296. https://doi.org/10.1007/s10457-019-00388-2

Ajit DSK, Handa AK, Newaj R, Chavan SB, Alam B, Prasad R, Ram A, Rizvi RH, Jain AK, Uma TD, Shakhela RR, Patel AG, Dalvi VV, Saxena AK, Parihar AKS, Backiyavathy MR, Sudhagar RJ, Bandeswaran C, Gunasekaran S (2017a) Estimating carbon sequestration potential of existing agroforestry systems in India. Agrofor Syst 91:1101–1118. https://doi.org/10.1007/s10457-016-9986-z

Ajit HAK, Dhyani SK, Bhat GM, Malik AR, Dutt V, Masoodi TH, Uma, Jain A (2017b) Quantification of carbon stocks and sequestration potential through existing agroforestry systems in the hilly Kupwara district of Kashmir valley in India. Curr Sci 113(4):782–785

Arora G, Chaturvedi S, Kaushal R, Nain A, Tewari S, Alam NM, Chaturvedi OP (2014) Growth, biomass, carbon stocks, and sequestration in an age series of *Populus deltoides* plantations in Tarai region of central Himalaya. Turk J Agric For 38:550–560. https://doi.org/10.3906/tar-1307-94

Brahma B, Pathak K, Lal R, Kurmi B, Das M, Nath PC, Nath AJ, Das AK (2018) Ecosystem carbon sequestration through restoration of degraded lands in Northeast India. Land Degrad Develop 29:15–25. https://doi.org/10.1002/ldr.2816

Brown S (1997) Estimating biomass and biomass change of tropical forests: a Primer. FAO Forestry paper – 134. ISBN 92–5–103955-0. Available on web site: http://www.fao.org/docrep/W4095E/w4095e00.htm#Contents

Chave J, Réjou-Méchain M, Búrquez A, Chidumayo E, Colgan MS, Delitti WB, Duque A, Eid T, Fearnside PM, Goodman RC, Henry M, Martínez-Yrízar A, Mugasha WA, Muller-Landau HC, Mencuccini M, Nelson BW, Ngomanda A, Nogueira EM, Ortiz-Malavassi E, Pélissier R, Ploton P, Ryan CM, Saldarriaga JG, Vieilledent G (2014) Improved allometric models to estimate the aboveground biomass of tropical trees. Glob Change Biol 20: 3177–3190. do: https://doi.org/10.1111/gcb.12629

Chinnamani S (1993) Agroforestry research in India: a brief review. Agrofor Syst 23:253–259. https://doi.org/10.1007/BF00704919

Chisanga K, Bhardwaj D, Pala N, Thakur CL (2018) Biomass production and carbon stock inventory of high-altitude dry temperate land use systems in North Western Himalaya. Ecol Proc 7(1). https://doi.org/10.1186/s13717-018-0134-8

Christensen BT (1996) Carbon in primary and secondary organomineral complexes. In: Carter MR, Stewart BA (eds) Structure and organic matter storage in agricultural soils. CRC Press, Boca Raton, pp 97–165

Devi B, Bhardwaj DR, Panwar P, Pal S, Gupta NK, Thakur CL (2013) Carbon allocation, sequestration and carbon dioxide mitigation under plantation forests of north western Himalaya, India. Ann For Res 56(1):123–135

Dhyani SK, Handa AK, Uma (2013) Area under agroforestry in India: an assessment for present status and future perspective. Indian J Agrofor 15(1):1–11

Dhyani SK, Ram A, Dev I (2016) Potential of agroforestry systems in carbon sequestration in India. Indian J Agric Sci 86(9):1103–1112

Dhyani SK, Ram A, Newaj R, Handa AK, Dev I (2020) Agroforestry for carbon sequestration in tropical India. In: Ghosh PK, Mahanta SK, Mandal D, Mandal B, Ramakrishnan S (eds) Carbon management in tropical and sub-tropical terrestrial systems, pp 313–331. https://doi.org/10.1007/978-981-13-9628-1_19

FSI, India State of Forest Report (2013) Forest Survey of India, (Ministry of Environment & Forests), Dehradun, India, pp. 71–80. Available at https://fsi.nic.in/forest-report-2013

Gera M, Mohan G, Bisht NS, Gera N (2011) Carbon potential of agroforestry under CDM in Punjab State of India. Indian J For 34:1–10

Guillerme S, Kumar BM, Menon A, Hinnewinkel C, Maire E, Santhoshkumar AV (2011) Impacts of public policies and farmers' preferences on agroforestry practices in Kerala, India. Environ Manag 48:351–364. https://doi.org/10.1007/s00267-011-9628-1

Gupta DK, Bhatt RK, Keerthika A, Mohamed MBN, Shukla AK, Jangid BL (2019) Carbon sequestration potential of *Hardwickia binata* Roxb. based agroforestry in hot semi-arid environment of India: an assessment of tree density impact. Curr Sci 116(1):112–116

Handa AK, Dev I, Rizvi RH, Kumar N, Ram A, Kumar D, Kumar A, Bhaskar S, Dhyani SK, Rizvi J (eds) (2019) Successful agroforestry models for different agro-ecological regions in India. Central Agroforestry Research Institute (CAFRI), Jhansi, and World Agroforestry (South Asia), New Delhi

ICAR-NAAS (Indian Council of Agricultural Research-National Academy of Agricultural Sciences) (2010) Degraded and wastelands of India: status and spatial distribution. ICAR, New Delhi

Intergovernmental Panel on Climate Change (IPCC) (2007) Climate change 2000: the scientific basis. Oxford University Press, Oxford

John AR, Raj AK, Kunhamu TK, Anoop EV, Jamaludheen V (2019) Forage yield and carbon dynamics of mulberry fodder banks under varying density and harvest interval in coconut garden. Indian J Agroforest 21(1):42–49

Joy J, Raj AK, Kunhamu TK, Jamaludheen V, Jayasree K (2019) Fodder production and carbon stock of Calliandra under coconut plantation. Range Manage Agroforest 40(1):109–117

Kanime N, Kaushal R, Tewari SK, Raverkar KP, Chaturvedi S, Chaturvedi OP (2013) Biomass production and carbon sequestration in different tree-based systems of Central Himalayan Tarai region. For Trees Livelihoods 22(1):38–50. https://doi.org/10.1080/14728028.2013.764073

Kaul M, Mohren GMJ, Dadhwal VK (2010) Carbon storage and sequestration potential of selected tree species in India. Mitig Adapt Strateg Glob Change 15:489–510. https://doi.org/10.1007/s11027-010-9230-5

Kaur B, Gupta SR, Singh G (2002) Carbon storage and nitrogen cycling in silvopastoral systems on a sodic soil in northwestern India. Agroforest Syst 54:21–29, https://doi.org/10.1023/A:1014269221934

Kirby KR, Potvin C (2007) Variation in carbon storage among tree species: implications for the management of a small-scale carbon sink project. For Ecol Manag 246:208–221. https://doi.org/10.1016/j.foreco.2007.03.072

Krishnan R, Sanjay J, Gnanaseelan C, Mujumdar M, Kulkarni A, Chakraborty S (eds) (2020) Assessment of climate change over the Indian Region: a report of the Ministry of Earth Sciences (MoES), Government of India. Springer. 226p

Kumar BM (2005) Land use in Kerala: changing scenarios and shifting paradigms. J Trop Agric 43(1–2):1–12

Kumar BM (2006) Carbon sequestration potential of tropical homegardens. In: Kumar BM, Nair PKR (eds) Tropical homegardens: a time-tested example of sustainable agroforestry, Advances in agroforestry, vol 3. Springer, The Netherlands, pp 185–204. https://doi.org/10.1007/978-1-4020-4948-4_11

Kumar BM (2011) Species richness and aboveground carbon stocks in the homegardens of central Kerala, India. Agric Ecosyst Environ 140:430–440. https://doi.org/10.1016/j.agee.2011.01.006

Kumar BM, George SJ, Jamaludheen V, Suresh TK (1998) Comparison of biomass production, tree allometry and nutrient use efficiency of multipurpose trees grown in woodlot and silvopastoral experiments in Kerala, India. For Ecol Manag 112(1–2):145–163. https://doi.org/10.1016/S0378-1127(98)00325-9

Kumar BM, Handa AK, Dhyani SK, Arunachalam A (2018) Agroforestry in the Indian Himalayan Region: an overview. In: Gordon AW, Newman SM, Coleman B (eds) Temperate agroforestry systems, 2nd edn. CABI Wallingford, UK, pp 153–172

Kumar BM, Singh AK, Dhyani SK (2012) South Asian agroforestry: traditions, transformations, and prospects. In: Nair PKR, Garrity D (eds) Agroforestry: the future of global land use, Advances in agroforestry, vol 9. Springer, pp 359–389. https://doi.org/10.1007/978-94-007-4676-3_19

Kunhamu TK, Kumar BM, Samuel S (2011) Does tree management affect biomass and soil carbon stocks of *Acacia mangium* Willd. stands in Kerala, India? In: Kumar BM, Nair PKR (eds) Carbon sequestration potential of agroforestry systems: opportunities and challenges, Advances in agroforestry, vol 8. Springer Science, The Netherlands, pp 217–230. https://doi.org/10.1007/978-94-007-1630-8_12

Kunhamu TK, Aneesh S, Kumar BM, Jamaludheen V, Raj AK, Niyas P (2018) Biomass production, carbon sequestration and nutrient characteristics of 22-year-old support trees in black pepper (*Piper nigrum* L.) production systems in Kerala, India. Agroforest Syst 92:1171–1183. https://doi.org/10.1007/s10457-016-0054-5

Kuyah S, Dietz J, Muthuri C, Jamnadass R, Mwangi P, Coe R, Neufeldt H (2012a) Allometric equations for estimating biomass in agricultural landscapes: I. Aboveground biomass. Agric Ecosyst Environ 158:216–224. https://doi.org/10.1016/j.agee.2012.05.011

Kuyah S, Dietz J, Muthuri C, Jamnadass R, Mwangi P, Coe R, Neufeldt H (2012b) Allometric equations for estimating biomass in agricultural landscapes: II. Belowground biomass. Agric Ecosyst Environ 158:225–234. https://doi.org/10.1016/j.agee.2012.05.010

Lal R (2008) Soil carbon stocks under present and future climate with specific reference to European ecoregions. Nutr Cycl Agroecosyst 81:113–127. https://doi.org/10.1007/s10705-007-9147-x

Mangalassery S, Dayal D, Meena SL, Ram B (2014) Carbon sequestration in agroforestry and pasture systems in arid north-western India. Curr Sci 107(8):1290–1293

Nair PKR (2012) Carbon sequestration studies in agroforestry systems: a reality-check. Agrofor Syst 86:243–253. https://doi.org/10.1007/s10457-011-9434-z

Nair PKR, Kumar BM, Nair VD (2009a) Agroforestry as a strategy for carbon sequestration. J Plant Nutr Soil Sci 172:10–23. https://doi.org/10.1002/jpln.200800030

Nair PKR, Nair VD, Kumar BM, Showalter J (2010) Carbon sequestration in agroforestry systems. Adv Agron 108:237–307. https://doi.org/10.1016/S0065-2113(10)08005-3

Nair PKR, Nair VD, Kumar BM, Haile SG (2009b) Soil carbon sequestration in tropical agroforestry systems: a feasibility appraisal. Environm Sci Policy 12(8):1099–1111. https://doi.org/10.1016/j.envsci.2009.01.010

Nath AJ, Das AK (2012) Carbon pool and sequestration potential of village bamboos in the agroforestry system of northeast India. Trop Ecol 53(3):287–293

National Poplar Commission of India (2012–2015) Country report on poplars and willows (2012–2015), Forest Research Institute, Indian Council of Forestry Research and Education, Dehradun, India., 40 p

Nayak AK, Rahman MM, Naidu R, Dhal B, Swain CK, Nayak AD, Tripathi R, Shahid M, Islam MR, Pathak H (2019) Current and emerging methodologies for estimating carbon sequestration in agricultural soils: a review. Sci Total Environ 665:890–912. https://doi.org/10.1016/j.scitotenv.2019.02.125

Nelson DW, Sommers LE (1996) Total carbon, organic carbon, and organic matter. In: Sparks DL, Page AL, Helmke PA, Loeppert RH, Soltanpour PN, Tabatabai MA, Johnston CT, Sumner ME (eds) Methods of soil analysis. Part 3. Chemical methods, Soil science Society of America Book Series no. 5. Soil Science of America and American Society of Agronomy, Madison, WI, pp 961–1010. https://doi.org/10.2136/sssabookser5.3.c34

Newaj R, Chaturvedi OP, Kumar D, Prasad R, Rizvi RH, Alam B, Handa AK, Chavan SB, Singh AS, Chaturvedi M, Karmakar PS, Maurya A, Saxena A, Gupta G, Singh K (2017) Soil organic carbon stock in agroforestry systems in western and southern plateau and hill regions of India. Curr Sci 112(11):2191–2193

Picard N, Saint André L, Henry M (2012) Manual for building tree allometric equations: from the field to the prediction. Food and Agriculture Organization of the United Nations, Rome, Centre de Coopération Internationale en Recherche Agronomique pour le Développement, Montpellier. 215 p

Pingale B, Bana OPS, Banga A, Chaturvedi S, Kaushal R, Tewari S (2014) Accounting biomass and carbon dynamics in *Populus deltoides* plantation under varying density in Tarai of central Himalaya. J Tree Sci 33:1–6

Puri S, Nair PKR (2004) Agroforestry research for development in India: 25 years of experiences of a national program. Agrofor Syst 61:437–452. https://doi.org/10.1023/B:AGFO.0000029014.66729.e0

Puri S, Panwar P (eds) (2007) Agroforestry systems and practices of India. New India Publishing Agency, Pitampura, New Delhi

Rajput BS, Bhardwaj DR, Pala NA (2015) Carbon dioxide mitigation potential and carbon density of different land use systems along an altitudinal gradient in north-western Himalayas. Agrofor Syst 89:525–536. https://doi.org/10.1007/s10457-015-9788-8

Randhawa MS (1980) A history of Indian agriculture, vol. 1 (Beginning to 12 Century) & vol 2, (Eighth to Eighteenth Century). Indian Council of Agricultural Research, New Delhi, India, vol 1: 414–415; vol 2: 67–68; 98–99

Rizvi RH, Dhyani SK, Newaj R, Karmakar PS, Saxena A (2014) Mapping agroforestry area in India through remote sensing and preliminary estimates. Indian Farm 63(11):62–64

Rocha D, Kunhamu TK, Santhoshkumar AV, Jamaludheen V, Raj AK (2017) Biomass production and carbon stocks in 12-year-old *Acacia mangium* managed at variable planting density and pruning regimes in central Kerala, India. Indian J Agroforest 19(1):69–74

Russell AE, Kumar BM (2019) Modeling experiments for evaluating the effects of trees, increasing temperature, and soil texture on carbon stocks in agroforestry systems in Kerala, India. Forests 10:803. https://doi.org/10.3390/f10090803

Saha SK, Nair PKR, Nair VD, Kumar BM (2010) Carbon storage in relation to soil size-fractions under some tropical tree-based land use systems. Plant Soil 328:433–446. https://doi.org/10.1007/s11104-009-0123-x

Saha SK, Nair PKR, Nair VD, Kumar BM (2009) Soil carbon stock in relation to plant diversity of homegardens in Kerala, India. Agrofor Syst 76:53–65. https://doi.org/10.1007/s10457-009-9228-8

Sathaye JA, Ravindranath NH (1998) Climate change mitigation in the energy and the forestry sectors of developing countries. Ann Rev Ener Env 23:387–437. https://doi.org/10.1146/annurev.energy.23.1.387

Sehgal J, Mandal DK, Mandal C, Vadivelu S (1992) Agro-ecological regions of India. 2nd ed., Tech. Bull. No. 24, National Bureau for Soil Survey and Land Use Planning, Nagpur, India. 130p

Singh AS, Singh JS (1999) Biomass, net primary production and impact of bamboo plantation on soil redevelopment in a dry tropical region. For Ecol Manag 119(1):195–207. https://doi.org/10.1016/S0378-1127(98)00523-4

Singh GB (1987) Agroforestry in the Indian subcontinent: past, present and future. In: Steppler HA, Nair PKR (eds) Agroforestry a decade of development. International Council for Research in Agroforestry, Nairobi, pp 117–138

Singh NR, Arunachalam A, Peetambari N (2019) Soil organic carbon stocks in different agroforestry systems of south Gujarat. Range Manage Agrofor 40(1):89–93

Singh SL, Sahoo UK (2018) Assessment of biomass, carbon stock and carbon sequestration potential of two major land uses of Mizoram, India. Internat J Ecol Environm Sci 44(3):293–306

Singnar P, Das MC, Sileshi GW, Brahma B, Nath AJ, Das AK (2017) Allometric scaling, biomass accumulation and carbon stocks in different aged stands of thin-walled bamboos *Schizostachyum dullooa*, *Pseudostachyum polymorphum* and *Melocanna baccifera*. For Ecol Manag 395:81–91. https://doi.org/10.1016/j.foreco.2017.04.001

Swamy SL, Mishra A, Puri S (2003) Biomass production and root distribution of *Gmelina arborea* under an agrisilviculture system in subhumid tropics of Central India. New Forests 26:167–186, https://doi.org/10.1023/A:1024478700645

Swamy SL, Puri S (2005) Biomass production and C-sequestration of *Gmelina arborea* in plantation and agroforestry system in India. Agrofor Syst 64:181–195. https://doi.org/10.1007/s10457-004-1999-3

Tanwar SPS, Kumar P, Verma A, Bhatt RK, Singh A, Lal K, Patidar M, Mathur BK (2019) Carbon sequestration potential of agroforestry systems in the Indian arid zone. Curr Sci 117(12):2014–2022. https://doi.org/10.18520/cs/v117/i12/2014-2022

Tejwani KG (1994) Agroforestry in India, Oxford & IBH, New Delhi. 233 p

Thakur S, Kumar BM, Kunhamu TK (2015) Coarse root biomass, carbon, and nutrient stock dynamics of different stem and crown classes of silver oak (*Grevillea robusta* A. Cunn. ex. R. Br.) plantation in Central Kerala, India. Agrofor Syst 89(5):869–883. https://doi.org/10.1007/s10457-015-9821-y

Tilman D, Lehman CL, Thomson KT (1997) Plant diversity and ecosystem productivity: theoretical considerations. Proc Natl Acad Sci U S A 94:1857–1861. https://doi.org/10.1073/pnas.94.5.1857

UNFCCC (2006) United Nations framework convention on climate change: handbook. Climate Change Secretariat, Bonn, Germany. 216p

UNFCCC (2007) Report on the second workshop on reducing emissions from deforestation in developing countries. Subsidiary body for scientific and technological advice. Item 5 of the provisional agenda: Reducing emissions from deforestation in developing countries. Twenty-sixth session, Bonn, 7–18 May 2007, 18p. http://www.rainforestcoalition.org/documents/UNFCCCSBSTA200703.pdf, accessed 15 October 2010

von Luetzow M, Kogel-Knabner I, Ludwig B, Matzner E, Flessa H, Ekschmitt K, Guggenberger G, Marschner B, Kalbitz K (2008) Stabilization mechanisms of organic matter in four temperate soils: Development and application of a conceptual model. J Plant Nutr Soil Sci 171:111–124. https://doi.org/10.1002/jpln.200700047

Walkley A, Black IA (1934) An examination of the Degtjareff method for determining soil organic matter and a proposed modification of the chromic acid titration method. Soil Sci 37:29–38

# Ecosystem Services from Agroforestry Systems in Australia

**John Doland Nichols, Samantha M. Lee, Rowan Reid, and John C. Grant**

## Abbreviations

| | |
|---|---|
| C | Carbon |
| $CO_2$ | Carbon dioxide |
| PES | Payments for ecosystem services |

## Forests and Soils in Australia

Australia is naturally known for its overall dryness, with forests, agriculture, and human population concentrated into a thin "skin" around the periphery from Queensland around to southern Australia and with a pocket in southwest Australia (Fig. 2). Approximately 17% of Australian landmass or 134 million ha is in "forest," with approximately 100 million of that in what could be better classified as "woodland"—short, widely spaced trees with grassy understories (Fig. 1). That leaves 83% of the continent in "non-forest," semiarid, or very arid areas (ABARES 2021) (Fig. 2).

Although the continent is mostly too dry to support trees, the forests that do exist contain massive diversity in a variety of forest types and tree species (Boland et al. 2006). A dominant genus of course is *Eucalyptus*, of which there are an estimated 700–900 species, which range from the edges of rainforests to the borders of grasslands and shrubby deserts. Australia is also a center of diversity for the genus *Acacia* with more than 900 species. The Casuarinaceae family plays an important role in

J. D. Nichols (✉) · S. M. Lee · J. C. Grant
Forest Research Centre, Southern Cross University, Lismore, NSW, Australia
e-mail: doland.nichols@scu.edu.au

R. Reid
Bambra Agroforestry Farm, Bambra, VIC, Australia
e-mail: rowan.reid@agroforestry.net.au

 
R. P. Udawatta, S. Jose (eds.), *Agroforestry and Ecosystem Services*,
https://doi.org/10.1007/978-3-030-80060-4_16

**Fig. 1** Eucalypt woodlands constitute approximately 9% of land surface area of Australia

many environments, particularly through the nitrogen-fixing associations developed within the genera of *Allocasuarina* and *Casuarina* and cones that support wildlife such as various species of cockatoos. Members of the important Southern Hemisphere Proteaceae family, with their ability to survive and thrive in low-phosphorus environments, are common. The rain forests at the time of European contact (1788) occupied approximately 2 million ha out of the 758 million ha on the continent, and hold high plant diversity commonly found in tropical rain forests, with dozens of tree species on 1 ha (ABARES 2021).

In general, the landforms of Australia are quite old and weathered, with a few exceptional areas that have either alluvial soils or experienced relatively recent volcanic activity (Fig. 3). Starting off with poor soils it has not taken long since European settlement began in 1788 for Australian soils to become even more degraded, through excessive (or too little) burning, cultivation, erosion by both wind and water, and practices that cause water tables to rise and increase salinity. Thus, there is great potential to use the restorative power of trees and improved agricultural management to increase productivity while simultaneously providing a multitude of ecosystem services.

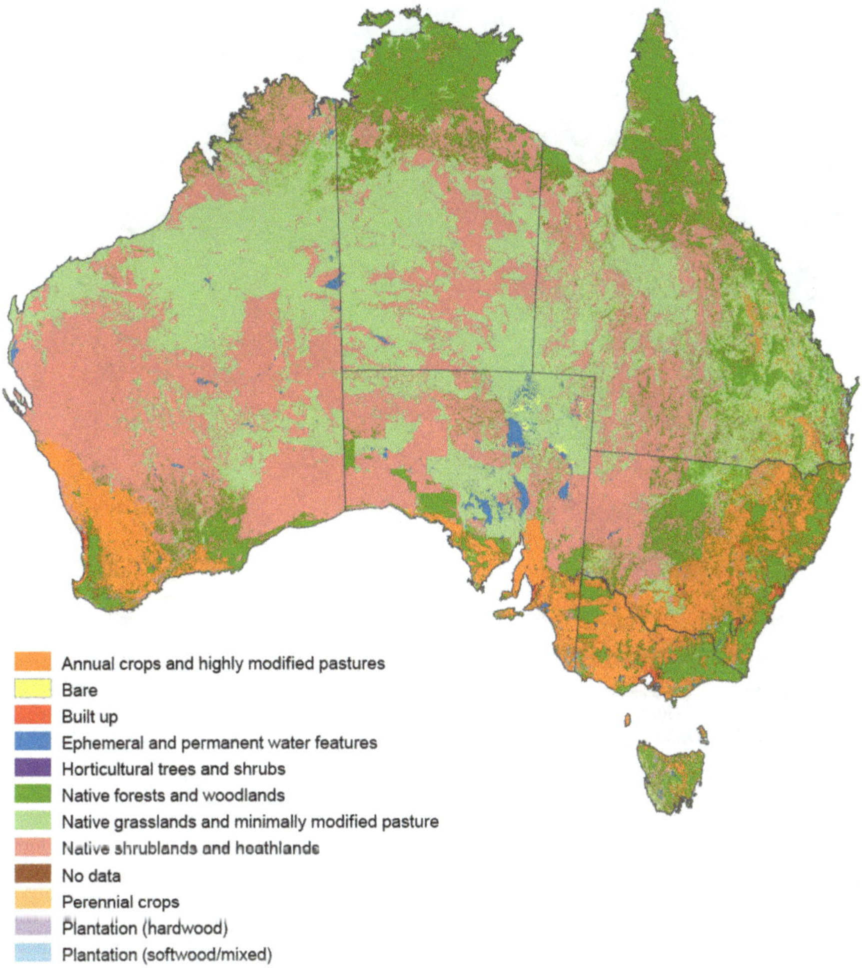


**Fig. 2** Vegetation types of Australia. From https://soe.environment.gov.au/sites/default/files/report2016/land/SoE2016-Land-web-resources/image/soe2016_lan_fig26_abares_vegetation_extent-01.png?acsf_files_redirect

## Ecosystem Services

In Web of Science database January 27, 2021, a search on "agroforestry" came up with 8610 references and on "agroforestry" combined with "ecosystem services" 810 publications, so the topic has been discussed!

Industrial agriculture has detrimental effects on many ecosystem services (Sandhu et al. 2012). However, the ecosystem benefits that can be derived from agroforestry within Australian agricultural landscapes are becoming more recognized (O'Grady and Mitchell 2017; Brown et al. 2018). Market drivers such as

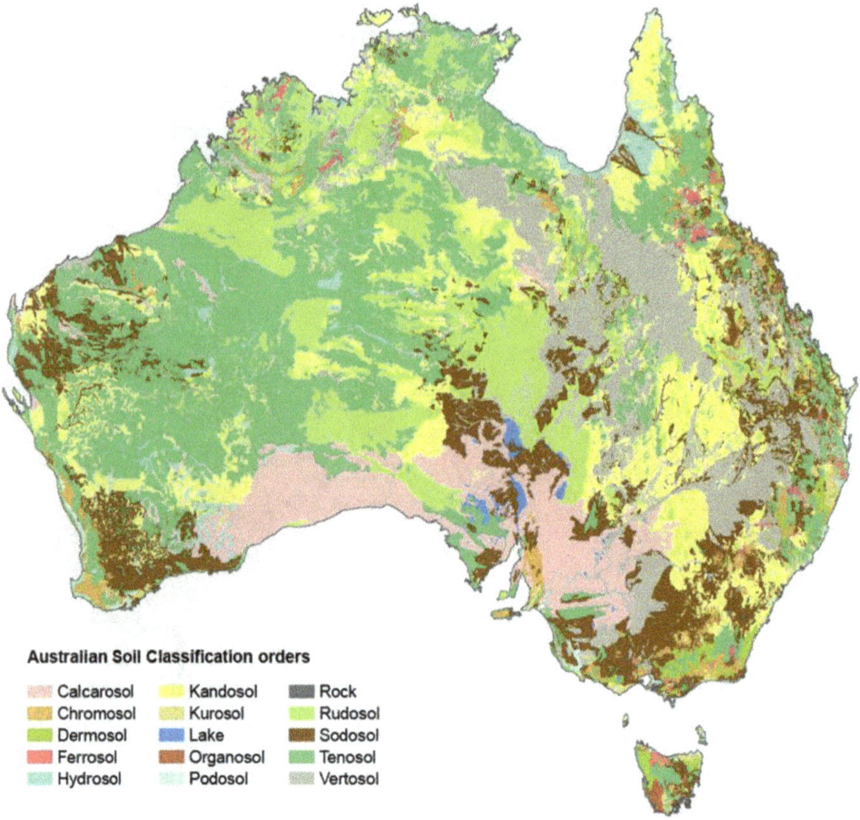


**Fig. 3** Australian soils. From https://soe.environment.gov.au/theme/land/topic/2016/soil-understanding

"Payments for Ecosystem Services" (PES) are well established in many regions around the globe (Do and NaRanong 2019). In Australia, however, it is a different matter—despite 30 years of talking about introducing PES in Australia, limited opportunities for participation exist. Previous trials have held rigid views about what constitutes an eligible system (for example, managed plantings do not qualify for conservation payments even if these areas deliver them) and have been unfair (due to a closed market), and returns have been mostly short-term.

## Agroforestry Solutions for Natural and Anthropogenic Issues

Moore and Bird (1997) summarized problems of land degradation in much of Australia and described development of agroforestry systems. Lefroy (2002) contributed a review paper on the use of forage trees and shrubs, which could be of major importance in Australia, where large areas are dedicated to production of cattle and sheep. While many definitions of agroforestry focus on the scale, appearance, or purpose of the system (alley farming, silvopastoral systems, forest farming, etc.) this approach has not worked well in Australia when describing why farmers engage in planting and managing trees. Reid and Stephen (2001) argued that what was important was the distinction between tree growing by farmers and that led or owned by corporations, industrial companies, nongovernment organizations (NGOs), and government agencies. They proposed an alternative definition that could be used to guide agroforestry research and development: "*Agroforestry is the commitment of resources by farmers, alone or in partnerships, towards the establishment or management of trees and forests on their land*" (Reid and Stephen 2001).

As such, agroforestry in Australia includes a wide variety of practices ranging from natural revegetation and protection of remnant trees on farmland to intensive management of trees on farms for timber. However, rather than managing trees and forests for a single purpose, most farmers seek multiple objectives from their forests with priorities commonly varying over time. Interviews with Australian farmers active in tree growing show that they appreciate the link between protecting and enhancing the environmental integrity of the farm—including its soil structure and fertility, biodiversity, and water quality—and the values important to them as family farmers (farm productivity, capital value, economic viability, and amenity) (Reid and Deans 2009).

What farmers choose to plant, the planting design, and their management reflect their aspirations and concerns. Drawing on a number of studies, Reid and Moore (2018) report that the provision of shelter for stock and crops, biodiversity enhancement, and land degradation control are the most important reasons given by farmers for planting or managing trees. Many farmers designed these plantings to also provide income, enhance the capital value of their farm, and improve the amenity or aesthetics of their properties. This approach has led to a diversity of agroforestry practices that reflect the diversity inherent within the farming community, the agricultural and forestry markets, and the land itself.

We would argue that there are few agroforestry systems in Australia as firmly established as the dominant systems in tropical countries, notably tropical home gardens and shade trees over perennial crops such as coffee and cacao. There have indeed been many research projects on agroforestry systems in Australia, particularly those done by the Joint Venture Agroforestry Program (JVAP) over 15 years, with results summarized by Powell (2009). The systems that do exist are mainly shelterbelts/windbreaks particularly in semiarid regions, and some on-farm tree planting for timber crops (Reid and Moore 2018). In the subtropical/tropical state of Queensland estimates are that there are 100,000 ha of planted *Leucaena leucocephala*, an exotic fodder species (Walton 2003) in large-scale cattle grazing operations. But there are also large areas "infested" with the species, and much of the literature about *Leucaena* treats it as a weed to be controlled or eradicated.

Some regions have detailed guides to the characteristics of tree species available for various purposes and the environmental requirements of those species. For example, Bird (2000) wrote a useful guide to growing high-quality sawlogs in South

Australia. Such guidebooks are yet to be developed for other regions, particularly for the subtropics and tropics which constitute highly variable landforms, soils, climates, and a wide array of native species that could be planted.

## Permaculture/Home Gardens

Australia has long been a center of development of permaculture (Mollison 1991) mainly in subtropical and temperate regions. This is somewhat similar as an agroforestry system to the tropical home garden (Kumar and Nair 2004). As in the USA, <2% of the Australian population are primary producers (directly involved in farming, forest harvesting, fishing) and live either in suburbs or in cities. Further, Australia suffers similar rates of "Western diseases" to the USA: this means a population with obesity rates approaching 40%, another third of the population overweight, high levels of heart disease, high blood pressure, diabetes, and so on.

Thus, a system like the home garden/permaculture plot, which is a small garden which integrates trees into a system of food production for one or several households, can address several problems simultaneously, providing healthful activities and fresh and beneficial food locally. A major environmental problem in Australia is the spread of property subdivisions with associated tree clearing and elimination of habitat. The implementation of many small-scale permaculture-based household gardens can have a positive impact on the whole suite of environmental services, including increasing landscape connectivity for wildlife.

## Grazing Systems and Shelterbelts in Australia

Bird et al. (1992) wrote an often-cited review of the benefits of shelter for protecting soil, plants, and animals, in broad-scale cropping and grazing areas. In terms of financial viability some 10–25% of a grazing property can be put into shelterbelts (Fig. 4) with no loss or increased income from wool, milk, or meat, because the tree/shrub cover actually increases production/lessens losses. The specific design and layout of these systems are important to maximize benefits and minimize costs (Fig. 4).

The enhanced productivity happens through a variety of mechanisms: lowering windspeeds of course—which can particularly lower destruction of seed crops by wind, increasing shade, water retention, and water availability, mitigating the effects of frosts, providing shelter where lamb survival can be higher, etc. As the impacts of climate change provide harsher conditions for agricultural stock and crops these benefits are likely to increase.

Windbreaks or shelterbelts can vary greatly in their design (Cleugh 2003; Basalt Bay Landcare Network 2014). In some cases, they are simple single rows of one species, such as exotic Lombardy poplars or one of the many native eucalypt species. In other instances, they can be quite complex and elegant (Fig. 5).

**Fig. 4** Example of engineered woodland compared with boundary shelterbelts, in terms of costs and areas sheltered, near Armidale, NSW. In addition to protecting paddocks from winds, engineered woodlands can also provide hydrological benefits such as slowing movement of water across a farm and biodiversity should improve through establishment of trees/shrubs as corridors for wildlife (source: Southern New England Land Care 2021)

**Fig. 5** A multilevel windbreak of mixed native and exotic species, at ~1300 m elevation Glen Innes, New South Wales, Australia, on a grazing property. Estimates are that woody vegetation can occupy up to 15–25% of a property with no loss of productivity by animals (in wool or meat), given the benefits provided by shrubs and trees (photo: J D Nichols 2007)

A well-designed shelterbelt can provide substantial production benefits to the landholder but also some or all of the usually cited ecosystem services, such as carbon sequestration, provision of habitat for wildlife and general biodiversity benefits, improvement or maintenance of soil quality, and improved water retention and quality (Jose 2009). Tree or shrub plantings can also be a source of forage for stock, providing improved productivity through increased nutrition as well as their extended growing period due to their deeper roots allowing access to water during dry periods (Bowen et al. 2015; Lefroy 2002).

## *Multipurpose Riparian Forests on Farms*

Australian research into the source of sediments and nutrients in waterways has highlighted the need to exclude stock from riparian areas and establish perennial vegetation to stabilize banks and trap nutrients (Harisine 1997; Abernethy and Rutherfurd 1999; Hubble et al. 2010). Hairsine (1997) reported that a riparian buffer strip composed of a 3 m strip of forest and 3 m of pasture grass was effective in trapping 98% of the sediments and 70% of the phosphorus flowing off a freshly cultivated farmland. Reid and Washusen (2001) and Smethurst (2004) explored the potential for these riparian forests on farmland to be managed for high-quality timber (Fig. 6).

## *Salinity*

Land cleared for pasture or annual crops is a major cause of land degradation and has led to increased soil salinity across much of Australia's agricultural land. This has resulted in major crop decline and toxicity, salinization of fresh river systems, soil erosion, increased flood risk, and denigration of riparian and terrestrial biodiversity, not to mention increasing aridity in already dry environments. Salinity represents a severe ecological and economic threat to Australia's rural communities and livelihoods.

Despite more than 60 years of documentation, salinity continues to be a threat to productive agriculture land. The most obvious answer in remediating saline soils is to reintroduce native vegetation into the landscape. Turner and Ward (2002) argue that it is not that simple—in areas that receive less than 600 mm of rainfall (i.e., most of Australia's agricultural land), broadscale planting of trees is not economic. Instead various land-use options must be combined and integrated (Turner and Ward 2002)—the very definition of agroforestry! Introducing salt-tolerant vegetation, combined with perennial pastures and trees that return some economic benefit to the farmer, has huge potential to mitigate degradation from salinity (Stirzaker et al. 2002). A document from the West Australian Government (Department of Primary Industries and Rural Development 2020) outlined case studies in

**Fig. 6** One of the authors (Rowan Reid) selectively logging his 32-year-old riparian strip on his farm in southern Victoria (photo courtesy: Rowan Reid)

addressing salinity issues in this state, which is one of the most heavily infected by salt issues (see https://www.agric.wa.gov.au/managing-dryland-salinity/managing-dryland-salinity-%E2%80%93-farmer-case-studies-western-australia). Cases are desired in pdfs available at the web site, and some are YouTube videos (https://www.youtube.com/watch?v=xtpZq5kOXu0&list=PLIRoVG3L9GNIj_IRVdhqg0OsprNCCwhxr&index=8). Most cases involve deploying shrubs in the genus *Atriplex* (saltbush) on which research has been conducted since the 1970s. If using woody shrubs may be considered a form of agroforestry then these are agroforestry systems, which help lower water tables, minimize rising salt, and enable continued economic uses of vast areas.

## *Carbon*

The present climate narrative has stressed the capability of increased atmospheric carbon dioxide ($CO_2$) to dismantle earth's current stability for human habitation (Ripple et al. 2020; Dinerstein et al. 2019; Wallace-Wells 2019). Addressing increasing greenhouse gas (GHG) emissions and capturing excess $CO_2$ have thus become urgent criteria for any farming or plantation enterprise, particularly given the great potential to convert $CO_2$ to biomass in vegetation systems (Schlamadinger et al. 2007). This mandate, coupled with calls to reverse ecosystem ruination (Wilson 2016), promotes agroforestry to the forefront with its potential to capture $CO_2$ across Australia's agricultural landscape conjunct with agricultural sustainability.

Albrecht and Kandji (2003) agree that agricultural lands have the potential to absorb large quantities of carbon (C) if trees are integrated across the landscape (as much as 228 Mg $ha^{-1}$). How then could this apply in Australian contexts? Zomer et al. (2016) estimate that of Australia's 597,000 ha of agricultural area the overall biomass is <10tC $ha^{-1}$, and it represents considerable potential in increasing biomass for C storage (though Florez et al. (2019) argue that biomass estimations are uncertain). Given that most Australian agricultural lands receive lower mean rainfall and higher variability per year, and on soils mostly considered substandard by global comparisons, the biophysical constraints of tree establishment within Australia's arid places cannot be ignored. However, even "simple" boundary plantings (such as windbreaks and shelterbelts discussed above) can sequester significant quantities of C (Albrecht and Kandji 2003). Reid (2017) notes that agroforests are not a secure carbon store, but are in unique standing as an opportunity for perpetual carbon sequestration through the repeated harvest of timber, cyclically freeing land to lock up more carbon (Reid 2017). Furthermore, increased diversity through strategic tree planting (e.g., N-fixing plants) would reduce net carbon loss through the reduction of fertilizer and pesticide input.

Therefore, a critical question must be asked: How willing are farmers to plant trees on farms to mitigate the effects of excess atmospheric $CO_2$? Despite the multitude of benefits in restoring deep-rooted vegetation to farmland (George et al. 2012; Smettem and Harper 2009), interviews conducted by Dumbrell et al. (2016) indicate that "tree planting" is an unlikely carbon farming practice that landholders would adopt. Though a large majority of those interviewed believed that climate change was occurring and their farms to be already experiencing the effects of climate change, carbon farming practices that were more favorable were those perceived by farmers to fit more readily into existing broadacre practices. Large-scale changes to the farming system—like tree planting—are just not enticing to most farmers. How then would this translate into motivation for farmers to adopt agroforestry practices? Dumbrell et al. (2016) make clear that those practices resulting in tangible, short-term production benefits are more alluring. Adding to this, agroforestry has been argued to be financially unviable—estimates by van Kooten et al. (2004) suggest that the cost of carbon sequestration through agroforestry is double the costs of forest conservation (Bryan et al. 2014). Though the study did not account for the potential value of wood products, nor consider the myriad of socio-economic benefits unique to agroforestry, if Australian farmers were to shift toward tree planting, substantial trial advancements must show significant economic advantages before widespread adoption will occur.

Poignantly, the Black Summer fires of 2019/2020 would not add confidence to Australian farmers to plant trees either. Why plant trees when the risk to live and livelihood is so great? While Damianidis et al. (2020) argue that agroforestry reduces wildfire risk, the report does not take into the account the flammability of

*Eucalyptus* and *Acacia* species—plants that dominate the Australian landscape and have adapted to thrive in fire-prone landscapes. Unconventional plantings might be a better option—many exotic deciduous species have been shown to be fire retardant. Silvicultural pruning techniques combined with grazing also reduce fire hazards, disconnecting ground fuel loads with the canopy. Fire management in agroforestry landscapes hence requires suitable species selection and strategic design. Previously dismissed cultural burning practices by Indigenous Australians must also come to the forefront in agroforestry design and fire management (Steffensen 2020).

Market-based incentives have the potential to encourage landholders to commit to sequestering C on agricultural lands. Despite the Australian Government introducing early legislation to promote farming projects that capture atmospheric $CO_2$, carbon markets in Australia remain confused (Dumbrell et al. 2016; Bryan et al. 2014). George et al. (2012) argue that robust carbon credit markets have the potential to expand agroforestry throughout arid landscapes. However, the additional capital, expertise, and labor required to implement broad changes may also lack, as well as the biophysical constraints of land prevent many landholders from adopting agroforestry practices (Bryan et al. 2014; Beardmore et al. 2019). Additionally, given market fluctuations as political cycles change, and shifting economic drivers, landholders must ask then what is the longevity of the carbon market in Australia, especially given the long-term nature of agroforestry (Bryan et al. 2014).

Furthermore, how can a "one-size-fits-all" model work given the biophysical diversity of the agricultural landscape (Beardmore et al. 2019). Until there is clear legislation, reflecting the diversity of biophysical constraints, and with long-term agreements in place, carbon markets will remain volatile, unsupportive, and confusing.

Estimates of soil carbon storage are variable: Dudek and LeBlanc (1990) and Lal (2008) estimate that two-thirds of terrestrial carbon is stored in soils; Dixon et al. (1994) argue that up to three times as much carbon is stored in soil or up to five times that of atmospheric and biomass carbon combined (Lal 2016). The variability in part is due to different soil depth parameters, and whether both organic and inorganic carbon estimates are included. Despite the variability, there are many farming and forestry practices that are suitable for increasing soil carbon (and thus have the potential to mitigate carbon emissions), particularly since SOC is a much bigger carbon sink than biomass. So why plant trees? Forests produce a soil carbon profile quite different to that of grasslands, accumulating organic material on the soil surface (material that is generally excluded from soil carbon calculations which focus on mineral soil carbon). Where forests are dense and highly productive, the tree biomass far exceeds soil carbon. In highly productive soils, the conversion from cropping to forest can lead to relatively large increases in carbon in a short period of time (Zomer et al. 2016; Osuri et al. 2019). Despite grasslands producing soils with higher carbon content in the upper soil mineral layers than forests generally do, deep-rooted vegetation stores carbon deeper into the soil than grasslands (Jobbagy and Jackson 2000). Additionally, the lignified material produced by trees tends to be longer lasting than the simpler compounds that dominate grass biomass.

But the carbon cycle is far from simple. Comparisons between pasture and plantations by Richards et al. (2007) suggest that hoop pine (*Araucaria cunninghamii*) plantations in northern New South Wales do not appear to be returning carbon to soil in a way that will return the soil carbon profile to that under the original forest in the short or medium term. Additionally, although Young et al. (2005) show the importance of forests for carbon storage in drier areas, how quickly carbon storage could be achieved from the conversion of pasture to planted forests within these regions remains unknown.

Metcalfe and Bui (2016) estimate that adding 0.4% of carbon into soil each year could neutralize Australia's GHG emissions. However, this 0.4% figure is virtually unattainable. Even in intensively managed soils where carbon increases can be achieved through a range of practices, the length of time that carbon is likely to be sequestered is unknown. Additionally, agricultural land covers only a small proportion of Australia's land area—any changes in management practices are only likely to impact relatively small areas (Lam et al. 2013).

Then there is the suggestion that fast-growing monoculture plantations sequester more C than other agroforestry options (George et al. 2012; Bonner et al. 2013; Huang et al. 2018). Though a prominent drawcard, carbon-based solutions are therefore not the answer. For most, carbon data is abstract and hard for the general population to comprehend and for farmers to link theoretical trends with on-ground repercussions. This is especially true in Australia, where wealth and infrastructure currently insulate most of the population from the tangible effects of climate change. Are scientific evidence-based predictions enough to convince farmers to adopt agroforestry within agricultural landscapes? If carbon sequestration is the ultimate definition of climate mitigation, then we run into problems. $CO_2$ reductionism and traditional notions of ecosystem services are really a brand of "nature trafficking" (Eisenstein 2018). Eisenstein (2018) further criticized the attempt to quantify "natural resources" which has gotten us into our present dire situation and justified projects that have devastating social and ecological effects (e.g., fracking, draining of wetlands).

## *Biodiversity Enhancement on Farms*

We believe that agroforestry systems and other forms of more regenerative practices should not be considered simply on the basis of one quantifiable benefit: "when dealing with a complex interconnected system, there is no such thing as an unbiased number" (Eisenstein 2018). The positive impact of a given agroforestry system, say a shelterbelt, may be quite minor if only judged by one number (decreased wind speed and therefore slightly higher crop yield) but the impacts overall may be much greater if biodiversity is also somehow added in (metrics of biodiversity assessment are challenging: see Jay et al. 2009, Baral et al. 2014, and Tennent and Lockie 2013), as well as the carbon benefits, improvement in timing and quantity of waterflows on a property and in a catchment, and so on. Also, as Reid (2017) explains, if we decide whether or not to plant a timber tree on a given farm solely on the basis of the traditional numbers of internal rate of return or net present value often the decision would be not to, as the "value" of the timber discounted over long time period would be quite low.

However, there is much more to consider other than simply the yield of money at rotation's end. In other words, agroforestry systems should be considered holistically.

For an example of an attempt to improve biodiversity, a small NGO in the Northern Rivers region of New South Wales, the Subtropical Farm Forestry Association (SFFA), received a $2.3 million grant to promote carbon sequestration and biodiversity on farms in a highly biodiverse subtropical region (SFFA 2020). Over 4 years, 2012–2016, it was able to oversee biodiversity-enhancing activities on some 600 hectares across more than 30 properties (SFFA 2020). Although the main activity was tree planting, there were other techniques employed, including weed management and installation of species-specific nest boxes—hollows to nest in are becoming rarer in native forests and are absolutely critical for the survival of many arboreal marsupials, as well as of birds and other fauna. Such efforts build corridors between reserves and parks and provide many ecosystem services and generally make a positive contribution, but they are small scale and large amounts of time and money often end up being squandered on one-off administration.

## *Socioeconomic/Policy Challenges*

Although the beneficial ecosystem services provided by trees in native forests, plantations, and agroforestry systems are widely recognized in Australia, it has been difficult for the society to develop policy and subsidy structures to support their implementation. As in the USA, Australia has a federal (Commonwealth) government and state government, and of course governments that change from election to election and invariably have differing priorities. This makes it difficult to develop long-term programs that promote the establishment of not only beneficial agroforestry systems, but also extension services to support them and funding for ongoing maintenance.

The biodiversity-on-farms program by Subtropical Farm Forestry Association was a "one-time-only" initiative mostly administered as a "legacy" program, that is, one inherited from a previous government. The building of new administrative structures to link between the Commonwealth government, the NGO, and the dozens of landholders consumed much of the energy and funding, so little on-ground benefit was seen after that.

Another example, a program called "Management Investment Schemes," was in place nationally, and up until about 2009 attracted approximately $3 billion in investments in tree plantations (Underwood 2007; Sydney Morning Herald 2009; Ferguson 2014). The final reckoning of this program has not been made public. Although some viable tree plantations were established, mainly blue gum plantings on short rotations for wood chip in the more temperate states, it is generally felt that the program became primarily focused on tax breaks and not on practicing good land management through tree planting. Further a general trend has been for traditional agricultural and forestry extension to be defunded, with technical assistance coming from a more fragmented sector of local groups and NGOs. This has made the regional or catchment-wide approach required to tackle broadscale problems such as salinity difficult.

Direct incentives, such as grants to cover the cost of seedlings and fences, have been one of the standard approaches used by state governments to encourage farmers to plant trees. In fact, many landholders now expect them as a prerequisite for planting trees. However, the farmers who need to adopt the favored agroforestry practices—in order to help achieve the desired public benefits—are either unable or unwilling to pay the full cost themselves. In our experiences in rural Victoria and northern subtropical New South Wales, landholders often expect "the government" to pay for ongoing maintenance as well as initial establishment costs. Reid (2017) argues that what makes the models being promoted by government unattractive to farmers is obvious: the high up-front costs; loss of existing production potential; long investment periods; high social, environmental, or market risk; and perception of low private benefit. Furthering this argument, Reid (2017) gives ten reasons as to why direct incentives are an ineffective tool for catalyzing spontaneous agroforestry adoption:

1. Direct incentives stifle, rather than drive, innovation and adaptation by only supporting technologies or methods approved by the funding agency.
2. They actively discourage farmers from implementing alternative practices that provide the same public benefits.
3. They support only a few of the potential recipients while alienating the majority who miss out because of the program's timeframes, conditions, preferred location, or eligibility criteria; people who are adept at paperwork, managing meetings, etc. will do much better than those who are focused on actual farming and agroforestry.
4. They can reward, rather than discourage, mismanagement, neglect, or inappropriate farming systems by supporting landholders who continue practices that are known to contribute to the problem.
5. They encourage farmers to overcommit by requiring a particular level of adoption, such as a minimum area of revegetation.
6. They undermine early adopters by not acknowledging or rewarding those farmers who have implemented similar technologies without having received public funds.
7. They discourage third-party investors who might have been willing to jointly fund multipurpose plantings that deliver both public and private benefit.
8. The fact that a farmer requires a grant or free inputs to implement the promoted practices provides a clear signal to other landholders that the innovation is not worth funding privately.

9. Over time a welfare mindset tends to develop amongst farmers to the point that many assume that investing in conservation practices is a public rather than a private responsibility.
10. When offered cost-share grants, farmers will see little point in engaging in community networks that build their own knowledge and confidence in tree growing and support the development of sophisticated and resilient agroforestry designs that are appropriate to each family and their landscapes.

Reid (2017) proposes three alternatives to direct incentives that governments and NGOs can use to support tree growing on farms:

1. Developing the physical and legal infrastructure required to support the industry, such as port facilities for exports or enabling legislation
2. Establishing fair and open markets for environmental, social, and commercial services that reward farmers for the public goods their trees provide, irrespective of how the outcomes are provided
3. Building the capacity of farmers, and those that service the sector, so that they are all able to make better decisions

## Conclusions

Australians are beginning to realize that traditional agricultural and forestry practices in this fragile continent, often imported from temperate regions in the Northern Hemisphere, have led to high levels of land degradation. Thus, we may hope in the struggle ahead that agroforestry and other regenerative forms of land use will become more widespread across the landscape.

What needs to be addressed is our failure to recognize that the common thread throughout each instance of ecological destruction and dismantling of ecosystem services—arguably the root of the problem—is the human factor. The flawed philosophy of reducing the complexity of natural systems to a set of preconceived human values within an economic framework is unlikely to make the difference needed to mitigate omnicide and ecocide. This may be partly attributed to the reduction of "nature" to a pragmatic, dispassionate concept labelled as "environment" and "services," separating most of humanity from the finite, natural world (Eisenstein 2018; Pretty et al. 2009). Through a rational framework, the nonlinear, multiscale, complex, intricate, and beautiful traits that define forests are pushed aside, further justifying overextraction and degradation. The environment—nature—is something to "use," "measure," and "manage"—and not relate to. This mindset has promoted the delusion that ecosystems can be "used" and "managed" at an ever-increasing rate while achieving exponential growth and development, without reciprocity (Ripple et al. 2018, 2020; Eisenstein 2018). Detachment from Australia's ecosystems has meant that much of society has proceeded in ignorance and failed to address the underlying issue of ecological devastation and biodiversity loss—ourselves.

## References

ABARES (2021) Australia's forests. Australian Government. https://www.agriculture.gov.au/abares/forestsaustralia/australias-forests. Accessed 29 January 2021

Abernethy B, Rutherfurd ID (1999) Guidelines for Stabilising streambanks with riparian vegetation, technical report 99/10, land and water resources Research and Development Corporation, Canberra. https://ewater.org.au/archive/crcch/ archive/pubs/pdfs/technical199910.Pdf. Accessed 5 February 2021

Albrecht A, Kandji ST (2003) Carbon sequestration in tropical agroforestry systems. Agric Ecosyst Environ 99(1–3):15–27

Baral H, Keenan RJ, Sharma SK, Stork NE, Kasel S (2014) Spatial assessment and mapping of biodiversity and conservation priorities in a heavily modified and fragmented production landscape in north-central Victoria, Australia. Ecol Indic 36:552–562

Basalt Bay Landcare Network (2014) Economic Benefits of Native Shelter Belts. https://www.stategrowth.tas.gov.au/__data/assets/pdf_file/0018/131913/Economic_Benefits_of_Native_Shelter_Belts.pdf

Beardmore L, Heagney E, Sullivan CA (2019) Complementary land use in the Richmond River catchment: evaluating economic and environmental benefits. Land Use Policy. https://doi.org/10.1016/j.landusepol.2019.104070

Bird PR (2000) Farm forestry in southern Australia: a focus on clearwood production of specialty timbers. Dept. of Natural Resources and Environment, Rutherglen

Bird PR, Bicknell D, Bulman PA, Burke SJA, Leys JF, Parker JN, Vandersommen FJ, Voller P (1992) The role of shelter in Australia for protecting soils, plants and livestock. Agrofor Syst 20(1–2):59–86

Boland DJ, Brooker MIH, Chippendale GM, Hall N, Hyland BPM, Johnston RD, Kleinig DA, McDonald MW, Turner JD (2006) Forest trees of Australia. CSIRO Publishing, Collingwood

Bonner MTL, Schmidt S, Shoo LP (2013) A meta-analytical global comparison of aboveground biomass accumulation between tropical secondary forests and monoculture plantations. Forest Ecol Manag 291:73–86

Bowen M, Chudleigh F, Buck S, Hopkins K, Brider J (2015) High-output forages for meeting beef markets—Phase 2. Meat and Livestock Australia. https://www.mla.com.au/Research-and-development/Search-RD-reports/final-report-details/Productivity-On-Farm/High-output-forage-systems-for-meeting-beef-markets-Phase-2/2910. Accessed 3 February 2021

Brown SE, Miller DC, Ordonez PJ, Baylis K (2018) Evidence for the impacts of agroforestry on agricultural productivity, ecosystem services, and human well-being in high-income countries: a systematic map protocol. Environ Evid. https://doi.org/10.1186/s13750-018-0136-0

Bryan BA, Nolan M, Harwood TD, Connor JD, Navarro-Garcia J, King D, Summers DM, Newth D, Cai Y, Grigg N, Harman I, Crossman ND, Grundy MJ, Finnigan JJ, Ferrier S, Williams KJ, Wilson KA, Law EA, Hatfield-Dodds S (2014) Supply of carbon sequestration and biodiversity services from Australia's agricultural land under global change. Glob Environ Change Hum Policy Dimen 28:166–181

Cleugh H (2003) Trees for shelter: a guide to using windbreaks on Australian farms. RIRDC, Barton, ACT

Damianidis C, Santiago-Freijanes JJ, den Herder M, Burgess P, Mosquera-Losada MR, Graves A, Papadopoulos A, Pisanelli A, Camilli F, Rois-Díaz M, Kay S, Palma JHN, Pantera A (2020) Agroforestry as a sustainable land use option to reduce wildfires risk in European Mediterranean areas. Agrofor Syst. https://doi.org/10.1007/s10457-020-00482-w

Department of Primary Industries and Regional Development: Agriculture and Food (2020) Managing dryland salinity – farmer case studies from Western Australia. Western Australian Government. 12/02/2021 at https://www.agric.wa.gov.au/managing-dryland-salinity/managing-dryland-salinity-%E2%80%93-farmer-case-studies-western-australia

Dinerstein E, Vynne C, Sala E, Joshi AR, Fernando S, Lovejoy TE et al (2019) A global deal for nature: guiding principles, milestones, and targets. Sci Adv. https://doi.org/10.1126/sciadv.aaw2869

Dixon RK, Winjum JK, Andrasko KJ, Lee JJ, Schroeder PE (1994) Integrated land-use systems: assessment of promising agroforest and alternative land-use practices to enhance carbon conservation and sequestration. Clim Chang 27(1):71–92

Do TD, NaRanong A (2019) Livelihood and environmental impacts of payments for forest environmental services: a case study in Vietnam. Sustain 11(15):22

Dudek D, LeBlanc A (1990) Offsetting new $CO_2$ emissions: a rational first greenhouse policy step. Contemp Econ Policy. https://doi.org/10.1111/j.1465-7287.1990.tb00642.x

Dumbrell NP, Kragt ME, Gibson FL (2016) What carbon farming activities are farmers likely to adopt? A best-worst scaling survey. Land Use Policy 54:29–37

Eisenstein C (2018) Climate: a new story. North Atlantic Books, Berkeley

Ferguson I (2014) Australian plantations: mixed signals ahead. Int Forestry Rev 16(2):160–171

Florez LG, Vanclay JK, Glencross K, Nichols JD (2019) Developing biomass estimation models for above-ground compartments in Eucalyptus dunnii and Corymbia citriodora plantations. Biomass Bioenergy. https://doi.org/10.1016/j.biombioe.2019.105353

George SJ, Harper RJ, Hobbs RJ, Tibbett M (2012) A sustainable agricultural landscape for Australia: a review of interlacing carbon sequestration, biodiversity and salinity management in agroforestry systems. Agric Ecosyst Environ 163:28–36

Hairsine P (1997) Controlling sediment and nutrient movement within catchments. Cooperative Research Centre for Catchment Hydrology, Clayton

Huang Y, Chen Y, Castro-Izaguirre N, Baruffol M, Brezzi M, Lang A, Li Y, Härdtle W, von Oheimb G, Yang W, Lui X, Pei K, Both S, Yang B, Eichenberg D, Assman T, Bauhus J, Behrens T, Buscot F, Chen X, Chesters D, Ding BY, Durka W, Erfmeier A, Fang J, Fischer M, Guo L, Guo D, Gutknecht HJ, He C, Hector A, Hönig L, Hu R, Klein A, Kühn P, Liang Y, Li S, Michalski S, Scherer-Lorenzen M, Schmidt K, Scholten T, Schuldt A, Shi X, Tan M, Tang Z, Trogisch S, Wang Z, Welk E, Wirth C, Wubet T, Xiang W, Yu M, Yu X, Zhang J, Zhang S, Zhang N, Zhou H, Zhu C, Zhu L, Bruelheide H, Ma K, Niklaus PA, Schmid B (2018) Impacts of species richness on productivity in a large-scale subtropical forest experiment. Science 362:80–83

Hubble TCT, Docker BB, Rutherfurd ID (2010) The role of riparian trees in maintaining riverbank stability: a review of Australian experience and practice. Ecol Eng 36(3):292–304

Jay VA, Sharpe D, Nichols D, Vanclay JK (2009) Sustainable private native forestry: a review of timber production, biodiversity and soil and water indicators, and their applicability to north-east NSW, ACT Australia RIRDC, Canberra

Jobbagy EG, Jackson RB (2000) The vertical distribution of soil organic carbon and its relation to climate and vegetation. Ecol Appl 10(2):423–436

Jose S (2009) Agroforestry for ecosystem services and environmental benefits: an overview. Agrofor Syst 76(1):1–10

Kumar BM, Nair PKR (2004) The enigma of tropical homegardens. Agrofor Syst 61:135–152

Lal R (2008) Carbon sequestration. Phil Trans R Soc B 363:815–830

Lal R (2016) Soil health and carbon management. Food Energy Security 5(4):212–222

Lam SK, Chen D, Mosier AR, Roush R (2013) The potential for carbon sequestration in Australian agricultural soils is technically and economically limited. Sci Rep. https://doi.org/10.1038/srep02179

Lefroy EC (2002) Forage trees and shrubs in Australia: their current use and future potential: a report for the RIRDC/L&W Australia/FWPRDC joint venture agroforestry program. RIRDC, Canberra

Metcalfe D, Bui E (2016) Australia state of the environment: land. Commonwealth of Australia. https://soe.environment.gov.au/theme/land/topic/soil-carbon-dynamics. Accessed 28 January 2021

Mollison B (1991) Introduction to permaculture. Tagari Publications, Tyalgum

Moore RW, Bird PR (1997) Temperate agroforestry in Australia. In: Gordon AM, Newman SM (eds) Temperate agroforestry systems. CAB International, Oxford, pp 119–148

O'Grady AP, Mitchell PJ (2017) Agroforestry: realising the triple bottom benefits of trees in the landscape. CSIRO, Australia

Osuri AM, Gopal A, Shankar Raman TR, DeFries R, Cook-Patton SC, Naeem S (2019) Greater stability of carbon capture in species-rich natural forests compared to species-poor plantations. Environ Res Lett. https://doi.org/10.1088/1748-9326/ab5f75

Powell J (2009) Fifteen years of the joint venture agroforestry program foundation research for Australia's tree crop revolution. RIRDC, Canberra

Pretty J, Adams B, Berkes F, Ferreira de Athayde S, Dudley N, Hunn E, Maffi L, Milton K, Rapport D, Robbins P, Sterling E, Stolton S, Tsing A, Vintinner E, Pilgrim S (2009) The intersections of biological diversity and cultural diversity: towards integration. Conserv Soc 7(2):100–112

Reid R (2017) Heartwood—the art and science of growing trees for conservation and profit. Melbourne Books, Melbourne

Reid R, Stephen P. (2001) The Farmer's Forest: Multipurpose Forestry for Australian Farmers. Australian Master Tree Growers Program. Joint Venture Agroforestry Program

Reid R, Deans W (2009) The Australian Master TreeGrower Program—providing a sound foundation for Australian agroforestry. JVAP Publication No. 09/157

Reid R, Moore R (2018) Agroforestry systems in temperate Australia. In: Gordon AM, Newman SM, Coleman BRW (eds) Temperate agroforestry systems. CABI, Oxfordshire

Reid R, Washusen R (2001) Sawn timber from 10-year-old pruned *Eucalyptus nitens* (Deane & Maiden) grown in an agricultural riparian buffer. Paper presented at the "Third Australian Stream Management Conference" Brisbane, 27–29 August 2001

Richards AE, Dalal RC, Schmidt S (2007) Soil carbon turnover and sequestration in native subtropical tree plantations. Soil Biol Biochem 39:2078–2090

Ripple WJ, Wolf C, Galetti M, Newsome TM, Green TL, Alamgir M, Crist E, Mahmoud MI, Laurance WF (2018) The role of scientists' warning in shifting policy from growth to conservation economy. Bios 68(4):239–240

Ripple WJ, Wolf C, Newsome TM, Barnard P, Moomaw WR (2020) World scientists' warning of a climate emergency. Bios 70(1):100

Sandhu HS, Crossman ND, Smith FP (2012) Ecosystem services and Australian agricultural enterprises. Ecol Econ 74:19–26

Schlamadinger B, Bird N, Johns T, Brown S, Canadell J, Ciccarese L, Dutschke M, Fiedler J, Fischlin A, Fearnside P, Forner C, Freibauer A, Frumhoff P, Hoehne N, Kirschbaum MUF, Labat A, Marland G, Michaelowa A, Montanarella L, Moutinho P, Murdiyarso D, Pena N, Pingoud K, Rakonczay Z, Rametsteiner E, Rock J, Sanz MJ, Schneider UA, Shivdenko A, Skutsch M, Smith P, Somogyi Z, Trines E, Ward M, Yamagata Y (2007) A synopsis of land use, land-use change and forestry (LULUCF) under the Kyoto protocol and Marrakech accords. Environ Sci Pol 10:271–282

Smethurst P (2004) Production forestry in riparian zones: Examples from Brazil, USA, Germany and Australia. 'Riparian Forestry' Gottstein Fellowship Report 2004

Smettem KRJ, Harper R (2009) Using trees to manage local and regional water balance. In: Nuberg I, George B, Reid R (eds) Agroforestry for natural resource management. CSIRO Publication, Collingwood, pp 37–52

Southern New England Land Care. (2021). Engineered woodlands, Information Sheet. https://snelandcare.org.au/linkedfiles/EWInformationSheet2web.pdf

Steffensen V (2020) Fire country—how indigenous fire management could save Australia. Explore Australia Publishing, Richmond

Stirzaker R, Vertessy R, Sarre R (2002) Trees, water and salt: an Australian guide to using trees for healthy catchments and productive farms. RIRDC, Canberra

Subtropical Farm Forestry Association. (2020) Connecting conservation areas through linked biodiverse carbon forestry systems; Case studies in integrated reforestation. Subtropical Farm

Forestry Association. https://www.subtropicalfarmforestryassociation.com/sffa-projects. Accessed 27 January 2021
Sydney Morning Herald. (2009, May 23). https://www.smh.com.au/business/stumped%2D%2Dthe-death-of-mis-20090522-bidn.html
Tennent R, Lockie S (2013) Market-based instruments and competitive stewardship funding for biodiversity conservation: the achievable reality. Aust J Environ Manag 20(1):6–20
Turner NC, Ward PR (2002) The role of agroforestry and perennial pasture in mitigating water logging and secondary salinity: summary. Agric Water Manag 53:271–275
Underwood R (2007) Assessing "management investment scheme" forestry projects: a best-practice template for commercial plantation development. Aust Forestry 70(4):269–274
van Kooten GC, Eagle AJ, Manley J, Smolak T (2004) How costly are carbon offsets? A meta-analysis of carbon forest sinks. Environ Sci Pol 7:239–251
Wallace-Wells D (2019) The uninhabitable earth. Crown Publishing Group, New York
Walton CS (2003) *Leucaena leucocephala* in Queensland. Q. G. Natural Resources and Mines, Brisbane
Wilson EO (2016) Half-earth: our planet's fight for life. Liveright Publishing, New York
Young R, Wilson BR, McLeod M, Alston C (2005) Carbon storage in soils and vegetation contrasting land uses in northern New South Wales, Australia. Aust J Soil Res 43:21–31
Zomer RJ, Neufeldt H, Xu J, Ahrends A, Bossio D, Trabucco A, van Noordwijk M, Wang M (2016) Global tree cover and biomass carbon on agricultural land: the contribution of agroforestry to global and national carbon budgets. Sci Rep. https://doi.org/10.1038/srep29987

# Agroforestry Integration and Multifunctional Landscape Planning for Enhanced Ecosystem Services from Treed Habitats

**Sarah T. Lovell, Erik Stanek, and Ronald Revord**

## Abbreviations

| | |
|---|---|
| CRP | Conservation Reserve Program |
| GIS | Geographic Information Systems |
| GM | Genetically modified |
| HEL | Highly erodible soils |
| LSA | Land suitability analysis |
| NRCS | Natural Resources Conservation Service |
| NWP | National Windbreaks Program |
| STE | Sediment trapping efficiency |
| TNSFP | Three-North Shelter Forest Program |
| US | United States |

## Introduction

Integrating trees and shrubs into the agricultural landscape could improve the provision of most ecosystem services, particularly compared with conventional monoculture crops. However, the establishment of agroforestry practices can represent a trade-off in terms of caloric yield and near-term profit for the farmer or landowner. In highly productive agricultural landscapes, the strategic placement of these treed habitats could optimize the benefits while allowing continued cropping on the most productive lands. Furthermore, the use of public funds to incentivize agroforestry

S. T. Lovell (✉) · R. Revord
The Center for Agroforestry, School of Natural Resources, University of Missouri, Columbia, MO, USA
e-mail: slovell@missouri.edu; r.revord@missouri.edu

E. Stanek
Balzac Brothers & Company, Charleston, SC, USA



R. P. Udawatta, S. Jose (eds.), *Agroforestry and Ecosystem Services*,
https://doi.org/10.1007/978-3-030-80060-4_17

practices might be best allocated based on optimized placement that offers the highest return and overall performance.

While interest in agroforestry is growing, little attention has focused on the potential for intentional planning for these treed habitats in the broader agricultural landscape. Regional planning, land-use planning, or simply "planning" as written hereafter relates to the purposeful disposition of land and resources, including the placement of permanent vegetation and habitat types such as agroforestry. At a finer resolution, the term "design" can be used to refer to the specific modification and spatial arrangement of landscape features, even down to the level of assigning individual species (Lovell et al. 2010).

We envision that this chapter will be a useful resource for many different *audiences*. However, we expect that it could have the most significant impact by informing four particular groups, which we define here. **Policy makers** are individuals determining how to allocate funds to support various agroforestry practices and research. **Regional planners** seek to prioritize zoning of land use and protection of natural resources, which could include integration of agroforestry practices. **Academics** (such as the authors of this chapter) define the focus of their research efforts to improve the functionality of agroforestry practices. **Agency personnel** assist those working directly with landowners/farmers to increase adoption conservation practices such as agroforestry. All of these groups have important roles to play in shaping a future transition to land-use types that offer a more extensive range of ecosystem services.

We define *stakeholders* separately from the audience, because we recognize that these groups will be impacted by the decision-making, even if they are less likely to engage directly with the material provided herein. **Landowners** are typically the ultimate decision makers. Without their approval, none of the practices will be implemented. Landowners can include individuals or groups (such as corporations) who may or may not be involved in the direct management of the land they own. Many landowners live on or near their agricultural properties, but increasingly they live far away (i.e., absentee landowners). **Farmers** are the individuals tasked with managing the farm operation. They may or may not be the landowners, and often they farm a combination of land they own and land they rent. Their decisions about future land-use changes are likely to be different for parcels they own versus those they rent. The **public** is the largest stakeholder group that will be impacted by land use as it relates to various ecosystem services or disservices from agroforestry practices. The provision of food, the protection of natural resources, and the creation of beauty in the landscape will be relevant to the broader public, even though the landowners are not compensated for many of these public benefits.

The placement of agroforestry practices has been studied at many different scales. We define two scales that cover the different aspects of decision-making and implementation. Much of the literature on the placement of various habitats such as agroforestry has focused on the watershed or regional scale. Guided by the discipline of landscape ecology, we will refer to these broader scales that consider spatial patterns and ecological processes as the "**landscape scale**" (Wu 2013a), unless a specific boundary is appropriate (e.g., a watershed for riparian buffer planning). As such, the "landscape" is defined as: "a geographic area in which variables of interest are spatially heterogeneous. The boundary of a landscape may be delineated based on geographic, ecological, or administrative units (e.g., a watershed, an urban area, or a county) which are relevant to the research questions and objectives" (Wu 2013b) (p. 5772).

While receiving less attention, the scale of the whole farm (hereafter referred to as the "**farm scale**") is also highly relevant for agroforestry, as this is most often the level at which decisions are made regarding land use and placement of treed habitats. Design for the farm scale considers all parcels owned and managed by an individual farmer or farm business unit and thus requires development in cooperation with these individuals. More resources are needed to support whole-farm planning, as there are currently few available (Lovell et al. 2010). This chapter covers both the landscape and the farm scales for the application of placement of agroforestry practices, recognizing that decisions about the efficient use of public funds for various practices would be more appropriate at the landscape scale, while decisions about benefits for the landowner or farmer would take place at the farm scale.

The following section of this chapter provides a historical perspective on the placement of agroforestry systems in the landscape, considering the contributions of landscape ecology, land suitability analysis, and landscape multifunctionality. The main portion of the chapter summarizes the literature on the planning and placement of each of the five agroforestry practices, considering scales from the landscape to the whole farm. For each practice, novel solutions are explored at the intersection of two approaches: (1) optimizing placement within the landscape for the public good and (2) matching the needs and preferences of the landowner. The final section of the chapter projects a path forward in which advanced technologies in landscape modeling and visualization are applied to the problem to link institutions with landowners, to inform land-use decisions and implementation.

## Background

Earlier work in landscape ecology, land suitability analysis, conservation planning, and landscape multifunctionality informed recent efforts to place agroforestry practices into the landscape. This section provides historical context for a topic that continues to evolve as new technology in spatial analysis becomes available.

### *Landscape Ecology*

The term "landscape ecology" was coined in 1939 by Carl Troll, a German geographer who contributed to the development of many of the concepts and terminology in the field (Turner et al. 2001). His work intended to characterize the distribution of ecological communities, considering the interaction between smaller landscape components in the broader landscape (Cushman et al. 2010; Wu 2013a). In the early 1980s, landscape ecology became recognized as a unique discipline, around the same time that the patch-corridor-matrix model became an accepted framework for landscape function (Forman and Godron 1981). This effort helped establish a language for describing the spatial pattern and arrangement of different components (habitats) of landscape structure. The model defined discrete areas consisting of specific vegetation as "patches"; linkages between patches as the "corridors"; and the "matrix" as the dominant landscape type (Forman and Godron 1986; Dramstad et al. 1996).

Fitting agroforestry into the patch-corridor-matrix model, however, creates a bit of a challenge depending on the extent to which the practices are truly integrated into the agricultural matrix. Are agroforestry practices considered to be distinct habitats that serve as the patches and corridors in an otherwise conventional agricultural system? Or are they part of the matrix? The answers may depend on the type of agroforestry practice (e.g., riparian buffer versus alley cropping system) and the scale at which the landscape is assessed. These questions may drive to a deeper point—that agroforestry offers opportunities to move beyond the focus on large, high-quality, so-called natural habitats, to instead focus on improving the overall quality of the matrix in agricultural landscapes. This approach would encourage thinking beyond a binary choice between "good" and "bad" habitats that either fit or do not fit the model, to consider opportunities for deeper integration of land-use types (Perfecto and Vandermeer 2002; Baudry et al. 2003; Bailey 2007) to intentionally design functional heterogeneity into the landscape pattern (Fischer et al. 2006).

## *Landscape Suitability Analysis*

Building on the work in landscape ecology, landscape suitability analysis (LSA) was developed as a method that "… focuses on the *fitness* of a given tract of land for a particular use…finding the optimal location for different uses of the landscape" (Ndubisi 2002) (pg 244). Soil scientists and landscape architects were among the first to develop LSA approaches to classify landscape zones and later to evaluate their potential for conservation or development, typically represented through a set of maps. One of the first approaches was the Gestalt method, developed by Lewis Hopkins. This method characterized the potential for a landscape to support human activity, primarily based on conditions of the existing landscape. Aerial imagery or remote-sensing data was used to record landscape patterns and in turn to recommend categories or land uses based on vegetative cover, wet areas, etc. (Ndubisi 2002).

Two land inventory systems developed in the early 1960s provided a base for land suitability analysis. The soil capability system developed by the Soil Conservation Service (now the Natural Resources Conservation Service, NRCS) was widely adopted at the time it was introduced and still informs landscape analysis today. The method was designed to classify soils based on a combination of properties, with the purpose of assigning a particular landscape (spatial area) with specific uses that its soils could support. The primary application was for determining the suitability of a parcel for production of particular crops. As such, this system has an important role to play in the determination of appropriate tree/shrub and alley crops for agroforestry (Soil Conservation Service 1961; Ndubisi 2002). Around the same time, Canadian forester G. Angus Hills developed a land inventory system referred to as the physiographic unit approach. Beyond the soil associations, this system integrated landforms and vegetation associations with the purpose of distinguishing landscapes for a variety of purposes (Ndubisi 2002).

Landscape and regional planning fields also contributed to the development of modern landscape suitability analysis. The resource pattern method was developed by Philip Lewis through his work on several planning projects, to identify unique landscape features or patterns that could become the "environmental corridors" integrated into plans and designs. The consideration of visual quality and recreational value was notable in this work (Ndubisi 2002). Publication of Ian McHarg's "Design with Nature" in 1969 substantially influenced the development of LSA for preserving nature. The method involved developing and mapping the ecological inventory of the landscape, and then determining appropriateness for various land uses to provide suitability maps. Key aspects were the development of holistic solutions based on multidisciplinary analysis and the quantification of landscape performance (Yang and Li 2016). The approach of creating a composite from a series of semitransparent overlays formed the basis for today's GIS outputs (Ndubisi 2002). McHarg's landscape suitability assessment framework aligned directly with another important framework focusing on landscape multifunctionality (Yang and Li 2016).

## *Multifunctional Landscape Framework*

Similar to the concept of "ecosystem services," landscape multifunctionality (or multifunctional agriculture) suggests that our agricultural systems have the potential to provide not only production functions but also ecological functions and cultural functions (Madureira et al. 2007). Conceptually and in landscape planning, multiple functions can be combined or stacked to increase the overall performance of agroecosystems (Lovell and Johnston 2009; Jordan and Warner 2010). The framework offers an opportunity for *redesigning the landscape* to support a "transition to sustainable food systems" (Gliessman 2010) that could expand the suite of goods and services from the landscape (Jordan and Warner 2010). Applications of this approach for agroforestry include (1) improving the quality of the agricultural matrix through the addition of treed habitats; (2) offering whole-farm designs that accommodate farmer preferences to integrate agroforestry; and (3) exploring the potential to use marginal lands for productive systems structured with trees and shrubs.

**Table 1** Landscape-level strategies for optimizing placement of the five agroforestry practices

| Type | Placement strategies | Justification |
|---|---|---|
| **Alley cropping** | | |
| | Prioritizing marginal lands | • Crop productivity is lower, so trees can be integrated with lower opportunity cost<br>• Areas are often prone to soil erosion or damage |
| | Optimizing tree productivity | • Zones suitable for given tree species increase success and profitability |
| | Identifying crops most compatible with trees and their relative distributions | • Low-maintenance crops decrease the risk of management-induced damage to the trees<br>• Crops with active growth periods that counter trees will minimize competition |
| **Riparian forest buffers** | | |
| | Buffering the zones beside streams at fixed width | • Offers standard protection for water resource and provides other ecological benefits |
| | Optimizing the buffer zone through variable width | • Accounts for site-specific physical conditions that impact buffer effectiveness |
| | Targeting placement based on soil units or terrain analysis | • Identifies areas and intercepts their runoff, based on soil properties or topographic data |
| **Windbreaks** | | |
| | Prioritizing regions where crop growth is constrained by strong winds | • Climatic conditions include high winds and rainfall<br>• High-value specialty crops are often most sensitive to winds or chemical drift |
| | Leveraging government programs and public support | • Visual quality and microclimate benefits create support from the public |
| **Silvopasture** | | |
| | Integrating trees in harsh climates | • Livestock in heat-stressed climates will benefit most from tree cover |
| | Targeting forested areas that could support livestock | • Open stands of trees would be appropriate and public lands might benefit |
| | Excluding livestock from sensitive areas | • Riparian zones in particular can be negatively impacted by livestock |
| **Multistory/forest farming** | | |
| | Prioritizing areas impacted by deforestation | • Degraded landscapes would benefit most from trees |
| | Identifying forested land that is underutilized | • The shady environment of a cleared understory can be utilized for specialty crops |
| | Connecting with communities to supply labor | • Many forest farming systems require intensive management |

## Placement of the Five Agroforestry Practices

The material in the following section is organized by the commonly recognized agroforestry practices, because much of the research and supporting literature are defined in that manner. Table 1 summarizes the landscape-level strategies for optimizing the placement of buffers based on these practices.

## *Alley Cropping*

Alley cropping is a literal integration of multiple crops at a fine resolution (intercropping), which fundamentally differentiates it from other agroforestry systems because production functions are emphasized. The focus on production creates an interesting opportunity for system placement, where the target areas might stretch beyond marginal/sensitive lands to include zones with productive soils. Introducing rows of trees into an alley cropping system can improve the provision of ecosystem services including carbon sequestration, nutrient cycling, water regulation, biodiversity, and overall productivity when compared with monocultures (Quinkenstein et al. 2009; Tsonkova et al. 2012).

### Landscape Scale

*Marginal Land.* A common approach for designating appropriate areas to transition to alley cropping is to target lands considered "marginal" for the production of relevant row crops (Tsonkova et al. 2012; Lovell et al. 2018). Highly erodible soils (HEL) can serve as good targets for land suitable for alley cropping, particularly to avoid loss of topsoil. In the USA alone, approximately 40 million ha are designated in this category (Garrett et al. 2009). Similar approaches have been used for determining regions that would be suitable for the production of bioenergy crops (Gelfand et al. 2013). The use of marginal lands for a productive purpose has primarily been promoted in response to the "food versus fuel" debate that raised concerns that increasing the production of bioenergy crops would come at a cost in terms of food produced on those areas (Cai et al. 2011; Shortall 2013).

Use of marginal lands for alley cropping is a logical approach from two perspectives. For one, conventional crops grown on marginal lands are often less productive; in fact, marginal land can be defined based on the low-yielding and high risk characteristics (Rhoads et al. 2016). In particular, zones that are prone to flooding may not produce consistent yields from one year to the next. As a result, the landowner might be more open to alternative production systems because these areas often result in net losses to the farm operation (Lovell et al. 2018). The second perspective considers the conservation/environmental benefits. Marginal land is often called sensitive land because it is prone to soil erosion or other damages. The integration of trees and shrubs on these limited areas could result in disproportionate benefits in terms of reduction of environmental impacts and economic risks, as these woody systems can stabilize soil and retain nutrients (Molnar et al. 2013).

*Optimizing Tree Productivity.* An alternative approach for placing alley cropping systems is to focus on optimizing the tree component. The same geospatial tools used to identify marginal lands for row crops can be used to find areas most suitable for various agroforestry species. While the approach may at first seem to run counter to the marginal land argument, optimized tree species placement is an additive approach that could increase opportunities for alley cropping.

A study by Wolz (2017) evaluated black walnut (*Juglans nigra*) systems, planted as either plantation forestry (trees alone) or rows in an alley cropping system with existing crops. These land-use alternatives were then compared based upon financial performance. The analysis focused on the Midwest USA, where corn-soybean rotation dominates and soils are considered to be among the most productive in the world. Based on long-term rates of return, alley cropping systems outcompeted the conventional corn-soybean rotation on 23.4% of the land. The identified lands included some areas that were marginal and some that were productive for row crops. These results suggest that opportunities for alley cropping would be missed by focusing on marginal lands alone (Wolz 2017). A unique aspect of this work is the merging of site suitability analysis with profitability (considering local cash rent rates) to provide a planning tool for landowners and investors. However, one current limitation of this approach is that high-resolution data for individual species is very limited. In this case, black walnut was the only species for which the complete data on soil suitability, timber prices, crop productivity, cash rents, and land cover were available.

*Compatibility of Alley Crops*. Another strategy for placing alley cropping systems is to identify the regions with crops that are compatible (i.e., minimal competition for resources) when intercropped with trees, and then map the suited distributions of those crops. The importance of appropriately pairing compatible tree species and alley crops cannot be understated (Garrett et al. 2009), and often the issues reach beyond the direct competition between trees and alley crop. For example, while select tree-row crop mixtures offer themselves as phenological or nutrient-use complements, contemporary management practices could severely limit the potential of such a transition in reality. With crops such as corn and soybeans, many farmers have come to rely heavily on genetically modified (GM) herbicide-resistant varieties that allow applications of various broad-spectrum herbicides late in the season. These applications would be extremely risky near valuable trees and shrubs that could be easily damaged by these herbicides (Garrett et al. 2009).

Fortunately, a broad range of alley crops could be managed in a way that is compatible with integrating trees. Wheat and other cool-season crops have been used successfully in long-term studies because less weed maintenance is required and less competition for light occurs during the active growing seasons between trees and alley crops. Long-term studies in France have demonstrated the success of growing winter wheat (which is actively growing from late fall to into early summer) in combination with hybrid walnut trees that do not leaf out until late spring (Lovell et al. 2018) (Fig. 1). Other cool-season crops could be good targets, as the weed management is also less of a factor. Forage as a hay crop shows good potential in the alley, although much more research is needed to understand productivity in systems with trees (Garrett et al. 2009).

Perhaps the greatest opportunity for integrating trees is with specialty crops in the alleys. Included in this category would be landscaping plants for nurseries; forest botanicals (florals, ginseng, mushrooms, etc.); landscaping plants for nurseries;

**Fig. 1** Alley cropping in France with hardwood trees and an alley crop of wheat

herbal ingredients for teas and other culinary uses; bioenergy crops; and decorative floral and handicraft items (Josiah et al. 2004; Palada et al. 2008; Garrett et al. 2009; Fletcher et al. 2012; Mori et al. 2017). Many of these crops were originally growing in forested habitats, so they are highly adapted to alley cropping situations. Fruit and vegetable cropping systems offer unique opportunities, as well. The integration of tree rows could help to reduce wind damage or sun scald on sensitive species. While orchards with fruit and nut trees may fit more into the "tree component" category, existing orchards could be targeted for introducing an alley crop or within-row shrub, particularly in the establishment phase when trees are smaller (Garrett et al. 2009).

Additionally, the temporal aspect of alley cropping systems is important to consider proactively, with a successional aspect that is driven by long-lived perennials. Initially, the tree crops might be established as rows in an existing crop field. As the system matures and trees become larger, they may outcompete the existing alley crop, so other more shade-tolerant species might become more appropriate. Compatibility between the trees and the alley crops is critically important, particularly considering the management practices that might include the use of pesticides. The successional plan needs to account for potential changes in the alley crop based on the maturation of the long-lived perennials as well as the landscape plan.

### Farm Scale

Effective integration of alley cropping at the farm scale depends upon local collaboration between landowners, regional planners, and government agencies (Tress and Tress 2003; Atwell et al. 2009; Arlettaz et al. 2010; Malezieux 2012). Regional planners may identify zones of high-impact placement for alley cropping practices and work with landowners to effectively have those practices adopted based upon constraints specific to individual farmers and their land. One such constraint is large-scale machinery used in conventional agriculture which creates a challenge in farming small, irregular shaped fields that may instead be used for suitable tree crops. For example, tracts of land that are curved may be difficult to harvest on the edges. To select the best possible crop for alley cropping on a particular farm, it is important to examine which crops grow well and become profitable based on local markets and landowner preferences, knowledge, and current agronomic practices. A handful of tree crops exist in any one region that may be selected for alley cropping, yet few meet all of the requirements to be a practical option. A best-fit system will vary from farm to farm because of the inherent complexity of designing and managing agroforestry systems. Farm-scale work that incorporates local knowledge with efficient designs is a necessity to build long-lasting alley cropping systems.

## *Riparian Forest Buffers*

Riparian buffers have received the most attention regarding placement of agroforestry practices. Their very name defines their position in the landscape (adjacent to a body of water), and their appropriate placement dictates the capacity to trap and filter sediment, nutrients, and pesticides. Without some form of targeted placement, these conservation practices are likely to be inefficient for improving water quality (Zaimes and Schultz 2011). For research purposes, some form of landscape modeling is most often used to determine areas that would benefit most from the installation of buffers. Typically, this research is conducted at the scale of the watershed, due to the linkage with hydrologic features. From a planning perspective, the identification of "hot spots" in the landscape is valuable for targeting conservation buffers to allocate funding (Qiu and Dosskey 2012), in which case political boundaries or jurisdictions may be more relevant. Different strategies have been used to target the placement of riparian buffers, but most fit into one of the three types: riparian buffer zone, soil erodibility, and terrain analysis (Tomer et al. 2009; Qiu and Dosskey 2012) (Table 2).

**Table 2** Three approaches to placing riparian buffers, showing how they each focuses on a different factor

| | Complexity | Target areas | Basis for placement | Methodology |
|---|---|---|---|---|
| Riparian buffer approach | Low | Areas adjacent to streams | Proximity to streams | Uses fixed or variable width buffers alongside streams, aided by GIS and remote-sensing technologies for location identification and analysis |
| Soil survey technique | High | Areas likely to intercept sediment from adjacent fields | Soil properties | Ranking of soil map units for how effectively a buffer, when placed there, would trap sediment carried by surface runoff |
| Terrain analysis | Medium | Areas which intercept water moving towards streams | Topographic and streamflow data | Uses topographic data to reveal the pathway of water movement and accumulation which are then classified and interpreted to reveal priority sites for buffers |

## Landscape Scale

*Riparian Buffer Approach.* The riparian buffer approach is probably the most straightforward, as it simply targets the zones beside streams—either a fixed width or a variable width. In the USA, many of the policies designated by state and local government can be written to protect a fixed width (e.g., the 30 m buffer on any side of the water body). Variable width buffers attempt to account for site-specific physical conditions that impact buffer effectiveness, often assisted by landscape analysis using GIS and remote sensing (Xiang 1993; Basnyat et al. 1999). The riparian buffer approach has been valuable in demonstrating the water quality benefits of prescribing perennial cover as the land use adjacent to water bodies (Basnyat et al. 1999). From the multifunctional landscape perspective, this approach would also contribute to the protection and expansion of many other functions and benefits of streams and wetlands, including wildlife habitat, biodiversity conservation, visual quality, and recreation (Lovell and Sullivan 2006).

*Soil Survey Technique.* As new spatial analysis tools and data become available, more complex approaches have been developed to target buffer placement for improving water quality specifically. The soil survey technique uses a ranking of soil map units, based on soil and slope information of the area, to estimate the sediment trapping efficiency (STE) of a buffer. The spatial data layers for soil map units are publically available for many regions (e.g., the entire USA). The approach seeks to target those areas where runoff could be intercepted by a buffer, based on known soil properties (Tomer et al. 2009).

*Terrain Analysis.* The terrain analysis technique utilizes topographic data (i.e., a digital elevation model) to identify potential buffer sites that are most likely to intercept water (Tomer et al. 2009; Dosskey et al. 2013). The approach is based on the hydrologic loading from contributing areas upland of the proposed buffer (Qiu and

Dosskey 2012). Where high-resolution data are available, the terrain analysis technique can be used to identify the specific locations where runoff can be intercepted, down to the scale of a farm or field (Tomer et al. 2009).

*Effectiveness*. Overall, when implemented broadly, the targeted placement strategies have been effective in improving water quality benefits to a greater extent than random placement (Zaimes and Schultz 2011). By determining the major contributing areas, along with the most effective receiving zones, conservation practices for buffers could be more effective and efficient. This type of holistic watershed approach has become much more accessible with broad use of GIS and other landscape models (Zaimes and Schultz 2011). Regarding cost-effectiveness, buffer zones constructed based on hydrological characteristics, not a fixed width, have been shown to be more cost effective because they include more areas of forests and wetlands that are not used for production (Tiwari et al. 2016). Qui and Dosskey (2012) found that strategies based on soil survey or topographic data were more cost effective than riparian focused approaches, primarily due to the high cost of installation with true riparian buffers containing trees (versus filter strips).

## Farm Scale

Despite the benefits of watershed-scale approaches, a number of constraints exist in working at that scale, including the following: mismatch in the locations of targeted hot spots and the allocation of funding, many diverse actors attempting to implement policies at various scales, and conflicts between the goals of various agencies (e.g., water quality versus enhancing habitat for riparian wildlife) (Wardropper et al. 2015). However, many of the broader findings from watershed-scale research are reflected at the farm scale. For example, a generalizable guideline is that the most effective locations for buffers are sites with large contributing upslope areas in which the riparian zone itself has relatively flat slopes that slow the flow of runoff to allow infiltration and sediment trapping (McGlynn and Seibert 2003; Tomer et al. 2003). Those features might be identified through site observation, particularly where the landowner is actively managing the land.

Regional agronomic practices may also impose unique challenges to riparian buffer placement. For example, in the Midwest USA, subsurface drainage tiles can significantly alter farm-scale hydrology. This alteration of the natural hydrology is often ignored in modeling approaches, yet tile inlets, outlets, and installation patterns can be determined using landowner records and site observation. In such situations, the tile flow might be redirected to the surface for a more intensive buffer treatment (Jaynes and Isenhart 2014). The issue of tile drainage need not be an excuse to avoid riparian buffers, considering all of the other benefits. Instead, multifunctional buffers can be paired with targeted technologies such as flow control structures for lateral distribution (saturated buffers) or filters/bioreactors to treat outflow directly. The buffers still manage the surface flow, including sediment containing phosphorus, and they play a critical role in reducing streambank erosion.

Farm-scale design can also connect more directly with local benefits, often beyond just environmental benefits. Landowner preferences for visual quality and recreational activities could be integrated into the placement of riparian buffers, in some cases encouraging buffer zones to be wider than prescribed by conventional targeting technologies. The needs of the landowner can also be matched with local government programs that provide funding through rent payments and establishment costs. Finally, at the farm scale, opportunities can be explored to integrate a production function into a multifunctional buffer, where riparian zones could be planted with trees and shrubs that produce fruits, nuts, florals, and other materials. This additional incentive could encourage further transition of lands near streams and wetlands.

## *Windbreaks*

Similar to other agroforestry practices, the placement of windbreaks can be prioritized based on the primary ecosystem services they are designed to supply. The most obvious ecosystem services from windbreaks are microclimate control and soil erosion reduction, which are indirectly related to production functions when trees serve to protect crops/livestock or the soil that supports productive systems. Windbreaks may also be one of the best agroforestry practices for supporting cultural services, particularly regarding visual quality, through increasing landscape heterogeneity or screening of undesirable features. Other cultural services include improving air quality by reducing the movement of unpleasant odors and restoring cultural heritage in the network of hedgerows that once covered the landscape.

### Landscape Scale

*Regional Specificity.* Regional characteristics related to climatic factors such as wind speeds and rainfall are often a strong determinant of the need for and use of windbreaks. When combined with the sensitivity and value of specific crops, the decision to install windbreaks can be driven by economics. For example, in Southern Patagonia, Argentina, production of certain agricultural commodities is constrained by the strong winds in the region. In that region, windbreaks are quite common, particularly in the valleys where the fertile soil supports high production but is prone to erosion (Peri and Bloomberg 2002). In Iran, desertification is a severe problem after years of exploitation of natural resources, so rows of trees are commonly planted to protect cultivated areas to allow crop production where it might otherwise be impossible (Amiraslani and Dragovich 2011).

*Government Programs.* In some regions, government programs have been developed to support the broad establishment of windbreaks that might better protect the entire landscape. The Three-North Shelter Forest Program (TNSFP) in Northeast China is one such example that was initiated in 1978 and intended to go through

2050, covering over 40% of China's total territory. In this semi-humid to semiarid climate, the windbreaks improve microclimate conditions for crops such as maize, particularly in years of high environmental stress due to drought and in zones that are most limited by moisture (Zheng et al. 2016). Similarly, in Australia, the National Windbreaks Program (NWP) was established in 1993 for regions that had high potential crop productivity but experienced moisture limitations during critical periods of the year. The high variability of rainfall in Australia further heightened the need for windbreaks. Studies demonstrated that certain crops benefited from the shelter (e.g., canola and potato), while others showed no yield gains (e.g., wheat and barley) (Cleugh et al. 2002). Targeting and mapping the areas with the most sensitive crops could be an effective strategy for prioritizing windbreak placement.

*Landscape-Scale Methods*. As with previous agroforestry practices, GIS can be a robust platform for assessing landscapes to prioritize windbreak placement. In addition to the basic potential to combine data on crop cover, soil type, and climatic conditions, GIS approaches used to place wind turbines might also guide windbreak placement. For example, Rodman and Meentemeyer (2006) used a GIS rule-based modeling framework to identify optimal locations for wind turbine network-based physical features that impact wind intensity (e.g., terrain), land use, environmental factors, and proximities to public resources. In the case of wind turbines, public opposition to these features drives decision-making for reducing the negative impact. Alternatively, the relationship with the public would be flipped for windbreaks, as most people would likely support them due to their visual quality and microclimate benefits (Rodman and Meentemeyer 2006).

## Farm Scale

While landscape factors can be useful for prioritizing regions for windbreak programs, farm-level placement connects windbreak functions with the priorities and preferences of landowners. At this scale, the relationship between all of the features of the farm can come into play, including the location of the homestead, livestock facilities, sensitive crops, etc.

*Protecting Soil and Crops*. The protection of soil and crops is a primary function of windbreaks and one that can improve profitability for the farmer. Windbreaks would ideally be oriented perpendicular to the problematic winds, with the sensitive crops on the leeward side (Brandle et al. 2009). If placed appropriately, the tree rows will minimize the physical crop damage from wind and wind-borne soil, in addition to reducing the desiccation from dry winds. Many vegetables, herbaceous fruit crops, young orchard trees, and other specialty crops are particularly sensitive to these conditions, whereas grains and forages better tolerate stresses that are a consequence of wind. Additionally, pollination can be improved for some of these crops, increasing the economic gains from establishing tree rows (Norton 1988). For the most sensitive crops, planting windbreaks at an interval of 6–8 times the height of trees is desirable, and fields may be divided into "compartments" to provide the greatest level of protection for the crops that benefit (or profit) the most (Finch 1988; Cleugh et al. 2002).

*Reducing Herbicide Drift.* In addition to protecting crops from direct wind damage, windbreaks can play an essential role in protection against the drift of herbicides to off-target species (Lazzaro et al. 2008). This issue is becoming increasingly important as genetically modified (GM) crops allow applications of herbicides during periods of the season when nontarget species can be highly sensitive (Bohnenblust et al. 2016). Regarding placement, windbreaks should be positioned at all boundaries between the crops receiving problematic herbicides and the off-target sensitive crops or features, regardless of prevailing winds, since wind patterns are often variable from one day to the next, and the application timing may be uncontrollable. In addition to protecting sensitive crops, windbreaks can also be used to protect aquatic habitats (Brown et al. 2004) and even ornamental landscape plantings from drift.

*Protecting Infrastructure.* Windbreaks have also gained popularity for their potential to protect the farmstead and infrastructure from cold winter winds and hot summer winds. In that role, they reduce energy needs for the household and barns, while also screening those living and working spaces from unpleasant views, odors, sounds, and dust (Brandle et al. 2009). Furthermore, windbreaks can protect the landscape plants and gardens of the farmstead, by blocking damaging winds and reducing the risk of herbicide drift. Management of drifting snow is an additional function of windbreaks in colder regions, where they can be placed to protect a sensitive crop, capture snow for moisture, or prevent accumulation on roads, driveways, or work areas (Scholten 1988; Shaw 1988; Brandle et al. 2009).

*Sheltering Livestock.* Appropriately placed windbreaks can also be used to reduce stress on livestock by providing more comfortable and safe conditions, particularly for animals that are most sensitive (Dronen 1988). To provide the best benefits for animal health and productivity, windbreaks should be established perpendicular to the prevailing winds that are most stressful, located 30–60 m from the animals (Dronen 1988). This placement might include blocking pastures, feedlots, and other livestock holding areas from cold winter winds, hot summer winds, or both. In addition to livestock protection, windbreaks can serve a function of mitigating odors and particulate matter when placed downwind of livestock areas, particularly those where animals are confined to small spaces (Tyndall and Colletti 2007; Willis et al. 2017). It should be noted that the optimal distance from the source (livestock) can vary depending on temperatures, but typically they are most effective when positioned nearby (15 m) the source (Lin et al. 2006).

## Silvopasture

Trees can be integrated into pastures, or livestock can be grazed in wooded areas to create silvopastoral systems. While livestock grazing is often practiced on land that is not suitable for cultivated crops (Sharrow et al. 2009), silvopasture principles create an opportunity to expand the land types for livestock grazing. Targets for optimizing placement of silvopastoral practices could focus on grazed areas that benefit from the addition of trees due to climatic conditions unfavorable for livestock, or where overgrazing has destabilized the ecosystem. Other "opportunity" lands could focus on open forests where the addition of livestock would provide an economic benefit without substantial consequences for the health of the ecosystem.

## Landscape Scale

*Adding Trees in Harsh Climates.* Many of the studies on silvopastoral systems have focused on their placement in areas where climatic conditions, mainly heat, place stress on livestock to the extent that the productivity of the animals declines (Fig 2). In Latin America, the goal to reduce heat stress was found to be a driver for integrating or retaining trees (Calle et al. 2017). Growth in meat consumption and production has resulted in overgrazing of many regions in Latin America, but the negative environmental impacts can be overcome through a transition to sustainable silvopastoral systems (Chara et al. 2017). In Southeast USA, where temperatures can be uncomfortably high for grazing animals, silvopasture has been shown to provide a milder microclimate compared to open pasture (Karki and Goodman 2015), and more consistent grazing across the pasture (Karki and Goodman 2010). A study in the arid climate of Israel demonstrated the potential of silvopasture to rehabilitate land that had been degraded due to unrestricted livestock grazing. The economic case to integrate trees is strong because the areas become less sensitive to drought-induced livestock loss (Mor-Mussery et al. 2013).

*Target "Opportunity" Lands.* Integration of livestock into woodland areas could be prioritized for "opportunity" lands where the benefits outweigh the risks to ecosystem health. The USA, for example, consists of 255 million ha in forest land, yet only 52 million acres ha (approximately 20%) are grazed. Areas with relatively open stands (compared with forests containing dense canopy cover) would be good targets for silvopasture because of higher pasture productivity (Bigelow and Borchers 2017). Another target, in the context of the USA, is to allow silvopasture

**Fig. 2** Silvopasture in France with trees scattered across the landscape to provide shade for livestock

practices to be eligible for payments from government support programs such as the Conservation Reserve Program (CRP) (Shrestha et al. 2004). In this program, the use of windbreaks contains low-growing grasses planted between rows of trees, presenting an opportunity for grazing.

*Exclusion Areas*. In identifying the best opportunities for placement for silvopasture, consideration should also be given to the avoidance of sensitive areas. For example, livestock grazing can lead to destruction and degradation of riparian habitats, impacting the habitat and resources for songbirds and other vulnerable species (Forrester et al. 2017). Fencing can be used to exclude livestock from riparian zones, providing benefits for wildlife and water quality of nearby streams (Line et al. 2016). Hillside forest ecosystems are also sensitive, and the exclusion of livestock can improve soil health and increase regeneration of trees (Etchebarne and Brazeiro 2016). In general, targeting areas for exclusion can be appropriate where habitat is degraded, wildlife is suffering, or water quality is declining (Forrester et al. 2017).

#### Farm Scale

In the case of silvopasture, farm-scale considerations are similar to the landscape scale, in identifying microclimates harsh for livestock and excluding animals from sensitive zones. But at the farm scale, the willingness of the landowner to introduce livestock is an important factor. The landowner will likely exhibit strong preference for the location of livestock grazing as it relates to proximity to the homestead, quality of pasture, and accessibility of infrastructure in barns, watering resources, etc. Visual quality preferences may come into play as well. Landowners may choose locations based on viewsheds open to livestock grazing in a silvopastoral system, which is often considered to be an ideal landscape aesthetic with canopy trees and open understory (Sharrow et al. 2009).

### *Forest Farming*

Forest farming, or multistory cropping, represents a remarkably diverse set of options, ranging from the cultivation of plants in the understory of an existing forest to the establishment of a new community of species in a multi-strata format (Fig. 3). One characteristic that many of these systems share is that they mimic the structure of a natural forest ecosystem that would have been endemic to the areas (Joffre et al. 1999; Malezieux 2012). As such, the placement of forest farming would be appropriate in nearly any area where the soil and climatic conditions can support tree establishment. Because these systems can require intensive management due to their complexity, prioritizing their placement at the landscape scale may be guided primarily by connections with the human resources needed to manage it.

**Fig. 3** Forest farming in Nicaragua with arabica coffee (*Coffea arabica*) planted beneath a mixture of native timber, citrus, and banana trees

## Landscape Scale

Regionally, priority for placement of forest farming might be given to landscapes that have experienced severe degradation due to deforestation. In the Caribbean country of Haiti, where deforestation has resulted in a degraded landscape, agroforestry has been proposed as a conservation strategy that would help to restore the ecosystem and provide valuable products (Zimmermann 1986). In tropical regions, forest farming can be established by adding additional canopy layers where plantations of understory tree crops such as coffee, banana, and cacao dominate the landscape. The use of leguminous trees in mixed agroforests is widely documented, and improvements in soil fertility have been noted (de Souza et al. 2010). Opportunities are likely to be greatest when efforts to implement forest farming are aligned with the regional markets for alternative crops and the local environmental needs (de Souza et al. 2012). Similarly, in temperate zones where annual crops dominate the landscape, forest farming could be targeted to marginal lands and areas of transition between different land uses (Bjorkund et al. 2018).

## Farm Scale

At the farm scale, decisions to integrate productive species into a forest system depend heavily on the preferences of the farmer and their desire to have the additional crop for personal use or sale (Rice 2011). The placement of various forest farming activities is often based on proximity to the homestead or other infrastructure. A study of homegardens in Nicaragua demonstrated that families organized

**Table 3** Key considerations for farm-scale placement of agroforestry practices based on landowner objectives and agroforestry functions

| Objectives | Agroforestry functions |
|---|---|
| Protect the soil and water resources | • Alley cropping systems and windbreaks to stabilize soil<br>• Riparian buffers to intercept and filter runoff<br>• Livestock exclusion reduces negative impacts |
| Create favorable microclimates | • Windbreaks for homestead and livestock structures<br>• Alley cropping or windbreaks to protect sensitive crops<br>• Silvopasture to provide shade |
| Utilize marginal lands | • Alley crops or other agroforestry systems for less productive or difficult-to-access areas<br>• Silvopasture for marginal areas not near water resources |
| Match landowner preferences | • Silvopasture to improve visual quality of livestock areas<br>• Windbreaks for comfort and wildlife habitat<br>• Riparian buffers to improve recreational value of property<br>• Forest farming for additional profit or items for personal use |

their outdoor plantings into different management zones based on plant requirements and management needs, in addition to soil suitability. Plants requiring regular watering and weeding were typically placed near the household, while trees needing less care are planted further away (Méndez et al. 2001). Additionally, a study on coffee agroforestry in Mexico revealed that farmers selectively promote the growth and management of preferred shade trees based on compatibility with understory crops as well as their potential ecosystem services. Farmers' management decisions were shown to be shaped not only by their knowledge generated from experience but also by outside groups such as governmental agencies and NGOs that can play a crucial role in shaping the structure of forest farming systems (Valencia et al. 2015). A summary of farm-scale objectives across various agroforestry practices is provided in Table 3.

## Moving Forward

For simplicity and consistency, this chapter has been organized based on the five commonly accepted agroforestry practices, following the format of much of the literature on agroforestry placement. Moving forward, however, a better approach might be to look at the agricultural landscape as a whole, considering where the greatest sensitivities (risks) exist and where the opportunities lie, and then prescribing a customized plan for establishment of perennial habitats. A clear example would be to design agroforestry systems at the boundaries of conservation zones or ecologically valuable habitats, to provide a buffer between the sensitive area and the agricultural landscape. The systems could be designed specifically to protect the area but also to be productive (Figs. 2 and 3).

For cultivated areas of large-scale grain crops, productive windbreaks could be a pragmatic solution (Fig. 4). The system could be designed somewhat like an alley

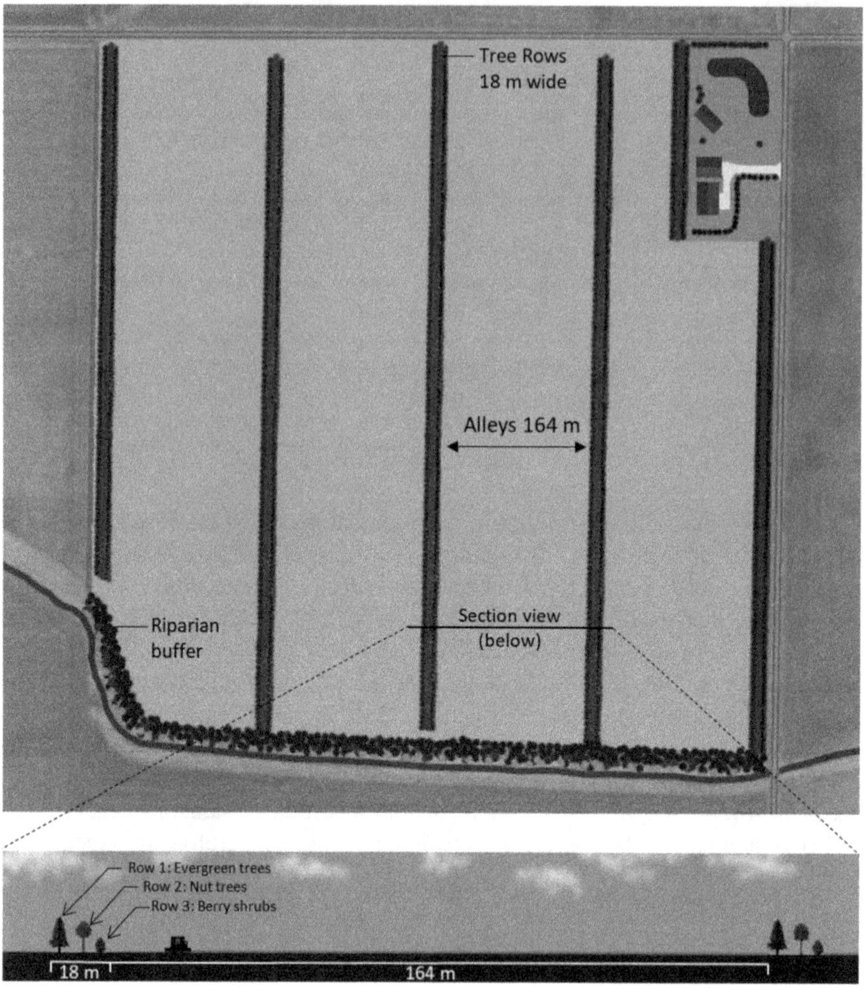


**Fig. 4** Plan view (top) and section view (bottom) of a design for integrating agroforestry on a farm to demonstrate the use of multiple productive windbreaks for protecting the crop and providing additional enterprises

cropping system, but with considerably wider alleys, where the alley distance is based on the protection zone provided by trees. While the tree zones would need to be wider to accommodate multiple rows, the overall land area taken out of production would be considerably lower, and the issues of managing the boundaries between crops would be lessened. Likewise, a system that combines alley cropping with pasture/livestock could be a solution for landowners interested in integrating animals. In situations that offer opportunities to include specialty crops, Mori et al. (2017) also suggest diversifying beyond the five practices to include a wide range of crops, woody plants, and animals that might become components of a more complex agroforestry system (Mori et al. 2017).

The Chinese TNSFP (described earlier in Windbreaks section) offers an opportunity for an innovative planning framework in which treed zones could be designated based on optimal benefits and then managed as public land. As a true regional planning effort, authors propose that "construction of shelterbelts should be planned in detail and should be combined with the layout and reconstruction of local roads, irrigation channels and other agricultural infrastructure" (Zheng et al. 2016). The addition of such a "green infrastructure" layer would be similar to approaches used in urban areas (Lovell and Taylor 2013), but in this case it is imposed onto an agricultural region.

Moving forward, multifunctional landscape planning approaches can be supported by new spatial analysis tools and methods. Precision agriculture, for example, can provide valuable data on yields, to identify less productive areas that might be good targets for conversion to agroforestry or other perennial crops (Dosskey et al. 2005). At a broader scale, marginal land can be spatially analyzed and mapped in GIS, using suitability models based on variables such as crop productivity, soil erosion, and other management-oriented land traits. Various scenarios that simulate the transition of marginal lands to agroforestry can then be tested or modeled to determine the environmental outcomes such as changes in landscape heterogeneity, soil erosion, and crop productivity (Mattia et al. 2018). To optimize the placement of agroforestry in a specific area (e.g., a target watershed), the "collaborative geodesign" approach could be implemented. This approach engages multiple stakeholders (i.e., landowners) in a collaborative planning process to optimize the production of agricultural commodities along with protection of natural resources (Stotterback et al. 2016). In addition to GIS tools, the use of visualization techniques, such as photorealistic images, can show agroforestry landscapes in situ to allow stakeholders to comprehend the makeup of long-term systems before adopting them (Tress and Tress 2003). If the goal is to improve the performance of the landscape, considering the transition of some portions of private lands, the value of involving stakeholders in the planning process cannot be understated (Landis 2017).

## Conclusion

As described nearly 100 years ago by J.R. Smith in his book *Tree Crops*, "[i]t is one thing to tell the farmer that here are good black walnuts or chestnuts or acorn-yielding oaks or honey locust trees, and it is quite another matter to organize these into an effective farm". The planning and placement of agroforestry can be complicated given the wide variety of species, combinations, and practices available. It is not unlike the complexity within landscape ecology, where a seemingly incredible plethora of biotic and abiotic factors are continually interacting. However, the development of spatial planning tools and methods, such as landscape suitability analysis, now allows researchers and planners to condense and understand the numerous factors involved in landscape planning. Science and technology are moving towards a new frontier in the ability to capture, compute, and model diverse agricultural systems. The variety of tools that now exist should be harnessed by landowners and planners alike for the most efficient planning and placement of multifunctional landscapes, especially when considering the unique aspects of each of the five common agroforestry practices. In future work, it is imperative that the development of agroforestry systems combines ecological, agronomic, economic, and social sciences. No one discipline is more important than another when considering the complex interactions of planning agroforestry systems, especially considering their use at larger scales. Each landscape offers its unique challenges, yet agroforestry provides the flexibility to meet those challenges and provides ecosystem services now and for generations to come.

**Acknowledgements** This material is based on the work that is supported in part by the National Institute of Food and Agriculture, U.S. Department of Agriculture, under Award Number 2014-68006-22041; the Illinois Nutrient Research and Education Council (NREC); and the University of Illinois Institute for Sustainability, Energy and Environment.

## References

Amiraslani F, Dragovich D (2011) Combating desertification in Iran over the last 50 years: an overview of changing approaches. J Environ Manag 92:1–13

Arlettaz R, Schaub M, Fournier J, Reichlin TS, Sierro A, Watson JEM, Braunisch V (2010) From publications to public actions: when conservation biologists bridge the gap between research and implementation. Bioscience 60:835–842

Atwell RC, Schulte LA, Westphal LM (2009) Linking resilience theory and diffusion of innovations theory to understand the potential for perennials in the US Corn Belt. Ecol Soc 14:17

Bailey S (2007) Increasing connectivity in fragmented landscapes: an investigation of evidence for biodiversity gain in woodlands. For Ecol Manag 238:7–23

Basnyat P, Teeter LD, Flynn KM, Lockaby BG (1999) Relationships between landscape characteristics and nonpoint source pollution inputs to coastal estuaries. Environ Manag 23:539–549

Baudry J, Burel F, Aviron S, Martin M, Ouin A, Pain G, Thenail C (2003) Temporal variability of connectivity in agricultural landscapes: do farming activities help? Landsc Ecol 18:303–314

Bigelow DP, Borchers A (2017) Major uses of land in the United States, 2012. In: Service, U.E.R. (Ed.), Washington DC

Bjorkund J, Eksvard K, Schaffer C (2018) Exploring the potential of edible forest gardens: experiences from a participatory action research project in Sweden. Agroforest Syst. 93:1107–1118

Bohnenblust EW, Vaudo AD, Egan JF, Mortensen DA, Tooker JF (2016) Effects of the herbicide dicamba on nontarget plants and pollinator visitation. Environ Toxicol Chem 35:144–151

Brandle JR, Hodges L, Tyndall J, Sudmeyer RA (2009) Windbreak practices. In: Garrett HE (ed) North American agroforestry: an integrated science and practice, 2nd edn. America Society of Agronomy, Inc., Madison, WI, pp 75–104

Brown RB, Carter MH, Stephenson GR (2004) Buffer zone and windbreak effects on spray drift deposition in a simulated wetland. Pest Manag Sci 60:1085–1090

Cai XM, Zhang XA, Wang DB (2011) Land availability for biofuel production. Environ Sci Technol 45:334–339

Calle Z, Giraldo AM, Cardozo A, Galindo A, Murgueitio E (2017) Enhancing biodiversity in neotropical silvopastoral systems: use of indigenous trees and palms. In: Montagnini F (ed)

Integrating landscapes: agroforestry for biodiversity conservation and food sovereignty. Springer, Cham, Switzerland, pp 417–438

Chara J, Rivera J, Barahona R, Murgueitio E, Deblitz C, Reyes E, Mauricio RM, Molina JJ, Flores M, Zuluaga A (2017) Intensive silvopastoral systems: economics and contribution to climate change mitigation and public policies. In: Montagnini F (ed) Integrating landscapes: agroforestry for biodiversity conservation and food sovereignty. Springer, Cham, Switzerland, pp 395–416

Cleugh H, Prinsley R, Bird RP, Brooks SJ, Carberry PS, Crawford MC, Jackson TT, Meinke H, Mylius SJ, Nuberg IK, Sudmeyer RA, Wright AJ (2002) The Australian National Windbreaks Program: overview and summary of results. Aust J Exp Agric 42:649–664

Cushman SA, Evans JS, McGarigal K (2010) Landscape ecology: past, present, and future. In: Cushman SA, Huettmann F (eds) Spatial complexity, informatics, and wildlife conservation. Springer, Tokyo, pp 65–82

de Souza HN, Cardoso IM, Fernandes JM, Garcia FCP, Bonfim VR, Santos AC, Carvalho AF, Mendonca ES (2010) Selection of native trees for intercropping with coffee in the Atlantic Rainforest biome. Agrofor Syst 80:1–16

de Souza HN, de Graaff J, Pulleman MM (2012) Strategies and economics of farming systems with coffee in the Atlantic Rainforest Biome. Agrofor Syst 84:227–242

Dosskey MG, Eisenhauer DE, Helmers MJ (2005) Establishing conservation buffers using precision information. J Soil Water Conserv 60:349–354

Dosskey MG, Qiu ZY, Kang Y (2013) A comparison of DEM-based indexes for targeting the placement of vegetative buffers in agricultural watersheds. J Am Water Resour Assoc 49:1270–1283

Dramstad WE, Olson JD, Forman RTT (1996) Landscape ecology principles in landscape architecture and land-use planning. Island Press, Washington, D.C.

Dronen SI (1988) Layout and design criteria for livestock windbreaks. Agric Ecosyst Environ 22-3:231–240

Etchebarne V, Brazeiro A (2016) Effects of livestock exclusion in forests of Uruguay: soil condition and tree regeneration. For Ecol Manag 362:120–129

Finch SJ (1988) Field windbreaks—design criteria. Agric Ecosyst Environ 22-3:215–228

Fischer J, Lindenmayer DB, Manning AD (2006) Biodiversity, ecosystem function, and resilience: ten guiding principles for commodity production landscapes. Front Ecol Environ 4:80–86

Fletcher EH, Thetford M, Sharma J, Jose S (2012) Effect of root competition and shade on survival and growth of nine woody plant taxa within a pecan *Carya illinoinensis* (Wangenh.) C. Koch alley cropping system. Agrofor Syst 86:49–60

Forman RTT, Godron M (1981) Patches and structural components for a landscape ecology. Bioscience 31:733–740

Forman RTT, Godron M (1986) Landscape ecology. John Wiley & Sons, New York

Forrester TR, Green DJ, McKibbin R, Bishop CA (2017) Evaluating the efficacy of seasonal grazing and livestock exclusion as restoration tools for birds in riparian habitat of the Okanagan Valley, British Columbia, Canada. Restor Ecol 25:768–777

Garrett HE, McGraw RL, Walter WD (2009) Alley cropping practices. In: Garrett HE (ed) North American agroforestry: an integrated science and practice, 2nd edn. America Society of Agronomy, Inc., Madison, WI, pp 133–162

Gelfand I, Sahajpal R, Zhang XS, Izaurralde RC, Gross KL, Robertson GP (2013) Sustainable bioenergy production from marginal lands in the US Midwest. Nature 493:514

Gliessman S (2010) Landscape multifunctionality and agriculture. J Sustain Agric 34:465–465

Jaynes DB, Isenhart TM (2014) Reconnecting tile drainage to riparian buffer technology for enhanced nitrate removal. J Environ Qual 43:631–638

Joffre R, Rambal S, Ratte JP (1999) The dehesa system of southern Spain and Portugal as a natural ecosystem mimic. Agrofor Syst 45:57–79

Jordan N, Warner KD (2010) Enhancing the multifunctionality of US Agriculture. Bioscience 60:60–66

Josiah SJ, St-Pierre R, Brott H, Brandle J (2004) Productive conservation: diversifying farm enterprises by producing specialty woody products in agroforestry systems. J Sustain Agric 23:93–108
Karki U, Goodman MS (2010) Cattle distribution and behavior in southern-pine silvopasture versus open-pasture. Agrofor Syst 78:159–168
Karki U, Goodman MS (2015) Microclimatic differences between mature loblolly-pine silvopasture and open-pasture. Agrofor Syst 89:319–325
Landis DA (2017) Designing agricultural landscapes for biodiversity-based ecosystem services. Basic Appl Ecol 18:1–12
Lazzaro L, Otto S, Zanin G (2008) Role of hedgerows in intercepting spray drift: Evaluation and modelling of the effects. Agric Ecosyst Environ 123:317–327
Lin XJ, Barrington S, Nicell J, Choiniere D, Vezina A (2006) Influence of windbreaks on livestock odour dispersion plume in the field. Agric Ecosyst Environ 116:263–272
Line DE, Osmond DL, Childres W (2016) Effectiveness of livestock exclusion in a pasture of Central North Carolina. J Environ Qual 45:1926–1932
Lovell ST, Johnston DM (2009) Designing landscapes for performance based on emerging principles in landscape ecology. Ecol Soc 14(1)
Lovell ST, Sullivan WC (2006) Environmental benefits of conservation buffers in the United States: evidence, promise, and open questions. Agric Ecosyst Environ 112:249–260
Lovell ST, Taylor JR (2013) Supplying urban ecosystem services through multifunctional green infrastructure. Ecol Soc 28:1447–1463
Lovell ST, DeSantis S, Nathan CA, Olson MB, Mendez VE, Kominami HC, Erickson DL, Morris KS, Morris WB (2010) Integrating agroecology and landscape multifunctionality in Vermont: an evolving framework to evaluate the design of agroecosystems. Agric Syst 103:327–341
Lovell ST, Dupraz C, Gold M, Jose S, Revord R, Stanek E, Wolz KJ (2018) Temperate agroforestry research: considering multifunctional woody polycultures and the design of long-term trials. Agrofor Syst 92:1397–145
Madureira L, Rambonilaza T, Karpinski I (2007) Review of methods and evidence for economic valuation of agricultural non-commodity outputs and suggestions to facilitate its application to broader decisional contexts. Agric Ecosyst Environ 120:5–20
Malezieux E (2012) Designing cropping systems from nature. Agron Sustain Dev 32:15–29
Mattia CM, Lovell ST, Fraterrigo JW (2018) Identifying marginal land for multifunctional perennial cropping systems in the Upper Sangamon River Watershed, Illinois. J Soil Water Conserv. 73(6):669–681
McGlynn BL, Seibert J (2003) Distributed assessment of contributing area and riparian buffering along stream networks. Water Resour Res 39:7
Méndez VE, Lok R, Somarriba E (2001) Interdisciplinary analysis of homegardens in Nicaragua: micro-zonation, plant use and socioeconomic importance. Agrofor Syst 51:85–96
Molnar TJ, Kahn PC, Ford TM, Funk CJ, Funk CR (2013) Tree crops, a permanent agriculture: concepts from the past for a sustainable future. Resources 2:457–488
Mori GO, Gold M, Jose S (2017) Specialty crops in temperate agroforestry systems: Sustainable management, marketing and promotion for the Midwest region of the U.S.A. In: Montagnini F (ed) Integrating landscapes: agroforestry for biodiversity conservation and food sovereignty. Springer, Cham, Switzerland, pp 331–366
Mor-Mussery A, Leu S, Budovsky A (2013) Modeling the optimal grazing regime of *Acacia victoriae* silvopasture in the Northern Negev. Israel J Arid Environ 94:27–36
Ndubisi F (2002) The first landscape-suitability approach. In: Ndubisi F (ed) Ecological planning: a historical and comparative synthesis. John Hopkins University Press, Baltimore, MD
Norton RL (1988) Windbreaks—benefits to orchard and vineyard crops. Agric Ecosyst Environ 22-3:205–213
Palada MC, Crossman SM, O'Donnell JJ (2008) Integrating high value horticultural crops into agroforestry systems in the tropics with focus on alley cropping. US Dept. Agr., Forest Serv So Forest Expt Stn, New Orleans

Perfecto I, Vandermeer J (2002) Quality of agroecological matrix in a tropical montane landscape: ants in coffee plantations in Southern Mexico. Conserv Biol 16:174–182
Peri PL, Bloomberg M (2002) Windbreaks in southern Patagonia, Argentina: a review of research on growth models, windspeed reduction, and effects on crops. Agrofor Syst 56:129–144
Qiu Z, Dosskey MG (2012) Multiple function benefit—cost comparison of conservation buffer placement strategies. Landsc Urban Plan 107:89–99
Quinkenstein A, Wollecke J, Bohm C, Grunewald H, Freese D, Schneider BU, Huttl RF (2009) Ecological benefits of the alley cropping agroforestry system in sensitive regions of Europe. Environ Sci Pol 12:1112–1121
Rhoads BL, Lewis QW, Andresen W (2016) Historical changes in channel network extent and channel planform in an intensively managed landscape: Natural versus human-induced effects. Geomorphology 252:17–31
Rice RA (2011) Fruits from shade trees in coffee: how important are they? Agrofor Syst 83:41–49
Rodman LC, Meentemeyer RK (2006) A geographic analysis of wind turbine placement in Northern California. Energy Policy 34:2137–2149
Scholten H (1988) Snow distribution on crop fields. Agric Ecosyst Environ 22(3):363–380
Sharrow SH, Brauer D, Clason TR (2009) Silvopastoral practices. In: Garrett HE (ed) North American agroforestry—an integrated science and practice, 2nd edn. America Society of Agronomy, Inc., Madison, WI, pp 105–131
Shaw DL (1988) The design and use of living snow fences in North America. Agric Ecosyst Environ 22(3):351–362
Shortall OK (2013) “Marginal land” for energy crops: exploring definitions and embedded assumptions. Energy Policy 62:19–27
Shrestha RK, Alavalapati JRR, Kalmbacher RS (2004) Exploring the potential for silvopasture adoption in south-central Florida: an application of SWOT-AHP method. Agric Syst 81:185–199
Soil Conservation Service, U.S.D.o.A (1961) Land-capability classification. Agricultural Handbook No. 210. U.S. Department of Agriculture, Washington, DC
Stotterback CS, Runck B, Pitt DG, Kne L, Jordan NR, Mulla DJ, Zerger C, Reichenbach M (2016) Collaborative geodesign to advance multifunctional landscapes. Landsc Urban Plan 156:71–80
Tiwari T, Lundstrom J, Kuglerova L, Laudon H, Ohman K, Agren AM (2016) Cost of riparian buffer zones: a comparison of hydrologically adapted site-specific riparian buffers with traditional fixed widths. Water Resour Res 52:1056–1069
Tomer MD, James DE, Isenhart TM (2003) Optimizing the placement of riparian practices in a watershed using terrain analysis. J Soil Water Conserv 58:198–206
Tomer MD, Dosskey MG, Burkart MR, James DE, Helmers MJ, Eisenhauer DE (2009) Methods to prioritize placement of riparian buffers for improved water quality. Agrofor Syst 75:17–25
Tress B, Tress G (2003) Scenario visualisation for participatory landscape planning—a study from Denmark. Landsc Urban Plan 64:161–178
Tsonkova P, Bohm C, Quinkenstein A, Freese D (2012) Ecological benefits provided by alley cropping systems for production of woody biomass in the temperate region: a review. Agrofor Syst 85:133–152
Turner MG, Gardner RH, O'Neill RV (2001) Landscape ecology in theory and practice—pattern and process. Springer, New York
Tyndall J, Colletti J (2007) Mitigating swine odor with strategically designed shelterbelt systems: a review. Agrofor Syst 69:45–65
Valencia V, West P, Sterling EJ, Garcia-Barrios L, Naeem S (2015) The use of farmers' knowledge in coffee agroforestry management: implications for the conservation of tree biodiversity. Ecosphere 6:17
Wardropper CB, Chang CY, Rissman AR (2015) Fragmented water quality governance: constraints to spatial targeting for nutrient reduction in a Midwestern USA watershed. Landsc Urban Plan 137:64–75

Willis WB, Eichinger WE, Prueger JH, Hapeman CJ, Li H, Buser MD, Hatfield JL, Wanjura JD, Holt GA, Torrents A, Plenner SJ, Clarida W, Browne SD, Downey PM, Yao Q (2017) Particulate capture efficiency of a vegetative environmental buffer surrounding an animal feeding operation. Agric Ecosyst Environ 240:101–108

Wolz KJ (2017) Frontiers in alley cropping: transformative solutions for temperate agriculture. Ecology, evolution, and conservation biology. University of Illinois at Urbana-Champaign, Urbana, IL, p 194

Wu J (2013a) Landscape ecology. In: Leemans R (ed) Ecological systems. Springer-Verlag, New York, pp 179–200

Wu J (2013b) Key concepts and research topics in landscape ecology revisited: 30 years after the Allerton Park workshop. Landsc Ecol. 28, 1–11

Xiang WN (1993) A GIS method for riparian water-quality buffer generation. Int J Geogr Inf Syst 7:57–70

Yang B, Li SJ (2016) Design with nature: Ian McHarg's ecological wisdom as actionable and practical knowledge. Landsc Urban Plan 155:21–32

Zaimes GN, Schultz RC (2011) Do randomly placed riparian conservation land-uses improve stream water quality in Iowa, USA? Pol J Environ Stud 20:1083–1092

Zheng X, Zhu JJ, Xing ZF (2016) Assessment of the effects of shelterbelts on crop yields at the regional scale in Northeast China. Agric Syst 143:49–60

Zimmermann T (1986) Agroforestry—a last hope for conservation in Haiti. Agrofor Syst 4:255–268

# Economic Valuation of Agroforestry Ecosystem Services

**Zhen Cai and Francisco X. Aguilar**

## Abbreviations

| | |
|---|---|
| BT | Benefit transfer |
| RP | Revealed preference |
| SP | Stated preference |
| TEV | Total economic value |
| WTA | Willingness-to-accept |
| WTP | Willingness-to-pay |

## Introduction

Agroforestry is an intensive land management system that combines trees and/or shrubs with crops and/or livestock for the increased social, economic, and environmental benefits (Center for Agroforestry, University of Missouri 2012). There are many types of agroforestry practices tailored to particular ecological conditions. For instance, common agroforestry practices include alley cropping, forest farming, homegardens, improved fallow rotations, riparian forest buffers, silvopasture, and windbreaks. Agroforestry land management practices provide functions similar to naturally occurring ecosystems, which can be categorized as provisioning (create ecosystem goods for human consumption), regulating (maintain ecosystem health),

Z. Cai (✉)
The Center for Agroforestry, The School of Natural Resources, University of Missouri, Columbia, MO, USA
e-mail: caiz@missouri.edu

F. X. Aguilar
Department of Forest Economics, Swedish University of Agricultural Sciences, Umea, Sweden
e-mail: Francisco.aguilar@slu.se

 
R. P. Udawatta, S. Jose (eds.), *Agroforestry and Ecosystem Services*,
https://doi.org/10.1007/978-3-030-80060-4_18

supporting (services that are fundamental in providing all other ecosystem services), and cultural (nonmaterial benefits including historical, spiritual, religious, recreation services among others) services (McAdam et al. 2009; Noel et al. 2009). Specific examples for these functions include crops, medicinal plants, fruits, and nuts (provision); clean water, air and soil, and carbon sequestration (regulation); wildlife habitat protection and biodiversity conservation (supporting); and recreation and ecotourism (cultural) (Jose 2009; Montagnini and Nair 2004).

The literature points to ecological gains from agroforestry over more intensive monocultural practices. Windbreaks can reduce wind velocity and particulate matter in the air, and riparian forest buffers reduce nonpoint source pollution from monoculture practices (Alavalapati and Nair 2001). Carbon sequestration rates for smallholder agroforestry systems in the tropics have been estimated to be ranging from 1.5 to 3.5 Mg C $ha^{-1}$ $year^{-1}$ (Watson et al. 2000). In temperate regions such as in the USA, the potential for carbon sequestration through alley cropping has been estimated to be around 73.8 Tg C/year (Montagnini and Nair 2004), and the potential for carbon sequestration in silvopastoral systems around 9.0 Tg C/year (Follett 2001). In terms of biodiversity conservation, agroforestry provides habitat for species that tolerate a certain level of disturbance and helps conserve floral and faunal species (Jose 2009). Obeng and Aguilar (2015), based on a review of 16 studies examining cacao farming systems in the tropics, conclude that multi-strata shaded cacao systems had significantly lower above- and belowground carbon than natural tropical forests but the loss was less dominant in shaded cacao over its monocultural management. Noticeably, there are reportedly positive marginal changes for mean species richness in soil and litter and some essential chemical and physical soil properties (calcium, magnesium, sand, and silt) of cacao agroforestry systems compared with a natural forest.

It is important to integrate the ecological changes between land management systems and quantify their economic value to ease comparative analyses and ultimately help with the optimal allocation of limited resources. Market goods and services have prices determined by societal values; however, not all ecosystem services provided by agroforestry have a market price. Non-rivalry and non-exclusion characteristics of some ecosystem services challenge the ability to set prices and trade them in market transactions. Benefits from the provision of ecosystem services cannot be limited to particular beneficiaries or their exclusion may not be prohibitively expensive (non-exclusionary condition) and the consumption of some services by one person may not affect the use by others (non-rivalry condition). These conditions can prevent the emergence of prices and market transactions. Nonetheless, although there might not be a financial value in markets, there is an inherent economic value to society by benefiting from their supply. Their respective value needs to be estimated because agroforestry competes for land, labor, and capital with all other land uses that might not provide the same level of nonmarket benefits. Neglecting the economic values of ecosystem services undervalues total economic returns from agroforestry practices and discourages the adoption of agroforestry at the expense of other practices that might seem more financially rewarding but may have a lower or even detrimental effect on societal well-being.

Economists have made substantial contributions to the values of marketable and nonmarketable products and services in other ecosystems (e.g., forestry, watershed); however, economic values of ecosystem services from agroforestry have been largely neglected (Pearce and Mourato 2004; Obeng et al. 2018). The quantification of ecosystem services from agroforestry helps better evaluate the social, economic, and environmental benefits among different land-use management practices (Hein et al. 2006). To address this need, this chapter discusses the values of ecosystem services provided by agroforestry and methods of valuating ecosystem services, and reviews the economic values of different agroforestry practices.

## Total Economic Value

The total economic value (TEV) is a framework that is widely used to quantify the value of ecosystem services. TEV captures the changes in utilities that individuals derive from a change in the provision of the ecosystem services (Pearce 2001). For instance, improved water, soil, and air quality brought by the establishment of a riparian forest buffer brings positive utilities to individuals. On the other hand, nutrient pollution caused by mismanagement of fertilizers can negatively impact drinking water sources and harm wildlife habitat. Moreover, TEV of ecosystem services provided by agroforestry captures both use and nonuse values. Use values capture current or future use of the ecosystem services. For instance, individuals can derive value directly by actual use of the goods or services (e.g., timber and crops), or can indirectly benefit from non-extractive use of ecosystem services (e.g., carbon sequestration, improved water quality). In addition, individuals may also have the option to preserve the services (e.g., biodiversity) for possible future use.

Nonuse values of services provided by agroforestry reflect the benefits derived from the knowledge that the service exists and is preserved, regardless of real or potential use of the particular good or service. Benefits from the existence of an ecosystem is the most important nonuse value. It reflects individuals' value placed on the existence of the agroforestry system, forests, and biodiversity, besides any possible present or future use (Attfield 1998; Marre et al. 2015). Bequest value is another nonuse value placed on the potential use of the agroforestry system by future generations (O'Garra 2009). Nonuse values might be rooted in altruistic (the value that individuals have for ensuring other people have the access to the ecosystem, forests and biodiversity) and biospheric (the value reflects a concern for the environment in itself) motivations (Lazo et al. 1997; Ojea and Loureiro 2007). Figure 1 summarizes agroforestry ecosystem functions, goods and services provided by agroforestry ecosystem, and TEV from the system.

Valuation of ecosystem services provided by agroforestry is a process by which various economic analyses are used to estimate TEV (Laurans et al. 2013). A central premise is that individuals are willing to exchange an ecosystem service's condition for a dollar amount—whether or not there is a price in traditional markets. Given the comprehensive nature of the TEV, it includes the monetary value of the costs or

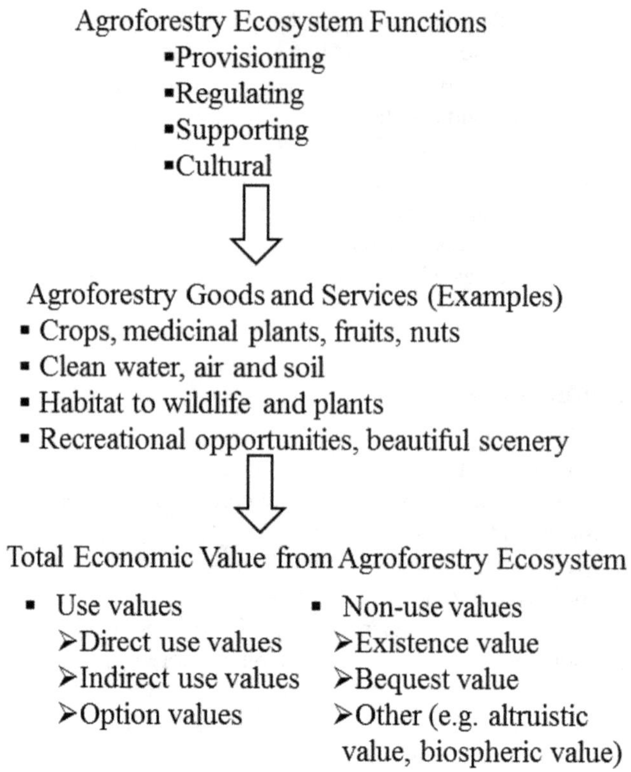


**Fig. 1** Agroforestry ecosystem functions, goods and services, and total economic value

benefits associated with changes in ecosystem services and is not limited to financial transactions. There are two concepts that can be used to measure the total value of ecosystem services and goods: willingness to pay (WTP) and willingness to accept (WTA). WTP is the maximum amount of money individuals are willing to pay for the change in ecosystem services where they obtain net gains, and WTA is the minimum amount of money individuals are willing to accept as compensation due to losses to ecosystem services or as payment to landowners to adopt agroforestry for ecosystem service provision. Ultimately, WTP and WTA estimates can be used to advise public policy and devise programs that can better reflect the true value that society places on agroforestry systems.

## Economic Valuation Approaches

Ecosystem services can be valued using different approaches including both primary and secondary methods. Primary methods capture individual WTP/WTA for changes in ecosystem services using either revealed preference (RP) or stated

**Table 1** Valuation methods and corresponding agroforestry goods and services valued

| Valuation approach | Valuation methods | Economic values | Agroforestry goods or services valued |
|---|---|---|---|
| Revealed preferences | Market prices | Use | Goods or services traded in markets (e.g., crops, timber, mushrooms) |
| | Cost-based approaches | Use | All goods and ecosystem services that prevented costs or expenses associated with replacing conditions (e.g., flood control, soil protection) |
| | Hedonic pricing | Use | Ecosystem services affect individuals' purchase behavior (e.g., proximity to riparian buffer) of a real estate |
| | Travel cost method | Use | Ecosystem services that contribute to recreational activities |
| Stated preferences | Contingent valuation method | Use and nonuse | All goods and ecosystem services |
| | Choice modeling | Use and nonuse | All goods and ecosystem services |
| Benefit transfer | Value transfer, function transfer | Use and nonuse | All goods and ecosystem services |

preference (SP) methods that observe individuals' actual or hypothetical choices or a combination of both (Freeman III et al. 2014). Benefit transfer is a secondary method that estimates the value of ecosystem services by adapting values estimated from past environmental valuation research of similar environmental characteristics and contexts. Table 1 describes valuation methods and corresponding agroforestry goods and services valued. Some valuation methods may only be appropriate to value certain agroforestry goods and services (e.g., market price method, hedonic pricing method, and travel cost method). Under certain contexts, RP and SP may be combined to value ecosystem services. For instance, the travel cost method can be used in conjunction with the contingent valuation method to better estimate recreational use values. Contingent valuation method is a method that is widely used to estimate nonmarket values of resources.

## *Revealed Preference Method*

RP methods value environmental goods and services by using real market information. Market price method is one of the RP methods and is frequently used in valuing marketable goods and services from agroforestry (Table 2). Economic values from agroforestry can be estimated by calculating the net present value of an agroforestry project over its duration or by estimating its average annual net income (Brandle et al. 1984; Robles-Diaz-de-Leon and Kangas 1998; Rasul and Thapa 2006). Other studies have emphasized that agroforestry can bring additional income

**Table 2** Past empirical studies on valuing ecosystem services provided by agroforestry

| Author(s) | Research region | Agroforestry practice | Economic values | Key findings |
|---|---|---|---|---|
| *Market price method* | | | | |
| Brandle et al. (1984) | Eastern Nebraska, USA | Windbreaks | Direct use value: Wheat production | Net present value of the wheat production from a 160-acre farm that has windbreaks was $22,184 based on a 50-year life of the windbreaks |
| Dangerfield and Harwell (1990) | Southeast USA | Silvopasture | Direct use value: Timber and livestock | Net present value of silvopasture was 71% higher than timber management only |
| Robles-Diaz-de-Leon and Kangas (1998) | Chesapeake Bay Region, USA | Riparian buffer | Direct use value: Non-timber forest products | Economic value of non-timber forest products from riparian forest buffer zones in the Chesapeake Bay region was estimated to be $60,934/ha/yr |
| Grado et al. (2001) | Southern Mississippi, USA | Silvopasture | Direct use value: Timber and livestock | Land expectation values of pine silvopastures were higher than those of pine plantations |
| Rasul and Thapa (2006) | Chittagong Hill Tracts, Bangladesh | Agroforestry | Direct use value: Timber and non-timber forest products | Agroforestry generated $388/ha/year income from timber and fruit sales |
| *Cost-based approach* | | | | |
| Niskanen (1998) | Thailand | Agroforestry | Indirect use value: Soil erosion control | The value of improved soil erosion control by agroforestry was estimated to be $1.2 per ton of soil erosion |
| *Hedonic pricing method* | | | | |
| Colby and Wishart (2002) | Tucson, AZ, USA | Riparian buffer | Use value: All services that provide use values | Total price premium for the 25,560 residential houses located within 1.5 miles from riparian corridors was $103.1 million |
| Shrestha and Alavalapati (2004b) | Florida, USA | Silvopasture | Direct use value: Recreation (hunting) | One percent increase in trees and other vegetation cover in ranchland led to a $0.03 per acre hunting lease price increase |

(continued)

**Table 2** (continued)

| Author(s) | Research region | Agroforestry practice | Economic values | Key findings |
|---|---|---|---|---|
| Bin et al. (2009) | North Carolina, USA | Riparian buffer | Use value: All services that provide use values | Average price premium for a house located in a riparian zone over a non-riparian house was $37,423 |
| *Contingent valuation* | | | | |
| Lant and Roberts (1990) | Iowa and Illinois River Basins, USA | Riparian buffer | Use value: Recreation, improved water quality, and aesthetics | Respondents were willing to pay $36.18, $48.65, and $49.47 for the improvement of water quality from poor to fair, fair to good, and good to excellent, respectively. These levels were determined by scores that were calculated based on the observable nature or aesthetics, level of adequacy for boating, fishing, and swimming |
| Cook and Cable (1990) | Kansas, USA | Windbreaks | Direct use value: Recreation (hunting) | The estimated aggregate value for hunting opportunities was $21,538,056 |
| Loomis et al. (2000) | South Platte River, USA | Riparian buffer | Use value: Improved water quality, soil erosion control, habitat for fish and wildlife, and recreation | Mean monthly WTP per household was $21 per month in a higher water bill for the increase in ecosystem services on a 45 mile stretch of the South Platte river |
| Lynch et al. (2002) | Maryland, USA | Riparian buffer | Indirect use value: Improved water quality, removal of chemicals from agricultural production | On average, respondents were willing to accept a $112/year payment for 15 years to install a buffer |
| Shrestha and Alavalapati (2003) | Florida, USA | Silvopasture | Use value: Improved water quality, soil conservation, carbon sequestration, wildlife habitat protection, and aesthetics | Ranchers were willing to accept a direct payment of $9.32/acre/year for silvopasture adoption on their farms. This led to a total annual payment of $56.45–$72.43 million for the adoption of silvopasture practices in Florida |

(continued)

**Table 2** (continued)

| Author(s) | Research region | Agroforestry practice | Economic values | Key findings |
|---|---|---|---|---|
| Holmes et al. (2004) | Little Tennessee River, Western North Carolina, USA | Riparian buffer | Use value: Improved soil and water quality, recreational uses, wildlife habitat, and ecosystem naturalness | Local residents were willing to pay $53.76 per household for a riparian restoration project |
| Qiu et al. (2006) | Dardenne Creek Watershed in St. Louis, USA | Riparian buffer | Use value: All services that provide use values | Respondents were willing to pay $1625 more for a house that is close to an openly accessible buffer along a creek compared to a house that is far from the buffer |
| Stone et al. (2008) | Indian | Windbreaks | Use value: All services that provide use values | Median household WTP for a windbreak restoration project was $11.65 per year |
| Grala et al. (2009) | North Central Iowa, USA | Riparian buffer | Direct use value: Recreation (hunting) | Agricultural landowners were willing to accept an average of US$30 per visit per party of four hunters to allow hunting of ring-necked pheasants |
| Sauer and Fischer (2010) | Northeim, Germany | Riparian buffer | Indirect use value: Soil erosion control, improved water quality, flood control | Respondents were willing to pay $46.64 to support the establishment of the buffer strips |
| Buckley et al. (2012) | Republic of Ireland | Riparian buffer | Use value: All services that provide use values | Approximately 53% of the respondents indicated no preference for establishing the buffer in their land due to concerns on potential interference with crop production. Among farmers who are willing to accept, their mean WTA for a 10 m riparian buffer zone was $1.72/meter |
| Grala et al. (2012) | Iowa, USA | Windbreaks | Indirect use value: Recreation (aesthetic) | A mean WTP ranged from US$4.77 to US$8.50 to support establishing windbreaks |
| *Choice modeling* | | | | |

(continued)

**Table 2** (continued)

| Author(s) | Research region | Agroforestry practice | Economic values | Key findings |
|---|---|---|---|---|
| Alavalapati et al. (2004) | Northern watershed of Lake Okeechobee, Florida, USA | Silvopasture | Indirect use value: Improved water quality and wildlife habitat, carbon sequestration | WTP estimates (per year for 5 years): water quality improvement: \$30.24 (from low to moderate) and \$71.17 (from low to high); carbon sequestration: \$58.05 (from low to moderate) and \$62.72 (from low to high); wildlife habitat improvements: \$49.68 (from low to moderate) and \$41.06 (from low to high). A total WTP for the environmental service: \$924.40 million over 5 years |
| Duke et al. (2012) | Delaware, USA | Riparian buffer | Use value: All services that provide use values | WTP for expanding riparian buffer in Delaware was estimated to be \$29.59 per household per year |
| Rolfe et al. (2006) | Fitzroy Basin in central Queensland, Australia | Riparian buffer | Indirect use value: Improved water quality | Landholders would require an average of \$3.75 per meter increase of riparian areas, and \$8/km per 1% increase in the level of minimum biomass condition |
| *Value transfer* | | | | |
| Kulshreshtha and Kort (2009) | Canadian Prairie Provinces, Canada | Riparian buffer | Use value: Reduced soil erosion, carbon sequestration, improved water and air quality, and recreation | Ecosystem services provided by riparian buffer were estimated to be over \$88.2 million, of which carbon sequestration accounted for \$46 million and reduced soil erosion services accounted for \$9.5 million |

(continued)

**Table 2** (continued)

| Author(s) | Research region | Agroforestry practice | Economic values | Key findings |
|---|---|---|---|---|
| Alam et al. (2014) | Southern Québec, Canada | Tree-based intercropping | Use value: Improved water, soil, and air quality; pollination; wildlife habitat; windbreaks; timber provisioning | The total value of the ecosystem services provided per year was estimated to be $2645/ha/year, of which the indirect use value was $1634/ha/year |
| *Function transfer* | | | | |
| Brenner et al. (2010) | Coastal zone of Catalonia, Spain | Riparian buffer | Use value: All ecosystem services that provide use values | A value of $8359/ha/year was provided by riparian buffer. Aesthetic and recreation ($3385/ha/year) and water supply ($4747/ha/year) values dominated values from all services |
| Bauer and Johnston (2017) | Great Bay watershed, USA | Riparian buffer | Indirect use value: Improved water quality | Economic values provided by riparian buffer were estimated to be up to $34 million when values were aggregated over all New Hampshire residents |

to farmers and estimated this income by comparing revenues generated from agroforestry practice with traditional agricultural or forestry practices (Dangerfield and Harwell 1990; Grado et al. 2001).

Cost-based approaches value environmental goods and services by considering the observable costs associated with the provision of them. Replacement cost method assumes that there are substitutes for environmental goods and ecosystem services, and the costs in providing these substitutes can be regarded as the value of the studied environmental goods and ecosystem services. Niskanen (1998) used the replacement cost method to estimate the value of soil protection brought by community- and agroforestry-based reforestation by using commercial fertilizers as substitutes (Table 2). Avoided cost method captures the actions or expenditures that individuals take to avoid a degradation of environmental quality. For example, individuals may purchase bottled water or water filters if a groundwater source is contaminated. The amount of money spent on treating the polluted water can be regarded as the value of improved water quality. The avoided cost method has been used to estimate the value of carbon sequestration under silvopastoral system in Galicia, Spain, by calculating the costs in avoiding reforestation (Fernández-Núñez et al. 2009). Mitigation or restoration cost method assesses the costs of mitigating the adverse effects caused by the absence of ecosystem services or the cost of restoring those services. Mitigation or restoration costs have been used widely in estimating greenhouse gas mitigation, nitrogen mitigation, waterfowl recreation from watershed, and wetland restoration (Heal 2000; Jenkins et al. 2010). But studies in agroforestry services using this method are limited.

Hedonic pricing method calculates a price for an environmental good or services by examining its effect on real estate prices using actual market data. This method is based on the assumption that real estate prices are determined by all characteristics of a property (e.g., construction area, number of rooms, distance to recreational areas or riparian forest buffers, distances to hog farms). Regression models can help discern the impacts of specific property characteristics on real estate prices. Model coefficients can be used to estimate the shadow prices of housing characteristics including goods or ecosystem services valued. For example, Herriges et al. (2005) suggest that the building of new livestock facilities resulted in a loss of 14–16% in home value when located directly downwind from a confined animal feeding operation. Isakson and Ecker (2008) report that in Iowa a house located within 2.5 miles of and directly downwind from a swine confined animal feeding operation lost about 15.3% of its value compared to real estate beyond this distance. The hedonic pricing method has been used to estimate the value of ecosystem services provided by agroforestry. The services provided by riparian forest buffer are the most frequently studied (e.g., Bin et al. 2009; Colby and Wishart 2002). Recreational value provided by silvopasture in Florida has also been estimated in the literature (Shrestha and Alavalapati 2004a).

The travel cost method is commonly used to evaluate recreational value provided by an ecosystem. Recreational value of an agroforestry site can be derived from individuals' travel costs and travel frequency by surveying site visitors. However, the literature using this method to estimate agroforestry recreational value (e.g., hunting, scenery) is very limited, probably because the sole ecotourism of an agroforestry site is uncommon but remains part of the multiple functions of farming (Barbieri and Valdivia 2010).

The major advantage of RP approach is that of estimating economic values of ecosystem services based on actual/observed behavior (Earnhart 2001). Data for the RP approach are often relatively inexpensive to collect compared to SP methods. However, there are some weaknesses of the RP approach (Table 3). Estimation bias is the main disadvantage of these approaches. For instance, values estimated from market price methods are determined by the market prices used; however, market failure or government interventions (e.g., subsidies from incentive programs) may distort these prices and lead to biased estimates. The hedonic pricing method may introduce estimation bias since it assumes that environmental attributes (e.g., wildlife protection, improved water and soil quality provided by a riparian forest buffer) can be reflected in real estate prices and people have perfect information about these attributes when making financial transactions. However, if real estate buyers/sellers were not familiar with the services provided by agroforestry, their values may not be reflected in real estate prices.

**Table 3** Revealed preference approach—advantages and limitations

| Valuation methods | Advantages | Limitations |
|---|---|---|
| Market prices | Simple to use<br>Data are easy to obtain | Can only be used to value goods and services with market prices<br>Prices of these goods and services have spatial and temporal variations<br>Market failure or government interventions may distort prices |
| Cost-based approaches | Costs are easier to estimate compared to the value of benefits | Costs to repair damages or to replace ecosystem services or goods may not be an accurate estimate of benefits (Bourlion 2015)<br>Man-made alternatives provide not only positive benefits but also negative benefits, which may not be captured in this method |
| Hedonic pricing | Data are relatively easy to obtain and valid | Implementation is complex<br>Estimates are determined by model specifications and interpretations<br>Assumes that environmental services can be reflected in real estate prices |
| Travel cost method | Data are relatively easy to obtain | Can only be used for recreational sites<br>Assumes that visiting the recreational area is travelers' only purpose<br>Estimating opportunity cost of time is difficult<br>Estimates are determined by model specifications and interpretations<br>Bringing sampling bias due to only travelers are surveyed (Bann 2002) |

Adapted from King and Mazzotta (2000)

## *Stated Preference Method*

SP methods can be used to estimate economic values of ecosystem services by eliciting individuals' utility change associated with a corresponding change in quality or quantity of an ecosystem services using hypothetical scenarios (Mitchell and Carson 1989). Compared to the RP method, SP can be used to estimate nonuse values for natural resources. SP methods often rely on a questionnaire to construct a hypothetical scenario, which specifies a change in a natural resource or environmental attribute compared to a baseline situation. The questionnaire in which respondents are asked to answer a series of questions can be used to estimate respondents' WTP for the environmental change (Mitchell and Carson 1989). The aggregation of each respondent's WTP over the population corresponds to the economic value of the change in ecosystem services. In cases where service providers are respondents (e.g., surveying farmers), WTA a certain amount of money to provide the services can be used as a proxy to value these services. Contingent valuation and choice modeling are two SP methods to value ecosystem services.

Contingent valuation method has been used in agroforestry to estimate the economic value of the services provided by riparian buffer, windbreaks, and silvopasture (Table 2). Choice modeling method can be used to estimate the values of specific attributes of environmental goods and services simultaneously. Rather than asking respondents' WTP directly, the choice modeling method generates several choice alternatives based on different WTP levels, and different levels of certain ecosystem services (e.g., low water quality level, moderate water quality level, and high water quality level). Several choice questions are generated by pairing these alternatives, and respondents are asked to choose the alternative that they prefer for each question. Statistical and regression models can be used to elicit individuals' WTP to each ecosystem service based on respondents' choices. In the agroforestry literature, there are only a few studies evaluating agroforestry ecosystem service values using choice modeling methods, and most of these studies focus on individuals' WTP for ecosystem service provided by silvopasture and riparian forest buffers.

The main advantage of the SP approach is that it can be used when RP data is lacking and can estimate nonuse values of ecosystem services. The hypothetical scenarios provide the flexibility to design survey questions and estimate the value of these services. There are also some limitations of the stated preference approach (Table 4). The accuracy of value estimates is greatly determined by the hypothetical scenario created and questions used to elicit WTP/WTA. The choice modeling method is difficult to implement. Choice questions used in the choice modeling method need to be clear and they are complicated to design; sometimes software may be used to simplify the generation of potential comparison groups (e.g., SAS, R). In the choice questions, specific ecosystem services should be chosen according to research objectives and corresponding levels should be introduced to respondents clearly. For instance, improved water quality can be measured using low, medium, and high levels, but these levels should be clearly defined sometimes with the assistance of pictures to provide visual impression (Aguilar et al. 2018).

**Table 4** Stated preference approach—advantages and limitations

| Valuation methods | Advantages | Limitations |
|---|---|---|
| Contingent valuation | Flexible and easy-to-design valuation questions<br>Can estimate nonuse value | Estimates contingent on the hypothetical scenario and questions used to elicit WTP |
| Choice modeling | Can estimate values for different environmental services simultaneously<br>Allows the estimation of overall preferences for any combination of attributes<br>Can estimate nonuse value | Complex-to-design choice modeling questions and analyze data<br>Repeated choice questions can result in respondent fatigue and bias results |

Adapted from The Regional Organization for the Conservation of the Environment of the Red Sea & Gulf of Aden (2015)

### *Benefit Transfer (BT) Methods*

BT methods estimate the economic values based on previous revealed and stated preference studies. The large number of economic valuation empirical studies currently available encourage the use of benefit transfer methods in evaluating ecosystem services. Benefit transfer has two methods: value transfer and function transfer. Value transfer identifies a benefit per unit estimate (e.g., per 1 acre riparian buffers, per 1 acre windbreaks) from a previous study and applies such effect to extrapolate it to sites of comparable conditions (Loomis and Richardson 2008). Function transfers value the ecosystem services by building a demand function or a meta-analysis/meta-regression to fit the characteristics of a new study site (e.g., social demographics, size of the new study site) (Rosenberger and Loomis 2000). The meta-analysis, a statistical based literature review tool, is one of the most popular methods when synthesizing scientific research and searching for ecosystem service values/patterns (Nijkamp et al. 2008). Based on the research objective of a benefit transfer study, meta-analysis selects comparable empirical case studies and derives relevant information using statistical methods. The results obtained can be used in providing the values for the benefit transfer study (Rosenberger and Loomis 2000). The benefit transfer method has been used in estimating total economic value of the ecosystem services provided by tree-based intercropping system and riparian forest buffers (Table 2).

By identifying essential values for ecosystem services from previous studies, the benefit transfer method can save time and expenses needed for primary valuation study. However, poor previous studies and publication bias may bring estimation bias. Estimation bias may also be introduced during the transfer of the values from the research site from the previous studies to the benefit transfer site due to many reasons such as the complex nature of ecosystems, context, and different socioeconomic factors (The Regional Organization for the Conservation of the Environment of the Red Sea & Gulf of Aden 2015).

## Summary and Conclusions

This chapter discussed economic value concepts and ecosystem service valuation approaches, and reviewed the existing empirical studies valuing ecosystem services provided by agroforestry. Economic value of ecosystem services provided by agroforestry is measured by social utility change (in dollars) caused by a change in service from agroforestry ecosystem. The stated preference revealed that preference and benefit transfer approaches are used to measure the economic value of ecosystem services provided by agroforestry. Some ecosystem service evaluation method may be more appropriate to value certain goods and services compared to other valuation methods. There are disadvantages of each estimation method and possible bias in their estimates, indicating a clear need to continue developing valuation tools to better understand the value of agroforestry ecosystem services (Nijkamp et al. 2008).

The literature has estimated the economic values of ecosystem services provided by agroforestry. On average, indirect use values (e.g., improved soil and water quality) individuals derive from agroforestry are estimated to be much higher than direct use values (e.g., crops, timber). However, the review suggested that the estimation is only focused on some use values (i.e., direct and indirect use values) of agroforestry ecosystems and seldom discusses the option value and nonuse values. The values of the existence of agroforestry, future use of agroforestry ecosystem services, and preservation of agroforestry ecosystem services for future generations are needed to be estimated.

Ecosystem services provided by agroforestry may depend on specific agroforestry practices applied and their economic values may be different. The literature focuses on the examination of economic values from riparian forest buffers and windbreaks, and has limited studies examining the economic values provided by silvopasture and alley cropping. The value of ecosystem services provided by forest farming is ignored in the literature according to our knowledge. More studies on valuing the agroforestry ecosystem services are needed to be conducted in the future to better understand the values of agroforestry and develop relevant policies to improve agroforestry adoption rate and further contribute to the earth's ecosystem.

## References

Aguilar FX, Obeng EA, Cai Z (2018) Water quality improvements elicit consistent willingness-to-pay for the enhancement of forested watershed ecosystem services. Ecosyst Serv 30:158–171

Alam M, Olivier A, Paquette A, Dupras J, Revéret JP, Messier C (2014) A general framework for the quantification and valuation of ecosystem services of tree-based intercropping systems. Agrofor Syst 88(4):679–691

Alavalapati JRR, Nair PKR (2001) Socioeconomics and institutional perspectives of agroforestry. In: Palo M, Uusivuori J (eds) World Forests, society and environment: markets and policies. Kluwer Academic Publishers, Dordrecht, The Netherlands, pp 71–81

Alavalapati JR, Shrestha RK, Stainback GA, Matta JR (2004) Agroforestry development: an environmental economic perspective. In: New Vistas in agroforestry. Springer, Dordrecht, pp 299–310

Attfield R (1998) Existence value and intrinsic value. Ecol Econ 24(2–3):163–168

Bann C (2002) An overview of valuation techniques: advantages and limitations. Asean Biodiversity 2(2):8–16

Barbieri C, Valdivia C (2010) Recreation and agroforestry: examining new dimensions of multifunctionality in family farms. J Rural Stud 26(4):465–473

Bauer DM, Johnston RJ (2017) Buffer options for the bay: economic valuation of water quality ecosystem services in New Hampshire's great bay watershed. http://www.bufferoptionsnh.org/wp-content/uploads/2017/12/BOB_Economic_Assessment.docx.pdf. Accessed 1 October 2018

Bin O, Landry CE, Meyer GF (2009) Riparian buffers and hedonic prices: a quasi-experimental analysis of residential property values in the Neuse River basin. Am J Agr Econ 91(4):1067–1079

Bourlion N. (2015) Sustainable Forest Management. https://planbleu.org/sites/default/files/upload/files/FactSheets_methods_EN.pdf. Accessed 1 October 2018

Brandle JR, Johnson BB, Dearmont DD (1984) Windbreak economics: the case of winter wheat production in eastern Nebraska. J Soil Water Conserv 39(5):339–343

Brenner J, Jimenez JA, Sarda R, Garola A (2010) An assessment of the non-market value of the ecosystem services provided by the Catalan coastal zone, Spain. Ocean Coast Manage 53(1):27–38

Buckley C, Hynes S, Mechan S (2012) Supply of an ecosystem service—Farmers' willingness to adopt riparian buffer zones in agricultural catchments. Environ Sci Pol 24:101–109

Center for Agroforestry, University of Missouri (2012) What is agroforestry? http://www.centerforagroforestry.org/practices/. Accessed 1 October 2018

Colby BG, Wishart S (2002) Quantifying the influence of desert riparian areas on residential property values. Appraisal J 70(3):304

Cook PS, Cable TT (1990) The economic value of windbreaks for hunting. Wildl Soc Bull. (1973–2006) 18(3):337–342

Dangerfield CW, Harwell RL (1990) An analysis of a silvopastoral system for the marginal land in the southeast United States. Agrofor Syst 10(3):187–197

Duke JM, Borchers AM, Johnston RJ, Absetz S (2012) Sustainable agricultural management contracts: using choice experiments to estimate the benefits of land preservation and conservation practices. Ecol Econ 74:95–103

Earnhart D (2001) Combining revealed and stated preference methods to value environmental amenities at residential locations. Land Econ 77(1):12–29

Fernández-Núñez E, Rigueiro-Rodríguez A, Mosquera-Losada MR (2009) Economic assessment of silvopastoral systems. Agroforestry Syst Tech Sustain Land Manage:303–315

Follett RF (2001) Soil management concepts and carbon sequestration in cropland soils. Soil Tillage Res 61(1–2):77–92

Freeman AM III, Herriges JA, Kling CL (2014) The measurement of environmental and resource values: theory and methods. Routledge

Grado SC, Hovermale CH, St Louis DG (2001) A financial analysis of a silvopasture system in southern Mississippi. Agrofor Syst 53(3):313–322

Grala RK, Colletti JP, Mize CW (2009) Willingness of Iowa agricultural landowners to allow fee hunting associated with in-field shelterbelts. Agrofor Syst 76(1):207–218

Grala RK, Tyndall JC, Mize CW (2012) Willingness to pay for aesthetics associated with field windbreaks in Iowa, United States. Landscape and urban planning, 108(2-4): pp.71–78

Heal G (2000) Valuing ecosystem services. Ecosystems 3(1):24–30

Hein L, Van Koppen K, De Groot RS, Van Ierland EC (2006) Spatial scales, stakeholders and the valuation of ecosystem services. Ecol Econ 57(2):209–228

Herriges JA, Secchi S, Babcock BA (2005) Living with hogs in Iowa: the impact of livestock facilities on rural residential property values. Land Econ 81(4):530–545

Holmes TP, Bergstrom JC, Huszar E, Kask SB, Orr F III (2004) Contingent valuation, net marginal benefits, and the scale of riparian ecosystem restoration. Ecol Econ 49(1):19–30

Isakson HR, Ecker MD (2008) An analysis of the impact of swine CAFOS on the value of nearby houses. Agric Econ 39(3):365–372

Jenkins WA, Murray BC, Kramer RA, Faulkner SP (2010) Valuing ecosystem services from wetlands restoration in the Mississippi Alluvial Valley. Ecol Econ 69(5):1051–1061

Jose S (2009) Agroforestry for ecosystem services and environmental benefits: an overview. Agrofor Syst 76:1–10

King DM, Mazzotta MJ (2000) Ecosystem Valuation. www.ecosystemvaluation.org. Accessed 1 October 2018

Kulshreshtha S, Kort J (2009) External economic benefits and social goods from prairie shelterbelts. Agrofor Syst 75(1):39–47

Lant CL, Roberts RS (1990) Greenbelts in the Cornbelt: riparian wetlands, intrinsic values, and market failure. Environ Plann A 22(10):1375–1388

Laurans Y, Rankovic A, Billé R, Pirard R, Mermet L (2013) Use of ecosystem services economic valuation for decision making: questioning a literature blindspot. J Environ Manag 119:208–219

Lazo JK, McClelland GH, Schulze WD (1997) Economic theory and psychology of non-use values. Land Econ:358–371

Loomis J, Richardson L (2008) Technical documentation of benefit transfer and visitor use estimating models of wildlife recreation, species and habitats. In: Report prepared for the National Council for Science and the Environment as part of the project "Development of an Operational Benefits Estimation Tool for the US". Defenders of Wildlife, Washington, DC

Loomis J, Kent P, Strange L, Fausch K, Covich A (2000) Measuring the total economic value of restoring ecosystem services in an impaired river basin: results from a contingent valuation survey. Ecol Econ 33(1):103–117

Lynch L, Hardie I, Parker D (2002) Analyzing agricultural landowners' willingness to install streamside buffers. Departmental Paper: Department of Agricultural and Resource Economics, University of Maryland

Marre JB, Brander L, Thebaud O, Boncoeur J, Pascoe S, Coglan L, Pascal N (2015) Non-market use and non-use values for preserving ecosystem services over time: A choice experiment application to coral reef ecosystems in New Caledonia. Ocean Coast Manage 105:1–14

McAdam JH, Burgess PJ, Graves AR, Rigueiro-Rodríguez A, Mosquera-Losada MR (2009) Classifications and functions of agroforestry systems in Europe. In: Agroforestry in Europe. Springer, Dordrecht, pp 21–41

Mitchell R, Carson R (1989) Using surveys to value public goods: the contingent valuation method. Resources for the Future, Washington, DC

Montagnini F, Nair PKR (2004) Carbon sequestration: an underexploited environmental benefit of agroforestry systems. Agrofor Syst 61:281–295

Nijkamp P, Vindigni G, Nunes PA (2008) Economic valuation of biodiversity: a comparative study. Ecol Econ 67(2):217–231

Niskanen A (1998) Value of external environmental impacts of reforestation in Thailand. Ecol Econ 26(3):287–297

Noel JE, Qenani-Petrela E, Mastin T (2009) A benefit transfer estimation of agro-ecosystems services. In: Western economics forum, Vol. 8, (1), pp. 18–28. Western Agricultural Economics Association

O'Garra T (2009) Bequest values for marine resources: how important for indigenous communities in less-developed economies? Environ Resour Econ 44(2):179

Obeng EA, Aguilar FX (2015) Marginal effects on biodiversity, carbon sequestration and nutrient cycling of transitions from tropical forests to cacao farming systems. Agrofor Syst 89(1):19–35

Obeng EA, Aguilar FX, McCann LM (2018) Payments for forest ecosystem services: a look at neglected existence values, the free-rider problem and beneficiaries' willingness to pay. Int For Rev 20(2):206–219

Ojea E, Loureiro ML (2007) Altruistic, egoistic and biospheric values in willingness to pay (WTP) for wildlife. Ecol Econ 63(4):807–814

Pearce DW (2001) The economic value of forest ecosystems. Ecosyst Health 7(4):284–296

Pearce D, Mourato S (2004) The economic valuation of agroforestry's environmental services. In: Agroforestry and biodiversity conservation in tropical landscapes. Island Press, Washington, pp 67–86

Qiu Z, Prato T. Boehrn G (2006) Economic Valuation of Riparian Buffer and Open Space in a Suburban Watershed1. JAWRA Journal of the American Water Resources Association, 42(6): pp.1583-1596.

Rasul G, Thapa GB (2006) Financial and economic suitability of agroforestry as an alternative to shifting cultivation: The case of the Chittagong Hill Tracts, Bangladesh. Agric Syst 91(1–2):29–50

Robles-Diaz-de-Leon LF, Kangas P (1998) Evaluation of potential gross income from non-timber products in a model riparian forest for the Chesapeake Bay watershed. Agrofor Syst 44(2–3):215–225

Rolfe J, Windle J, Reeson A, Whitten S (2006) Assessing the incentives needed to improve riparian management in grazing systems: comparing experimental auctions and choice modelling approaches. In 50th Annual Conference of the Australian Agricultural and Resource Economics, Sydney

Rosenberger RS, Loomis JB (2000) Using meta-analysis for benefit transfer: In-sample convergent validity tests of an outdoor recreation database. Water Resour Res 36(4):1097–1107

Sauer U, Fischer A (2010) Willingness to pay, attitudes and fundamental values—On the cognitive context of public preferences for diversity in agricultural landscapes. Ecol Econ 70(1):1–9

Shrestha RK, Alavalapati JR (2003) Florida Ranchers' Willingness to adopt silvopasture practices: a dichotomous choice contingent valuation approach. Bugs, Budgets, Mergers, and Fire: Disturbance Economics 317

Shrestha RK, Alavalapati JR (2004a) Effect of ranchland attributes on recreational hunting in Florida: a hedonic price analysis. J Agric Appl Econ 36(3):763–772

Shrestha RK, Alavalapati JR (2004b) Estimating Ranchers' cost of agroforestry adoption. In: Valuing agroforestry systems. Springer, Dordrecht, pp 183–199

Stone K, Bhat M, Bhatta R, Mathews A (2008) Factors influencing community participation in mangroves restoration: a contingent valuation analysis. Ocean Coast Manage 51(6):476–484

The Regional Organization for the Conservation of the Environment of the Red Sea & Gulf of Aden (2015). https://pdfs.semanticscholar.org/presentation/8c20/ee3503f90be27829457e31e3654a602e3063.pdf. Accessed 1 October 2018

Watson RT, Noble IR, Bolin B, Ravindranath NH, Verardo DJ, Dokken DJ (eds) (2000) Land Use, land-use change, and forestry. Intergovernmental Panel on Climate Change (IPCC), Special report. Cambridge Univ. Press, New York

# Correction to: The Role of Temperate Agroforestry Practices in Supporting Pollinators

**Gary Bentrup, Jennifer Hopwood, Nancy Lee Adamson, Rae Powers, and Mace Vaughan**

**Correction to:**
**Chapter 11 in: R. P. Udawatta, S. Jose (eds.),**
***Agroforestry and Ecosystem Services,***
**https://doi.org/10.1007/978-3-030-80060-4_11**

An error in the production process unfortunately led to publication of the book before incorporating the below corrections. This has now been corrected and updated throughout the book:

On page 279

In Table 1, the intext figures in the last column have been duplicated as the same in the second column.

The intext figures in the last column of the Table 1 have been corrected.

The updated online version of this chapter can be found at
https://doi.org/10.1007/978-3-030-80060-4_11



R. P. Udawatta, S. Jose (eds.), *Agroforestry and Ecosystem Services*,
https://doi.org/10.1007/978-3-030-80060-4_19

**Table 1** Common insect pollinator groups

| | | |
|---|---|---|
|  |  |  |
| **Honey bee** | **Bumble bees** | **Ground-nesting bees** |
| Order: Hymenoptera<br>Family: Apidae<br>Genus and species: *Apis mellifera* | Order: Hymenoptera<br>Family: Apidae<br>Genus: *Bombus* | Order: Hymenoptera<br>Families: Andrenidae, Apidae, Colletidae, Halictidae |
| The European honey bee (native to Europe, Africa, and Asia) is a domesticated species that lives in large perennial social colonies (hives), with division of labor within the colony. Only the queen reproduces, while others gather nectar and pollen to feed brood (larvae) and store food (honey) for the winter. Feral colonies in the United States are somewhat rare; most hives are managed by beekeepers | Bumble bees form annual social colonies. Queen bumble bees that mated the previous fall start nests in spring and by mid-summer colonies can have dozens or hundreds of workers. They nest in insulated cavities such as under clumps of bunch grass or in old rodent nests. There are 46 recognized bumble bee species in North America | Most native bees live solitary lives, with each female working alone to build her nests and collect and provide food for her offspring. About 70% of our solitary bee species nest underground, digging slender tunnels in which they build individual cells for each egg and its provisions |
|  |  |  |
| **Tunnel-nesting bees** | **Flower-visiting flies** | **Flower-visiting beetles** |
| Order: Hymenoptera<br>Families: Apidae, Colletidae, Halictidae, Megachilidae | Order: Diptera<br>Families: Anthomyiidae, Bombyliidae, Syrphidae, Tachinidae, others | Order: Coleoptera<br>Families: Cantharidae, Coccinellidae, Scarabaeidae, others |
| Approximately 30% of solitary bee species nest in tunnels, inside already hollow stems or by chewing into the pithy center of stems, or in existing holes in wood, sometimes man-made. Most tunnel-nesting bees are solitary species | Flower-visiting flies consume nectar and sometimes pollen. Many hover flies (family Syrphidae) resemble bees or wasps in coloration. Larvae of some species are voracious predators of small insects, like aphids | Flower-visiting beetles consume nectar and pollen, and may also chew on flower parts. Larvae of some species are predatory, hunting other insects (including crop pests) as food, while others are herbivorous or are decomposers |

(continued)

**Table 1** (continued)

| | | |
|---|---|---|
|  |  |  |
| **Flower-visiting wasps** | **Flower-visiting moths** | **Butterflies** |
| Order: Hymenoptera<br>Families: Sphecidae, Vespidae, Tiphiidae, Scoliidae, others | Order: Lepidoptera<br>Families: Sphingidae, Noctuidae, Arctiidae | Order: Lepidoptera<br>Families: Papilionidae, Hesperiidae, Pieridae, Lycaenidae, Nymphalidae |
| Predatory wasps, most of which are solitary, hunt for prey to bring back to their nest as food for their young. They build nests in cavities or in the ground, and may utilize pieces of grass, mud, or resin in construction of their nest. Adults maintain their energy by consuming nectar and/or pollen, and in the process may also transfer pollen between flowers | Moths, which are often subdued in color and tend to fly at dusk or night, are less visible than other groups, but many are important specialist pollinators of wild plants, while some also pollinate crops. Moths as a group form a critical food source for other wildlife | With their striking transformation from a chubby plant-chewing caterpillar to a delicate pupa to a graceful nectar-drinking adult, butterflies are some of the most beloved insects. Some species have narrow host plant needs for their caterpillars while others feed on a wide variety of plants |

Source: Flower-visiting beetle image by Jennifer Hopwood and remaining images by Nancy Lee Adamson

# Index





R. P. Udawatta, S. Jose (eds.), *Agroforestry and Ecosystem Services*,
https://doi.org/10.1007/978-3-030-80060-4

CPSIA information can be obtained
at www.ICGtesting.com
Printed in the USA
LVHW082252100522
718467LV00004B/22

9 783030 800598